The Justification of War and International Order

THE HISTORY AND THEORY OF INTERNATIONAL LAW

General Editors

In the past few decades the understanding of the relationship between nations has undergone a radical transformation. The role of the traditional nation-state is diminishing, along with many of the traditional vocabularies which were once used to describe what has been called, ever since Jeremy Bentham coined the phrase in 1780, 'international law'. The older boundaries between states are growing ever more fluid, new conceptions and new languages have emerged which are slowly coming to replace the image of a world of sovereign independent nation-states which has dominated the study of international relations since the early nineteenth century. This redefinition of the international arena demands a new understanding of classical and contemporary questions in international and legal theory. It is the editors' conviction that the best way to achieve this is by bridging the traditional divide between international legal theory, intellectual history, and legal and political history. The aim of the series, therefore, is to provide a forum for historical studies, from classical antiquity to the twenty-first century, that are theoretically informed and for philosophical work that is historically conscious, in the hope that a new vision of the rapidly evolving international world, its past and its possible future, may emerge.

PREVIOUSLY PUBLISHED IN THIS SERIES

The Battle for International Law
South-North Perspectives on the Decolonization Era
Edited by Jochen von Bernstorff and Philipp Dann

Rewriting the History of the Law of Nations
How James Brown Scott Made Francisco de Vitoria the Founder of International Law
Paolo Amorosa

To Reform the World
International Organizations and the Making of Modern States
Guy Fiti Sinclair

The New Histories of International Criminal Law
Retrials
Edited by Immi Tallgren and Thomas Skouteris

Sovereignty
A Contribution to the Theory of Public and International Law
Hermann Heller, edited and introduced by David Dyzenhaus

Law and the Political Economy of Hunger
Anna Chadwick

The Justification of War and International Order

From Past to Present

Edited by
LOTHAR BROCK AND HENDRIK SIMON

OXFORD
UNIVERSITY PRESS

OXFORD
UNIVERSITY PRESS

Great Clarendon Street, Oxford, OX2 6DP,
United Kingdom

Oxford University Press is a department of the University of Oxford.
It furthers the University's objective of excellence in research, scholarship,
and education by publishing worldwide. Oxford is a registered trade mark of
Oxford University Press in the UK and in certain other countries

First Edition published in 2021
Impression: 5

Published in the United States of America by Oxford University Press
198 Madison Avenue, New York, NY 10016, United States of America

British Library Cataloguing in Publication Data
Data available

Library of Congress Control Number: 2020942234

ISBN 978–0–19–886530–8

DOI: 10.1093/oso/9780198865308.001.0001

Printed and bound by
CPI Group (UK) Ltd, Croydon, CR0 4YY

Preface by the Editors

The idea for this book was born in the summer of 2017. But even before, the observation that the use of force throughout history needs justification and that the justification of war interacts with the construction of international order had become a recurring and increasingly central topic in our research projects at the Peace Research Institute Frankfurt and in our seminars taught at Frankfurt's Goethe-University. Lothar encountered the justification of war as a theoretical endeavour and a political practice in the context of his engagement with the *modern* project of achieving peace through law, which in our view still is to be defended, by being critically examined, against the 'hard facts' of international anarchy which *Realists* of all walks of academic life and life in general like to refer to. At the same time, Hendrik was engaged in writing well-placed articles and a historic study on the 'myth of the "free right to go to war"' (which, of course, sometimes suffered under the workload of producing the present volume). Our discussions led to the idea of engaging with the role of norms in political discourses on the legitimation of violence under a broader geographical, historical, and disciplinary perspective—from past(s) to present(s), as the book's subtitle indicates. For it seemed conspicuous that war has never been and probably will never be waged without recourse to norms, as we and our authors emphasize in this volume. So the contributions of the book follow, in different ways, the assumption that the history of war is also a history of its justification and interacts with what is recognized as international order.

With regard to the relevant literature, we soon realized that, although there is an intense philosophical debate on the justification of war, in particular in connection with the 'just war' tradition, empirical research on the nexus of war justifications and international order in theory and political practice is still rare to date. In this respect, the present volume brings together different approaches and promotes new dialogues across the disciplinary boundaries of the History and Theory of International Law, International Relations, International Political History, Political Theory, and Sociology. Of course, we make no claim to present the final or second to final word in this matter. Hopefully, however, the present volume contributes to the series editors' project of 'bridging the traditional divide between international legal theory, intellectual history, and legal and political history'.

The fact that our project could be implemented is primarily, of course, thanks to our authors, whom we are deeply indebted to for their inspiring contributions which soon turned *our* project into a project of all participating. We owe them great thanks. In addition, a number of individuals have contributed to the genesis of the book, to whom we would like to express our sincere thanks: first of all, Beate Jahn, Anuschka Tischer, and particularly Miloš Vec, who all enthusiastically supported the idea for this book from the very beginning and who helped us with advice and support. Anna Geis, Thilo Marauhn, Gert Krell, and Benno Teschke have critically read the introduction and made important suggestions. The same applies to the anonymous peer reviewers, who, in an exemplary manner, provided a stimulating critique of what we were about to do. The project was presented for the first time at a small panel at the EISA Conference 2018 in Prague, attended by Mustafa Aksakal (who came all the way from Washington!), Chris Brown, Beate Jahn, and Benno Teschke. We are grateful for their encouragement in pursuing the project.

We also would like to thank the Peace Research Institute Frankfurt for providing a stimulating environment and the necessary working space. Special thanks go to Cornelia Hess, who helped us with formatting some of the chapters. Last not least we thank the OUP team, especially Merel Alstein, Jordan Burke, and Jack McNichol, and the Newgen team for their extremely helpful way in which they mentored the publication. The willingness of Nehal Bhuta, Anthony Pagden, and Benjamin Straumann to suggest the project for inclusion in the 'History and Theory of International Law' Series of Oxford University Press, was, of course, decisive to get the project going in the present context.

Finally, a very special thanks goes to Ingrid Krüger from Tübingen, who provided the cover image. The original of the aquarelle is hanging in Lothar's office at the Peace Research Institute Frankfurt. What it tells us as cover of this volume is, of course, in the eye of the beholder. We ourselves felt that, in analogy to a famous book on the history and theory of international law, it may be seen as offering an image of how the apologetic and the utopian side of (international) law coalesce.

Frankfurt, June 2020

Lothar Brock and Hendrik Simon

Series Editors' Preface

The present volume seeks to examine both the historical emergence of the modern international order and its normative foundation. This foundation is excavated via the normative justifications that have been put forward by the historical actors partaking in international conflicts. Lothar Brock and Hendrik Simon, profitably avoiding the pitfalls of scholarly parochialism, have gathered a global cast of contributors. The chapters cover a very wide range of historical periods and geographical locations indeed and range from the early modern period to the current international legal order, discussing a great number of legal and normative arguments and practices from places in Europe, indigenous Spanish America, India, the Ottoman Empire, to China and Russia. All this historical depth and geographical and scholarly range is disciplined by a definitive focus that lies throughout on the justification of war and on the crucial interaction between normative justification and conflict on the ground, between legal scholarship and political practice.

This approach successfully overturns unexamined assumptions that have taken a hold in scholarship and yields often counterintuitive results. It turns out that far from lending support to a *realist* view of international order the investigation of the historical record of political practice often shows a surprisingly strong hold of normative justifications on the imagination and argumentation of political actors, even in contexts such as the nineteenth-century European 'Concert' of Great Powers. Depending on historical and geographical context, norms could at times and under certain conditions exercise even more exogenous pull in political practice than in legal theory. Norms and the conceptual materials they are built from, that is, managed to have, as a matter of historical fact, causal effects on political practice and social reality.

Overall, we can see that the fine-grained analysis of both normative justification and political practice over the long historical term leads to many necessary revisions of longstanding orthodoxies. These revisions cannot be had cheaply, but must be shown in the painstaking way that is under display in the well-integrated scholarship assembled here. The vision emerging from this rich volume cannot be captured in a few paragraphs, but if pressed to name the most salient upshot, one might point to the striking historical datum that the need for justification is never really absent throughout the regions and time periods under scrutiny and that this justificatory need shows 'on the ground' and in practice no less than it does in learned treatises.

29 April 2020

Benjamin Straumann

Table of Contents

PART III THE NINETEENTH CENTURY AS THE BIRTH ERA OF THE MODERN WAR DISCOURSE

PART IV FROM THE LEAGUE TO THE UN: THE UNIVERSE OF WESTERN INTERNATIONAL LEGAL ORDER REVEALING ITS SELF-CONTRADICTIONS

PART V 'DEMOCRATIC WARS' AND THE POST-COLD WAR INTERNATIONAL ORDER: RISE AND DECLINE OF THE 'LIBERAL PEACE'

AN ATTEMPT AT A SYNTHESIS

List of Contributors

Mustafa Aksakal is Associate Professor of History and Nesuhi Ertegün Chair of Modern Turkish Studies at Georgetown University, Washington, DC, where he teaches Ottoman and Middle Eastern History. He is the author of *The Ottoman Road to War in 1914* (2006); 'Holy War Made in Germany? Ottoman Origins of the 1914 Jihad', *War in History* (2011); and 'The Ottoman Empire', in Robert Gerwarth and Erez Manela (eds), *Empires at War. 1911–1923* (2014).

Mikhail Antonov is Professor of Law associated with the Law Faculty at the National Research University 'Higher School of Economics' (Saint Petersburg) and practising member of the Saint Petersburg Bar Association. Among his recent publications are 'History of Russian Law and Its Interpretations', *Review of Central and East European Law* (2020); 'Religion, Sexual Minorities, and the Rule of Law in Russia', *Journal of Law, Religion and State* (2019); 'Legal Realism in Soviet and Russian Jurisprudence', *Review of Central and East European Law* (2018); 'Conservative Philosophy and Doctrine of Sovereignty: A Necessary Connection?', *Archiv für Rechts- und Sozialphilosophie* (2017).

Beate Jahn is Professor of International Relations at the University of Sussex, UK. Her publications include *The Cultural Construction of International Relations* (2000); *Classical Theory in International Relations* (2006); *Liberal Internationalism* (2013); 'Kant, Mill, and Illiberal Legacies in International Affairs', *International Organization* (2005); 'Theorizing the Political Relevance of IR Theory', *International Studies Quarterly* (2017); and 'Liberal Internationalism: Historical Trajectory and Current Prospects', *International Affairs* (2018).

Arnulf Becker Lorca is a researcher at the Pontificia Universidad Católica de Valparaíso, Henry Steiner Visiting Professor at Harvard Law School, and a Lecturer at Brandeis University. His book *Mestizo International Law: A Global Intellectual History, 1842–1933* (2015) was the winner of the 2016 Book Prize of the European Society of International Law.

Lauren Benton is Barton M. Biggs Professor of History at Yale University. Her publications include, with Lisa Ford, *Rage for Order: The British Empire and the Origins of International Law, 1800–1850* (2016); *A Search for Sovereignty: Law and Geography in European Empires, 1400–1900* (2010); and *Law and Colonial Cultures: Legal Regimes in World History, 1400–1900* (2002), which received the World History Association's Bentley Book Prize and the J. Willard Hurst Book Prize of the Law and Society Association.

Lothar Brock is Senior Professor of Political Science at Goethe-University Frankfurt and at the Peace Research Institute Frankfurt. His English publications on the topic of this book include 'The Use of Force in the Post-Cold War Era. From Collective Action back to Pre-Charter Self Defense?', in Michael Bothe, Mary Ellen O'Connell, and Natalino Ronzitti (eds), *Redefining Sovereignty. The Use of Force After the Cold War* (2005); *Democratic Wars. Looking at the Dark Side of Democratic Peace* (2006, co-ed. with Anna Geis and Harald Müller); and 'Between Sovereign Judgement and the International Rule of Law', in Anthony Lang, Jr and Mathias Albert (eds), *The Politics of International Political Theory* (2019).

Chris Brown is Emeritus Professor of International Relations at the LSE and the author of *International Society, Global Politics* (2015); *Practical Judgement in International Political Theory* (2010); *Sovereignty, Rights and Justice* (2002); *International Relations Theory: New Normative Approaches* (1992). He is co-editor (with Terry Nardin and N.J. Rengger) of *International Relations in Political Thought* (2002) and (with Robyn Eckersley) of *The Oxford Handbook of International Political Theory* (2018). His textbook *Understanding International Relations* (2019) is now in its fifth edition.

Manjiao Chi is Professor and Founding Director at the Center for International Economic Law and Policy (CIELP), Law School, University of International Business and Economics (UIBE), China. His research fields cover international law, especially international trade and investment policy and law, dispute settlement, and global governance. He is a founding editor-in-chief of Asian Yearbook of International Economic Law, and author of *Integrating Sustainable Development in International Investment Law: Normative Incompatibility, System Integration and Governance Implications* (2018).

B.S. Chimni is Professor of International Law at School of International Studies, Jawaharlal Nehru University. Among his publications are 'Third World Approaches to International Law & Individual Responsibility in Internal Conflict', 2 *Chinese Journal of International Law* (2003, with Antony Anghie); and *International Law and World Order. A Critique of Contemporary Approaches* (2nd ed., 2017).

Christopher Daase is Professor of International Organizations at Goethe-University Frankfurt and Deputy Director of the Peace Research Institute Frankfurt. Among his publications are *Transformations of Security Studies. Dialogues, Diversity and Discipline* (2015, with Gabi Schlag und Julian Junk); *Clausewitz on Small War* (2015, with James Davis); and *Recognition in International Relations. Rethinking a Political Concept in a Global Context* (2015, with Caroline Fehl, Anna Geis, and Georgios Kolliarakis).

Nicole Deitelhoff is Professor of International Relations and Theories of International Order at Goethe-University Frankfurt and Director of the Peace Research Institute Frankfurt. Among her publications are *Internationalization and the State. Sovereignty as the External Side of Modern Statehood* (with Michael Zürn), in Stephan Leibfried et al. (eds), *The Oxford Handbook of Transformations of the State* (2015) and 'The Discursive Process of Legalization. Charting Islands of Persuasion in the ICC Case', *International Organization* (2009).

Oliver Eberl is Lecturer in Political Theory and the History of Ideas at Leibniz University Hannover. Among his publications are 'Kant on Race and Barbarism: Towards a more complex view on racism and anti-colonialism in Kant', in *Kantian Review* (2019); 'The Metaphysics of International Law. Kant's "Unjust Enemy" and the Limitation of

Self-Authorization', in Sorin Baiasu, Sami Philström, and Howard Williams (eds), *Politics and Metaphysics in Kant* (2011).

Anna Geis is Professor of International Security and Peace Studies at the Institute of International Politics at Helmut-Schmidt-University/University of the Federal Armed Forces in Hamburg (Germany). Among her publications are *Recognition in International Relations. Rethinking a Political Concept in a Global Context* (2015, with Christopher Daase, Caroline Fehl, and Georgios Kolliarakis) and *The Janus Face of Liberal Democracies. Militant 'Forces for Good'* (2013, with Harald Müller and Niklas Schörnig).

Aimee Genell is an Assistant Professor of History at the University of West Georgia. She is the author of 'The Well-defended Domains: Eurocentric International Law and the Making of the Ottoman Office of Legal Counsel', *Journal of Ottoman and Turkish Studies* (2016) and 'Ottoman Autonomous Provinces and the Problem of "Semi-Sovereignty" in International Law', *Journal of Balkan and Near Eastern Studies* (2016). She is completing her manuscript, *Empire by Law: The Ottoman Origins of the Mandates System in the Middle East.*

Sohail H. Hashmi is Professor of International Relations on the Alumnae Foundation and Professor of Politics at Mount Holyoke College, South Hadley, Massachusetts. He is the editor of *Just Wars, Holy Wars, and Jihads: Christian, Jewish, and Muslim Encounters and Exchanges* (2012).

Axel Heck is Senior Lecturer in International Relations at the Institute of Political Science at the University of Kiel. He has published respectively in the *European Journal of International Relations*, *International Studies Perspectives*, and the *Zeitschrift für Internationale Beziehungen.*

Thomas Hippler is Professor of Modern and Contemporary History at University of Normandy in Caen. His publications include *Citizens, Soldiers, and National Armies: Military Service in France and Germany, 1789–1830* (2007); *Bombing the People: Giulio Douhet and the Foundations of Air-Power Strategy, 1884–1939* (2013); and *Governing from the Skies: A Global History of Aerial Bombing* (2017).

Isabel V. Hull is the John Stambaugh Professor of History (retired) at Cornell University. A German historian, she is the author of *The Entourage of Kaiser Wilhelm II* (1982); *Sexuality, State and Civil Society in Germany, 1700–1815* (1996); *Absolute Destruction: Military Culture and the Practices of War in Imperial Germany* (2004); and most recently, *A Scrap of Paper: Breaking and Making International Law in the First World War* (2014), which was awarded the certificate of merit from the American Society of International Law.

Anthony F. Lang, Jr is a Professor of International Political Theory at the University of St Andrews, where he has been since 2004. He writes on global constitutionalism, the just war tradition, and international political theory more widely. He has published the books *Agency and Ethics: The Politics of Military Intervention* (2002); *Punishment, Justice and International Relations: Ethics and Order after the Cold War* (2008); and *International Political Theory: An Introduction* (2014), along with eight edited volumes and numerous articles and chapters.

Felix Lange is a research fellow at the Kollegforschungsgruppe 'The International Rule of Law—Rise or Decline?'. He published on the history of international law, inter alia, in the European Journal of International Law and the Heidelberg Journal of International Law. Among his publications on the topic of the book are 'The Multifaceted Emergence of the Ius Cogens-Doctrine—A Multicausal Historical Account', *Leiden Journal of International Law* (2018) and 'The Dream of a Völkisch Colonial Empire—International Law and Colonial Law during the National Socialist Era', *London Review of International Law* (2017).

Siddharth Mallavarapu is Professor at the Department of International Relations and Governance Studies, School of Humanities and Social Sciences at Shiv Nadar University in India. He has been featured on Theory Talks and E-International Relations. He is the author of *Banning the Bomb: The Politics of Norm Creation* (2007); *International Relations in India: Bringing Theory Back Home* (2005, with Kanti P. Bajpai); and *International Relations: Perspectives for the Global South* (2012, with B.S. Chimni). His most recent contribution is forthcoming in a special issue of the journal *Global Constitutionalism* (2020).

Thilo Marauhn is Professor of Public and International Law, Justus Liebig University Gießen, and Head of Research Group Public International Law, Peace Research Institute Frankfurt (PRIF). He serves as President of the International Humanitarian Fact-Finding Commission. Among his most recent publications is 'The International Rule of Law in Light of Legitimacy Claims', in Heike Krieger, Georg Nolte, and Andreas Zimmermann (eds), *The International Rule of Law. Rise or Decline?* (2019).

Paul Robinson is a Professor in the Graduate School of Public and International Affairs at the University of Ottawa. He has also served as an officer in both the British and the Canadian armies. His publications include *The White Russian Army in Exile* (2002); *Just War in Comparative Perspective* (2003); *Ethics Education in the Military* (2017, with Nigel De Lee and Don Carrick); and *Russian Conservatism* (2019).

Gabi Schlag is Senior Lecturer at the Eberhard Karls-University Tübingen, Germany. She is the co-editor of *Transformations of Security Studies: Dialogues, Diversity and Discipline* (2016) and the special issue *Visualizing Violence: Aesthetics and Ethics in International Politics* (Global Discourse, 2017). Results of her research on visual global politics are published in *Media, War & Conflict*, *Critical Studies on Terrorism*, and *European Journal of International Relations*.

Hendrik Simon is Research Associate at the Peace Research Institute Frankfurt and Lecturer at Goethe-University Frankfurt. His English publications on the topic of this volume include 'The Myth of *Liberum Ius ad Bellum*. Justifying War in 19th-century Legal Theory and Political Practice', *European Journal of International Law* (2018) and 'Theorising Order in the Shadow of War. The Politics of International Legal Knowledge and the Justification of Force in Modernity', *Journal of the History of International Law, Special Issue: Politics and the Histories of International Law* (2020).

Michael Stohl, Professor of Communication, Political Science and Global Studies, University of California, Santa Barbara. He is the author, co-author, editor, or co-editor of

16 books and more than 100 scholarly journal articles and book chapters. His most recent volume is *Constructions of Terrorism* (2017, co-edited with R. Burchill and S. Englund).

Benno Teschke is a Professor in the Department of International Relations and former Director of the Centre for Advanced International Theory (2016–2019) at the University of Sussex. He is the author of *The Myth of 1648: Class, Geopolitics and the Making of Modern International Relations* (2003), awarded the 2004 Isaac and Tamara Deutscher Memorial Prize. His recent publications include a debate with Gopal Balakrishnan in the *New Left Review* on Marxism and Carl Schmitt and 'Carl Schmitt's Concepts of War: A Categorical Failure', in Jens Meierhenrich and Oliver Simons (eds.), *The Oxford Handbook of Carl Schmitt* (2016).

Anuschka Tischer is a Full Professor in Early Modern History at the University of Würzburg (Germany). Among her publications are books on the French Diplomacy at the Congress of Westphalia (1999), on Justifications of War in Early Modern Europe (2012), and (together with Derek Croxton) a Dictionary on the Peace of Westphalia (2002).

Miloš Vec is Professor for European Legal and Constitutional History at the University of Vienna and Permanent Fellow at the Institute for Human Sciences (IWM). Among his publications on the history of international law are *The Transformation of Foreign Policy: Drawing and Managing Boundaries* (2016, with Andreas Fahrmeir and Gunther Hellmann); *Paradoxes of Peace in 19th Century Europe* (2015, with Thomas Hippler); and 'From the Congress of Vienna to the Paris Peace Treaties of 1919', in Bardo Fassbender and Anne Peters (eds), *The Oxford Handbook of the History of International Law* (2012).

Wolfgang Wagner is Professor of International Security at the Department of Political Science and Public Administration, Vrije Universiteit Amsterdam. Relevant publications include *The Democratic Politics of Military Interventions. Political Parties, Contestation, and Decisions to Use Force Abroad* (2020) and 'War and Punitivity under Anarchy', *European Journal of International Security* (2018, with Wouter Werner).

Nina Wilén is Director for the Africa Programme at the Egmont Institute for International Relations and Assistant Professor at the Department of Political Science at Lund University. Dr Wilén's areas of expertise include peacekeeping, peacebuilding, security sector reform, and gender and the military. She has published extensively on these topics and is also the author of *Justifying Interventions in Africa: (De)Stabilizing Sovereignty in Liberia, Burundi and the Congo* (2012).

INTRODUCTION

1
The Justification of War and International Order: From Past to Present

Hendrik Simon and Lothar Brock

1. A Genealogical Approach: The Justification of War and the Historical Evolution of International Order

This book departs from a simple but momentous observation: the history of war is also a history of its justification. The use of force in international relations has always been accompanied by political, scholarly, and public discourses on its appropriateness. This is to say that the justification of the use of force is tied inextricably to its contestation because there would be no need for justifying the use of force if the latter were not regarded as basically problematic. Accordingly, the justification of certain wars entails a critique of war in general. We understand the unity of the justification and the critique of the use of force both in political practice and academic theory-building as constitutive for the emergence of international order.[1]

Since the justification and the critique of the use of force involve normative judgement, international order rests on a paradoxical, perhaps even dialectical[2] relationship between war and normativity: war challenges and drives the formation of international order as an 'order of justification'.[3] Therefore, the history of the modern international order first and foremost can be told as a genealogy of endeavours to facilitate the use of force *and* to hedge it. This way the book addresses the interaction between the justification of specific wars and the formation of international order as offering a frame of reference for the justification and critique of war as such.

As we pursue this issue, we proceed on the assumption that contrary to *realist* claims in International Law (IL), International Relations (IR), or International History, the justification of war rarely happens simply as empty 'propaganda'.[4] Throughout history even powerful actors tend to refer to the language of normativity to justify their forceful acts. By doing so, they (voluntarily or involuntarily) contribute to the emergence of

[1] R. Tuck, *The Rights of War and Peace: Political Thought and the International Order from Grotius to Kant* (1999); S.C. Neff, *War and the Law of Nations: A General History* (2005); O'Connell, 'Peace and War', in B. Fassbender and A. Peters (eds), *The Oxford Handbook of the History of International Law* (2nd ed., 2014) 272; Lesaffer, 'Too Much History: From War as Sanction to the Sanctioning of War', in M. Weller (ed.), *The Oxford Handbook of the Use of Force in International Law* (2015) 35.

[2] Benjamin, 'Critique of Violence', reprinted in P. Demetz (ed.), *Walter Benjamin, Reflections: Essays, Aphorisms, Autobiographical Writings* (1978) 277; M. Foucault, *Society Must Be Defended: Lectures at the Collège de France 1975–1976*, ed. by A.I. Davidson (2003); H. Joas and W. Knöbl, *War in Social Thought: Hobbes to the Present* (2013); J. Bartelson, *War in International Thought* (2018).

[3] R. Forst, *Normativity and Power. Analyzing Social Orders of Justification* (2017).

[4] W.G. Grewe, *The Epochs of International Law*, translated and revised by M. Byers (2000), at 531.

Hendrik Simon and Lothar Brock, *The Justification of War and International Order: From Past to Present* In: *The Justification of War and International Order*. Edited by: Lothar Brock and Hendrik Simon, Oxford University Press (2021).
 DOI: 10.1093/oso/9780198865308.003.0001

'communication communities' (*Kommunikationsgemeinschaften*).[5] These communication communities are based on and shape a common understanding of what war in general is about and when it is justified. Since the early modern period such communities constitute an increasingly informed audience towards which justifications of the use of force are directed.[6] The resulting public sphere has been and continues to be a major focus of dealing with the use of force. While the structure of the public sphere was transformed (from dynastic to inter-national discourse) and the technique of its representation changed (from written to printed war manifestos to mass media), the general discursive principles of justifying war remained the same. Thus, discourses on the use of force construct 'international order' as a normative frame of reference for politics and theory alike.[7] As such a frame of reference, international order is to be understood as a 'normative order', in which norms function as an instrument of politics (including the politics of theory-building) and at the same time structure the practice of justifying and practising the use of force.[8]

Our fundamental thesis is that in their justifications of war, states and other political actors refer to (existing or presumed) norms of the international order to depict their own violence as legitimate, that is 'appropriate behaviour'.[9] According to this practice, a state's effort to justify violence is to be interpreted as an expression of this state's awareness of the fact that the use of force may damage not only its war opponent, but also his own standing in the normative order constituting the existing communication community.[10] With this, the discourse of justifying war interacts with the international order. As norms of the international order shape the justification practices of states, the practice of justification in specific cases shapes the general normative order. Accordingly, we claim that the communicative practices of justifying war and international order- building are to be understood as co-constitutive.

This discursive co-constitutionalization of the justification of war and international order is of course not to be understood as static. We can only understand it

[5] A. Tischer, *Offizielle Kriegsbegründungen in der Frühen Neuzeit. Herrscherkommunikation in Europa zwischen Souveränität und korporativem Selbstverständnis* (2012), at 22, 220 f.; see also the contribution by Anuschka Tischer in this volume. For the concept of '*Kommunikationsgemeinschaften*' from a philosophical perspective of discourse ethics see K.-O. Apel, *Transformation der Philosophie—Band II: Das Apriori der Kommunikationsgemeinschaft* (1999).

[6] See also Tischer in this volume.

[7] H. Bull, *The Anarchical Society. A Study of Order in World Politics* (2nd ed., 1995), at 180–83; Neff, above note 1; C. Peevers, *The Politics of Justifying Force: the Suez Crisis, the Iraq War, and International Law* (2013); Brock and Simon, 'Die Selbstbehauptung und Selbstgefährdung des Friedens als Herrschaft des Rechts. Eine endlose Karussellfahrt? (The Self-Assertion and Self-Destruction of Peace as Rule of Law. Riding a Merry-Go-Round?)', 59 *Politische Vierteljahresschrift* (2018) 269 ; H. Simon, *Der Mythos vom 'freien Recht zum Krieg'. Zu einer Genealogie der modernen Kriegslegitimation* (forthcoming).

[8] R. Forst and K. Günther (eds), *Die Herausbildung normativer Ordnungen: Interdisziplinäre Perspektiven* (2011); see also Forst, above note 3.

[9] Finnemore and Sikkink, 'International Norm Dynamics and Political Change', 52 *International Organization* (1998) 887, at 891; Jepperson, Wendt, and Katzenstein, 'Norms, Identity, and Culture in National Security', in P.J. Katzenstein (ed.), *The Culture of National Security: Norms and Identity in World Politics* (1995), at 54.

[10] On norms in International Relations, see ibid., and F. Kratochwil, *Rules, Norms and Decisions, On the Conditions of Practical and Legal Reasoning in International Relations and Domestic Society* (1991); Wendt, 'Anarchy Is What States Make of It: The Social Construction of Power Politics', 46 *International Organization* (1992) 391; Checkel, 'The Constructive Turn in International Relations Theory', 50 *World Politics* 2 (1998) 324; A. Wiener, *The Invisible Constitution of Politics. Contested Norms* and International Encounters (2008).

scientifically in its historical context. Due to the ambiguity of norms as political instruments and determining factors structuring political discourse, international order is always contested and thus changes over time. The conceptualization of this change depends on the basic theoretical or ideological preferences of the observer. From a *realist* point of view, the change of international orders understood as orders of the justification of the use of force can be expected to be confined to the *form* of international politics while its *substance* as politics under anarchy persists.[11] From a rationalist-institutionalist viewpoint, normative change responds to the necessities of reducing transaction costs in a world of ever-increasing complexity.[12]

Critical scholars rather emphasize the ideological bias accompanying both these approaches. While the materialists within their ranks may address normative change as an issue of re-arranging the political economy of international relations (see Benno Teschke in this volume),[13] those building on basic ideas of the European Enlightenment on the one hand,[14] its postmodern and postcolonial critics on the other,[15] would rather look for the emergence of an international rule of law as an international order[16] reducing but also reproducing the arbitrary use of force. Under this third perspective, international law functions as a 'gentle civilizer of nations',[17] *and* at

[11] C. von Clausewitz, *On War* (1832, reprinted 1976), at 605; Lueder, 'Krieg und Kriegsrecht im Allgemeinen', in F. von Holtzendorff (ed.), *Handbuch des Völkerrechts* (1889) 169; Schmitt, 'The Turn to the Discriminating Concept of War' (1938), in C. Schmitt, *Writings on War*, trans. by T. Nunan (2011), 30; Morgenthau, 'Positivism, Functionalism, and International Law', 34 *American Journal of International Law* (1940) 260; K.N. Waltz, *Theory of International Politics* (1979); W.G. Grewe, *Friede Durch Recht?* (1985); see also Jütersonke, 'Realist Approaches to International Law', in A. Orford and F. Hoffmann (eds), *The Oxford Handbook of the Theory of International Law* (2016) 327.

[12] R.O. Keohane, *Complexity* (1980); Ikenberry, 'The End of Liberal International Order?', 94 *International Affairs* (2018) 7.

[13] R.W. Cox with T.J. Sinclair, *Approaches to World Order* (1999); Knox, 'Marxist Approaches to International Law', in Orford and Hoffmann, above note 11, 306; B.S. Chimni, *International Law and World Order. A Critique of Contemporary Approaches* (2nd ed., 2017), at 440; see also the contributions by Mallavarapu, Teschke, and Chimni in this volume.

[14] I. Kant, *Project for a Perpetual Peace: A Philosophical Essay* (1795, reprinted 1891), at 99; H. Wehberg, *Die Aechtung des Krieges. Eine Vorlesung an der Haager Völkerrechtsakademie und am 'Institut Universitaire de Hautes Etudes Internationales'(Genf)* (1930); H. Kelsen, *Peace through Law* (1944); Habermas, 'Kant's Idea of Perpetual Peace, with the Benefit of Two Hundred Years' Hindsight', in J. Bohman and M. Lutz-Bachmann (eds), *Perpetual Peace: Essays on Kant's Cosmopolitan Ideal* (1997) 113; H. Williams, *Kant and the End of War: A Critique of Just War Theory* (2012); see also the contribution by Oliver Eberl in this volume.

[15] Benjamin, above note 2; Foucault, above note 2; Derrida, 'Force de loi: Le "fondement mystique de l'autorité"/Force of Law: The "Mytical Foundation of Authority"', 11 *Cardozo Law Review* (1990) 919; G. Agamben, *Homo Sacer: Sovereign Power and Bare Life* (1998); A. Anghie, *Imperialism, Sovereignty and the Making of International Law* (2004); Pahuja, 'The Postcoloniality of International Law', 46 *Harvard International Law Journal* (2006) 459; C. Menke, *Recht und Gewalt* (2012); D. Loick, *Juridismus. Konturen einer kritischen Theorie des Rechts* (2017); see also the contributions by Mallavarapu, Becker Lorca, Benton, Chimni, Hippler, and Lange in this volume.

[16] Hobbes developed the idea that the domestic state of nature could be overcome through the legalization of the relationship between the ruler and the ruled. Kant took up this idea but realized that the Hobbesian approach could not simply be projected onto the system of states. His contribution to establishing a critical tradition in international law consists in taking up the need for legalization of international relations but linking it to domestic reform and the universal rights of people. Legalization in this understanding does not result in offending all war but in approaching peace as a piecemeal process. See also Habermas, above note 14, and Eberl in this volume.

[17] M. Koskenniemi, *The Gentle Civilizer of Nations: The Rise and Fall of International Law. 1870–1960* (2002).

the same time as an instrument of power and domination.[18] The question of how the enabling and the restraining functions of international law interact in changing historical contexts and how this impacts on international order is the central issue of the present volume.

As to this issue, today everything seems to be in flux. The cosmopolitan version of international order which relies on the idea of progressive (peace-enhancing) legalization has come under increasing pressure with war and violence nourishing the appetite of politicians and ordinary people (including academics) alike for reinforcing existing borders or reconstructing them in order to protect themselves against the vicissitudes of a world seemingly falling apart. But the world has always been falling apart for someone, even in 'good times'. Whether the rule of law is in rise or decline cannot be decided at this point of time. Perhaps it does not have to be decided either.[19] In any case we should be careful not to declare its premature or long-awaited death (see also the contribution by Thilo Marauhn).[20] If 'words are politics' as Martti Koskenniemi reminds us,[21] it would be all the more important to understand theorizing on adequate normative standards as practice of shaping them in specific historical contexts.

There is no better time for analysing the importance of law as a normative force than these troubled years. In terms of *longue durée*, reference to positive international law as a normative frame of reference for addressing the use of force has spread out particularly since the mid-nineteenth century when international humanitarian law became canonized.[22] It acquired a concrete form when the age-old critique of the arbitrary use of force finally[23] resulted in its formal prohibition first through the Kellogg-Briand Pact and later through the UN Charter and respective regional arrangements.[24] The fact that this happened between and after two world wars and certainly did *not* result in the abolishment of war, makes it difficult to determine what really has changed.[25]

[18] I. Hurd, *How to Do Things with International Law* (2017); Brock and Simon, above note 7; Reus-Smit, 'The Politics of International Law', in C. Reus-Smit (ed.), *The Politics of International Law* (2004) 14.

[19] Brock, 'The Use of Force by Democracies in the Post-Cold War Era. From Collective Action Back to Pre-Charter Self Defense?', in M. Bothe, M.E. O'Connell, and N. Ronzitti (eds), *Redefining Sovereignty. The Use of Force after the End of the Cold War: New Options, Lawful and Legitimate?* (2005) 21; see also Krieger and Nolte, 'The International Rule of Law—Rise or Decline? Points of Departure', 1 *KFG Working Paper Series* (2016); for a legal historical perspective, see Koskenniemi, 'Imagining the Rule of Law: Rereading the Grotian "Tradition"', 30 *European Journal of International Law* (2019) 17; see also Marauhn's chapter and our concluding chapter.

[20] Franck, 'Who Killed Article 2(4)? or: Changing Norms Governing the Use of Force by States', 64 *American Journal of International Law* (1970), 809; see also Marauhn's chapter in this volume.

[21] Koskenniemi, 'Miserable Comforters: International Relations as New Natural Law', 15 *European Journal of International Relations* (2009) 395, at 395.

[22] M. Abbenhuis, *The Hague Conferences and International Politics. 1898–1915* (2018); M. Abbenhuis, C.E. Barber, and A.R. Higgins (eds), *War, Peace and International Order? The Legacies of the Hague Conferences 1899–1907* (2017); K. von Lingen, *Crimes against Humanity. Eine Ideengeschichte der Zivilisierung von Kriegsgewalt. 1864–1945* (2018).

[23] For its historical roots, see the contributions by Tischer, Simon, and Hull in this volume.

[24] O.A. Hathaway and S.J. Shapiro, *The Internationalists: How a Radical Plan to Outlaw War Remade the World* (2017); M.M. Payk, *Frieden durch Recht? Der Aufstieg des modernen Völkerrechts und der Friedensschluss nach dem Ersten Weltkrieg* (2018).

[25] See B.S. Chimni's critique in his contribution on 'Peace through Law' (chapter 13) in this volume. See also our concluding chapter.

All heroic endeavours to reach a breakthrough towards the establishment of an international rule of law have produced fragile results. Argumentative reference to positive law is being challenged (increasingly, it seems) not only by power-politically inspired practitioners and academics, but also by narratives of other extra-legal normative spheres (like ethics, morality, honour in connection with religion, social custom, military necessity, technological standards etc.). Thus, war discourses may not only be seen as the sum of disputes over 'justifying reasons',[26] but also as the sum of disputes between different narratives and traditions of justification.[27] The simultaneity of different normative frameworks of reference may be summarized as 'multi-normativity'—a descriptive concept which helps to underline the complexity of the justification of war.[28] The crucial question is how these different normative spheres relate to each other. They may partly match, partly co-exist, and partly collide.

Ideally, the existence of multiple normativities would provide more flexibility in dealing with war and conflict. But this is not necessarily the case in practice as the debate on 'humanitarian intervention' has shown (see also the contribution by Chimni in Part VII). For instance, instead of bridging the gap between the commitments of the international community to protect people from mass atrocities (R2P) on the one hand, the limited ability of the UN Security Council to act on the other, reference to the just-war tradition may provide a shortcut to action but opens up new arenas of conflict as it pits morality against procedural rules of the UN Charter. Likewise, efforts to redefine sovereignty in order to provide a new legal basis for the international protection of people from mass atrocities may result in a deepening rift between considerations of legality and legitimacy of the use of force thus impeding any progress towards hedging the arbitrary use of force.[29]

So would it be helpful, as Chris Brown suggests in his contribution, to return to the just-war tradition all together and be done with the analytical focus on an international rule of law? Or should we stick to the latter in order to be able to navigate the tension between the legality and the morality of the use of force?[30] We leave this question open at this point but will return to it in our conclusions to the present volume.

[26] Forst and Günther, above note 8, at 11; Forst, above note 3.

[27] The resulting complexity is addressed by von Bernstorff, 'The Use of Force in International Law before World War I. On Imperial Ordering and the Ontology of the Nation-State', 29 *European Journal of International Law* (2018) 233, at 260; see also D. Kennedy, *The Dark Sides of Virtue: Reassessing International Humanitarianism* (2005).

[28] Vec, 'Multinormativität in der Rechtsgeschichte', in Berlin-Brandenburgische Akademie der Wissenschaften (vormals Preußische Akademie der Wissenschaften) (ed.), *Jahrbuch 2008* (2009) 155, at 162–65; M. Stolleis, *Geschichte des öffentlichen Rechts in Deutschland. Staats—und Verwaltungsrechtswissenschaft in West und Ost 1945–1990* (2017), at 684, 696; Duve, 'Was ist Multinormativität?—Einführende Bemerkungen', 25 *Rechtsgeschichte—Legal History Rg* (2017) 88; see also the contribution by Miloš Vec to this volume.

[29] Brock, 'The Politics of International Law. On the Transformation of Violence into Governance', *ISA Working Paper* (2005), online: <www.peacepalacelibrary.nl/ebooks/files/37552505X.pdf> (last visited 10 January 2020); Bothe et al., above note 19; see also our concluding chapter in this book.

[30] See also Habermas, 'Bestiality and Humanity: A War on the Border between Legality and Morality', 6 *Constellations* (1999) 263.

2. 'Mind the Gap!'—Navigating between Theoretical and Practical Discourses on the Legitimacy of War and International Order

The preceding observations call attention to the fact that we have an insufficient understanding of the role of international norms in past and present discourses on war justifications. From our point of view, this has something to do with the fact that, in comparison to dogmatic history, there is a lack of historical studies which examine the political practice of the justification of war and which address the linkages between justifying war and modelling international order. For instance, there are countless monographs, anthologies, and articles on the just-war tradition.[31] But most of the writing concentrates on theoretical discourse.[32]

We posit that there is a need to confront theoretical considerations with the actual practice of justifying war. Trying to do this is not to degrade theory-driven observations to second place but rather to achieve a better understanding of how scholarly discourse and political decision-making interact and relate to each other. From early modern (European) discourses on international law,[33] over the professionalization of international legal scholarship in the late nineteenth century[34] to the present,[35] there has been a special discursive connection between political practice and legal scholarship in the highly political field of justifying and hedging war.[36] In retrospect, this relationship manifested itself in the emergence of a body of customary law.[37] As the *opinio iuris* of a state might be hard to identify without full insights into its motives to

[31] M. Farrell, *Modern Just War Theory: A Guide to Research* (2013).

[32] Tuck, above note 1; F. Allhoff, N.G. Evans, and A. Henschke (eds), *Routledge Handbook of Ethics and War. Just War Theory in the 21st Century* (2013); A. Lang, Jr., C. O'Driscoll, and J. Williams (eds), *Just War: Authority, Tradition, and Practice* (2013); S. Lazar and H. Frowe (eds), *The Oxford Handbook of Ethics of War* (2015); D. Brunstetter and C. O'Driscoll (eds), *Just War Thinkers from Cicero to the 21st Century* (2018).

[33] See B. Kingsbury and B. Straumann (eds), *The Roman Foundations of the Law of Nations. Alberico Gentili and the Justice of Empire* (2011); S. Kadelbach, T. Kleinlein, and D. Roth-Isigkeit (eds), *System, Order, and International Law. The Early History of International Legal Thought from Machiavelli to Hegel* (2017); and the contributions by Tischer and Becker Lorca to this volume.

[34] Koskenniemi, above note 17; L. Nuzzo and M. Vec (eds), *Constructing International Law. The Birth of a Discipline* (2012); Payk, above note 24; see also Simon and Vec in this volume.

[35] See von Bernstorff, 'International Legal Scholarship as a Cooling Medium in International Law and Politics', 25 *European Journal of International Law* (2015) 977; Peters, 'International Legal Scholarship Under Challenge', in A. Nollkaemper, J. d'Aspremont, T. Gazzini, and W. Werner (eds), *International Law as a Profession* (2017) 117.

[36] On the politicization of law(yers) in modern war discourse see Simon, 'Theorising Order in the Shadow of War. The Politics of International Legal Knowledge and the Justification of Force in Modernity', 22 *Journal of the History of International Law* (2020) 218; see also Alston, 'The Myopia of the Handmaidens: International Law and Globalization', 8 *European Journal of International Law* (1997) 435; and Peters, above note 35.

[37] See also von Bernstorff, above note 27, at 240; Vec, 'Sources in the 19th Century European Tradition. The Myth of Positivism', in S. Besson and J. d'Aspremont (eds), *Oxford Handbook on the Sources of International Law* (2017) 121; on the history and theory of customary international law see Kelly, 'Customary International Law in Historical Context. The Exercise of Power without General Acceptance', in B.D. Lepard (ed.), *Reexamining Customary International Law* (2017) 47; and Chimni 'Customary International Law: A Third World Perspective', 112 *American Journal of International Law* (2018) 1; see also Hull and Vec in this volume.

go to war,[38] 'state practice' is always in need of interpretation. Legal scholars provide such interpretation by (re-)constructing 'practice'[39] and thereby formulate norms on the legitimacy of war in conjunction with state practice. With this, lawyers (inside and outside of courts) contribute to the shaping of what is accepted by the respective political publics as international law. Historical and critical reconstruction of the respective scholarly discourses (*Wissenschaftsgeschichte*)[40] rightly represents a core issue of the History and Theory of International Law.

However, it is analytically problematic to confine oneself to the observation of scholarly discourse.[41] A rather obvious objection here is that those contemporary legal scholars, who for good reasons kept a distance to political practitioners, might have misinterpreted a state's *opinio iuris* due to the lack of political insider knowledge.[42] Another objection which goes into the opposite direction is that legal scholars, who made themselves available to the service of states—only think of Grotius and his commissioned works for the Dutch East India Company[43]—might have been partial, or they may even have played 'an *instrumental* role in developing and cementing new justifications for the use of force'.[44] These lawyers would not correspond to Kelsen's ideal type of an 'objective lawyer',[45] but rather to his real type of a mouthpiece for the powerful by presenting political interests 'as what is objectively right'.[46]

What is striking on a methodological basis, is that international scholarship, while providing a set of questions and issues to be engaged with when states use force, has been for long dominated by *a-historical* theoretical writings.[47] By analysing the work of Hugo Grotius, Michael Walzer, and Cécile Fabre from a historical perspective, Anthony Lang, Jr concludes (in his chapter to the present volume), that the just-war tradition, in order to be a truly political instead of a moral theory, must take into account the historical narratives that structure it. Much in this spirit, German historian Konrad Repgen (1923–2017) has noted already in his seminal article on *Kriegslegitimationen in Alteuropa* published in 1985, that historical *exempla* and *dicta* in works of early modern scholars offered largely 'ornamentations' but not historical foundations for their argument.[48] But, as Stephen C. Neff has put it convincingly,

[38] C. Parry, *The Sources and Evidences of International Law* (1965), at 67 f.; see also Carty, 'Doctrine versus State Practice', in B. Fassbender and A. Peters, above note 1, 972, at 982f.

[39] Carty, above note 38, at 974.

[40] M. Stolleis, *Rechtsgeschichte schreiben. Rekonstruktion, Erzählung, Fiktion?* (2008); Hueck, 'The Discipline of the History of International Law. New Trends and Methods on the History of International Law', 3 *Journal of the History of International Law* (2001) 194.

[41] Anuschka Tischer and Isabel V. Hull in their contributions.

[42] Parry, above note 38; Anthony Carty even goes so far as to call legal scholars' constructions of political practice 'unscientific', Carty, above note 38, at 995.

[43] van Ittersum, 'The Long Goodbye: Hugo Grotius' Justification of Dutch Expansion Overseas, 1615–1645', 36 *History of European Ideas* (2010) 386; Hathaway and Shapiro, above note 24, part I; see also the contributions by Lang, Jr, Mallavarapu, and Tischer in this volume.

[44] von Bernstorff, above note 27, at 260. Emphasis added.

[45] Kelsen, 'Science and Politics', 45 *The American Political Science Review* (1951) 641, at 641.

[46] Kelsen, 'Juristischer Formalismus und Reine Rechtslehre', 58 *Juristische Wochenschrift* (1929), at 1723; quoted from von Bernstorff, above note 27, at 977; see also Simon, above note 36.

[47] See also Carty, above note 38, at 996.

[48] '*Exempla und dicta sind bei Gelehrten der frühen Neuzeit weitgehend Ornamentik, aber nicht tragende Mauer oder Fundament*', see Repgen, 'Kriegslegitimationen in Alteuropa. Entwurf einer historischen Typologie', in 9 *Schriften des Historischen Kollegs. Vorträge* (1985), at 15.

'history is no subject for purists'.[49] Theoretical arguments need historical and political contextualization as much as historical and political discourses require theory-driven interpretations.[50] Thus, both legal and political discourses have to be addressed. In this respect, the historical approaches to normative war discourses and international order have so far been mainly shaped by studies in legal theory.

This argument is not new: the lack of studies on the politics of justifying war[51] and on the linkages between theory and practice was criticized at an early stage, particularly in research on the history of international law and international relations.[52] But efforts to fill this gap are still rare.[53] In his article of 1985, Repgen referred to the phenomenon that since the thirteenth century political actors were under pressure to justify war which resulted in numerous war manifestos.[54] But to his surprise, Repgen found that nobody, neither at home nor abroad, had engaged with this material.[55] As Repgen himself speculated, the lack of interest in official pronouncements may have been due to the ruling scientific discourses at the time of his writing: *Realism* was dominant not only in the History of International Relations and IR, but also in the History of International Law. Wilhelm Grewe's standard book on the history of international law as a history of shifting hegemonies,[56] in which Grewe, as already quoted, spoke of the justification of war as empty 'propaganda', was first published in 1984 and thus almost simultaneously with Repgen's article of 1985 (originally a lecture held in 1984). The *realist* 'empty propaganda' approach trumped Repgen's plea for taking political declarations of war more seriously for a long time.

[49] Neff, above note 1, at 164.

[50] R. Koselleck, *Vergangene Zukunft: Zur Semantik geschichtlicher Zeiten* (10th ed., 2017).

[51] Some books offer interesting insights into the practice of domestic political propaganda and media discourse, see J.A. Kuypers, *Bush's War: Media Bias and Justifications for War in a Terrorist Age* (2005); D. Welch and J. Fox (eds), *Justifying War. Propaganda, Politics and the Modern Age* (2012). Research on the justification of the 'war on terror', 'democratic warfare', and 'humanitarian interventions', with a relatively new focus on the 'Responsibility to Protect', comes closer to addressing the linkage between justification and international order-thinking, see also A. Bellamy and T. Dunne (eds), *The Oxford Handbook of the Responsibility to Protect* (2016). However, the respective studies mostly concentrate on the past thirty years and fall short of addressing the theory and practice of justifying force as historical processes of constructing international order.

[52] W. Preiser, *Die Völkerrechtsgeschichte, ihre Aufgaben und ihre Methode* (1964); W. Janssen, *Die Anfänge des modernen Völkerrechts und der neuzeitlichen Diplomatie: Ein Forschungsbericht* (1965); Repgen, above note 48; Tischer, above note 5; Fassbender and Peters, 'Introduction: Towards a Global History of International Law', in Fassbender and Peters, above note 1, 1; Simon, 'The Myth of *Liberum Ius ad Bellum*. Justifying War in 19th-Century Legal Theory and Political Practice', in 29 *European Journal of International Law* (2018) 113. See also on the 'practice turn' in International Relations C. Bueger and F. Gadinger, *International Practice Theory* (2018).

[53] Lesaffer, 'Defensive Warfare, Prevention and Hegemony: The Justifications for the Franco-Spanish War of 1635', 8 *Journal of the History of International Law* (2006) 91; A. Geis, L. Brock, and H. Müller (eds), *Democratic Wars. Looking at the Dark Side of Democratic Peace* (2006); Tischer, above note 5; N. Rengger, *Just War and International Order. The Uncivil Condition in World Politics* (2013); I.V. Hull, *A Scrap of Paper. Making and Breaking International Law during the Great War* (2014); E.J. Kolla, *Sovereignty, International Law and the French Revolution* (2017); Peevers, above note 7; Simon, above note 52; Hathaway et al., 'War Manifestos', 85 *University of Chicago Law Review* (2018) 1139.

[54] Repgen, above note 48, at 17.

[55] Ibid., at 18.

[56] Grewe, above note 4; Fassbender, 'Stories of War and Peace: On Writing the History of International Law in the "Third Reich" and After', 13 *European Journal of International Law* (2002) 479, at 511ff.

A dialogue between Repgen and a US-American colleague highlights what, in Repgen's view, was (and still is in our view) at stake: as he was searching the files of the National Library in Munich week after week in the winter of 1984, Repgen told the colleague how excited he was about war declarations as objects of historical studies. His colleague responded, 'with the sympathetic, sad look of a neurologist': 'What on earth do you want to do with these texts?' Repgen replied: 'Read them.' His colleague: 'But these texts contain nothing but lies.' Repgen: 'For exactly that reason I want to study them.' He then went on:

> These texts amount to pleas which supposedly use a number of recurrent arguments for the justification of the use of force. If this proves to be correct then we should be able to generate a fairly complete compilation of these arguments. This way we could advance our understanding of the onset of wars in Europe from the fall of Constantinople to the French Revolution. Nobody can force us to continue to look at international relations as an issue of the balance of power, ... which has been the dominant approach for the past 250 years. Presumably we would learn more by looking into the specific reasons given by the war-parties for the legitimacy of their use of force. We would thus be able to construct a pattern which would enable us to discover and describe the particular as it relates to the general without taking refuge to whatever is fashionable at a certain time.[57]

In this response, Repgen was able to refute a central objection against reading government proclamations voiced by his colleague. Nevertheless, this objection still appears in current methodological debates. Proponents of the heuristic emptiness of government proclamations tend to ignore that turning to normative discourses in political practice does not necessarily also mean to recognize these observed justifications as 'true'/'false' or 'good'/'bad' and thus succumb to an (alleged[58]) *Rankean* seduction—instead of trying 'to show what really happened' *('wie es eigentlich gewesen')*.[59] Although it is truly helpful to know the 'motives' behind war justifications in order to reconstruct a state's *opinio iuris*—as Clive Parry highlighted it[60]—Repgen's and our approach is not about motives. Neither is it our intention to 'reduce normative orders to practices'.[61] Rather, our aim is to emphasize the discursive construction of normative orders through the communicative reference to norms in political practice on the one hand and legal theory on the other (see below). While war discourses in legal theory

[57] The dialogue can be found in Repgen, above note 48, at 19 f.; our translation.

[58] Lutz Raphael has argued that already in the contemporary reception, the idealistic-philosophical part of Ranke's method had largely been ignored. Thus, Ranke's methodology had generally been misinterpreted as a 'trivial positivist ideal of objectivity', L. Raphael, *Geschichtswissenschaft im Zeitalter der Extreme. Theorien, Methoden, Tendenzen von 1900 bis zur Gegenwart* (2003), at 67f.; see also Iggers, 'The Image of Ranke in American and German Historical Thought', 2 *History and Theory* (1962) 17.

[59] F. L. von Ranke, *Sämtliche Werke*, vol. 33/34 (1885), at 7; see also Carty, above note 38, at 978f.

[60] C. Parry, *The Sources and Evidences of International Law* (1965), at 67 f.; see Carty, above note 38, at 982f.

[61] See Möllers, 'Ernst Rudolf Hubers letzte Fußnote. Die normative Ordnung des Nationalsozialismus und die Grenzen der Kulturgeschichte', 10 *Zeitschrift für Ideengeschichte* (2016) 47, at 56: '*Aber eine normative Ordnung auf diese [Praktiken] zu reduzieren bedeutet, ihr das Aspirative, das Kontrafaktische, das Normative abzunehmen.*'

are firmly established in the History and Theory of International Law, normative discourses in political practice (from past to present) still have to be reconstructed.

Repgen's early critique of dogmatic reservations about the role of norms in international politics and his plea for an analytical turn in dealing with this topic more recently have been followed up by an impressive study on belligerent princes' public declarations of war by Anuschka Tischer.[62] By analysing how belligerent princes in early modern Europe actually justified war, which norms were accepted, and to what extent the emergence of modern international law resulted from a process involving elaborate communication in the context of power politics, Tischer reveals the heuristic productivity of an empirical approach to the study of the politics of justification. She also shows to what extent the politics of justification in early modernity contributed to the shaping of modern international law. This way she helps to narrow the gap in our knowledge about the interplay between theoretical discourses on international law and political practice.

Along this line, the present anthology should help to emphasize the importance of minding and bridging the gap between political and theoretical war discourses. For instance, Isabel V. Hull shows in her studies on German justifications of war and international law in the historical context of the First World War, that there was considerable controversy inside the German government about the content of official declarations, and even about whether they were necessary at all. The international reception of Germany's arguments by other powers underline the argument of other authors in this volume—that is Anuschka Tischer for early modernity, Hendrik Simon for the nineteenth-century Great Powers, Paul Robinson and Mikhail Antonov for recent debates between Russia and the West, or Michael Stohl for constructions of 'terrorism': within different historical communication communities or communities of practice, the justification of the use of force became a political discourse about normativity. This at least in part conditioned political leaders to believe themselves to be restrained in their decisions for or against war by a common understanding of when the use of force was permissible.

Again, this re-orientation of research towards the inclusion of the practice of justification does not at all devalue the analysis of doctrinal discourses on the justification of war and international order as historical sources and philosophical reflections. Thus, some of the contributions in this volume may have a more 'classical', theory-driven focus (see also the contributions by Chris Brown, Anthony Lang, Jr, Oliver Eberl, and B.S. Chimni) than those chapters that aim at writing a history of international order on the basis of political practice. The interesting question is to what extent these different approaches speak to each other. As already pointed out, we assume that there is a close interplay between political practice and theoretical reflection. This interplay of theory and practice comes to the fore in the role of the international lawyer who observes legal practices and by doing so intervenes in the formation of law.[63] For instance, the contributions by Miloš Vec, Thilo Marauhn, Bhupinder S. Chimni, Siddharth Mallavarapu, and Arnulf Becker Lorca show that in discourses addressing the normative side of war lawyers themselves operate as visionaries[64] or critics of political practice[65] and

[62] Tischer, above note 5; see also Anuschka Tischer's contribution to this volume.
[63] See also Janssen, above note 52, our translation; Simon, above note 36.
[64] See also Koskenniemi, above note 17; Hathaway and Shapiro, above note 24.
[65] See the contribution by Chimni on 'Peace through Law' in this volume.

this way get involved in the shaping of international order. Bridging the gap between practice and theory might then be 'the most difficult problem of the scientific treatment of international law in its history'.[66] Nevertheless, we deem it both necessary and worthwhile to work on this issue, preferably from different disciplinary perspectives and methodological angles.

3. The Politics of Justification: War Discourses as a Struggle for Order/Authority

As mentioned above, this book's core thesis is that the history of war justifications also refers to the history of international order(s). Thus, the subject of war discourses touches on the quest for ordering the international, which has always been at the core of scholarly debates in (the History of) International Law and International Relations.[67] Or, as Nicholas J. Rengger has put it,

> the very fact of the existence of multiple and often widely diverse 'communities' coupled with the fact of their interactions and interrelations makes the problem of order at the international level inescapable. Accounts of 'international' or 'world' order are the inevitable result as is the fact that accounts of international relations cannot but try and deal with the problem of order.[68]

The way in which order is thought about is thus the subject of historically changeable worldviews. This is of central importance for a genealogical treatment of the role of international norms in war discourses. However, there are no commonly shared definitions of 'international order', especially with regard to one of its most important fields, the order of war and violence. It depends on one's theoretical point of view whether international order is understood as 'a pluralist and limited society of sovereign states',[69] or as the rule of the powerful[70] and rich,[71] or—in Kantian terms—as a legal order that will eventually overcome war.

While *Realists* have no problem with defining war as an aspect of international order, normative theory tends to juxtapose the two. Richard Falk defines international order as 'the distribution [both] of power and authority among the political actors on the global stage'.[72] With this he points to the double feature of order as based on

[66] Steiger, 'Ius belli in der Völkerrechtsgeschichte—universelle Geltung oder Beschränkung auf "anerkannte Kulturvölker"?', in T. Bruha, S. Heselhaus, and T. Marauhn (eds), *Legalität, Legitimität und Moral. Können Gerechtigkeitspostulate Kriege rechtfertigen?* (2008) 59, at 62.

[67] Tuck, above note 1; Kadelbach et al., above note 33; G. Krell, *Weltbilder und Weltordnung. Einführung in die Theorie der internationalen Beziehungen* (4th ed., 2009); G. Hellmann (ed.), *Theorizing Global Order: The International, Culture and Governance* (2018); see also the special issue G.J. Ikenberry, I. Parmar and D. Stokes (eds), 'Ordering the World? Liberal Internationalism in Theory and Practice', 94 *Foreign Affairs* (2018).

[68] N.J. Rengger, *International Relations, Political Theory, and the Problem of Order* (2000), at 2; see also Rengger, above note 53.

[69] A. Hurrell, *On Global Order: Power, Values, and the Constitution of International Society* (2013), at 3.

[70] See also Grewe, above note 4.

[71] See also Chimni, above note 13.

[72] Falk, 'World Orders, Old and New', 98 *Current History* (1999) 29.

'power' and 'authority' (which encompasses political, moral, or legal authority).[73] On the one hand, international order—like any order—is based on political, social, and economic power (asymmetries). On the other hand, however, international orders are also 'normative orders' since *any* political order needs justification.[74] Order not only offers stability (first and foremost of expectations), but also involves the issue of legitimacy and authority. Society and order are not possible without the formation of generally shared convictions[75] and accepted norms: *ubi societas, ibi ius*.[76] Furthermore, 'no political society, national or international, can exist unless people submit to certain rules of conduct'.[77] Where no norms, rules, or standards are considered binding, anarchy prevails.

This is to say that international order is constituted by power *and* norms.[78] (The Histories of) International Law and International Relations have usually treated power and normativity as a dichotomy. However, both spheres are historically closely interwoven: norms rely on political acceptance, and even the crudest power politics refer to justification. With this the analytical focus shifts from dealing exclusively with the distribution of power to the inclusion of the processes through which legitimacy[79] and authority in an international order are established.

Thus, the formation of order at the international level is not only a political but also a social process of criticism and conflict, as critical theorists like Walter Benjamin or Michel Foucault emphasize.[80] Furthermore, the emergence of normative orders, especially at the international level and particularly in the highly politicized field of justifying war, is shaped by power struggles *and* struggles over the authority of fundamental normative standards concerning the use of force. Rudolf von Jhering (1818–1892) put this aptly with reference to legal norms, 'Law is a struggle' of mankind (in the shape of 'nations, of the state, of classes, of individuals') to tame and, thus, to civilize itself.[81]

[73] M. McDougal (ed.), *Studies in World Public Order* (1960); on 'legal authority' see Besson 'The Authority of International Law: Lifting the State Veil', 31 *Sydney Law Review* (2009) 343 ; Zürn, 'From Constitutional Rule to Loosely Coupled Spheres of Liquid Authority. A Reflexive Approach', 9 *International Theory—A Journal of International Politics, Law and Philosophy* (2017) 261.

[74] Forst, above note 3; Forst and Günther, above note 8; A. Hirsch, *Recht auf Gewalt. Spuren philosophischer Gewaltrechtfertigung nach Hobbes* (2004); Brock and Simon, above note 7.

[75] Deitelhoff, 'The Discursive Process of Legalization: Charting Islands of Persuasion in the ICC Case', 63 *International Organization* (2009) 33.

[76] This is an assumption that is widespread both in IL and IR, even if different schools and authors disagree on the binding power of these rules in international or world order, see J.J. Brierly, *The Law of Nations: An Introduction to the International Law of Peace* (1963), at 1; K.D. Wolf, *Internationale Verrechtlichung* (1993), at 39, 51; Koskenniemi, 'The Politics of International Law', 1 *European Journal of International Law* (1990) 1, at 1; Kolla, above note 53.

[77] E.H. Carr, *The Twenty Years' Crisis: 1919–1939: An Introduction to the Study of International Relations* (1962), at 41.

[78] Ikenberry, 'The Logic of Order: Westphalia, Liberalism, and the Evolution of International Order in the Modern Era', in G.J. Ikenberry (ed.), *Power, Order, and Change in World Politics* (2014) 83, at 83f.; Forst, above note 3.

[79] T. Franck, *The Power of Legitimacy Among Nations* (1990).

[80] Benjamin, above note 2; Foucault, above note 2. For other relevant theorists see above note 15.

[81] R. von Jhering, *The Struggle for Law*, translated by J.J. Lalor (1915/1997), at 1; See also Cotterrell, 'The Struggle for Law: Some Dilemmas of Cultural Legality', 4 *International Journal of Law in Context* (2009) 373; J. von Bernstorff and P. Dann (eds), *The Battle for International Law South-North Perspectives on the Decolonization Era* (2019).

This notion of law (and other spheres of normativity) as a social struggle refers to the procedural character of the emergence of normativity: as outlined above, discourses on norms *socialize* political actors with regard to domestic[82] as well as international politics.[83] Within these discourses, norms continuously are the subject of social and political debates. This points to *a fundamental dialectic of the historical discourse of the justification of war and international order*: by referring to order in their justification of war, political actors or theorists not only attempt to identify their own use of force as 'appropriate behaviour' in accord with the norms of the international order,[84] they also claim to restore an order recognized as legitimate. However, in the asserted restoration of old normativity, political actors mix old and new forms, concepts and ideas of ordering the international. In other words: In their war justifications, political actors claim to restore an old order; but they actually create something new precisely with this claim.[85]

Processes of building international order always have involved the justification of the use of force as a means to provide for peace or at least as a way of restoring civil order.[86] To quote Jhering again: 'The end of the law is peace. The means to that end is war.'[87] This entanglement of international ordering and the use of force Walter Benjamin would have described as *schicksalhafte Gewalt* ('fateful violence'). In his famous essay *Kritik der Gewalt* ('Critique of Violence'), Benjamin defined violence as the origin of law, which appeared either in the form of law-making or law-preserving violence: 'Lawmaking is power making, and, to that extent, an immediate manifestation of violence.'[88] This violence cannot be completely overcome even in a purely legal system. Or, in the words of Hans Kelsen: a normative order always is a 'coercive order' (see also Daase and Deitelhoff in this volume).[89]

Of course, it makes a big difference whether the use of force is embedded in the rule of law or in some authoritarian set-up in which *submission* prevails. However, even under the rule of law, law remains a double-edged sword because it operates under the condition of an uneven distribution of power and vulnerability. The rule of law therefore tends to nudge the citizen into an *uneasy acceptance* of rules typical for a hegemonic constellation.[90] The point is that the dialectic relationship between war and order is perpetuated not only by the exertion of power as such but in a continuous discourse on the legitimacy of the use of force and the authority to exercise it. This calls into question *realist* narratives of anarchy, and also *liberal* narratives of progress in the

[82] Peevers, above note 7.

[83] Kratochwil, above note 10; Wendt, above note 10; Checkel, above note 10; Wiener, above note 10.

[84] Finnemore and Sikkink, above note 9, at 891; Jepperson, Wendt, and Katzenstein, above note 9, at 54.

[85] See also Tischer, above note 5; and our concluding chapter, which deals with Koselleck's *Wiederholungsstrukturen*.

[86] See also Joas and Knöbl, above note 2; see also the contributions by Anna Geis/Wolfgang Wagner, Arnulf Becker Lorca, and Beate Jahn to this volume.

[87] von Jhering, above note 81.

[88] Benjamin, above note 2; in a similar vein, Michel Foucault reversed Clausewitz' famous formula of war as 'the continuation of politics with other means'; Foucault spoke of politics as the 'continuation of war with other means', see Foucault, above note 2.

[89] Kelsen, above note 14.

[90] Buckel and Fischer-Lescano, 'Gramsci Reconsidered: Hegemony in Global Law', 22 *Leiden Journal of International Law* (2009) 437.

discourse on violence in international relations.[91] Instead, the relationship between war and order underlines the need for a systematic historical contextualization of war discourses as complex struggles for normativity in the face of power politics. As mentioned above, neither power nor normativity prevail in these discourses. They rather interact continuously.

Discourses on the use of force are therefore understood here as discourses in which the emerging or existing normative frame of reference for the use of force at the same time is claimed to be valid, contested, modified, and rejected.[92] To define norms only on the basis of their written codification as a fixed 'set of rules' is therefore a clear analytical mistake. International law is about the social practice of negotiating acceptance of norms in historical communication communities,[93] or, in a modified version of a formulation by Randall Lesaffer, it is about '*multiple normativities* in action'.[94] This also underlines, once again, the importance of turning not only to theory but also to the *political practice* of justifying war.

In the late nineteenth century, in the midst of the European imperial world order, the struggle for order involved strategic efforts by Japan and some of the existing political entities in the South to study European international law in order to be able to meet the requirements for being accepted as subjects of international law.[95] For many states, belonging to the normative order of European international law became the yardstick of 'civilization' in the nineteenth century. For other socio-political entities, this combination of international law and 'civilization' meant that European standards were imposed on them by unequal treaties, interventions, colonialism, and 'civilizing missions' (see the contributions by Mallavarapu, Becker, Benton, Genell and Aksakal, Chimni, Hippler, Lange, Wilén, Jahn, and Hashmi to the present volume).[96] The discourse on the use of force was clearly Western-dominated as it distinguished between legitimate reasons and means for war among major powers or 'civilized' states on the one hand, enforcement between European states and the non-European '(semi-)periphery' on the other.[97] This order worked on the basis of simultaneous *submission*, *uneasy acceptance*, traces of *mutual recognition*, and an *international rule of law*. What we experience today is a cacophony of the basic working principles of international order which reveals that there is a lot of tension between them.

[91] On the narratives of 'anarchy' and 'progress' in modern historiographies of the justification of war and international order, see Hendrik Simon's chapter in this volume.

[92] A. Wiener, *Contestation and Constitution of Norms in Global International Relations* (2018); see also Wiener, above note 10.

[93] J. Brunnée und S.J. Toope, *Legitimacy and Legality in International Law: An Interactional Account* (2010); Brunnée und Toope, 'Norm Robustness and Contestation in International Law: Self-Defense against Nonstate Actors', in 4 *Journal of Global Security Studies* (2019) 73.

[94] 'Law in action', see Lesaffer, 'Introductory Note' to R. Kubben, *Regeneration and Hegemony: Franco-Batavian Relations in the Revolutionary. 1795–1803* (2011), at xii.

[95] A. Becker Lorca, *Mestizo International Law: A Global Intellectual History 1842–1933* (2015); Payk, above note 24, at 42; G. Simpson, *Great Powers and Outlaw States. Unequal Sovereigns in the International Legal Order* (2009).

[96] Ibid.

[97] I.V. Hull, *Absolute Destruction: Military Culture and the Practices of War in Imperial Germany* (2004); H. Kleinschmidt, *Diskriminierung durch Vertrag und Krieg. Zwischenstaatliche Verträge und der Begriff des Kolonialkriegs im 19. und frühen 20. Jahrhundert* (2013); von Bernstorff, above note 27.

Furthermore, as Arnulf Becker has shown in his book on *Mestizo International Law*, during the first three decades of the twentieth century and in particular during the interwar period, non-Western lawyers, politicians, and activists referred to international law to support the demand for self-government. In this process, the semi-periphery of international order appropriated and transformed the international law discourse and with this pushed for a non-Western imprint on the legal order of the world. This shows that the universalization of international law was marked by contributions from quite distinct sources.[98] Playing on Becker's conclusion, both from his book and his contribution to this volume, we assume that rather than being purely European, the discourse on the use of force is increasingly *mestizo* (see also the contributions by Siddharth Mallavarapu and B.S. Chimni).

A genealogical perspective on these discourses inevitably raises questions relating to change and continuity, for instance in the form of the fall of an 'old' and the emergence of a 'new' international order.[99] It is tempting to construct a narrative of progress in dealing with conflict on the basis of such a dichotomy. But does it do justice to the complex histories of the justification of war and international order? To which extent and how does reference to norms in war discourses change (on the part of observers and actors) from early modernity to the present? Is there a changing balance between power-political, legal, and moral references in favour of legal references designed to cope with norm collisions (rule of law)? Has the dialectic relation between peace (as a goal) and war (as a means of achieving it) changed since it was invoked for the first time in antiquity? Or are we riding a merry-go-round in our efforts to think peace in a world of war?[100]

Narrativity, as Jürgen Osterhammel expressed it with reference to Reinhart Koselleck's concept of 'repetition structures' *(Wiederholungsstrukturen)*, is not limited to linear narratives that can be distinguished easily from one another.[101] Following this observation, the aim of the volume is to provide material for a genealogy of different histories of striving for international order and resulting setbacks. As some of the contributions in this anthology show, change in terms of an increasing weight of legal arguments (in comparison to moral or political ones) is always precarious and repeatedly marked by a relapse into mere power politics combined with moral reasoning.[102] Thus, the aim of the book is to highlight the fragility *and* the persistence of shared normative orientations in the struggle for international order.

Continuity and contingency also have to be addressed with a view to the categorization of the use of force both in political practice[103] and in normative theory.[104] In the

[98] Generally on Universalism and Particularism as paradigms of international law see G. Teubner, *Constitutional Fragments: Societal Constitutionalism and Globalization* (2012); von Bogdandy and Dellavalle, 'Universalism and Particularism as Paradigms of International Law', *IILJ Working Paper* (2008).

[99] Hathaway and Shapiro, above note 24.

[100] Brock and Simon, above note 7.

[101] Osterhammel, 'Über die Periodisierung der neueren Geschichte', 10 *Berichte und Abhandlungen der Berlin-Brandenburgischen Akademie der Wissenschaften* (2006) 45.

[102] See also the contribution by Vec to this volume.

[103] Clausewitz, above note 11; A. Herberg-Rothe, *Der Krieg. Geschichte und Gegenwart. Eine Einführung* (2nd ed., 2017); A. Geis (ed.), *Den Krieg denken. Kriegsbegriffe und Kriegstheorien in der Kontroverse* (2006); J. Lindley-French and Y. Boyer (eds), *The Oxford Handbook of War* (2012).

[104] Neff, above note 1; T. Hippler and M. Vec (eds), *Paradoxes of Peace in Nineteenth Century Europe* (2015); C. Chinkin and M. Kaldor, *International Law and New Wars* (2017).

title of this book, we speak of 'war' as this might be the most common socio-political term of collective violence. However, the contributions in this volume are by no means limited to interstate wars in the 'classic' *Clausewitzian* sense. Rather, they engage with and reflect upon different forms and justifications of force, such as 'wars of aggression', 'preventive' and 'pre-emptive war', 'colonial war', 'humanitarian intervention', 'self-defence', 'measures short of war', or 'coercion'. Though all of these terms are not really new, the way they relate to each other is getting ever more complex: under a largely legalized order for the justification of the use of force, 'war' turns into 'collective security', 'intervention' turns into 'the Responsibility to Protect', the meddling with domestic wars into 'just policing', etc. Even 'just war' may give way to 'justified war' (see Chris Brown's chapter in this book).

Is this a mere play of words? We think not. The change in the language points to changing requirements for the justification of the use of force. As such it reflects the basic predicament of an order based on the use of force and the limits of solving the problem once and for all.

4. Riding a Merry-Go-Round? Power Politics, Multiple Normativities, and the History of War Discourses—The Contributions to this Book

In what follows we introduce the contributions to the volume. The concluding chapter will attempt to synthesize the different perspectives with a view to the interrelationship between the justification of war and the transformation of international order.

In Part I, this volume starts out with basic considerations on the relationship between the justification of war and international order from two different theoretical perspectives. Academic approaches to war discourses usually depart from the theory of 'just war' (*bellum iustum*), the oldest and thus perhaps most influential ensemble of criteria for the justification of war in the European (and, according to Chris Brown's contribution, *not only* European) tradition. We take up this classical approach because the fundamental problems of justifying the use of force can be summarized on the basis of this theory. As Anthony Lang, Jr shows in his contribution, just-war theory limits the range of justifications of war by establishing the much discussed normative criteria and thus opens up the use of force to inter-subjective critique. Its central focus is on the restoration of an injured right without causing new injustice. In the absence of *effective* centralized decision-making, international order is hence based on specific moral considerations of equal political entities, that is on political judgement of sovereign states (see also Chris Brown's contribution).

By arguing for a historicization of decisions on the use of force, Lang, Jr addresses the contingency of the underlying judgement under precarious conditions and the difficulties in both theory and practice to determine what is just(ified).We will return to this issue in our concluding remarks. While Anthony Lang, Jr engages with the European roots of the discourse on war, normativity, and international order, Siddharth Mallavarapu examines the relationship between 'war' and 'empire' as a means to articulate a critique of the domination of international order by European or Western positions. Engaging with various critical perspectives in the disciplines of

International Relations and Third World Approaches to International Law (TWAIL), Mallavarapu argues that the history of international law is closely linked to the justificatory rationales of empire. With this, Mallavarapu pleads for a stronger and more critical historicization of war discourses. Juxtaposing the texts by Mallavarapu and Lang, Jr, it becomes clear that the legitimization of the use of force involves both authority[105] and rule (*Herrschaft*)[106] as possible starting points of the analysis of international order.[107]

On this background, Parts II, III, and IV of the book are dedicated to the justification of war and international order from the fifteenth century to 1945. Part II begins with the historicization of early modern war discourse. The common theme of the contributions gathered here is that of historical *transformations*—a constitutive topic for the history of the (early) modern war discourse: with the increasing monopolization of domestic violence, the notion of *raison d'état* and positive law (*bellum legale*) developed alongside the doctrine of just war as a frame of reference for justifying violent coercion. But how far did the resulting transformation of international order go? On the basis of her systematic analysis of the vast majority of declarations of war between the fifteenth and eighteenth centuries,[108] Anuschka Tischer shows in her chapter that contrary to the assumptions of mainstream theoretical opinion in (the History of) International Relations and International Law, war was constantly in need of justification in early modern political practice: nearly all belligerent princes in Europe justified their wars in public declarations in order to demonstrate that they acted in accordance to the customs of international law. These observations call into question Carl Schmitt's claim that a 'non-discriminating concept of war' (*nicht-diskriminierender Kriegsbegriff*)[109] emerged in the early modern era which was to say that since then war as such was no longer in need of justification.

In his contribution to reading (legal) history anew, Arnulf Becker departs from the observation that the Spanish and Portuguese conquests and raids in Latin America have so far mainly been told from the perspective of European international legal theorists (reaffirming Mallavarapu's critique). In contrast, Arnulf Becker shows that the colonized were engaging in the discourse on war justifications, too, thus offering a new source for grasping the role of law in European imperialism. Becker argues that 'conquest' as a military phenomenon was not outside law, but rather a violent enterprise governed by *legal mechanics*.

The most important historical break in early modernity for the formation of a modern international order under which intra-European conflicts were fought out and extra-European ambitions (of the Europeans) prospered is classically found in the Westphalian Peace of 1648 constituting the 'Westphalian (state) system' based on sovereign statehood, territoriality, and equal rights. This assumption has been questioned

105 Zürn, above note 73; see also the contribution by Anthony Lang, Jr in this volume.

106 Daase and Deitelhoff, 'Jenseits der Anarchie: Widerstand und Herrschaft im internationalen System', 56 *Politische Vierteljahresschrift* (2015) 299.

107 See also Forst, above note 3.

108 Tischer's sources are 343 declarations of war, 1 of the fifteenth century, 83 (~24 per cent) of the sixteenth century, 156 (~46 per cent) of the seventeenth century, and 103 (~30 per cent) of the eighteenth century, see Tischer, above note 5, at 29.

109 Schmitt, above note 11.

by Benno Teschke early on.[110] By bringing diplomacy and a unique type of British grand strategy ('blue water policy') back into the scientific discourse, Teschke argues in his present contribution that early modern international order was re-organized in the Peace of Utrecht (1713) which concluded the War of the Spanish Succession by imposing a new pro-British 'normative' set of rules for post-Utrecht early modern international relations. The question then is, to which extent or in which way this 'unique type of British grand strategy' came to bear in the context of the revolutionary developments on the Continent and the *Concert of Powers* that provided an answer to them.

The European discourse on war and international order which unfolded in the upcoming Enlightenment was shaken by the French Revolutionary Wars. This was commented on in particular detail by Immanuel Kant. As it turned out, he was one of the most influential philosophers of the Enlightenment with a view to the topic discussed here. As Oliver Eberl explains in his chapter, Immanuel Kant's famous proposal for 'eternal peace' can only be understood in the historical context of the French Revolutionary Wars involving the entire continent. According to Eberl, Kant repudiated both the idea of 'just war' and of war as a function of necessity. More specifically, Kant's idea was directed against the claim of the revolutionaries to defend the revolution against an 'unjust enemy'. With this, Kant anticipated an important project of modernity—the attempt to hedge collective violence as the normative core of international order.

The realization of this project did not only start in the twentieth century, but rather finds its birth era in the 'long nineteenth century' as the contributions in Part III of this book argue. Here, Hendrik Simon engages with 'one of the most prominent and provocative doctrines in the history of international law',[111] the 'free right to go to war' proposition (*liberum ius ad bellum*). According to an overwhelming majority of scholars, war in the nineteenth century was seen as legitimate whenever it served the sovereign's will to power. As a consequence, the nineteenth-century international order functions as an anarchic background for the construction of progress in the outlawry of war in the twentieth century. Simon's chapter challenges these assumptions by analysing both political and theoretical discourses on the justification of war. He deconstructs the thesis of an unlimited 'right to go to war' in the nineteenth century and identifies the historical roots of this myth.

With a view to the justification of the use of force in the nineteenth century, Simon's underlying assumptions are in basic accord with Lauren Benton's contribution which addresses the issue with regard to the intra-imperial use of force. Benton analyses normative plurality in the context of the rise of intervention in the nineteenth century. She highlights the complex political and legal *interpolity relations* between European captains and 'representatives' of non-European entities. As she shows with a fresh

[110] B. Teschke, *The Myth of 1648. Class, Geopolitics, and the Making of Modern International Relations* (2003); see also Duchhardt, '"Westphalian System". Zur Problematik einer Denkfigur', 269 *Historische Zeitschrift* (1999) 305; Osiander, 'Sovereignty, International Relations, and the Westphalian Myth', 55 *International Organization* (2001) 251; J. Burkhardt, *Vollendung und Neuorientierung des frühmodernen Reiches 1648–1763* (10th ed., 2006), at 28; F. Dhondt, *Balance of Power and Norm Hierarchy: Franco-British Diplomacy after the Peace of Utrecht* (2015).

[111] Weiler, 'Editorial', 29 *European Journal of International Law* (2018) 1, at 6; See also Simon, above note 52.

analytical perspective on local political practice, Navy captains were often acting as diplomats, but also justified wars and made contracts. Benton highlights that the discourse on the justification of war and international order was rather complex not only in Europe, but within the global-local relations all over the nineteenth-century world. Thus, her contribution cautions against simple lessons to be drawn for instance on present debates on the relationship between the global and the local.

Simon's and Benton's chapters are followed by three contributions dealing with the decay of the nineteenth-century international order and 'the seminal catastrophe' of the twentieth century (George Kennan): the First World War. Isabel Virginia Hull picks up the issue of the 'free right to go to war' again, and contextualizes the Imperial German war justifications of 1914 in the broader context of nineteenth-century international order. Engaging with the German war discourse on the one hand and nineteenth-century customary international law of self-defence, pre-emption, prevention, and aggression on the other, Hull concludes that the German decision to wage war in 1914 was not only a serious breach of international moral norms, but a violation of generally recognized international law.

Not only the German Empire claimed (pre-emptive) 'self-defence' as a narrative of justification for waging war in 1914: Aimee Genell and Mustafa Aksakal show in their chapter that the Ottoman Empire had perhaps one of the most sumptuous schemes to make participation in the First World War appear as a defensive move. Genell and Aksakal argue that the Ottomans' justification for war is best understood in terms of the deep sense of defeat and resentment at the relentless violation of Ottoman international rights and sovereignty by the Europeans. To unfold this argument, Genell and Aksakal place the Ottoman justifications of 1914 into the broader historical context of discrimination despite of, but also with the help of international law.

The legalization of international relations (which still lacked a *general* legal prohibition of *all* types of wars)[112] contributed little to preventing the 'Great War' and did not even strengthen the authority of law as *ius in bello*. Miloš Vec displays with his analysis of the legal debates on poison gas that scholars of all European nations were highly politicized during the First World War. Instead of criticizing the use of poison gas by referring to law, multiple concepts and theories were used to temper the law and to refer to it only selectively. Although the German gas attack of 22 April 1915 took place in the immediate aftermath of intensive legal efforts to civilize war and limit the means of warfare, international criticism was surprisingly weak. As Vec maintains, international law nevertheless was alive in the war—but so were power politics and other modes of justifying the use of force.

In line with these findings, Part IV of the book is devoted to the increasing universalization of the international legal discourse on the use of force from the League to the UN, which is also addressed from non-European perspectives: Bhupinder S. Chimni's chapter critically revisits the rich debate on the causes of the First World War and takes up the question whether more international law could have prevented the war. Chimni doubts Francis Boyle's thesis that even slight changes in the behaviour of central actors before 1914 together with the then-abandoned US legal positivist project of

[112] In the nineteenth century, legal regulations against war were discussed in theory and practice, but it was not possible to agree on a *general* ban on war, see also Simon and Hull in this volume.

war prevention could have effectively counteracted the outbreak of a general systemic war in Europe. He—like Mallavarapu—sees the decisive shortcoming of these debates in the underestimated entanglement of international law with imperialism and war.

Moving from a broad historical approach to the invention and evolution of air warfare in the early twentieth century, Thomas Hippler's contribution is dedicated to the justification of aerial warfare especially, but not exclusively, in the realm of strategic thinking. His main aim is to point out the conceptions of international order behind the different justifications of aerial warfare and air strikes, in particular with regard to the strategic choice to target civil populations first implemented as colonial 'police bombing'. With a view to the non-discriminatory violence of air strikes, Hippler again underlines the self-contradictions of the universe of the Western international legal order. This observation is of particular importance for Part IV.

In the following chapter, the book turns to one of the darkest moments in the history of the justification of war and international order, National Socialism. While the idea of securing a 'living space' (*Lebensraum*) in East Europe through an aggressive war was central to National Socialist Germany's foreign policy, the Nazis also strove to restore German colonies. Felix Lange's chapter takes up this topic and engages with National Socialists' dreams, concepts, and justifications concerning the regaining of the German colonies in Africa. Lange explains how the dictatorial and racist regime used the justificatory potential of international law for the political purpose of regaining the African colonies. He demonstrates the role of international law as an instrument of power politics but also as a potential force shaping it.

Parts V, VI, and VII of the book mainly deal with histories of the justification of war under the current international legal order as laid down in the UN Charter. One important question here is whether the Charter's prohibition of the unilateral use of force in Article 2(4) brought an end to multi-normativity and the entanglement of law with imperialism. Neither during the Cold War[113] nor afterwards, did this struggle end. Part V focuses on liberal justifications of war: Anna Geis and Wolfgang Wagner observe that following the end of the Cold War, democratic leaders have often attempted to legitimize the use of military force by referring to 'liberal' justifications such as the protection of human rights, the promotion of democracy, and the advancement of an international rule of law. Their chapter provides a critical review of the theory and practice of 'democratic wars' after 1990. The authors argue that 'democratic wars' are a specific manifestation of liberal world ordering practices. Whether these wars are becoming obsolete in the current international setting remains on open question.

In his chapter, Michael Stohl addresses one particular liberal narrative of justification. He explores, how '9/11' altered the construction of threat and how this, in turn, affected arguments for the use of force in the context of counterterrorism. According to Stohl, new efforts are required to rein in or prevent the next set of abuses when the inevitable next 'existential' threat occurs. He concludes that it is important to recognize the mistakes of the global response to terrorism during the past two decades as *not* inevitable, and sees this as an opportune moment to begin 'the hard work that will

[113] Peevers, above note 7; Marauhn in this volume.

be required' to come up with more constructive approaches to dealing with comparable threats.[114]

Much in this sense, but with a different analytical focus, Axel Heck and Gabi Schlag argue that discourses on the legitimate use of military force do not only take place in the UN, in national parliaments, ministries, and expert circles—they also take place on the ground among those who are involved in war fighting. In their chapter, they take a closer look at cultural representations of the Afghanistan war, especially in US films and documentaries, showing how the selected material refers to and contests an international normative order. With this, they add an important perspective on non-governmental communities in the discourse on the justification of war and international order.

Nina Wilén's and Beate Jahn's contributions deal with a close relative of war, 'interventions'. Wilén explores the paradoxical relationship between sovereignty, intervention, and the international normative order, by examining how the Economic Community of West African States (ECOWAS) justified its intervention in Liberia's civil war. Based on critical discourse analysis, Wilén argues that ECOWAS' intervention as well as its justification strategy were clearly influenced by the international normative frame of reference. She refers to the tension between global and local orders of the use of force, which underlines that the dialectical unity of the justification of war and the transformation of international order is even more complex than often assumed.

Beate Jahn offers a critical contribution on the use of force as 'humanitarian intervention'. Jahn shows that classical authors—Francisco de Vitoria, Edmund Burke, Thomas Paine, and John Stuart Mill—as well as state practice have regularly contemplated and often justified the use of force as protection of people in other states or regions. The rise of the concept of humanitarian intervention during the 1990s thus did not signify the advent of a new type of war. Instead, Jahn argues, it functioned as a 'doctrinal advance guard' for a new international political order as a (short-lived?) hegemonic liberal order.

While the previous contributions of Part V mainly deal with dominant actors of the West, Sohail Hashmi, Paul Robinson, Mikhail Antonov, and Manjiao Chi turn to non-Western approaches to war justifications and the ensuing controversies between East and West (Part VI). Sohail H. Hashmi engages with the advent of liberal international law in, and its impact on, Muslim articulations of *jihad*. As Muslim states acceded to the international legal regime governing war and peace, the very relevance of classical notions of *jihad* was called into question. Hashmi's chapter analyses how Muslims have sought to reconcile the Islamic law of war and peace with international humanitarian law during the past century. He underlines that the discourse on the justification of war and international order cannot be confined to the narrative of the West. There are various culturally divergent histories of justifying and criticizing violence.

Following an interdisciplinary approach (moral theory and law), Robinson and Antonov analyse Russian conceptions of justifying war in history and today. In doing so, the authors refer to the centrality and paradoxes of sovereignty as the central norm

[114] Stohl in this volume, at 325.

in Russian war discourse. They argue that since the collapse of the Soviet Union, political leaders have tended to adopt a *realist* approach and to frame war in terms of security rather than justice. At the same time, Russia claims to uphold international law and its core norm of sovereignty against Western ambitions .

Chi's chapter addresses the Chinese take on these issues. He identifies a 'shift of China's approach to the use of force' from a more traditional reasoning, based on the 'five principles of co-existence', to a concept more open to universalist claims. Nevertheless, according to Chi, China strongly adheres to a restrictive understanding of the principles of the use of force in the UN Charter. While traditionally having been an opponent to Western conceptions of 'humanitarian intervention' for a long time, according to Chi, China has in recent years become a conditional supporter of the concept. Chi argues that this may signal a deviation from China's traditional state-centric perspective on the global normative order. Thus, the contributions by Robinson/ Antonov and Chi both suggest that the tension between states' rights and human rights[115] may gain in importance as a factor shaping a post-liberal international order.

With this, and taking Jahn's observation into account, we ask: Are the same old questions still being asked today? Are we riding a merry-go-around of history where reiteration clearly outruns innovation? The final Part VII of the volume critically examines the impact of legal claims in war discourses in the face of classical and new challenges to international order. A contribution, which may agitate all those who believe in the possibility and importance of an international rule of law, is provided by Chris Brown. Brown fundamentally questions the relevance of international law in the decision-making on the use of force. He argues that international law has completely failed as a frame of reference for the justification and limitation of war. Brown suggests a universalized reading of the just-war tradition, defending it against both its 'overt enemies' and its 'supposed friends'.

In contrast to Brown, Thilo Marauhn's chapter can be read as a defence of the role of international law in the discourse of the international use of force. Starting from Thomas M. Franck's famous 1970 essay 'Who killed Article 2(4)?', Marauhn addresses the question to what extent the authority of international law as a vocabulary for framing the use of force is in rise or decline. Marauhn warns against weakening the prohibition of the use of force. Instead, he argues that Article 2(4) UN Charter should still be understood as a cornerstone of public international order in support of international law as a 'gentle civilizer' (Koskenniemi).

Bhupinder S. Chimni throws light on these issues from a different angle than Brown and Marauhn. In accordance with his earlier chapter in the book on 'peace through law', in this second contribution to our book he provides an overview over the meaning and practice of humanitarianism since the sixteenth century, focusing on its connection with the justification and critique of colonialism and imperialism. Reflection on this connection according to Chimni is a precondition for writing a global history of humanitarianism, which would be crucial for developing an understanding of humanitarianism that is not internally linked to the vision of empire.

[115] See also Brock, above note 19.

Along this line, Christopher Daase and Nicole Deitelhoff argue in their contribution that the current crisis of the liberal international order requires a reorientation or at least a re-evaluation of its normative foundations. According to Daase and Deitelhoff, however, the focus should be on the concept of 'legitimate coercion' rather than 'legitimate force' since in most institutions and also in international law the enforcement of norms—as the very basis of order—does not only or even primarily rely on military or police force but on various forms of political and economic coercion. Thus, their approach calls for a shift of the debate from the emergence of 'authority' to the working of everyday 'rule' (*Herrschaft*) at the international level.

The final chapter of the volume, an attempt at a synthesis, addresses the key issue of whether historical changes in the justification of war point to new possibilities of furthering peace or whether the respective claims would only open up ever more rounds of masquerading self-serving politics as offering new opportunities for improving the world. Discussing the findings of the contributions to the book, we assess to what extent historical changes in the justification of the use of force point to the emergence, stagnation, or the decline of a normative order which would serve not only the renaming of war but also the promotion of peace.

PART I
BASIC THEORETICAL CONSIDERATIONS
On War and Order(s)

2

Politics, Ethics, and History in Just War

*Anthony F. Lang, Jr**

1. Introduction

In an interview in 2006, Jean Bethke Elshtain, a prominent theorist of just war, argued that the 2003 Iraq war was morally justified. In part, this built on her 2002 book, *Just War Against Terror*, in which she argued that the United States had a responsibility to order the world in accordance with its role as the most powerful state in the system.[1] In the interview, she states the following:

> Another piece that I think is very, very important and that didn't get stressed as I would have liked it to have been, actually, was the question of what we loosely call humanitarian intervention. In the classic just war teaching, it's protecting the innocent from certain harm. The data on what was going on in Iraq was horrid and overwhelming, about the numbers of lives lost in the years of Baathist rule and on an ongoing basis. That is, it wasn't just stuff that happened in the past, but there were continuing horrors: the children's prisons, the number of Iraqi children by UN figures lost because of Saddam's gaming of the food and medicine for oil program, and so forth. Estimates ran to 60,000 kids a year. So the damage was ongoing, so I think that added to the moral justification.[2]

Elshtain's argument was invoked by others seeking to justify the Iraq war in terms of the just war tradition, such as James Turner Johnson.[3] The key line for the purposes of this chapter is Elshtain's claim that 'it wasn't just what happened in the past'. What she is referring to in making this point is probably Michael Walzer's critique of the Iraq war; his argument was that if Saddam Hussein deserved to be removed from office, the policies that merit this response have been going on since he came to power and are not necessarily new right now. For Walzer, putting Iraq into a historical narrative was important for understanding why the intervention might <u>not</u> have been justified.[4]

There are, of course, other reasons to contest the Iraq war; perhaps the most common one comes from figures such as Noam Chomsky, who pointed to the purely

* Thanks to Chris Brown, Cynthia Enloe, Cian O'Driscoll, James Pattison, Rory Cox, and the editors for helpful comments.

[1] J.B. Elshtain, *Just War Against Terror* (2002).

[2] Kim Lawton interviews Jean Bethke Elshtain, Religion and Ethics Newsweekly (24 March 2006), available at <http://www.pbs.org/wnet/religionandethics/2006/03/24/march-24-2006-jean-bethke-elshtain-extended-interview/12480/> (last visited 29 November 2019).

[3] J.T. Johnson, *The War to Oust Saddam Hussein: Just War and the New Face of Conflict* (2005).

[4] M. Walzer, *Arguing about War* (2004).

Anthony F. Lang, Jr, *Politics, Ethics, and History in Just War* In: *The Justification of War and International Order*. Edited by: Lothar Brock and Hendrik Simon, Oxford University Press (2021). © the many contributors.
DOI: 10.1093/oso/9780198865308.003.0002

economic reasons for the US intervention, that is oil wealth.[5] Rather than explore this more traditional critique of the intervention, this chapter will take the disagreement between Elshtain and Walzer on the relevance of 'stuff that happened in the past' to think critically about the past and its relevance for the politics and ethics of war. Walzer argues that putting the conflict in a wider historical ambit provides more context than Elshtain's focus on the present. Instead of focusing on that recent past, I argue in this chapter that Walzer and many other theorists of just war use history to justify and, perhaps, enable war.

The chapter proceeds as follows: the first section provides a brief overview of contemporary just war theory, focusing on the different strands and how they do and do not reflect a historical orientation. The section that follows looks to three thinkers: Hugo Grotius, Michael Walzer, and Cécile Fabre. Grotius famously drew upon and developed the just war tradition (and international law) by incorporating historical examples in ways that previous theological writers did not. But Grotius' efforts here did not reflect a historical narrative, which was not really available to him at that time. Rather, he drew upon examples plucked from classical and Biblical history to justify his principles. Michael Walzer, in his classic *Just and Unjust Wars*, undertakes a similar approach. The book, subtitled 'A Moral Argument with Historical Illustrations', sees history as a wealth of comparative cases which can be used to justify particular principles. So, while he can draw on more historical examples, both chronologically and geographically, Walzer undertakes precisely the same approach as does Grotius. Finally, Fabre argues in her book *Cosmopolitan War* in a very similar fashion, though extended further. Fabre, unlike other analytic just war theorists (such as David Rodin or Jeff McMahan) argues that she will draw on history to make her arguments. In addition, though, as with McMahan and others in this tradition, she turns to hypotheticals to make her argument. This extends and essentializes the approach of Walzer and Grotius; that is, if historical examples can be plucked out of the wider record without any specific context, why not simply create fictional examples without any context to make moral points?

Having reviewed these three theorists, the final section proposes how historical narratives might be employed in just war theorizing. I first connect the works of Grotius, Walzer, and Fabre to the idea of casuistry, a theoretical approach that focuses on judgment. But, following the lead of Nicholas Rengger, I suggest that using judgment cannot be divorced from some deeper structures, structures which often emerge from historical narratives. I then turn to two alternative narratives about war, a grand postcolonial narrative and a more personalized feminist narrative. While neither of these is perfect, they suggest how taking a narrative approach can point us towards alternative ways of seeing the judgments we make about war. This leads to some reflections on the politics of writing historical narratives and the importance of understanding power within those narratives. The chapter concludes less with a critique that solves this problem but more an awareness that ensuring the just war tradition is one of political theory with attention to history rather than abstract moral theory is perhaps the only answer we have.

[5] 'On the War in Iraq', Noam Chomsky interviewed by David McNeil, (31 January 2005), available at <https://chomsky.info/20050131/> (last visited 29 November 2019).

2. Just War: Ethics, History, and Politics

For the purposes of this chapter, I categorize three types of work on just war: international law, just war theory, and just war tradition. In particular, I look to the way in which the three approaches use history in making their evaluative claims. While all three provide important insights into how we evaluate war, they use history in different ways and, as a result, do not always connect with each other in their evaluations.

In the nineteenth century, formalized rules emerged concerning the use of force (see Part III in this book). These sought to counter the more traditional approaches which either relied on natural law or a *raison d'état* approach that enabled the use of force at the whim of political leaders. This newly emerging tradition intersected with liberal theories of the state and a professionalization of international law by those within such liberal states.[6] As the power of these liberal states increased, particularly Great Britain and the United States, they sought to transpose some of their liberal constitutional thought onto the wider international community.[7] Some of these efforts were supplemented by humanitarians whose witness to the horrors of war led to the creation of institutions such as the International Committee of the Red Cross, which today serves as the 'guardian' of many of the international legal instruments concerning the use of force.[8] There is no single historiographical approach within the wide field of international law (though certainly good histories; see Stephen Neff's historical overview[9]), evidence for which is that a recent handbook on the history of international law is over 1000 pages long.[10] Indeed, within that handbook can be found a diversity of methods, theoretical assumptions, and regional variations. Anthony Carty's contribution in particular is helpful in critically highlighting how assumptions about theory and practice were shaped by nineteenth-century legal theorists. Carty exposes the ways in which a category called 'practice' which differed from theory resulted from positivists who wanted to better ground their insights in the supposed real world. But in so doing, they unconsciously reinforced the positivist assumptions that helped to create a progressive international legal history.[11]

Hence, what often emerges from the international law approach is a progressivism which suggests that the laws which govern state uses of force have improved over the years and will continue to do so. While efforts to find in the past evidence for respect for rights or the existence of treaties do appear at times, the more common approach among international lawyers is to focus on how the international legal system continues to develop and improve. This progressivism is rarely acknowledged but it does structure the international legal tradition in some critical ways.[12]

[6] M. Koskenniemi, *The Gentle Civilizer of Nations: The Rise and Fall of International Law, 1870–1960* (2001).

[7] J.T. Shotwell, *War as an Instrument of National Policy and Its Renunciation in the Pact of Paris* (1929).

[8] D. Forsythe, *The Humanitarians: The International Committee of the Red Cross* (2005).

[9] S. Neff, *Justice among Nations: A History of International Law* (2014).

[10] B. Fassbender and A. Peters (eds), *The Oxford Handbook of the History of International Law* (2012).

[11] Carty, 'Doctrine Versus State Practice', in ibid., at 972.

[12] H. Haflidadottir. *Progressive Development of International Enforcement: Public International Law and Compliance with International Environmental Obligations* (2017) (PhD thesis on file at the University of St Andrews).

A second category of thought can be labelled just war theory. This approach is also sometimes called the revisionist just war theory, as it has become a form of theorizing about war and peace among political philosophers in the last thirty or forty years. This approach tends to avoid using figures in the history of political thought and concrete historical examples in making its claims about the use of force. The history of political thought can be disregarded, on this view, because the assumptions and background conditions of those who formulated the rules and principles concerning the use of force in ages past do not correspond to the conditions of today. For instance, the largely Christian background which informs the thinking of figures such as Augustine and Aquinas no longer functions in largely secular contexts. The historical record of war and violence certainly informs the questions being asked by those working within the just war theory approach. At the same time, in order to achieve analytical clarity, these philosophers often leave out the many conditions and causal factors that inform any particular use of military force. By creating hypothetical examples, these efforts can better capture the fundamental ethical assumptions underlying such decisions.

Of course, this categorization does not capture the approach of any one of these theorists. Jeff McMahan, one of the leading figures within this approach, notes the importance of Hugo Grotius in framing the way he understands the use of force.[13] C Cécile Fabre, to whom I return below, begins many of her insights with reference to historical events and introduces her book by noting that she wishes to draw on history more directly.[14] So, there is some variety here. But, the dominant tendency in this work tends to draw on hypothetical examples rather than real historical events.

A third category can be labelled the just war tradition, within which there is a more prominent place for historical examples and the history of political thought. This approach to just war is to look to the practices and historical interpretations of war over the years to establish and reinforce the norms which today shape our understanding. Perhaps the leading figure here is James Turner Johnson, who has long argued that understanding the past is crucial to understanding the moral standards we use to evaluate particular wars. He states this in response to a series of papers that explored his work in the *Journal of Military Ethics*:

> There is a great deal of use of the term 'just war' in contemporary literature on the ethics of war, but sadly, much of what appears under that heading is consciously or unconsciously unconnected to the broad cultural and moral tradition that historically has defined just war as a conception and body of practice. My work, by contrast, has been focused on the tradition that has developed and carried this idea historically and the implications to be drawn from this tradition of just war for present-day reflection and, maybe, practical decision-making.[15]

Johnson's extensive historical works have built up this approach to the study of just war, and many others have adopted his historical narrative and his methodological approach. Indeed, my own work, especially my use of Grotius as a key figure in

[13] J. McMahan, *Killing in War* (2009), at 16–17.
[14] C. Fabre, *Cosmopolitan War* (2009), at 15.
[15] Johnson, 'Thinking Historically about Just War', 8 *Journal of Military Ethics* (JME) (2009) 246, at 246.

the tradition, has borrowed from Johnson's method and particular narrative.[16] This chapter will return to this idea of constructing a narrative in the final conclusion, but for now, it is important to emphasize that the just war tradition approach is perhaps the only one of the three examined here which truly engages with history in a narrative sense. Johnson's narrative approach constructs a direct line from the history of the tradition to the modern day efforts by international legal theorists to limit war.[17] Indeed, one might argue that built into his tradition of thought there is something like the progressivism of the international legal approach.

These three approaches to the evaluation of war—international law, just war theory, and just war tradition—each have a different way of engaging with history. As noted, only the just war tradition really uses historical narrative to make a case for its moral evaluation. Interestingly, however, it is this approach that is perhaps most often used to justify war. As noted above, both Johnson and Jean Elshtain, who also fall into the just war tradition category, argued in favour of the Iraq War. Those writing from the just war theory and international legal perspectives were strongly against this use of military force. In the conclusion of this chapter, I will suggest that it is not simply the idea of seeing history as a narrative which shapes moral evaluation of war; perhaps, once we recognize the importance of such narratives, we need to think critically about what kind of narratives we use and how those narratives might predispose us to the justifications offered by states for their wars. That is, by understanding how history works, both narrative and otherwise, we can be more clear about how politics enters into our evaluations of war.

3. Three Theorists of Just War

In this section, I want to ground my argument by looking to three different theorists of just war: Hugo Grotius, Michael Walzer, and Cécile Fabre. Each one of these theorists has contributed to our understanding of just war. Moreover, each one of them has been greatly influential in how I evaluate war. That is, I do not see their works as fundamentally flawed. Rather, in researching and writing this chapter, I have come to see how each one of them draws on history in a way that can more easily justify war. In grouping them together, I also recognize that they write from very different theoretical starting points and with very different political milieus as their backgrounds. How do they relate to each other? I will return to this below, but I would argue that each of them participates in a kind of 'naturalist' just war. I purposefully use this rather ambiguous term rather than the more standard 'natural law' term, because Walzer and Fabre would not necessarily consider themselves part of a natural law tradition. Indeed, Grotius, considered one of the central figures in the development of the natural law approach to just war and someone who consciously drew upon natural law

[16] Though, I do have quibbles with Johnson on some matters concerning the authority of that tradition and how we should understand religious practice within it; see Lang, Jr., 'The Just War Tradition and the Question of Authority' 8 *JME* (2009) 202.

[17] Johnson, 'From Moral Norm to Criminal Code: The Law of Armed Conflicts and the Restraint of Contemporary Warfare', in A.F. Lang, Jr, A.C. Pierce, and J. Rosenthal (eds), *Ethics and the Future of Conflict: Lessons from the 1990s* (2003).

theorizing, also saw himself as departing in some important ways from the natural law tradition he inherited from the ancients and medieval Catholic theologians. Instead, I use naturalist here because it seems to me that all three theorists assume that the human condition has not fundamentally changed over the decades, centuries, and perhaps millennium, to justify drawing historical examples from random times and places. Only Grotius tries to connect his naturalism to nature itself, especially when he cites the behaviour of animals as justifications for self-defence.[18] Yet, with each one of these figures, they assume a certain human nature which they may not articulate but which makes sense of the way they use history. That is, they do not see history as a narrative which constructs our moral frameworks for evaluating war, but instead see history as evidence for how the human person behaves over time.

Hugo Grotius is well known for his use of history in making the claims he does about war and peace. Hugo Grotius, or Huigh de Groot as he was named at birth, was born on Easter Sunday, 10 April 1583.[19] He grew up in an educated and cultured Dutch household and studied philosophy, classical philology, and Oriental languages (Hebrew and Arabic) at Leiden, and went on to obtain a law degree from the University of Orleans in 1599. Through the patronage of Johann van Oldenbarneveldt, the Land's Advocate for Holland, Grotius became involved in domestic and international politics in the Dutch Republic. He also worked for the Dutch East Indies Company, a role in which he defended the company's use of force in a conflict in the East Indies. This defence led to the publication of *De Jure Pradae*, or *The Rights of Prize and Booty*, which set out his ideas about natural law and the use of force.[20] A small section of this longer book was published in 1608 as *Mare Liberum*, or *The Free Seas*, which defended the right of free trade and movement on the high seas.[21] In 1607, he was appointed advocate general, or what one might call attorney general, in Holland, along with being named the official historian of the Netherlands. As a result of some of his political activities, he was put in prison where he laid out his arguments for *The Rights of War and Peace*, his most well-known book, published in 1625. He died in 1645 while serving as a diplomat for Sweden.

There is an extensive secondary literature on Grotius, both critical and commendatory. Rather than engage with this work, or with the substantive claims Grotius makes about war and peace, I want to briefly here examine his use of history. As argued above, Grotius uses history as a form of evidence, along with claims about the animal world, in order to justify the use of force. Rather than a narrative approach to history, however, Grotius finds moments in time as evidence for the different positions he seeks to make. Grotius does not provide any direct historiographical reflections in his work, though he does comment on his use of history. In this chapter, I will only look to his comments in *The Rights of War and Peace*, where he does offer a few thoughts on history. Moreover, I will provide a quick account of how he uses history in Book 1 of that work.

[18] H. Grotius, *The Rights of War and Peace* (2005 [1625]), at 182 f.

[19] For biographical details, see Lang, Jr, 'Hugo Grotius (1583–1645)', in D.R. Brunstetter and C. O'Driscoll (eds), *Just War Thinkers: From Cicero to the 21st Century* (2018) 128.

[20] H. Grotius, *The Rights of Prize and Booty* (2006 [1604]).

[21] H. Grotius, *The Free Seas* (2004 [1608]).

There are some places in the text where Grotius makes a point about history, though these comments require some interpretation. Early on in the text, when he comments on some French authors, he states that they

> introduce history into the study of law. Their decisions and reasons will often furnish us with Matter for the Search for Truth ... For I profess truly, that as Mathematicians consider Figures abstracted from Bodies, so I, in treating of Right, have withdrawn my Mind from all particular Facts.[22]

Withdrawing from the particulars to the general sounds more like a philosopher's use of hypotheticals than it does the provision of a narrative historical account which makes sense of the use of force over the ages. At other points, Grotius will claim that 'historians' support the claims he is making.[23] He also notes at another point that 'both sacred and profane History do testify ... ' suggesting that he sees the two forms of history as distinct, but that both reinforce the claims he is making.[24]

What historical examples does he actually employ to justify the use of force? As one might imagine from a late Renaissance figure such as Grotius, his references come from two sources: the classical tradition of Greece and Rome and the Biblical tradition. In Book 1, those references are more tilted towards the classical tradition than the Bible, but he does draw on both. Calculating them is difficult, as many of the references are to authors who reference historical events, especially in the Classical era. In Book 1, I counted at least ten references to events from Ancient Greece and Rome, some of which are direct historical references, and others are to authors (Plutarch, Thucydides) who are referencing historical events. References from the Ancient Israelites or pre-Israelite Canaan number at least five, with two references to the New Testament and one to the early Church Fathers. Again, the last is not a historical reference directly, but is an example of how individual authors considered the use of force. There are two references to Ancient Egypt, two to Norway (the timeline on this reference is unclear, but probably refers to what we would call the Viking era). There are scattered references to Germany and France, though these are not to the contemporary era in which Grotius lived, but rather to the times when Rome was expanding and fighting various tribes from these regions.

From this brief summary it does not appear that Grotius saw himself as writing a narrative of any particular war, nor did he see his ideas in the context of any wider world historical narrative. Interestingly, he does make one reference to historical evidence from outside of the largely European and Biblical contexts that dominate his text; this comes in a footnote, where he states that 'We find even at this Day amongst the Arabians and Africans several Nations of Savages and Vagrants, without Laws, Magistrates or any form of Government.'[25] So, for Grotius, the truth of his arguments are not confined to any one place or time, but rather are universal. This results, of course, from his views on natural law, a body of thought to which he made some

22 Grotius, above note 18, at 132.
23 Ibid., at 103.
24 Ibid., at 267.
25 Ibid., at 133, fn. 2.

important contributions.[26] According to natural law, the human condition results in certain forms of politics and law which are attainable by any person or community operating in accordance with right reason. Historical evidence does not alter this but rather reinforces it.

However, it is likely that Grotius understood himself as writing within a particular historical narrative, one that perhaps he did not see the need to specifically identify. As a Protestant thinker during the late Renaissance, we could say that Grotius saw his work within a broadly defined Christian universal narrative. Importantly, he contested the existing Scholastic narrative which combined Christian thought with classical antiquity. But, it is likely that he saw himself as correcting that narrative and reshaping it, rather than resisting it. So, for Grotius, we can perhaps see a narrative of sorts, but one that is not explicitly identified as such. And, this narrative, with both classical justifications of war in Rome and Biblical justifications of force by the Ancient Israelites, easily leads to justifying the use of force in the many contexts he cites in his texts.

Michael Walzer adopts a similar approach to the study of war, though without this Christian background narrative. Walzer was born in 1935 in New York City and has been both an activist and philosopher. For many years, he published not only academic analyses of the world but also edited and wrote for *Dissent* magazine, a social democratic publication focused on the United States. He has written extensively about communitarian political theory, political obligation, the intersection of religion and politics (both the Protestant Reformation and Jewish political thought), and the ethics of war. For Walzer, some of these topics are tied together through his early engagement with the American participation in the Vietnam War; as Chris Brown notes,

> to participate in American politics in the 1960s necessarily involved adopting a position on the Vietnam war; Walzer, by then a teacher at Princeton and then Harvard was, predictably, an active member of the anti-war movement, and his academic focus shifted as a result of this engagement.[27]

Walzer himself notes in the Preface to the first edition of *Just and Unjust Wars* that 'I did not begin by thinking about war in general but about particular wars, and above all about the American intervention in Vietnam'.[28] At the same time, Walzer is not a pacifist. His critique of the American intervention draws on the just war tradition, though rather lightly at points. He certainly does not defend a religious account of the tradition, something which differentiated his work from almost all of the previous writing about just war when his book first appeared in 1977. Walzer argued, for instance, that the Israeli attack on Arab states in 1967 was justified as an act of self-defence.[29]

Walzer is more explicit than Grotius on how he understands the relationship between history and the ethics of war. First, he argues that his position reflects that of 'common morality'. That common morality is located not in particular activities,

[26] K. Haakonssen, *Natural Law and Moral Philosophy: From Grotius to the Scottish Enlightenment* (1996).
[27] Brown, 'Michael Walzer (1935–present)', in D.R. Brunstetter and C. O'Driscoll (eds) *Just War Thinkers: From Cicero to the 21st Century* (2018) 205
[28] M. Walzer, *Just and Unjust Wars: A Moral Argument with Historical Illustrations,* (5th ed., 2015), at xxiii.
[29] Ibid., at xxiv.

however, but in the 'opinions of mankind'.[30] He then goes on to state quite clearly that he does not accept what he calls 'historical relativism':

> Now change and variation are certainly real enough, and they make for a tale that is complex in the telling. But the importance of that tale for ordinary moral life and, above all, for the judgment of moral conduct is easily exaggerated. Between radically separate and dissimilar cultures, one can expect to find radical dichotomies in perception and understanding. No doubt the moral reality of war is not the same for us as it was for Genghis Khan; nor is the strategic reality. But even fundamental social and political transformation within a particular culture may well leave the moral world intact or at least sufficiently whole so that we can still be said to share it our ancestors. It is rare indeed that we do not share it with our contemporaries, and by and large we learn how to act among our contemporaries by studying the actions of those who have preceded us. The assumption of that study is that they see the world much as we do.[31]

Walzer, in other words, confronts more clearly what Grotius only stated in a roundabout way: historical differences do not necessarily represent different moral realities, but, instead, provide evidence for the ways in which the ethics of war does not differ much over the centuries.

In Part One of *Just and Unjust Wars* (tellingly entitled 'The Moral Reality of War'), this use of history becomes clear. Walzer's examples stretch across the following moments in time: the Peloponnesian War; the Battle of Waterloo; the medieval battle between France and England at Agincourt (as described by both historians and Shakespeare); Napoleon's conquest of Europe; the American Civil War (Sherman's burning of Atlanta in particular); and the German campaign in North Africa in the Second World War (focusing on Erwin Rommel). Throughout the text, numerous other historical examples are deployed by Walzer in order to find evidence of the moral reality of war, found primarily in the judgments upon which not only historians but literary figures have drawn. Walzer is less interested than Grotius in finding a natural law foundation for the claims he is making; rather, as with the subtitle of his book, he is looking for 'moral illustrations' to reinforce this 'common morality' that he argues structures our lives. Over the course of the book, these references reinforce and support the wider just war tradition of thought of justice in going to war and justice in the conduct of war.

This suggests that Walzer's work falls into the category of naturalist thinking in that it seeks to reflect a universal language of war. Walzer does not seem to have much of a grand narrative in mind in his work, and he rarely cites the tradition of thought that informs the just war tradition. So, unlike Grotius, there is no single narrative that seems to underpin Walzer's work.

A final figure to explore is Cécile Fabre. Fabre, born in France in 1971, is a political philosopher at Oxford University. Her work stretches across a number of important areas. Her early work focused on the way in which rights can function within a

[30] Ibid., at 15.
[31] Ibid., at 16.

constitutional order, with special attention to social and economic rights. She then turned to the body as the subject of her next body of work, looking to the ways in which theories of justice can help us to think about matters such as reproductive rights and prostitution. After this, perhaps because war is a practice that harms and destroys bodies, she spent the next part of her career working on issues related to war and peace. This culminated in two important books, *Cosmopolitan War* (2013) and *Cosmopolitan Peace* (2016). In both cases, she draws upon the core insight of cosmopolitan theory to orient her reflections. She summarizes that position here:

> Cosmopolitanism is the view that human beings are the fundamental and primary loci for the moral concern and respect and have equal moral worth. It is individualist, egalitarian, and universal, and insists that political borders are arbitrary from a moral point of view, and more precisely ought not to have a bearing on individuals' prospects for a flourishing life.[32]

This orientation locates her work within a wider analytical tradition of political philosophy, which includes figures such as Jeff McMahan and David Rodin. For these figures, reflections on war and peace originate in liberal individualism, which means that the authority structures of the state and the rules of international law are less important than the rights of and respect for individuals.

For some, this liberal cosmopolitan approach is problematic because it fails to confront certain features of the tradition, including the authority of the political community and the rules of international law. Indeed, some have argued that it might well undermine the purpose of war, which is to defeat the other side.[33] Others have argued that a distinction between what is sometimes called 'traditionalists' versus 'revisionists' is somewhat overblown.[34] While Fabre uses a few historical examples, she also defends the use of hypothetical examples in theorizing about war, something that takes the selective use of history even further. Moreover, she sees those hypotheticals and the historical examples that she uses as contributing to what she calls 'pre-theoretic intuitions'.[35] Interestingly, these intuitions seem very close to what Walzer calls 'common morality'. She defends her use of such intuitions and her construction of them in the following way:

> However, if used sparingly and judiciously, hypotheticals do help us isolate morally relevant features of particular cases and thereby uncover our intuitions. This is not to imply that intuitions are the only sources of moral knowledge: far from; rather they are one such source—the raw data, if you will—of our moral thinking. Moreover, as we shall see at various points in this book, some of the intuitions which underpin my considered account of the morality of war are rather radical and immune from the charge of undue conservatism.[36]

32 C. Fabre, *Cosmopolitan War* (2013), at 16.

33 Brown, 'Revisionist Just War Theory and the Impossibility of Victory', in A. Hom, K. Mills, and C. O'Driscoll (eds), *Moral Victory: The Ethics of Winning War* (2017) 85.

34 Pattison, 'The Case for Non-Ideal Morality of War: Beyond Revisionism versus Traditionalism in Just War Theory', 46 *Political Theory (PT)* (2016) 242.

35 Fabre, above note 32, at 12.

36 Ibid., at 13

She later goes on to argue that her approach to using historical examples is 'resolutely Walzerian' rather than analytic.[37]

I read Fabre, then, as taking the naturalists position as far as possible, by combining historical examples with hypotheticals. In the same way that Walzer and Grotius extract from history principles of warfare or conventional wisdom about warfare, Fabre's use of history and hypotheticals takes that moral abstractness a step further. Her work does differ from political philosophers who avoid history altogether, but it takes the approach I have identified here to its limits. She also does not write within any single narrative approach, though perhaps her cosmopolitan orientation would link her more closely with the progressivism of the international legal tradition. In a sense, then, Grotius, Walzer, and Fabre demonstrate, from what would appear to be very different theoretical positions, a similar methodological style in arguing about war. In the next section, I suggest some alternative ways in which we might see theorizing and arguing about war, which can rely on history, but which require a different approach to how we think, one that emphasizes the importance of narrative over example.

4. History and Just War

I have located in the different approaches to the study of war and in three theorists in particular some historiographical assumptions that inform their evaluations of war. I have labelled Grotius, Walzer, and Fabre naturalist thinkers in that they assume a certain truth across time and place about the justifications for using military force. For all three of them, there is a need to make judgments about war on the basis of these historical or hypothetical examples. Before I turn to two alternative ways of narrating war, let me briefly say something about making judgments about war.

Nicholas J. Rengger argued that judgment about war often takes place through the medieval model of casuistry, or the use of examples rather than rules to morally evaluate practices. While sympathetic to this approach, Rengger also criticizes those who used it for not carefully understanding the background conditions that made such judgments possible. This background condition might arise out of a cultural context or historical narratives, but it is often not articulated. Rengger highlights this in two separate but overlapping essays on Michael Walzer's defence of the supreme emergency. In *Just and Unjust Wars*, Walzer argues that the rules of war may be set aside in moments of emergency, but that the limits on such exceptions must be very carefully policed. Walzer explores two cases, both from the Second World War: the British bombing of German cities and the dropping of the atomic bombs on Japan. In both cases, he argues that we cannot rely on utilitarian calculations alone but on confronting our competing moral commitments. He concludes that the British bombings were justified but the dropping of the atomic bombs was not. This is an interesting, and perhaps counterintuitive conclusion, one that opens up many interesting questions on the limits imposed by just war and how to keep to those limits in situations of conflict.

[37] Ibid., at 15.

Rengger highlights the fact that throughout Walzer's book, and especially in this section, he is defending the state and the state-based system. For Rengger, however, this kind of defence is problematic. It puts the existence of this particular political construct before the ultimate purpose of the just war tradition, which is to restrain rather than enable the use of force. Walzer fails to recognize that the just war tradition is

> in a very important sense, *constituted by and through* its rules, and they cannot be 'trumped' or overridden by any particular social or contextual circumstance. That, indeed, is what it means, I think, to see the just war tradition as Walzer at least claims to do as a casuistic tradition. Any given action, the bombing of German cities, for example would have to be examined in the context of the 'rules' which are not merely laws, and nor are they fixed, but they certainly have a centre of gravity around a common set of precepts. Political communities may have a putative right of defence under these precepts, for example, but such a right is always conditional on other precepts being met and is never absolute.[38]

The idea that while the rules are not fixed laws but that they have some sort of 'centre of gravity' is elusive but important. Rengger develops this point in a slightly different context when he examines the idea of political judgment. Judgment played an important role in the way in which Rengger understood politics, but he also brought a certain scepticism to claims that judgment would be the way in which we can avoid becoming too rule-governed when it comes to evaluating war. In a chapter on the work of Chris Brown, who has developed the idea of judgment quite extensively in his own work (see Brown's contribution to this volume), Rengger brings to bear this critical perspective. While he appreciates Brown's idea of judgment, Rengger argues that it lacks what the medieval theorists of casuistry and judgment had; a shared cultural, political, and ethical framework within which making judgments made sense. The ability to make judgments in the contemporary global order would require a much greater consensus about what counts as legitimate evaluative standards than we currently have; in fact, to suggest that such standards can be found and cultivated in the world today is, perhaps, 'wishful thinking'.[39]

How does this relate to the argument of this chapter? What Rengger points us to is the difficultly of drawing historical examples from a range of eras without any context. Walzer's work is very much in the tradition of political judgment; rather than set out specific rules, he argues for a contextual interpretation of those rules. Walzer identifies his context in the preface to *Just and Unjust Wars*, when he locates his impetus for writing the book in his concerns about the US war in Vietnam. And, in the prefaces to the four editions that followed, along with his book of essays on war,[40] he situates his thinking in particular contexts. However, that context is a particularly American liberal one, one that may not work in Europe, the Middle East, East Asia, or Africa. Moreover, how long will the context of the sovereign state make sense? This is not

[38] N.J. Rengger, *Just War and International Order: The Uncivil Condition in World Politics* (2013), at 154 f.

[39] Rengger, 'Practical Judgment: Inconsistent—or Incoherent?', in A.F Lang, Jr and M. Albert (eds), *The Politics of International Political Theory: Reflections on the Work of Chris Brown* (2019) 55, at 63.

[40] Walzer, above note 4.

to say that every theorist has to think about every single context. But, when making claims that are putatively universal, as the just war tradition often does, perhaps more reflections on how the context generates our ability to make judgments is important to consider.

With this idea of the context in mind, and recognizing that the context is often shaped by the historical narratives we employ to understand that context, let me suggest two alternative narrative forms. These narratives are perhaps more conscious and explicit about the politics that inform them. Let me make clear that I do not necessarily think these are the only valid narratives. Rather, the point to emphasize here is that understanding and articulating the narrative context within which we evaluate war is central to the judgments we make about it.

The first is what we might roughly call the postcolonial narrative form, one we can find by critically engaging the background conditions found in the work of Grotius. As noted above, Grotius served for a time as a lawyer for the Dutch East Indies Company, a quasi-public corporate body that played a crucial role in the expansion of Dutch power into the East Indies. In 1602, he wrote a legal defence of a Dutch merchant who had captured a Portuguese ship carrying a significant amount of goods from the East Indies, a short version of which appeared in 1608. This defence formed the basis of Grotius' ideas about the freedom of the seas, arguing that no state could control them as the Portuguese claimed along with important insights into the justification of military force. In addition, he developed many of his ideas about the use of force and its relationship to natural law. This study, in many ways, can be seen as the origin of his more famous work on the rights of war peace which solidified his name as a leading theorist of both international law and the just war tradition.

What needs highlighting here is how the ideas for which Grotius became famous arose from a colonial enterprise. Edward Keene pointed this out in his study of Grotius and world order.[41] In that study, which positioned itself against figures such as Hedley Bull who had argued that Grotius provides us with a way to think about international affairs through institutions and law, Keene argues that those laws and institutions arose from political agents seeking to establish the dominance of Europe in relation to the rest of the world. This critical perspective was further developed by Martine van Ittersum's historical study of Grotius and his role in the colonial project in modern day Indonesia.[42] This study provides an alternative historical narrative, one that provides important details on Grotius and the germination of his ideas. In addition, van Ittersum locates this in relation to the ways in which Grotius' reputation within the Netherlands and Europe more widely created the image of the 'father of international law' by whitewashing this colonial context. Her work here reinforces efforts by international legal theorists and historians of ideas, such as Anthony Anghie and Anthony Pagden, both of whom, in different ways, have given us narratives of international law which contrast starkly with the progressive story of international law and international affairs.[43] By demonstrating that the history of the use of force is often interconnected

[41] E. Keene, *Beyond the Anarchical Society: Grotius, Colonialism, and Order in World Politics* (2002).

[42] M. van Ittersum, *Profit and Principle: Hugo Grotius, Natural Rights Theories, and the Rise of Dutch Power in the East Indies, 1595–1615* (2006).

[43] A. Anghie, *Imperialism, Sovereignty, and the Making of International Law* (2005) and A. Pagden *The Burdens of Empire: 1539 to the Present* (2015).

to the history of colonialism,[44] the stories of a slow emergence of more 'civilized' behaviours about warfare becomes a story of how ideas of civilized behaviour are used to justify force for economic and imperial reasons. The recent account by Hathaway and Shapiro on outlawing war connects to this critique, though not directly; they claim that Grotius represents the 'old world order' in which international law could be used to justify war after the fact. While not highlighting the colonial dimension of his career in the way that others have, their critique reinforces it.[45]

To explore such works in more detail is beyond the space allowed in this chapter. However, what is important to highlight here is that the postcolonial narrative explicitly highlights the background conditions that allow powerful states to justify their use of force. It suggests that without the existence of these efforts to 'universalize' their justifications for using force, the powerful actors in these contexts would not be able to make such claims. By clarifying this alternative background through this critical narrative, we are able to see some of the tensions found in the naturalist style of not only Grotius, but many other thinkers who constitute the just war tradition (see the following chapter by Siddharth Mallavarapu).

A second alternative I wish to mention comes from a very different perspective. In fact, it moves away from all three theorists identified above. The first critique I made of the three theorists is that they fail to address historical narratives, but rather use history as a form of examples that reinforces existing assumptions about common morality. The second critique is that our ability to draw on history and make judgments about that history requires a set of background assumptions that are difficult to find at the global level. This last point brings those two together by suggesting how we might construct parallel narratives about war which might link together in ways we do not expect.

The feminist scholar Cynthia Enloe has long analysed the ways in which global politics is constituted by forms of violence and militarization that rely on assumptions about gender.[46] In some of her work, she parallels the types of historiographical assumptions found in figures such as Walzer; perhaps her training as a political scientist has led her to see history as a form of examples that can be drawn upon in order to better understand these gender dynamics. In one of her works, however, she adopts a slightly different approach to studying war and violence. In 2010, Enloe published *Nimo's War, Emma's War: Making Feminist Sense of the Iraq War.* In that book, Enloe constructs a series of overlapping narratives which explore the experiences of different women in the United States and Iraq. These narratives arose from her interviews and readings with many different people, both men and women, but the resulting text seeks to make 'feminist sense' of the Iraq War. In the introduction to the work, Enloe reflects on how the stories of the eight women she explores in the book challenged her analytical abilities. As the four Iraqi and four American women emerge from her

[44] See the chapter by Siddharth Mallavarapu for a critical overview. Furthermore, see the contributions by BS Chimni, Arnulf Becker Lorca, and Thomas Hippler to this volume.

[45] O.A. Hathaway and S.J. Shapiro, *The Internationalist: How a Radical Plan to Outlaw War Remade the World* (2017), at 3–30; for a critique of the 'old order'/'new order'-dichotomy see Simon's and Brock's introduction and Simon's chapter in this volume.

[46] C. Enloe, *Does Khaki Become You? The Militarization of Women's Lives* (1989) and *Morning After: Sexual Politics at the End of the Cold War* (1993).

pages, she worried that her narratives about them might reduce them to simple characters of 'wife', 'widow', 'working woman', or 'mother'. In trying to avoid doing this, she wanted to reveal their particular realities. At the same time, she wanted to find in their stories something general or perhaps even universal:

> My own inclination is toward comparison and generalization, to delve into one to shed light on the other, to use the particular to reveal the general. That's the teacher in me. At the same time, though, more perhaps than in any of my previous writings, I have worried about doing justice to each woman individually. Emma is more a Latina, more than an American mother of a teenage son. Safah is more than Iraqi teenage girl, more than a massacre survivor. I realize now that I scarcely know how much more.[47]

Enloe here is struggling with her ability to translate these particular historical narratives into something from which she can generalize about war. What keeps these narratives focused, of course, is that they centre on the experiences of women. In so doing, she can draw out some general lessons on gender, war, and justice in new ways. Importantly, Enloe here is finding a middle ground between the type of postcolonial narrative identified in the work of figures such as van Ittersum and the hypothetical or historical examples found in Fabre and Walzer.

In so doing, though, Enloe takes seriously the difficulties of using this first person narrative style to understand and make judgments about war. She realizes that she can never quite know the inner landscape of any one of these women, and that there is a real danger in creating fictional characters:

> I have kept a tight rein on my imagination, however. I have not drifted into fictionalizing. Thus I have made no attempts here to tease out psychological inner voices, to attribute emotions beyond what each woman herself has described.[48]

This balance is difficult to achieve, but at least Enloe here is honest about the challenges of avoiding fiction while writing an alternative historical account.

Enloe does not understand her work as directly about ethics of war, nor does she see herself as writing within the just war tradition. Rather, I invoke her work here to demonstrate a different methodological approach to the use of history in understanding and, eventually, evaluating war. Enloe creates a narrative (or series of narratives) that brings out a very different perspective on war and justice. It holds together through her focus on women and their lived experiences, from which she can draw some broad lessons about how militarization works in different cultures but seems to link back to underlying assumptions about gender roles. At the same time, her particular narratives ensure that the lessons drawn are not too simple or simplistic. Instead, they suggest how we might construct an ethics of warfare that finds a middle ground between the easy use of historical examples and the use of fully constructed, airtight narratives from which too simplistic conclusions can be drawn.

[47] C. Enloe, *Nimo's War, Emma's War: Making Feminist Sense of the Iraq War* (2010), at 3.
[48] Ibid., at 12.

5. Conclusion

In their introduction to this volume, Hendrik Simon and Lothar Brock argue that the messy relationship between theory and practice, particularly when it comes to justifying war, needs to be more carefully explored. In this chapter, I have sought to respond to their challenge by looking at how history is used as a form of 'practice', the function of which needs to be understood more clearly. As Carty does in his critical reading of how international legal theorists have constructed the idea of 'practice',[49] so I have tried to critically read how 'history' functions for those writing on the ethics of war.

The chapter began with a reflection on the justifications offered for the Iraq War. The final section ended with some reflections on an alternative approach to the Iraq War, one that leads much less easily to justifications for it. It should be noted that some of the figures drawn upon by Enloe do indeed justify the war; the wives of American soldiers and the female soldiers she discusses have qualms but do not see the war as an imperial enterprise. And, some of the Iraqi women see the toppling of Saddam Hussein as not necessarily a bad thing. But, their narratives complicate our ability to justify war. They make it a much more complex exercise by bringing out the multiplicity of background conditions that inform our understanding of that particular war. And, by looking to individuals both in the United States and Iraq, they take us beyond the postcolonial narrative found in the previous section.

The chapter set out to explore the ways in which history functions in our evaluations of war and peace. It began by finding in three figures—Hugo Grotius, Michael Walzer, and Cécile Fabre—a particular approach to the use of history which revolves around using particular historical moments or vignettes to draw general lessons about the ethics of warfare. The chapter proposed an alternative way of seeing history by drawing out the importance of judgment and narrative in the justifications for war. This chapter has pointed to how history plays a role in how just war theorists make judgments about war. Hugo Grotius, Michael Walzer, and Cécile Fabre have all played important roles in helping us to understand the ethical dilemmas that war and the use of force create. By using their work as a starting point, I have tried to suggest some alternative frameworks through which we might understand war better. By purposefully injecting politics into what is often seen as a simple set of moral judgments about war, we can begin to think more carefully and critically about how we all make those evaluations.

[49] Carty, above note 11.

3

Imperialism, International Law, and War: Enduring Legacies and Curious Entanglements

Siddharth Mallavarapu

1. Introduction

The objective of this intervention is straightforward. It seeks to examine the nexus between imperialism and international law in the closely allied realms of the use of force, the international discourse surrounding responsibility, and ultimately war itself. None of these dimensions make for strange bedfellows. Imperialism has demanded a robust and nimble-footed yet ambitious global legal edifice that props it up, explains its invocation in diverse historical milieus, and makes room for exceptions whenever warranted in the actual application of the law.[1] It is premised on two other important elements. The first relates to 'standards of civilization'.[2] From the historical record, it is evident that not everybody is invited to the high table. It demands a certain great power etiquette and equipoise to be welcomed. And second it is premised on conquest. High achievers are invited and there is disdain for those at the receiving end of history. It is a world of victors with little empathy for losers.

If one were to get straight to the heart of the matter in political parlance, the nature of the beast is *sovereignty*. To those at the helm, there is a clear sense of what this means. To those who are labouring under the rigours of 'differential sovereignty'[3] or 'racial sovereignty',[4] what the term means to whom and under what conditions is more deeply contested. However, what is less disputed is the rhetoric of sovereign equality in a world of deep structural inequalities. In reality, sovereign equality may sound like a worthwhile aspirational tenet, though in practice every actor in history who has been denied that status knows for a fact that there is something elusive and chimerical about it.

In terms of a road map, the piece observes the following rhythms. It begins with rehearsing key claims positing a connection between imperialism and the nature of international law. In this context, legal facets relating to colonialism and 'racial

[1] Anghie, 'Imperialism and International Legal Theory', in A. Orford and F. Hoffman (eds), *The Oxford Handbook of the Theory of International Law* (2016) 156.
[2] Obregón, 'The Civilized and The Uncivilized', in B. Fassbender and A. Peters (eds), *The Oxford Handbook of the History of International Law* (2012) 917.
[3] A.A. Azoulay, *Potential History: Unlearning Imperialism* (2019), at 380–443.
[4] Nisancioglu, 'Racial Sovereignty', *European Journal of International Relations* (online first, 2019) 1, <https://journals.sagepub.com/doi/abs/10.1177/1354066119882991>

Siddharth Mallavarapu, *Imperialism, International Law, and War: Enduring Legacies and Curious Entanglements* In: *The Justification of War and International Order*. Edited by: Lothar Brock and Hendrik Simon, Oxford University Press (2021).
 DOI: 10.1093/oso/9780198865308.003.0003

sovereignty' merit further scrutiny. Second, it closely explicates two strands of argument that seek to decipher the workings of imperialism in the twenty-first century. Third, it also raises the inevitable question as to what might constitute a response to contemporary 'global imperialism'.[5]

There are a couple of other critical elements that this piece seeks to flag patiently. Questions relating to the use of force, the rhetoric of responsibility, and the actual practice of war warrant a closer examination. To illustrate the doublespeak of contemporary international law, the focus here shifts to 'human shields' and 'drones' as a recent technology of modern warfare.

A final aspect relates to institutional design. Are formal international organizations equipped to deal with the nature of crises faced by the world today? What other forums (if any) are better equipped to step up to the plate? Are international bureaucracies and their accompanying rationality more of a problem than a solution when it comes to addressing complex issues that remain germane to the very texture of world politics? These questions are aimed at unsettling any complacence about simplistic solutions that might be advanced in the light of the many contradictions that visit international law, imperial intent, and war in the twenty-first century.

2. International Law and Imperialism: Preliminary Observations

The most explicit argument forging a connection between imperialism and international law is advanced by Antony Anghie. His claim is that 'imperialism generates a new epistemology and a new set of questions when viewed from the perspective of those who have been the object of imperialism'.[6] More poignantly he asks, '[h]ow is international law, its operations and character, experienced and understood by peoples who have been systematically disadvantaged by its applications?'[7] Similar to the lament Robert Vitalis expresses of the American variant of the discipline of International Relations, premised as it is on a historiography that reveals how it papered over imperialism when there was a rich tradition within the discipline of such engagement in the early twentieth century particularly among a series of African-American scholars in the Howard School,[8] Anghie suggests that '[b]y the 1980s, then imperialism had essentially disappeared as a concern for scholars engaged in the theory of international law'.[9]

It is perhaps now as good a time as any other to bring imperialism squarely back into the discussion. What is worth noting is that 'the subjects and objects of imperialism are continuously shifting, just as imperialism takes very different forms—and always has'.[10] Distinguishing between a sovereign Europe and a non-sovereign 'Other',

[5] B.S. Chimni, *International Law and World Order: A Critique of Contemporary Approaches* (2nd ed., 2017), at 499–524.
[6] Anghie, above note 1, at 159 f.
[7] Ibid.
[8] R. Vitalis, *White World Order, Black Power Politics: The Birth of American International Relations* (2017).
[9] Anghie, above note 1, at 162.
[10] Ibid.

Anghie exposes the differential treatment meted out within the world of international law in the manner in which the non-European world was characterized vis-à-vis the European gold standard (used metaphorically, here).[11] The former colonies never measured up to their European colonizers and legal technologies of rule were actively deployed to remind them of their inferior status. To make matters worse, a metric of 'civilized' and 'uncivilized' was integral to the differential DNA of international law. As Liliana Obregón asserts, '[c]ivilization is a key concept to understand the imagined values of the 19th-century political communities and their relation to international law.'[12] The apparently benign language of 'civilization' masked exploitation and reproduced the fundamental lop-sidedness of power. There was a proliferation of diverse sources that re-enforced notions of 'civilized' and 'uncivilized'.[13] Obregón confirms that '[t]he 19th and early 20th centuries epitomized the use of the civilized/uncivilized in its relation to international law, in particular with the doctrine of recognition and the promotion of an implied standard for Europe's "civilizing mission".'[14]

What is of special relevance to us here is the manifestation of this binary when it comes to questions of peace and war. Undoubtedly a substantial part of future scholarly endeavours in this domain will have to unpack what transpired particularly in relation to questions of war and peace.[15] The untold story of international law might reveal a great deal more about the process of instantiation/imposition of a conventional European template and simultaneous attempts to obscure it with an apparently neutral, benign, and earnest language. Advancing a research agenda for interested scholars, Obregón argues that '[y]et a comprehensive history of the concept of civilization, and its accompanying adjectives civilized/uncivilized in international law has to be written.'[16]

If we treat colonialism as a particular phase of imperialism, international law again acquires a different tonality. Mathew Craven for instance suggests that

> [i]t is no longer possible to read Grotius without attending to the fact that much of his work seemed to be written as an apology for the whole Dutch commercial expansion into the Indies, or engage with the historic formation of notions of war, sovereignty and territory, and not notice the role they assumed in violent expansion of European empires.[17]

How often is Grotius taught against this backdrop (see also the chapter by Lang, Jr)? More emphatically, Craven observes that

> . . . colonialism has remained a significant background theme, providing the setting for doctrinal debates over freedom of the high seas, the use of force, title to territory, recognition, and statehood. Dealing with its legacy, of course, was also a central

11 A. Anghie, *Imperialism, Sovereignty and the Making of International Law* (2005).
12 Obregón, above note 2, at 917.
13 Ibid.
14 Ibid., at 921.
15 Ibid.
16 Ibid., at 938.
17 Craven, 'Colonialism and Domination', in Fassbender and Peters, above note 2, 862, at 862.

> preoccupation in the 20th century both informing institutional initiatives (mandates and trusteeships) and emergent doctrines such as that relating to self-determination, sovereignty over natural resources, human rights, the law of armed conflict, state succession, and scarcely a single arena of international law that has not, in some manner or other, been informed by this history.[18]

It is virtually impossible therefore to conceive of the international as scripted on a *tabula rasa*. Any nook or cranny of international legal jurisprudence you turn to there are more than evident traces of past lineages and legacies that lodge themselves in myriad forms in the present.

Not unconnected with the colonial quest was a special interest in the appropriation of land and territory. International law was an indispensable ally here as well. Andrew Fitzmaurice argues

> Europeans enjoyed a spectrum of legal arguments to justify empire and colonization over the 500 years from the first voyage to Columbus to the collapse of empire in the 20th century. There was a remarkable stability in these doctrines over those 500 years, even while they were subjected to ceaseless reinterpretation. The dominant doctrines were discovery, conquest, cession, and occupation.[19]

There were peculiar complications though. The first challenge was to either use or jettison the language of 'rights'. It was decided to go along with a warped interpretation of rights as opposed to doing away with it altogether. Fitzmaurice captures this move well when he suggests that

> [t]he idea that subjects, or individuals, possess rights was central to European political consciousness from late medieval times. Rights theories were based on natural law and therefore made claims to universal application. It was the universal notion of rights which impeded the arguments of discovery and conquest. Europeans therefore sought a justification of the encroachments on other people's lands that would be proofed against the rights-based critiques. In such circumstances, the best justification would itself have to come from the rights tradition and this is precisely why apologists for European expansion seized upon the argument of occupation.[20]

Occupation paved the way for land grabs. Treaties were part of the compact, whether in Africa or elsewhere like Australia. Historically,

> [b]y the time the first professional body of international jurists was established in 1873, the *Institut de droit international*, the effort to codify international law led to a taxonomy of occupation which included not only *res nullius*, but also *territorium nullius* and *terra nullius*. Each of these terms was employed in order to distinguish

[18] Ibid., at 863.

[19] Fitzmaurice, 'Discovery, Conquest, and Occupation of Territory', in Fassbender and Peters, above note 2, 840, at 840.

[20] Ibid., at 851.

> the particular conditions in which occupation may be possible and also to distinguish different possible degrees of occupation.[21]

In a nutshell, what emerges clearly from these accounts is the realization that

> . . . international law surged in the 19th century as a discipline constituted by the tension in defining its inner and outer limits between the civilized and uncivilized. The events that mark the dates of international law's historical narrative and origins are defined around these spheres, with the final utopia of international law as the ultimate achievement of civilized progress. To be *in* international law was to be part of a civilization while to be *out* of international law was to be lawless and savage.[22]

This is something which comes up with unfailing regularity and any observer of the scene cannot miss in our times the deeper import of this framing.

3. Sustaining Imperialism: Three Accounts of Sovereignty

It might be worth our while to dwell on an integral meta-facet of the link between international law and imperialism, namely, sovereignty. We have already alluded to the manner in which sovereignty was crystallized in its European self while all other selves were rendered as pale shadows inadequate to the task of measuring up to similar standards. Perhaps as a start, we must all take Audre Lorde (1932–1992) more seriously as Ariella Aïsha Azoulay recognizes.[23] When invited to speak at a feminist collective, Lorde bemoaned '[t]he absence of any consideration of lesbian consciousness or the consciousness of Third World Women . . .'.[24] She went on to argue that '[w]ithout community there is no liberation, only the most vulnerable and temporary armistice between an individual and her oppression. But community must not mean a shedding of our differences, not the pathetic pretence that these differences do not exist.'[25] What matters '. . . is learning to take our differences and make them strengths. For the master's tools will never dismantle the master's house. They may allow us temporarily to beat him at his own game, but they will never enable us to bring about genuine change.'[26] Decolonized states may need to ask whether it is even possible to win the sovereignty game in the manner in which the dice has been cast. Can the 'master's tools' be subverted and rebooted all in one lifetime? Or is this an endeavour condemned to fail and remain a transient triumph, if any?

Three arguments are well worth considering here. The first is an argument advanced by Azoulay ultimately aimed at the re-imagination of a forgotten notion of sovereignty

[21] Ibid., at 856.
[22] Anghie, above note 1, at 938.
[23] A.A. Azoulay, *Potential History: Unlearning Imperialism* (2019).
[24] A. Lorde, *The Master's Tools Will Never Dismantle The Master's House* (1984), available at <https://collectiveliberation.org/wp-content/uploads/2013/01/Lorde_The_Masters_Tools.pdf> (last visited 13 March 2020).
[25] Ibid.
[26] Ibid.

trampled as we write by more virulent dominant statist forms. Reflecting on her own conflicting identities, French Jewish mother and Algerian father, Azoulay claims

> that prior to the destruction of worlds and the imposition of imperial sovereignty, different types of sovereignty existed and still exist and can be reconstructed through rehearsals with others. What are these worldly sovereignties that they perform? What were the sovereignties that were destroyed?[27]

Distinguishing between two kinds of sovereignty, the first referred to as 'imperial sovereignty' and the second labelled as 'worldly sovereignty', Azoulay advances a case for the latter. She clarifies what she means by these terms in some detail by configuring

> . . . sovereignty as a story of struggle between two forms of sovereignty, imperial and worldly. Imperial sovereignty consists of and manifests itself through ruling apparatuses, with minimum connection to worldly activities, which it tends to destroy or replace by extractive, productive, and computational activities required for its own operations. *Worldly sovereignty* refers to the persisting and repressed forms and formations of being in the world, shaped by and through intimate knowledge of the world and its secrets, of its multiple natural, spiritual, political, and cosmological taxonomies preserved and transmitted over generations and shared among those entitled and invested to protect them. Imperial sovereignty consists of the massive expropriation of people's skills so as to transform them into governable subjects in a differential body politic. Worldly sovereignty consists of care for the common world in which one's place among others is part of the world's texture.[28]

Azoulay is not entirely at ease with Anghie's critique of European sovereignty. She argues that Anghie's account does not sufficiently 'undermine the univocality of the concept of sovereignty itself . . .' beholden as it is to particular notions of 'power' as well as 'authority'.[29]

Anghie himself excavates other idioms of thinking about sovereignty. This brings me to the second account of sovereignty that I wish to place here for further consideration. A historical moment that is conducive to this project is the Bandung Conference in 1955. Anghie suggests that '[i]t is surely tempting to treat the Bandung Conference, the historical meeting of Asian and African states in 1955, as formulating a Third World vision of sovereignty in much the same way as Westphalia articulated an idea of sovereignty that dominated in relations among Western states thereafter.'[30] There were attempts to carve out other meanings of sovereignty. They stemmed from the recognition that 'virtually all the states of Africa and Asia, including China and Japan, were not regarded as fully sovereign and thus incapable of participating on completely equal terms in the international system and in the making of international law.'[31]

27 Azoulay, above note 3, at 383.
28 Ibid., at 387 f.
29 Ibid.
30 Anghie, 'Bandung and the Origins of Third World Sovereignty', in L. Eslava, M. Fakhri, and V. Nesiah (eds), *Bandung, Global History, and International Law* (2017) 535, at 535.
31 Ibid., at 537.

Against this background, *Panchsheel* as conceptualized in Bandung articulated five distinct principles. This includes '(1) mutual respect for sovereignty and territorial integrity, (2) nonaggression, (3) non-interference in internal affairs, (4) equality and mutual benefit, and (5) peaceful coexistence.' However, as Anghie points out, these principles resonate well with the UN Charter. For instance,

> [t]he ideas of non-interference and mutual respect for sovereignty and territorial integrity correspond with principles found in Article 2(4) of the Charter, which prohibits the use of force against the territorial sovereignty of a state; Article 2(7), which prohibits interference by the United Nations in matters within the domestic jurisdiction of a state; and Article 2(1), which holds that all sovereign states are equal.[32]

However, it is precisely the UN model of privileging states over all else that provokes Azoulay to stridently argue that

> the UN was founded with the double agenda of protecting imperial interests through national bodies founded on the imperial principle of differential rule and of making sure that appeals against sovereign states could be addressed only to institutions in which states' interests are represented. The UN, founded by imperial sovereignties such as the United States and France, cannot provide protection against differential rule since their structure was based on legitimating differential rule. Under the triumphant call of ending World War II campaigns of partition and occupation were pursued—in Palestine, Korea, India, Germany, Japan—in order to prepare the establishment of new sovereign states through which the principle of imperial sovereignty would be preserved. More than the League of Nations after World War I, the United Nations after World War II was instrumental in transforming differential sovereignty into the one agreed-upon form of living with others and consolidating authoritative nation-states as its constituent and component parts.[33]

A third variant of critically coming to terms with the legacy of sovereignty is best embodied in Kerem Nisancioglu's thoughtful introduction of the category 'racial sovereignty'.[34] Suspicious of claims rendering race and sovereignty as 'analytically and historically discrete' categories, Nisancioglu argues that 'a focus on sovereignty might help us better understand racism'.[35] His claims tie up well with our earlier thread of interests in the connections between colonialism, land, and the workings of international law. Nisancioglu argues that

> [i]n drawing together authority, territory and population as its defining components, sovereignty is arguably a distinctive racializing practice in its capacity to affect how we understand the classifications of, relationships between, and the technologies of control over lands and peoples.[36]

32 Ibid., at 539.
33 Azoulay, above note 3, at 415.
34 Nisancioglu, above note 4.
35 Ibid., at 2 f.
36 Ibid., at 3.

One specific site to study the 'materiality of racial sovereignty' is the 'increasing visibility of the border'.[37] Returning to Vitalis, if we regard International Relations as essentially not about inter-state relations but about 'race-relations', Nisancioglu's account reiterates 'the coarticulation of sovereignty and racialisation through practices of colonial dispossession'.[38] Most critically, 'racial sovereignty' is not a done and dusted bygone but 'structural and ongoing'.[39]

It is perhaps worth asking what implications these arguments carry for thinking about imperialism as we proceed. Both Azoulay and Anghie are alerting us to other modes of envisaging sovereignty that go beyond the 'Westphalian' historical benchmark. Azoulay privileges a 'worldly sovereignty' to the prevalent 'imperial sovereignty' and Anghie conveys Bandung sensibilities when it comes to recasting sovereignty. Nisancioglu inserts the consciousness of race and how it operates as a structuring device when it comes both to the cognition and subsequent instantiation of sovereignty. None of these models come without their riders and caveats. To Azoulay, it is worth asking what are the political conditions that might inhibit or encourage the realization of 'worldly sovereignty'. To Anghie, the question of the extent and success or failure of translation of non-Westphalian principles of sovereign world-making in modern crucibles like the UN merit re-examination. Finally, to Nisancioglu, we need to methodologically process trace what forms of racial assumptions inform the workings of sovereignty manifest not merely in international law but also in the articulation of politics between the powerful and the weak in the contemporary world. I suggest here that the answers to these questions hinge fundamentally on our assessment and reading of imperialism both historically and in terms of its twenty-first century manifestations.

4. Deciphering Imperialism: Two Theses

Drawing on a critical lineage, Emanuele Saccarelli and Latha Vardarajan make a persuasive case that if we wish to understand the use of force and justificatory rationales surrounding war in the contemporary epoch, there is really no escape from 'a serious engagement with imperialism as an enduring historical reality'.[40] Two distinctions are in order here. First, the authors argue we must avoid conflating imperialism with empire. Second, a similar conflation must be avoided with colonialism. With regard to the former, they observe that

> [o]nce placed in a proper historical framework, imperialism is not reducible to empire, and attempts to blur or conceal the differences between the two behind the undeniable historical continuities linking them are not just problematic at the level of theory but raise a number of political issues as well.[41]

[37] Ibid., at 20.

[38] Ibid., at 3. See also R. Vitalis, *White World Order, Black Power Politics: The Birth of American International Relations* (2017).

[39] Ibid., at 4.

[40] E. Saccarelli and L. Vardarajan, *Imperialism: Past and Present* (2015), at 6.

[41] Ibid., at 23.

Similarly while distinguishing first of all between species of colonialism, Saccarelli and Vardarajan note that while Spanish colonialism 'led to the accumulation of wealth', British colonialism of an 'imperialist' variety 'led to the accumulation of capital'.[42] In other words, '[f]ar from being an issue of specific policies, imperialism was a systematic feature of capitalism.' In order to understand the workings of imperialism, we need to recognize its role 'in the world arena'.[43]

Tracing the origins of imperialism to the 'late nineteenth century', Saccarelli and Vardarajan go on to argue that this specific historical phase generated 'profound changes in the working of capitalism, including the rise of monopolies, the dominance of finance capital, and the recurring of economic crises'.[44] None of this transpired without a heavy political fallout. To them 'war and violence have been with us for a long time. But the political and economic parameters through which they are systematically produced and reproduced today are part of a definite historical epoch of which the present is regrettably still part.'[45]

If we bring this analysis to speed with ongoing conflicts raging in different parts of the world, it continues to be essential to invoke the analytics of imperialism. Three heuristic moves are essential here. The first involves a reminder of classic geopolitics to explain contemporary conflicts, second an acknowledgment of the high noon of its manifestation during the Cold War years, and third its resilient unwillingness to recede from its more virulent manifestations even in the post-Cold War world. With regard to the first phase, Saccarelli and Vardarajan suggest that

> [t]he recent events in Syria have a series of important precedents, from the early twentieth century British protectorates in the Middle East, to the more recent overthrow and brutal killing of Muammar Qaddafi by some of the same elements now active against Assad, aided and abetted by the same foreign powers.[46]

Afghanistan also harks back to an earlier era 'with its own distinctive and brutal history'.[47] Saccarelli and Vardarajan ask if what we are witnessing is a continuation of 'the classic "great game" pitting rival powers against each other in the region since the nineteenth century'.[48] With regard to the second dimension, the Cold War years, they confirm what by now is well established that 'this was a period in which wars, rebellions, covert and overt involvement of the imperialist powers, the toppling of elected governments, and establishment of dictatorial regimes were to become routine'.[49] Calling the bluff on 'humanitarianism' (see B.S. Chimni's chapter in Part VII), Saccarelli and Vardarajan conclude that it was merely 'a fig leaf to justify imperialist political manoeuvrings … not in and of itself a novel phenomenon'.[50] Reading this

[42] Ibid.
[43] Ibid., at 54.
[44] Ibid., at 17.
[45] Ibid., at 5.
[46] Ibid., at 4.
[47] Ibid.
[48] Ibid.
[49] Ibid., at 171.
[50] Ibid., at 192.

account, you may ask why at all we must be any longer taken aback by political developments in Libya, Syria, or Ukraine.

An equally scathing appraisal of 'global imperialism' and its specific manifestation in contemporary international law (the third dimension alluded to above) as 'global imperial international law' is offered by B.S. Chimni in his book *International Law and World Order: A Critique of Contemporary Approaches*. For our purposes here, I shall merely flag the ten facets identified in this context:[51] the first aspect relates to finance capital. Chimni argues that '[t]he hegemony of finance capital is a function of worldwide liberalization of finance controls encouraged by state and non-state actors and embodied in bilateral, regional and multilateral agreements.'[52] Further, '[b]oth trade and investment treaties are being used by powerful nations to seek capital account liberalization without appropriate safeguards.'[53] The second dimension relates to transnational corporations or TNCs. According to Chimni, '[t]hese include hundreds of Bilateral Investment Protection Treaties (BITS), an increasing number of Free Trade Agreements (FTAs) with investment protection provisions' among other instruments.[54] A third dimension alludes to David Harvey's notion of 'accumulation by dispossession',[55] Chimni argues that international economic law is complicit in perpetuating this phenomenon. A fourth aspect relates to claims of free trade and specifically the counsel 'to developing countries ... to rapidly liberalize international trade'.[56] This is particularly ironical considering that 'developed countries industrialized behind tariff walls, and also the range of protectionist measures in place even today (e.g. in agricultural commodities)'.[57] Fifth, intellectual property rights regimes disadvantage the Global South when there is evidently 'the control of advanced capitalist countries over the knowledge economy'.[58] Sixth, Chimni argues that a feature of global imperial international law is 'the undermining of labour rights, both at the national and global levels'. Seventh, a manifestation of global imperialism is evident in the domain of the environment and 'the continued exploitation of Nature'.[59] An eighth aspect and particularly pertinent here is 'the displacement of violence to the third world'. This is illustrated through the cases of 'Afghanistan, Iraq, Libya and Syria [that] symbolize this displacement of violence to the third world; there are also a vast number of other ongoing internal conflicts'. A ninth dimension relates to the 'efforts of the advanced developed countries to limit legal migration leading to the growth of a global smuggling and trafficking industry'.[60] This is particularly visible also in attitudes towards both refugees and asylum seekers. A final aspect relates to the proliferation of 'international human rights law' as well as 'the ascendance of international

[51] Chimni, above note 5.
[52] Ibid., at 509.
[53] Ibid.
[54] Ibid., at 510.
[55] D. Harvey, *The New Imperialism* (2003).
[56] Chimni, above note 5, at 511.
[57] Ibid
[58] Ibid., at 512.
[59] Ibid., at 513.
[60] Ibid., at 514.

institutions established by global law that are coming to constitute a nascent global state'.[61]

Is there a way out of this impasse? Chimni is firmly of the view that

> [t]here are ongoing social movements the world over led by a range of forces that include left and democratic parties, civil society organizations, and a coalition of third world states that seek the reform of contemporary international law.[62]

His hopes are pinned on the need to make 'the narrative of resistance ... an integral part of the analysis of international law and institutions'.[63] Elsewhere reflecting on the Bandung legacy, Chimni claims that '[t]he essential point is that the struggle against new imperialism requires a certain deftness that combines peoples struggles with a focus on national solutions and strengthening national sovereignty with non-national movements against global capital.'[64] We might ask from this account what might it mean 'to have an anti-imperialist jurisprudence'?[65] Without further ado, let us proceed to consider a couple of proposed epistemic strategies to resist imperialism by recasting responsibility in the twenty-first century with implications for the manner in which doctrines surrounding the use of force, legitimation strategies coalescing around claims to responsibility, and the possibility of war itself may be more circumscribed.

5. Recasting 'Responsibility': Two Theses and Illustrations of 'Responsibility' Gone Awry

An interesting fresh set of claims are advanced by Robin Dunford and Michael Neu in a recent intervention that contextualizes Responsibility to Protect (R2P) afresh. Rejecting a naive *tabula rasa* approach to international conflicts that imagines 'crises like Syria simply crop up', they remind us of 'the deeper structural causes of mass atrocity, the way in which purported humanitarian saviours are heavily implicated in the constant reproduction of these causes'.[66] What emerges as particularly salient in this discussion is an invitation to pry open the category of 'responsibility' and recognize its differential moral valences both as idea and as practice.

Let us begin by examining what Dunford and Neu regard as irresponsible and what constitutes bad citizenship internationally speaking. Irresponsibility is premised on 'the arbitrary and distorting narrowness of the calculations commonly used to justify military interventions'.[67] Bad citizens act in bad faith when 'the states often called on to intervene are not fit for purpose'.[68] Some examples may further clear the fog and it may be worthwhile spelling this out in some detail. According to Dunford and Neu,

[61] Ibid., at 515.
[62] Ibid., at 516.
[63] Ibid.
[64] Chimni, 'Anti-Imperialism: Then and Now', in Eslava, Fakhri, and Nesiah, above note 30, 35, at 48.
[65] Ibid., at 41.
[66] Dunford and Neu, 'The Responsibility to Protect in a World of Already Existing Intervention', 25 *European Journal of International Relations* (2019) 1080, at 1081–82.
[67] Ibid., at 1095.
[68] Ibid., at 1093.

> [w]hen states are bad international citizens on account of the way in which they contribute to (and fail to address) everyday atrocity, refuse to protect people fleeing from conflict and help produce mass atrocities, they are not fit to engage in military intervention in order to protect people. This does not only mean, for instance, that France was not fit to intervene in Rwanda on account of its arms sales to Rwanda, or that the UK is not fit to intervene in Syria on account of its failure to adequately support refugees displaced by the Syrian conflict. It is to say that such actors are not fit for the purpose of humanitarian military intervention full stop. Their moral character is such that they cannot be expected to protect responsibly. Rather, their military interventions can be expected, for reasons outlined earlier, to have inhumanitarian consequences.[69]

Returning to the question of responsibility, Dunford and Neu highlight 'the large extent of intervention that is already occurring, including monetary intervention, activity that degrades the environment and attempts to stoke ethnic tensions'.[70] These assorted actions are encouraged by intervening external powers with both resources, oftentimes self-serving sectarian political agendas and capacity. Besides this there are a huge number of unaddressed forms of structural violence, 'say, the extreme hunger experienced by 795 million of the world's people or the avoidable death of six million children under the age of five every year'.[71] There are deeper inconsistencies in what forms of intervention are viewed as essential by intervening powers, to what extent they tend to be tethered to more short-term geopolitical interests narrowly construed and their eventual efficacy in terms of intended mandates. None of these answers come easy as amply demonstrated by the existing historical record. The larger point Dunford and Neu advance is that

> R2P is concerned only with protecting a pool of people from a particular form of violation, namely, large-scale assault that is direct and physical. Other 'pools' that could be protected from other forms of violation through non-military means simply do not make it to the picture. This is an arbitrary choice; it discriminates against the millions who suffer greatly but find themselves in the 'wrong pool'.[72]

This brings us back to root causes that need to be addressed. A piecemeal approach to global conflicts will only work with symptoms while failing to acknowledge the deeper causes of international churning. The imperial international order is a huge, and nevertheless oftentimes overlooked, detail of this structural backdrop. Absent here is a larger discussion of the structural changes that are warranted to redress various forms of global inequality and the institutional wherewithal and international political resolve that might be required to actually make these changes feasible.

[69] Ibid., at 1093.
[70] Ibid., at 1081.
[71] Ibid., at 1087.
[72] Ibid., at 1088.

Another voice I would like to add here to help us think through the politics of resistance is Vasuki Nesiah who encourages us 'to challenge the injustices of empire'.[73] In her piece titled 'Resistance in the age of Empire: Occupied Discourse Pending Investigation', Nesiah compels us to take a close look at the question of torture in the context of the excesses committed by the United States at Abu Ghraib and the scant disregard Israel demonstrated for the International Court of Justice (ICJ) Advisory Opinion on the Wall of 9 July 2004.

To begin with, torture in this narrative is 'inserted into the logic of empire in complex ways'.[74] Nesiah argues that

> torture normalises resource exploitation and structural domination by presenting [it] as hyperbolically 'abnormal'—but torture may also enable dominion through the spectacle of dehumanisation, as it were, of how a subject population is the object of empire's will and whims.[75]

Torture, she goes on to suggest, is only a tiny part of a larger arsenal of technologies of rule evident for instance in Iraq. Nesiah reveals that apart from torture, 'there are also uniforms and flags at checkpoints, the awarding of contracts and the toppling of statues, the dropping of cluster bombs and driving of armoured vehicles'.[76] The paraphernalia of imperial formations is not to be trivialized. More than a decade later while Iraq is faced with other crises including those from its immediate neighbour Iran, it does not absolve intervening powers of their blood-stained culpability in contributing to Iraq and the region's many unresolved predicaments including an unnecessary loss of life of thousands of ordinary civilians and the destruction of world heritage—both intended and unintended.

Similarly, the ICJ Advisory Opinion rendering illegal Israel's construction of a wall on the West Bank, may be viewed

> as both a material and symbolic manifestation of the colonial occupation through its recounting of the systematic betrayal of the Palestinian people from the time of the mandate period. In effect, the ICJ's account of this entrenched history also suggests that the construction of the wall was enabled in many ways by the relationship between an occupying force and a subject people.[77]

Israeli non-compliance demonstrated that 'the ICJ opinion was a fly to be swatted away by an emboldened construction team pouring cement into the scar tissue of the laws and norms of self-determination'.[78]

Nesiah's overall verdict is that '[t]oday, self-determination discourse is a terrain whose critical energies are spent—spent by a politics whose terms were limited to the

[73] Nesiah, 'Resistance in the Age of Empire: Occupied Discourse Pending Investigation', 27 *Third World Quarterly* (2006) 903, at 903.
[74] Ibid.
[75] Ibid.
[76] Ibid., at 907.
[77] Ibid., at 915.
[78] Ibid.

nation-state; it has few resources to take on the more difficult internationalised politics of empire'.[79] We will need then to look elsewhere and eventually 'develop vocabularies of resistance that engage with this historical habitus'.[80]

6. Illustrating Imperialism in the Realm of War

A telling illustration of the salience of colonial continuities in the law of war is evident in the discourse surrounding human shields. Nesiah turns the tables on purported legal and normative standards espoused when it comes to human shields by highlighting the fact that

> [t]he colonizer/colonized distinction trumped the civilian/combatant distinction and exposed the radical instability of the principles defining the notion of human shields; the colonizer seldom thought he had reached the threshold of disproportionality in violence against the colonized. Instead, the 'civilizing mission' rendered the colonized body perennially vulnerable.[81]

A huge part of the problem lies with the 'strategic imperatives' narrowly defined rendering some humans over others as fair game for an avowed larger cause—inescapable 'collateral damage' at best.[82]

What is also worth paying attention to is the manner in which the human figure is invoked in these accounts. This is particularly conspicuous when it comes to an assessment of 'human shields'. Article 8(2)b(xxiii) of the International Criminal Court Statute states that 'utilizing the presence of a civilian or other protected person to render certain points, areas or military forces immune from military operations' is a 'war crime' from the perspective of international law.[83] Nesiah is categorical in her claim

> that today, the language of 'humanity' and 'counterterrorism' converge in the discourse of 'human shields' to facilitate a parallel dynamic in countries such as Afghanistan, Yemen, Syria [which] are rendered perennially vulnerable when framed as victims of belligerent forces deploying 'human shields'.

Nesiah points us to the contradictions that are rife in the deployment of human shields. She suggests 'while the language of human shields is advanced in the name of protecting civilians, often, it is also doing the opposite; namely, it is translating

[79] Ibid., at 918.
[80] Ibid.
[81] Nesiah, 'Human Shields/Human Crosshairs: Colonial Legacies and Contemporary Wars' (Symposium on Critical Perspectives on Human Shields) (2016), available at <https://www.cambridge.org/core/services/aop-cambridge-core/content/view/183CC05C31768B39BC727E5BAA9BBBF1/S2398772316000064a.pdf/human_shieldshuman_crosshairs_colonial_legacies_and_contemporary_wars.pdf> (last visited 13 March 2020).
[82] Ibid.
[83] ICRC, *Customary IHL Database*, available at <https://ihl-databases.icrc.org/customary-ihl/eng/docs/v1_rul_rule97#Fn_7198635_00002> (last visited 13 March 2020).

contemporary iterations of colonized or precarious bodies into "legitimate" targets'.[84] Thus, it comes as no surprise that 'Mosul, Aleppo, Gaza, Kandahar land on our headlines as places with stakes not only for humans living in those territories but for all humanity'.[85] When civilians are used to protect military targets as human shields, then the use of force to wipe those military targets themselves constitute a 'war crime' given the inevitable civilian casualties that would result from any such effort.[86] The ultimate outcome is that 'the law of war is itself weaponized'.[87] What could be worse?

Similar questions have also arisen in the context of drones. Traditional concerns about sovereignty and the 'non-use of force' appear to be the real casualty in this narrative. Markus Gunneflo for instance points out that

> [h]ere the problem is one of treating targeted killings as a human rights and humanitarian law problem while staying silent about the aggression against sovereign states they also constitute. This leads to a restructuring of legitimate forms of warfare with tremendous effects for post-colonial states and for international law more generally.[88]

Reminding us of an earlier set of claims advanced in relation to 'extraterritorial assassinations' which invariably invoked core interrogations of the violation of sovereignty, Gunneflo observes that 'we have, today, almost the opposite relationship, questions of sovereignty being consumed by considerations of the rights or status of the individuals affected'.[89]

7. Are International Bureaucracies Culpable?

I wish to briefly dwell here on institutional design questions and the impact it has on conflict prevention. The case that really stands out here is Rwanda. While there are many arguments that surface about major power apathy in Rwanda, it is perhaps also worth recounting that the UN Secretariat may have also been particularly myopic due to the tyranny of 'bureaucratic rationalization'.[90] Touko Piiparinen's scholarship makes this point rather sharply especially when it comes to a reliance on hierarchy when the warning signs of a pending genocide were clearly communicated well in advance to the UN Secretariat. As Piiparinen suggests,

> [t]he mechanisms of modern bureaucracies are notorious for their inflexible rules and procedures, whose function is to attain enhanced efficiency, accuracy and stability in

[84] Nesiah, above note 81.
[85] Nesiah, above note 81.
[86] Ibid.
[87] Ibid.
[88] Gunneflo, 'Drones and the Decolonization of International Law' (2018), available at <https://mei.nus.edu.sg/wp-content/uploads/2018/10/Markus-gunneflo.pdf> (last visited 13 March 2020).
[89] Ibid.
[90] Piiparinen, 'The Rise and Fall of Bureaucratic Rationalization: Exploring the Possibilities and Limitations of the UN Secretariat in Conflict Prevention', 14 *European Journal of International Relations* (2008) 697.

> decision-making. At times, however, this comes as a Pyrrhic victory. The downside, or dysfunction, of bureaucratization is routinized and mechanic behaviour.[91]

While Rwanda also brought to the fore prevarication by major western powers, it also revealed that a part of the problem lay in how 'the UN visualized the world through rigid, self-constructed bureaucratic categories which proved misleading and obfuscating'.[92] Readers will find an interesting account of the contradictory accounts of two UN officials, Dallaire and Booh-Booh particularly relevant to gauge the 'structural' impediments within the Secretariat in the context of Rwanda. This combined with 'intelligence gathering capacities of the UN still fettered by the uncompromising principle of impartiality' did not culminate in a proper acknowledgment at all of the nature of the crisis then brewing.[93]

8. In Lieu of a Conclusion

Consistent with the objectives articulated in my introduction, the piece seeks to consciously reside in the connections forged between imperialism, international law, and war. The desire to map enduring legacies compels us to revisit the big tropes—imperialism and the constitution of modern international law. The move to curious entanglements reveals how this regime manifests in an assorted contemporary legal regime around questions relating to the use of force, the discourse around international responsibility, and war. I do not intend rehearsing all these arguments again here but it might be worth reiterating some central contentions. The manner in which I have structured this piece and the literature I have summoned in order to establish scholarly claims in this domain are a conscious choice. Since these arguments are critical of the mainstream assumptions in the manner in which these questions are normally approached, I have taken more time to chronicle these voices as well as try to dip into the theoretical and empirical slices of evidence they draw on to advance their claims. Without further ado, let me distil in a nutshell these core claims and try to gesture to some possible implications this might have for anybody interested in rehabilitating international law as well as bringing back attention to the conditions that inform war and peace in the light of the burgeoning intersections of international law and world politics.

First and foremost there is an undeniable link between imperialism and the make-up of international law. One could extend the argument to suggest that the materiality of law manifest in formal international organizations also bears the weight of this history. The vantage point from which we observe international law and the evolving legal regimes around the laws of war in particular makes all the difference in how we view the process. Anghie's urgent reminder of turning the gaze to 'those who have been the objects of imperialism' reveals a new look at the furniture in the room and most critically the 'epistemology' that informed the arrangement of the constituent

91 Ibid., at 705.
92 Ibid., at 718.
93 Ibid., at 714 and 716.

parts.[94] Closely allied to this history of international law are the justificatory rationales of the 'civilizing mission', the logic of colonial conquest, the annexation of territories, and the distortion enforced on the life-worlds of the colonized. It is important not to view the colonized as passive putty in the hands of the colonial master as they fought back to articulate their own agency as the Bandung moment most clearly testifies to.

Second, conceptions of sovereignty are vital to understanding the links between imperialism, international law, and war. Three variants are presented in this piece. Azoulay makes a fervent plea for a 'worldly sovereignty' in contrast to 'imperial sovereignty', Anghie commendably resurrects a Bandung conceptualization of sovereignty and Nisancioglu's account reveals an ongoing project of 'racial sovereignty'. The reason why I juxtapose these different readings of sovereignty is to also partake of debates surrounding the re-negotiation of the enduring legacies of the histories of imperialism. Can we at all jettison some of these constructs or are we compelled to rehabilitate them in the light of their dominant reflexes which have been deeply unsettling particularly to 'those who have been the objects of imperialism'.[95]

Third, imperialism is a thread that runs through the entire piece. What is evident is its resilience as a meta-assemblage that seriously inflects both international law and world politics in more ways than one. Establishing the urgency and conceptual validity of the category to understand contemporary raging political conflicts and global political economy developments is integral to this narrative. Saccarelli, Vardarajan, and Chimni make a valuable set of contributions here that are well worth taking on board. To complicate the moral valences of resistance, I have drawn in this piece on some more intellectual allies who have tried to patiently think through how imperialism could be resisted against the backdrop of the dominant *Zeitgeist*. Dunford and Neu unravel 'international responsibility' for us as does Nesiah when it comes to singling out important moments in the hypocrisies surrounding the term. These range from torture as practice not peripheral but central to the perpetuation of hegemonizing some human bodies over others to the cold reception an ICJ Advisory Opinion receives when it seeks to render justice in the face of imperial support. These examples can be multiplied. Here the discourse surrounding 'human shields' and 'drones' come in for critical scrutiny both in Nesiah's and Gunneflo's individual accounts. What is interesting here is to observe the implications it carries ultimately for sovereignty where classic conceptions of sovereignty fall by the wayside and newer individual rights are privileged.

Finally, I also delve into international institutional bureaucratic politics with unfortunate consequences as in a failure to prevent genocide in Rwanda. If there is one thing all these various elements in my narrative suggest, it is that imperialism, international law, and war are here to stay and to hope that the 'better angels of our nature' prevail is oftentimes a naïve expectation in the light of a disturbing historical record.[96] However, this does not inevitably imply that we throw our hands in despair

[94] Anghie, above note 1.

[95] Ibid.

[96] This is a usage from Abraham Lincoln and also serves as the title of a book by S. Pinker, *The Better Angels of Our Nature: Why Violence Has Declined* (2011).

and cease trying to rehabilitate our record of collective violence, our differential treatment of fellow human beings, and a law struggling to be more just and persuasive especially to 'those who have been objects of imperialism'.[97] Only then can we begin to resuscitate claims both in the ideational and material realms to a genuine international 'responsibility'.

[97] Anghie, above note 1.

PART II
THE EARLY MODERN WAR DISCOURSE
A Process of Transformation?

4

Princes' Justifications of War in Early Modern Europe: The Constitution of an International Community by Communication

Anuschka Tischer

1. Introduction

The history of international law and of law of war usually focuses on their theory. Historians analyse the sophisticated treatises on just war that were written in particular in the sixteenth and seventeenth centuries by Francisco de Vitoria, Domingo de Soto, Hugo Grotius, and others. However, the decision about what was accepted by the international community was not taken by the theorists. It was taken by the princes and republics together with their councils and chancelleries, by the outcome of wars, and not in the end by the judgement of the public. Francisco de Vitoria for example argued in one of his lectures in the 1530s that the glory of a prince would be no reason for war, because it was of no common interest.[1] No other legal theory put that into question. Nevertheless, in 1672 Louis XIV declared war on the Netherlands and accused them of staining his glory with their behaviour, without even further explaining what this behaviour was.[2] The cause seemed to be weak, but no one questioned openly that the glory of the king of France was worth waging war for. The Netherlands refused the conjoined French and English declaration as 'contrary to all right and reason' which was a vague and quite common rejection.[3] They did not start a debate on glory. Yet, other princes hereafter did not follow Louis' XIV example. The argument of glory remained an exception. With the declaration of Louis XIV, however, it was accepted

[1] F. de Vitoria, 'De iure belli' 3,3, in F. de Vitoria, *Vorlesungen II (Relectiones): Völkerrecht—Politik—Kirche*, ed. by U. Horst, H.-G. Justenhoven, and J. Stüben (1997), at 656–58.

[2] 'La mauvaise satisfaction que Sa Majesté a de la conduite que les Estats Géneraux des Provinces-Unies des Païs–Bas tiennent depuis quelques années à son égard, estant parvenuë jusques à vn tel point, que Sa Majesté ne peut plus, qu'aux dépens de sa gloire, dissimuler l'indignation que luy cause vne maniére d'agir si peu conforme aux grandes obligations dont Sa Majesté, & les Roys ses Prédecesseurs, les ont si liberalement comblez.', in: *Ordonnance Du Roy: Par laquelle Sa Majesté, aprés avoir resolu de faire la Guerre aux Estats de Hollande, deffend à ses Sujets d'y avoir aucune communication ny Commerce. Du 6. Avril 1672.* (1672), 3. Bibliothèque Nationale de France (BNF): F-23613 (289).

[3] *A Placat of the High and Mighty Lords the States General of the United Netherlands; Conteining a sharp Prohibition to the Inhabitants of these Lands, and also concerning the Navigation and Commerce of the Neutrals to the Havens of France and England, and also Goods of Contrabande, and Aßurances of Ships and Goods belonging to the Subjects of the Kings of France and Great Brittany./Placcaet, Van de Hooge Mogende Heeren Staten Generael der Vereenichde Nederlanden, houdende scherp verboodt jegens de Ingesetenen der selver Landen; . . .* (1672), 1. BNF: FOL-NC-1082 (29).

Anuschka Tischer, *Princes' Justifications of War in Early Modern Europe: The Constitution of an International Community by Communication* In: *The Justification of War and International Order*. Edited by: Lothar Brock and Hendrik Simon, Oxford University Press (2021). © the many contributors. DOI: 10.1093/oso/9780198865308.003.0004

that for him and his successors on the throne, glory was a reason for war.[4] The practice of international law had caught up with theory.

The scholars who wrote on the theory of international law stated more or less their own academic view. They were Catholics or Protestants, had various national backgrounds, wrote in different epochs, were influenced by other authors or current events, had disparate opinions, and came from different schools.[5] Scholars had no general authority and were sometimes clearly partial: in his famous work *Mare Liberum* Hugo Grotius argued in 1609 in favour of the freedom of the sea and in particular of the trade routes, attacking thereby the Spanish-Portuguese monopoly and supporting the interests of his homeland, the Netherlands. John Selden on the other hand underlined in *Mare Clausum* England's dominion over its adjacent waters. When in 1652, during an Anglo-Dutch war for maritime and commercial interests, an English translation of the book was published, the translator praised this coincidence: 'It is a gallant fight to see the Sword and Pen in victorious Equipage together.'[6]

Many belligerent princes read treatises on international law and on just war. When King Gustavus II Adolphus of Sweden prepared his invasion of the Holy Roman Empire during the Thirty Years' War, he studied thoroughly Hugo Grotius' recently published *De iure belli ac pacis.*[7] Although, even if he did, neither his enemies nor his allies necessarily did the same. Moreover, no official war declaration or manifesto ever referred to a legal theorist. Only thorough analysis can sometimes give an idea of by whom a prince was influenced. Emperor Charles V for example was probably the first ruler in the sixteenth century who adapted the theory of mutual just war (*bellum iustum ex utraque parte*) invented by Francisco de Vitoria and Domingo de Soto. They suggested that a prince might be mistaken about the justice of his cause and actually be convinced that he was right. Then indeed both belligerents believed that they were in a just war.[8] It is, however, no real surprise that the two theorists had some influence on the Emperor as they were his confessors. Even the fact that Gustavus II Adolphus read Grotius should be seen in its context: the author wrote *De iure belli ac pacis* in exile in France and later became a Swedish resident. According to his biography, Grotius stood for an anti-Habsburg direction and like the Swedish king he was no Catholic.

If a prince considered legal theory for his politics, he chose a theory that was close to him: Gustavus' II Adolphus Catholic Habsburg enemies will hardly have read Grotius, whereas the Spanish School of Salamanca was no general authority for the Protestants of their time. The theorists evidently influenced the development of international law and they also compiled current views and practices, although they usually gave examples from history or from the Bible and not from their own time. Most notably, the

[4] For further examples see A. Tischer, *Offizielle Kriegsbegründungen in der Frühen Neuzeit: Herrscherkommunikation in Europa zwischen Souveränität und korporativem Selbstverständnis* (2012), at 157f.

[5] S. Besson and J. d'Aspremont (eds), *The Oxford Handbook of the Sources of International Law* (2017); O. Asbach and P. Schröder (eds), *War, the State and International Law in Seventeenth-Century Europe* (2010).

[6] J. Selden, *Of the Dominion, Or, Ownership of the SEA* ... (1652), at b.

[7] Piirimäe, 'Just War in Theory and Practice: the Legitimation of Swedish Intervention in the Thirty Years War', 45 *The Historical Journal* (2002) 499.

[8] Tischer, 'Der Wandel politischer Kommunikation im Kriegsfall: Formen, Inhalte und Funktionen von Kriegsbegründungen der Kaiser Maximilian I. und Karl V.', 9 *Militär und Gesellschaft in der Frühen Neuzeit* (2005) 7.

treatises do not indicate whether their statements were generally accepted, how norms were applied or what role they played in the wars of a single prince or republic.

2. Justifications of War—The General Framework

Such questions must be answered from practice. Peace treaties are therefore analysed for quite some time as sources of international law,[9] but their normative character should be regarded in light of the fact that there was enough room for interpretation.[10] In many wars the parties argued with different interpretations of a peace treaty. In particular the 1648 Peace of Westphalia, which became a kind of role model for international law, was at the same time a popular argument for war. It was used in this sense by the king of Poland only eight years after the settlement and still in 1792 during the French Revolution.[11]

The numerous public justifications for war are still an underrated type of source. Due to the invention of the printing press, nearly every war was accompanied by public justifications of sometimes more than 100 pages in length.[12] They were directed to the enemy in the form of a war declaration or counter-declaration, to other princes in order to explain why a war was regarded as just and necessary, and finally to an unspecified public.

A justification could not, of course, really reach everyone or even a majority of the people. The justification which was probably spread with more effort than any other before, namely the pamphlet promoting the invasion of England by William III of Orange in 1688, numbered about 60.000 copies.[13] It helped William to expel James II, but the number printed was marginal compared to the number of inhabitants of Europe or even of England. The public debate about the justifications was in fact limited to those who could understand them and who had access to a copy or at least to the content—besides the political interest in the printed justifications, the commercial aspect was already well known and calculated in the early modern period.[14] The growing market for news and information created a base for debates on war and international law. This audience was much broader than the group of princes, their counsellors, and scholars. Nevertheless, a recent study has shown that the comprehensive early modern publications of international treaties and other sources on international

[9] See for a general overview R. Lesaffer (ed.), *Peace Treaties and International Law in European History: From the Late Middle Ages to World War One* (2004).

[10] For the problems caused alone by the French translations of the Latin Peace of Westphalia see Braun, 'Les traductions françaises des traités de Westphalie (de 1648 à la fin de l'Ancien Régime)', 190 *XVII*e *siècle* (1996) 131.

[11] See also Tischer, above note 4, at 139–42.

[12] For a detailed list see ibid., at 223–61. Recently a wide transepochal and global analysis of public war justifications was made by Hathaway, Holste, Shapiro, van de Velde, and Lachowicz, 'War Manifestos', 85 *The University of Chicago Law Review* (2018) 1139.

[13] *The Declaration of His Highnes William Henry, by the Grace of God Prince of Orange, &c. Of the Reasons Inducing him, to appear in Armes in the Kingdome of England, for Preserving of the Protestant Religion, and for Restoring the Lawes and Liberties of England, Scotland and Ireland* (1688), see also Claydon, 'William III's *Declaration of Reasons* and the Glorious Revolution', 39 *The Historical Journal* (1996) 87. For further references see Tischer, above note 4, at 86.

[14] See Tischer, above note 4, at 86–7.

law were influenced by a complex mixture of political, academic, and commercial interests.[15] Thus, the source base for the history of international law and for the debate on it in the early modern period should always be critically reviewed.

The belligerents who justified their wars usually addressed no specific small group, but either formally their enemy or a wider public. Thus, their willingness to communicate and the new means of communication together constituted an open international community. Illiteracy was not regarded as an obstacle: as the state of war was a legal one, its declaration could be like other legal acts proclaimed at public places and thus be directed 'To all, who these Presents shall see or hear read.'[16] Unlike the works of legal theory, the justifications were rarely published in Latin but in more common languages like German, English, French, or Dutch. The wider the audience a prince wanted to reach—his own people, the enemy's subjects, or a wider European audience—the more copies and translations existed and were distributed in different countries. The authors sometimes insisted that their argument was clear enough to be understood by everyone or even by people of a simple mind. Critics complained indeed that official justifications and the reasons for war were discussed by ordinary people at public places. If the authors mentioned addressees, the circle could be very wide. Literally they directed their arguments to the world. Sometimes they claimed that it was obvious to everyone and everywhere, even in the extra-Christian part of the world, that they were right.[17]

This implies that political justifications of war were less academic than scholarly treatises, and they often have been regarded as sources of propaganda and not of international law. One can argue to the contrary, however, that international law was shaped by political interest, military means, and public debates together with the more academic theories. Justifications of war should be taken seriously as sources of international law, because they argued with it and were part of its practical consequences. Unlike modern propaganda, they did not demonize the enemy.[18] They relied on reason, on the justice of their cause, and pretended that their explications could convince everyone, including the enemy himself who usually was not accused of being a villain, but regarded as being mistaken, perhaps due to poor counsellors. The view was underlined by the academic theory that a war could be just on both sides, when all belligerents believed themselves to have a just cause and the legal situation was not really evident.[19] Thus, belligerents used public justifications to explain the whole situation conforming to their understanding of law.

Every war justification referred more or less to the three criteria established for a just war by Thomas Aquinas in the thirteenth century: the princely authority of the

[15] B. Durst, *Archive des Völkerrechts. Gedruckte Sammlungen europäischer Mächteverträge in der Frühen Neuzeit* (2016).

[16] *A Declaration of War of the States General of the United Provinces against the King of Sweden. Translated Out of Dutch* (1675), 3. BNF: FOL-NC-1082 (46).

[17] For the justifications in the context of the public sphere see Tischer, above note 4, at 79–107.

[18] See also Tischer, 'Den Gegner bekämpfen, aber nicht beleidigen: Friedensorientierte Rhetorik in frühneuzeitlichen Konflikten', in M. Espenhorst (ed.), *Frieden durch Sprache? Studien zum kommunikativen Umgang mit Konflikten und Konfliktlösungen* (2012) 97.

[19] See also Tischer, above note 8.

belligerent, his just cause, and his upright intention which meant that he (or she) went to war only for the just cause and was always ready for peace.[20]

The first criterion was crucial, because other princes or sovereigns in the first place had to accept a war declaration or manifesto. To be at war with someone meant to recognize his princely authority. Thus, a justification of war could be a means to demonstrate authority or even sovereignty: William I of Orange did so during the Dutch Revolt, when he published an apology in the form of a war manifesto in 1581 and demonstrated his just cause against Philip II of Spain. William pretended to be a sovereign prince and underlined his view together with the apology in an open letter to all Christian rulers.[21] Philip II on the other hand had put William's authority in question with a proclamation that authorized a possible assassination of the Prince of Orange for being 'a publique plague'.[22] Such a call would have been inconceivable against an enemy in an open, lawful war. The Spanish king underlined his own sovereignty which included the authority over William of Orange as his subject. A subject was not able to lead a just war against his overlord, because he did not have the authority to do so.

At the beginning of the Thirty Years' War the Bohemian estates pursued a strategy similar to that of the Prince of Orange and published apologies which made the revolt look like an act of political authority.[23] Unsurprisingly, Ferdinand II, Emperor and King of Bohemia, refused this view. He did not deal in detail with their arguments for the revolt, but declared in public that whatever their reason was, they lacked the legal basis to use force. For him as their sovereign prince their actions were insolence and rebellion.[24] Both the Dutch and the Bohemian Revolt had a different outcome: the first one was the onset of a new state, the second one ended with total suppression. In retrospect, the apology of William I of Orange is a justification of war, whereas the apologies of the Bohemian estates are sources of an internal conflict. In both cases however, the apologies were attempts by the authors to have their princely authority accepted, that is to be accepted for being able to lead a just war. It was the enemy, but in dubious cases even more so the international community, who judged this.

When a revolutionary government claimed authority, the use of a traditional and well-established international form like a public war declaration or manifesto was a suitable means to demonstrate that it was not the intention to overthrow the whole states system. The revolutionary government under Oliver Cromwell was quickly accepted by the European princes and referred in its war declarations of the 1650s to the 'Law of Nations', to its own 'Just and most Reasonable Causes', and to the international

[20] Thomas Aquinas, *Summa theologiae*, book 2,2, quaest. 40, Art. 1: Sancti Thomae Aquinitatis doctoris Angelici Opera omnia Leonis XIII P. M. edita 8, (1895), at 312.

[21] The Letter and the Apology are edited in H. Wansink, *The Apologie of Prince William of Orange against the Proclamation of the King of Spaine*, edited after the English edition of 1581 (1969), at 1–6, 12–147. See also ibid., at 17: 'I am so assured of the iustice of my cause.'

[22] The proclamation is edited ibid., at 150–69.

[23] *Apologia Oder Protestation aller Dreyer des Königreichs Böhaimb den Leib/vnd Bluth vnsers Herrn Jesu Christi vnder baider gestalt empfangender Ständt/Auß was für Vrsachen Sie haben ein gewisse Defension, vnd beraitschafft anordnen müssen. Erstlich Gedruckt zu Prag 1618.* (VD 17: 23:285025L); *Die Grosse Oder Andere Apologia Der Stände deß Königreichs Böheimb/so den Leib vnd das Blut vnsers HErren vnd Heylands JEsu Christi vnter Beyder Gestallt empfahen....* (1619), VD 17: 12:109639U.

[24] *Der Röm. Käy. May. Ferdinandi II. Edictal Cassation vnd Annullation, mit ahngehefter Reseruation, wider die angemaste newe nichtige Wahl vnd Crönung inn Böhaimb/etc.* (1620), VD 17: 23:685067L.

treaties concluded by James I and Charles I.[25] Even after the French Revolution of 1789 the means of justifications was used. When the revolutionary government after the execution of Louis XVI called its citizens to war against the 'tyrants' of Europe and to a holy crusade against the kings ('*croisade sainte contre les rois*'), it stated at the same time that its wars were purely defensive and mainly caused by the fact that its enemies did not recognize the French Republic as a sovereign state and would not foster a normal relationship with it.[26]

Princely authority, which was the basic criterion for a just war, was still a vague category in the early modern period, but it was more and more equated with sovereignty so that only a sovereign state or prince could lead a just war. One might argue that with the idea of sovereignty the just cause and upright intention even became irrelevant for a just war and that war became merely a political instrument, because a sovereign was responsible to no one and all his international actions were legal by definition.[27] The continuing practice of justifications, however, contradicts this view.[28] Even Louis XIV, the role model of a sovereign king, published a long manifesto in 1688 to demonstrate his 'reasons' and 'upright intention' to the 'whole of Christendom'.[29] The Sun King's language was now amazingly close to that of Thomas Aquinas, whereas he had commenced his government with a much more autocratic way of declaring war, for example on the Netherlands in 1672.[30] The Prussian invasion of Silesia in 1740 or the gradual partition of Poland suggest that the traditional just war theory was taken less seriously in the eighteenth than in the seventeenth century. Remarkably, precisely in the age of Enlightenment the difference between elaborate justifications and power

[25] Cromwell's government released its most elaborate declaration of war, probably written by John Milton, as an ally of France against Spain: *A Declaration of His Highness, By the Advice of His Council, Setting forth, On Behalf of this Commonwealth, the Justice of their Cause against Spain. Friday the 26th of October, 1655. Ordered by His Highness the Lord Protector, and the Council, That this declaration be forthwith printed and published* (1655), British Library: E.1065. (1.). Reprint in W.G. Grewe (ed.), *Fontes Historiae Iuris Gentium/ Quellen zur Geschichte des Völkerrechts/Sources Relating to the History of the Law of Nations*, vol. 2: 1493–1815 (1988), at 457–63. The above citations are ibid., at 457. Further justifications for the war against the Netherlands in 1652 are listed in Tischer, above note 4, at 241.

[26] *Exposé de la conduite de la Nation Française, envers le peuple Anglais, Et des motifs qui ont amené la rupture entre la REPUBLIQUE FRANÇAISE & le Roi d'Angleterre ... Envoyé aux Départmens et aux Armées* (1793), citation ibid., 29. BNF: 8-LE38-184; *Décret de la Convention Nationale, Du 7 Mars 1793, l'an second de la république Françoise, Qui déclare que la République Françoise est en guerre avec l'Espagne.—Rapport Sur les hostilités du Gouvernement Espagnol, & sur la nécessité de déclarer que la république Françoise est en guerre avec le Roi d'Espagne* (1793), BNF: 4-LA32-1094 (60); see also Eberl's and Simon's chapters in this volume.

[27] See e.g. Lesaffer, 'War, Peace, Interstate Friendship and the Emergence of the *ius publicum Europaeum*', in R. G. Asch, W. E. Voß, and M. Wrede (eds), *Frieden und Krieg in der Frühen Neuzeit. Die europäische Staatenordnung und die außereuropäische Welt* (2001) 87.

[28] For further arguments see Steiger, 'Ius bändigt Mars. Das klassische Völkerrecht und seine Wissenschaft als frühneuzeitliche Kulturerscheinung', in ibid., 59. The idea that sovereigns were free to go to war without restrictions is now also criticized for the 19th century: Simon, 'The Myth of *Liberum Ius ad Bellum*: Justifying War in 19th-Century Legal Theory and Political Practice', 29 *The European Journal of International Law* (2018) 113. See also the contribution by Hendrik Simon to this volume.

[29] *Mémoire des Raisons, qui ont obligé le Roy de France Louis XIV. à reprendre les Armes & qui doivent persuader toute la Chrétienté des sinceres Intentions de Sa Majesté, pour l'affermissement de la tranquilité publique* (1688), BNF: 4-Lb-37-3923.

[30] See also above note 2. For this change of Louis' XIV justifications see Tischer, 'Mars oder Jupiter? Konkurrierende Legitimationsstrategien im Kriegsfall', in C. Kampmann, K. Krause, E. Krems, and A. Tischer (eds), *Bourbon—Habsburg—Oranien um 1700. Konkurrierende Modelle im dynastischen Europa* (2008) 196.

politics was growing. Nevertheless, belligerents did continue to present their just cause and upright intention to the public.

For the time from the late fifteenth up to the late eighteenth century, there are more than 340 such printed justifications and replies.[31] Some of them have been analysed thoroughly, for example the Swedish manifesto of 1630 which was not in the end the result of Gustavus' II Adolphus and his councillors' reading of Hugo Grotius,[32] or the French declaration of war on Spain of 1635 which reflects the extensive previous discussions in the French government.[33] There are, however, only a few attempts to take a more general and continuous look at the justifications of war, although they reflect the pre-modern international community with its values, norms, and interaction.[34]

The community shaped by the justifications was a European and a Christian one, not only in respect of common values, but also because different means of communication emerged in Europe: the printing press, public postal systems, modern diplomacy, etc. Princes and scholars perceived the Ottoman Empire's laws and customs of war,[35] but a legal community and regular political interaction began cautiously only in the eighteenth century.[36] Still in that time the implementation of the printing press and public debates in the Ottoman Empire failed.[37] On the other hand, many forged Ottoman justifications circulated in Europe, whereas due to public control no other war declarations or manifestos were forged systematically.[38] Up to the eighteenth century, the community of European princes and states was moreover limited to western Christianity. When opening up to Europe, Tsar Peter I adapted its forms of international law, including public war manifestos and other standard documents.[39] He also invented public information by means of the printing press, although with little success.[40] By using traditional forms, the Tsar achieved, however, a smooth integration of Russia into the international system as it was established by the European powers.

31 See also above note 12.

32 See Piirimäe, above note 7.

33 Lesaffer, 'Defensive Warfare, Prevention and Hegemony. The Justifications for the Franco-Spanish War of 1635', 8 *Journal of the History of International Law* (2006) 91, at 141 (with reference to the exhaustive earlier research).

34 See Repgen, 'Kriegslegitimationen in Alteuropa. Entwurf einer historischen Typologie', 241 *Historische Zeitschrift* (1985) 27; Tischer, above note 8.

35 See e.g. A. Reland, *Türckisches Krieges-Recht/Oder kurtze Doch ausführliche Beschreibung der Weise/wie die Türcken gegen die Christen Krieg führen, und was sie dabey für Gesetze haben. ... Und nun zum erstenmahl in die Teutsche Sprache übersetzet, Zwey Bücher von der Türckischen oder Mohammedischen Religion*, vol. 2 (1717).

36 For a general overview see Komatsu, 'Die Türkei und das europäische Staatensystem', in C. Roll (ed.), *Recht und Reich im Zeitalter der Reformation. Festschrift für Horst Rabe* (1996) 121; for the nineteenth century, see Genell and Aksakal in this volume.

37 Kunt, 'Reading Elite, Elite reading', in P. Sadgrove (ed.), *Printing and Publishing in the Middle East* (2008) 89.

38 D. C. Waugh, *The Great Turkes Defiance. On the History of the Apocryphal Correspondance of the Ottoman Sultan in its Muscovite and Russian Variants* (1978); Maier, '"Ontsegh-brief van den Turckschen Keyser ... " Ein fiktiver Brief des türkischen Sultans an den König von Polen in russischer Übersetzung (1621)', in P. Ambrosiani et al. (eds), *Jako blagopesnivaja ptica: Hyllningsskrift till Lars Steensland* (2006) 135.

39 Piirimäe, 'Russia, the Turks and Europe: Legitimations of War and the Formation of European Identity in the Early Modern Period', 11 *Journal of Early Modern History* (2007) 63. For the use of the standard forms of capitulation see Ungerns-Šternbergs (Ungern-Sternberg), 'Kā notiek kapitulācija? Baltiešu kapitulācijas Pēterim I Eiropas kontekstā', 48 *Latvijas Vesture* (2002) 11.

40 G. Marker, *Publishing, Printing, and the Origins of Intellectual Life in Russia, 1700–1800* (1985), at 17–40.

The law of nations to which the war justifications of the European Christian powers referred in the early modern period was meant to be a global system and indeed a law of all nations. Due to the authority of classical pre-Christian authors, of Roman history, and of natural law, international law extended from the very beginning beyond the limits of Christianity. The authors of the justifications argued that in general their norms were accepted also by Turks and Tatars ('*Türcken vnd Tartarn*'),[41] by honourable pagans ('*bey den Ehrbaren Heyden*'),[42] even between the most hostile nations and the most barbaric ('*entre les Nations mesmes les plus ennemies & les plus barbares*'),[43] or even among the wildest nations and infidel pagans who have no knowledge of God ('*selbst unter denen wildesten Nationen und ungläubigen Heyden die von GOtt keine Kenntniß haben*').[44] Thus, on the one hand, the community was, at least theoretically, open to all nations which were ready to accept the rules.

On the other hand, however, these rules were defined by the European powers. Their concept was global, but international law was first of all developed for European relations.[45] When the European powers acted on a global stage they extended their own legal sphere to other continents and also used international law as a means for expansion. Since medieval times the Christian powers had developed legal strategies to justify wars of conquest against Moslems in the Mediterranean or 'pagans' in Eastern Europe.[46] Spreading the Christian faith was mandatory, thus every military defence of non-Christians could be delegitimized as resistance against the natural order. This was also the basis for the further development of international law in times of the overseas expansion since the late fifteenth century.[47] The theory of just war was affirmed, but it left hardly any option for non-Christians to wage a just war.[48] Eventually this concept worked even without the Christian argument, because the expanding European powers and many scholars regarded others (and in particular the American natives) as uncivilized and claimed to have a natural authority over them.[49]

[41] *Der Röm. Käy. May. Ferdinandi II. Edictal Cassation* (1620), above note 24, at Bii.

[42] *Abdruck Zweyer PATENTEN, Deren eins im Nahmen Ihrer Königl^{n}. Maytt. zu Dännemarck/Norwegen/ etc. am 7. Sept: jetzt-lauffenden 1658. Jahrs publiciret; Worin der gantzen Welt der Schwedischer Friedensbruch* [!] *kund gethan/und alle/in Dero Reichen/Fürstenthümbern und Landen in Schwedischen Kriegsdiensten/zu Wasser uñ Lande begriffenen/Landsassen/Vnterthanen und Eingebohrnen/ernstlich und bey Vermeydung schwerer Straffe anbefohlen wird/aus Schwedischen Diensten zu tretten. Das andere/von Ihrer Churfůrstl^{n}. Durchleuchtigkeit zu Brandenburg/etc. Anzeigend die Vrsachen/so die Römische/Kayserliche Mayt./vnd Ihre Churfůrstliche Durchl^{t}. bewogen/mit Dero Kriegs-Heer in Holstein zu rücken* (1658), [A3^{r}]. VD 17: 14:080778C.

[43] *Manifeste dv Roy de Pologne, povr servir de response av Manifeste pvblie par le Roy de Svede, Touchant la guerre qu'il fait à la Pologne* (1656), 3. BNF: M-6348.

[44] *Rußisches Manifest wieder Schweden. Petersburg 1741.* Printed in: *Sammlung XXIII. Curiöser Zur allerneuesten Staats-Historie gehöriger Kleinen Schriften* (1742), No. VII. Herzog August Bibliothek Wolfenbüttel (HAB): AB: Ge 643.

[45] See also Steiger, 'Rechtliche Strukturen der Europäischen Staatenordnung 1648–1792', 59 *Zeitschrift für ausländisches öffentliches Recht und Völkerrecht (Heidelberg Journal of International Law)* (1999) 609.

[46] R.A. Williams, *The American Indian in Western Legal Thought. The Discourses of Conquest* (1990), at 29–41, 59–74. See also S. Brand-Pierach. *Ungläubige im Kirchenrecht. Die kanonistische Behandlung der Nichtchristen als symbolische Manifestation politischen Machtwillens* (2004): <https://d-nb.info/972051961/34> (last visited 7 September 2019).

[47] See also Williams, above note 46.

[48] J. Muldoon, *The Americas in the Spanish World Order. The Justification for Conquest in the Seventeenth Century* (1994); A. Anghie, *Imperialism, Sovereignty and the Making of International Law* (2004), 13–31; Asbach and Schröder, above note 5.

[49] See also Williams, above note 46, at 100–08.

The Christian European powers were not the only ones who confronted others with their specific view of international relations. The Ottoman Empire regarded all its relations with non-Muslim powers as basically hostile. Accordingly, it did not conclude any enduring peace treaty with non-Muslims before the end of the seventeenth century and attacked its Christian neighbours without any legal concerns.[50] Finally it was the successful combatant who defined the rules of international law. These were, in particular in the global context, Christian European powers. The inner-European competition was a part of this process and the European conflicts, including the struggle for international law, were transferred to other continents: Spain and Portugal negotiated over their competition already in the fifteenth century and divided the world outside Europe among themselves. The Pope mediated this conflict and with its outcome, he approved the Portuguese and Spanish claims.[51] When Bartolomeo Diaz sailed along Africa in 1488 and Vasco da Gama proceeded on this route to India in 1497, both men erected huge pillars (*padrões*) ashore that were engraved with the arms of the king of Portugal and made his claim visible.[52] Other nations rejected this claim. In 1615 the English ambassador to the Great Mogul, Sir Thomas Roe, chose a specific European way to underline his different view: he formally declared war on the Portuguese vice-king at Goa.[53] As the scenario seemed odd and no military encounter took place, it has been questioned if this declaration really happened, but actually it made perfect sense in the way European powers competed overseas.

Despite their rivalry the European powers shared a common legal background and also an attitude to not taking overseas actors seriously as sovereign members of an international community.[54] Their specific policies varied. Evidently they had to deal with the particular conditions they came upon and which were completely different in Africa, Asia, or the Americas. In his chapter to this volume, Arnulf Becker shows how the Spanish treated the American natives like natural subjects: the *requerimiento* that was read to them seemed to meet the standards of a declaration of war, but its demand that they accept the king of Spain as an overlord and to allow the propaganda of the Christian faith made clear who was sovereign and who not. Even the Inca princes were treated like subjects whom the Spanish did not hesitate to kill. Their appeal to the authority of a natural law of nations that guaranteed a peaceful coexistence of different communities was in vain.[55]

[50] For the Ottoman–European relations see Komatsu, above note 36. See also above note 35. For the various Ottoman attacks on Venice see Poumarède, 'L'Entrée en guerre de Venise contre les Turcs (XVIe-XVIIe siècle)', in É. Schnakenbourg (ed.), *Les entrées en guerre à l'époque moderne. XVIe–XVIIIe siècle* (2018) 37.

[51] See also Williams, above note 46, at 78–81.

[52] An unknown member of Vasco da Gama's first expedition reported how several of those *padrões* were erected. He also noticed one of those put up by Bartolomeo Diaz a decade before: D. Kopke and A. da Costa Paiva (eds), *Roteiro da viagem que em descobrimento da India pelo Cabo de Boa Esperança fez dom Vasco da Gama em 1497* (1838), at 14–16, 90–91. Some of the *padrões* still exist today.

[53] The declaration, dated 20 October 1615, is edited in W. Foster (ed.), *The Embassy of Sir Thomas Roe to the Court of the Great Mogul, 1615–1619. As narrated in his Journal and Correspondence*, vol. 1 (1899), at 76f. An English representative in Japan, Richard Cocks, wrote about the declaration in a letter to Richard Wickham on 14 July 1616: K. Murakami and J. Murakawa (eds), *Letters Written by the English Residents in Japan. 1611–1623* (1900), at 181–82.

[54] For the long-term development of this attitude see Anghie, above note 48.

[55] See also the chapter by Becker Lorca in this volume. For the *requerimiento* see also Williams, above note 46, at 91–93; Muldoon, above note 48. The *requerimiento* is printed in Grewe, above note 25, at 68–70.

In comparison, the Portuguese in Asia were not in the position to conquer large territories and subdue people in a way the Spanish did in South America. Nevertheless, when Vasco da Gama travelled to India again in 1502 he brought with him a strong fleet and acted as a violent conqueror. He presented harsh demands to the prince of Calicut (modern Kozhikode) who had always shown hospitality to the foreigners. Although the prince behaved in a diplomatic manner, da Gama combined his demands with the use of force: he let thirty-two men hang in front of his ships and bombed the city of Calicut. Da Gama made clear that he regarded his claims as legal.[56] The way he confronted the prince of Calicut was, however, definitely not in the manner one might have addressed a European prince. During the following centuries various European powers fought numerous wars overseas and they had different ways of doing so. Nevertheless, there was a clear tendency in the early modern period to define overseas wars as being merely a kind of legitimate violence and not war in the full sense.[57] The concept of open just war which required a public justification and included the acceptance of the enemy as another sovereign was still restricted to Europe.

In general, the early modern law of nations had different facets. It was the basis for a regulated international interaction and war was an essential part of this interaction. The early modern law of nations was meant to be a law of all nations and actually it was the basis for our modern global international law. Apart from that, European powers still had double standards. Thus, the justifications of war which this chapter focuses on were from Europeans for Europeans.

3. The Way of Reasoning

The justifications of war further developed the idea of just war. The just cause which every just war needed remained nevertheless vague, and it would be difficult to make a list of such causes in public justifications. In general, a war was regarded as just when a sovereign or his ally were attacked or when their rights were violated. Among scholars of international law there was a consensus that fear alone was no just reason for a war.[58] Nevertheless, many justifications did not mention an attack but listed evidences that seemed to indicate a preparation of war. At the beginning of the Seven Years' War in 1756, both Frederick II of Brandenburg-Prussia, who in fact started the war, and Maria Theresa of Austria accused each other of having increased their troops and thus prepared for war.[59]

In times without precise boundary marks and control, it was not easy to judge what was an offence and what a defence. The concept that a strike back would be allowed

[56] C. v. Rohr (ed.), *Neue Quellen zur zweiten Indienfahrt Vasco da Gamas* (1939), at 20–23.

[57] The French e.g. also avoided any formalities of an open war when they attacked the Iroquese in 1609. See Lignereux, 'Et Champlain tira (juillet 1609). De l'hostilité à la guerre. Modalités d'une entrée en guerre aux lendemains de la fondation de Québec', in Schnakenbourg, above note 50, at 19.

[58] Oschmann, 'Der *metus iustus* in der deutschen Kriegsrechtslehre des 17. Jahrhunderts', in F. Bosbach (ed.), *Angst und Politik in der europäischen Geschichte* (2000) 101.

[59] *Déclaration du Roi sur les Motifs, qui obligent Sa Majesté, d'entrer avec son armée dans les états héréditaires du Roi de Pologne électeur de Saxe* (1756), BNF: M-13042.—*Anmerkungen über die von Anbeginn des gegenwärtigen Kriegs bis anhero zum öffentlichen Druck gediehene königl. Preußische Kriegs-Manifesten, Circularien und Memoires* (1756), 5r-8r. BNF: Manuscrits occidentaux, Ms. Allem. 27, fol. 5r-32v.

without further investigation was justified as a basic human right of natural law. The pre-modern society conceded the use of violence to all of its members in a way that was far beyond the current concept of self-defence and included not only a protection of the body and the house but also of honour.[60] Thus, everyone could basically follow the argument of Gustavus II Adolphus who, when invading the Holy Roman Empire in 1630 without having been attacked or his rights violated, stated that in everyday life people would go and extinguish the neighbour's fire without asking and waiting until their own house began to burn.[61] More than a century later, in 1756, Frederick II of Brandenburg-Prussia used quite a similar argument when he stated that his war was self-defence ('*Nothwehr*') and something that in accordance with all rights in the world is allowed even to the most humble human being for his defence and self-preservation ('*so nach allen Rechten in der Welt, auch dem geringsten unter den Menschen zu seiner Vertheidigung und Selbsterhaltung erlaubet ist*').[62]

It is even more difficult to give a list of precise causes when it comes to the violation of rights. Like in every other legal conflict there was a clash of different laws and various interpretations, and somehow every case was individual. The various wars of succession in the early modern period show this very well. The unclear legal situations and the conflicts they caused were the reason why many sovereigns concluded treaties in order to establish clear rules, but this was still a drop in the ocean. Still worse, many treaties became a reason of quarrel or an argument, too. Belligerents accused each other of having breached an agreement or having failed in their obligations.[63] In general, the list of just causes ranges from more familiar ones like a dispute over a hereditary right to more particular ones like having captivated the Pope or quite unusual ones like owing money to the belligerent[64] or disturbing commerce and violating nature by the deviation of a river.[65] Even in the last three issues the accused took the causes seriously in their responses and did not reject them immediately as a pretext.

[60] S. Backmann, H.-J. Künast, S. Ullmann, and B. A. Tlusty (eds), *Ehrkonzepte in der Frühen Neuzeit. Identitäten und Abgrenzungen* (1998); K. Schreiner and G. Schwerhoff (eds), *Verletzte Ehre. Ehrkonflikte in Gesellschaften des Mittelalters und der Frühen Neuzeit* (1995).

[61] The manifesto is edited in S. Goetze, *Die Politik des schwedischen Reichskanzlers Axel Oxenstierna gegenüber Kaiser und Reich* (1971), at 349–65, reference at 350.

[62] *Pro-Memoria, Oder Sr. Königl. Majest. in Preußen gründliche Beantwortung des zu Regenspurg von dem Kayserl. Reichs-Hofrath in Wien eingegebenen Commission-Decrets*. Dated Berlin, 9 October 1756, at [3] and 5. HAB: Gl 4° Sammelbd. 10 (2).

[63] See also Tischer, above note 4, at 139–42.

[64] Both mentioned in the declaration of Henry VIII of England against Emperor Charles V in 1528, printed in *Abclag beder Königen von Franckreych vnd Engelandt/Auch Römischer Kay. May. mündtlich vnd schrifftlich antwort zu Burgos. 22. vnd 27. Januarij/gehandelt/Im. 1528. Jar*, Aiii'''-B'. VD 16: M 583. Charles' response is in the same pamphlet.

[65] Frederick III of Denmark against Charles X Gustavus of Sweden: *Jus feciale Armatæ Daniæ Nebenst kurtzem Entwurff Vieler/hocherheblichen Bewegnüssen/dadurch die zu Dennemarck/Norwegen/der Wenden und Gothen/etc. Königliche Maytt: hochgenötiget worden/König Carll Gustaffen und der Cron Schweden/nach zugefügten/unerträglichen Beschwerden/und von Ihnen zerschlagener Güte/offenbahre Fehde zu Wasser und Lande durch dero Herold gebührlich anzukündigen/und alle nachbarliche Freundschafft auffzuheben* (1657), [Bii']. VD 17: 14:081843E.—Counter-justification of Charles X for the German estates: *Königlicher Majestät von Schweden Sendschreiben An Churfürstl. Durchl. von Maintz/Wie auch An andere des Heiligen Römischen Reichs Chur-Fürsten und Stände/Worinnen die zwischen beyderseits Könige in Schweden und Dennemarck gepflogene Streitsache benebens dero Umbständen erleutert/und das Reich ümb Hülfe gegen Dennemarck ersuchet wird* (1657), VD 17: 14:081493R.

The concept of a just cause becomes still more incalculable when we take into consideration that a ruler's main purpose was not to defend himself, but to protect his subjects and their rights. This obligation did not end at his country's borders. War justifications referred to actions that were committed against subjects in other countries or at sea: stealing or confiscating their commercial goods and ships, killings, and other kinds of injustice. Even excessive tolls were mentioned. In the time of emerging globalization and of competition in commerce and expansion, the causes for war evidently increased. On the other hand, the war justifications conveyed the idea that a prince represented in the first place his country's and subjects' interests and security, not his own. This was an important step towards the modern state.[66]

War justifications might not give a clear (or maybe better: might give an overly extended) answer to the question of what was regarded as a just cause, but there was at least a consensus on other ambiguous aspects of international law. First of all, everyone agreed that the reasons for every war had to be explained to the enemy and to others, and most belligerents offered such an explanation. This should not be underestimated. It indicates that they accepted, at least in theory, a kind of control, contrary to the popular view that an early modern prince was a ruler who could do more or less what he wanted and did not care for others' opinions. In fact, there came even criticism from below as from the Lutheran pastor Johann Ammersbach who in 1658 blamed the princes for their justifications, which in turn made the people judge and discuss the justice of wars, although they had neither the knowledge nor the authority to do so.[67] For Ammersbach this was against the order given by God, and he warned that it would finally lead to uprisings. The princes, who invited the audience to judge on their cause, did obviously not share this view. They affirmed themselves to be part of an international community and therefore not only willing to accept common rules and norms, but also to demonstrate that they observed them. The Declaration of Independence of the United States of America in 1776 followed exactly the pattern of a European war justification when the authors not only referred to the 'Law of Nature and of Nature's God', but also stated that 'a decent respect to the opinions of mankind requires that they should declare the causes which impel them to the separation'.[68] Finally, the European concept became a global one.

There was also a consensus among early modern belligerents that a war had not only to be justified, but to be declared in advance. Even in the few cases that did not follow this rule, the consensus itself was not put into question. When Gustavus II Adolphus invaded the Holy Roman Empire in 1630 without any previous warning, he avoided the term 'war'. Having studied international law and in particular Hugo Grotius, he may have judged that his cause for an invasion was weak. In the manifesto the king and his councillors tried to suggest that Sweden undertook 'a sort of "police-action"' (Pärtel Piirimäe), because the Emperor was not able or not willing

[66] See also Tischer, above note 4, at 148–51.

[67] H. Ammersbach, *Præpostervm. Mvndi. Jvdicivm. Oder Christliche Betrachtung der allgemeinen Kriegs-Discursen/welche heutigs Tags in der Welt geführet warden/Darauß zu ersehen/Ob und wie man zu diesen Zeiten im gemeinen Leben und Wandel von Krieg und andern Stats-Sachen redden/richten und urtheilen sol.* (1658), VD 17: 23:285292U.

[68] Facsimile and transcript of the Declaration with further information on the homepage of the US National Archives: <www.archives.gov/founding-docs/declaration> (last visited 25 August 2018).

to defend the Imperial Estates against Wallenstein's uncontrollable army.[69] Frederick II of Brandenburg-Prussia, once he had occupied Silesia in 1740, refused explicitly to have encouraged any idea that a declaration of war was not necessary.[70] He argued that he had negotiated unsuccessfully his demands and had occupied Silesia for security reasons during a particular situation.

It is also remarkable that all justifications of war were argued on a legal base. The whole explanation was strongly connected to the early modern European society, but the basic argument was always a legal one. This applies also to matters of religion.[71] Although, there were wars of religion in the sixteenth and seventeenth centuries, war justifications did not refer to religion as a just cause, but to the legal situation. A ruler could blame deviants as disturbers of public peace and order, but with the Reformation special laws like the Peace of Augsburg of 1555 were passed in many countries and regulated the relation between Catholics and Protestants. Subsequently, it was those laws that were disputed, not religion itself.

At the outbreak of the Thirty Years' War, probably one of the best-known wars with a religious background, the opponents quarrelled about different interpretations of the so called Letter of Majesty of 1609 which tolerated the Bohemian Protestants. When the rebels in their apologies called themselves those 'estates of the Kingdom of Bohemia who take the Body and Blood of our Lord Jesus Christ in both forms'[72] and thus underlined the religious dimension of the conflict, Ferdinand II insisted that the dismissal of the king and the election of a new one would be in the same way forbidden to estates who took the Holy Communion in only one form.[73] When later Gustavus II Adolphus did intervene in the Thirty Years' War in support of the German Protestants, he listed several reasons for his justification. He referred to religion only in so far as he admitted to interfering for the defence of Germans with whom he had much in common, including religion.[74]

The number of anonymous or even of semi-official pamphlets shows the real importance of religion in many early modern conflicts. Religion played, of course, a crucial role for the identity, government, and policy of early modern princes, and they mentioned their religious beliefs in their war justifications. However, religion was never mentioned as the essential reason for war, it was rather linked to defence, to laws, and treaties. Even in the wars against the Ottoman Empire the official argument was defence, not religion.

The aspect of religion demonstrates very well that on the one hand early modern justifications of war had a legal framework that was supposed to be global and adaptable to all nations, religions, and epochs, but on the other hand this framework was interpreted in accordance with the further political and social values of that period.

[69] Piirimäe, above note 7, at 517.
[70] 'Beantwortung der sogenannten actenmässigen und rechtlichen Gegeninformation über das rechtsbegründete Eigenthum des königlichen Churhauses Preussen und Brandenburg auf die schlesische [!]' Herzogthümer Jägerndorf, 'Liegnitz, Brieg, Wohlau etc. Anno 1741', in R. Koser (ed.), *Preussische Staatsschriften aus der Regierungszeit König Friedrichs II.*, vol. 1. (1877) 136, at 147f.
[71] See Tischer, above note 4, at 165–71.
[72] See also above note 23.
[73] *Der Röm. Käy. May. Ferdinandi II. Edictal Cassation* (1620), above note 24, at Aii.
[74] Goetze, above note 61, at 349–65; Piirimäe, above note 7.

This provided a unique mixture, because many justifications did indeed not focus on the just cause. They fulfilled much more the third criterion established by Thomas Aquinas which is often neglected in the history of international law, that is the upright intention.

In the early modern period, a type of war justification was developed that combined legal thinking with propaganda. It was a result of the demand for an upright intention, but it made perfect sense when the printing press and other improved means of communication allowed communication with a wider audience. For this audience the belligerent did not so much point out one single just cause, but he showed in a much wider frame that they shared similar views and values. The justifications thus ensured identification between a sovereign and his country, but also between other sovereigns in an abstract idea of Europe.[75] Even those who were at war against each other regarded themselves as members of the same community, because mutual respect and the affirmation of common features—often actually blood relationship—were basic. Enemies accused each other of having broken the rules, but both affirmed these rules' validity. Theoretically, the community which was outlined in war justifications was regarded as global, as a universal Republic of the World ('*universelle Republique du Monde*') as a Portuguese declaration of independence stated in 1641.[76] In fact, however, the community was the Christian society of Europe with its social order.

Essentially the argumentations of the war justifications were developed for and from the community of the belligerents themselves, the 'Society of Princes' (Lucien Bély), the political and social elite of Europe which shared the same ethos and way of behaving and which by numerous marriages was nearly a Family of Princes.[77] Many of their values were generally accepted. War justifications argued for honour, friendship, common features, or gratefulness, and they rejected ambition. These concepts and ideas were essential for the entire society.[78] They were values and virtues of personal interaction, because war was still regarded as a matter between two princes, not between nations or countries. With the further development of states, however, this concept slowly changed. The values of personal interaction were also transferred to the country or nation. When for example Elizabeth I supported the Netherlands against Spain in 1585, she stated that there was a 'speciall mutuall amitie to bee obserued betwixt the people and inhabitants of both parties', that is England and the Netherlands, and she confirmed that this friendship always had been stronger than any hostilities between the princes.[79] Later in the seventeenth century Oliver Cromwell and after him Charles II and William III of Orange transferred the term of honour from the personal to a national context, when they spoke of the honour of the English nation.[80]

[75] Tischer, above note 4, at 209–18.

[76] *Manifeste Du Royaume de Portugal. Par le quel se declare le droit, les causes, & le moyen qu'il a eu pour se soustraire de l'Obeissance du Roy de Castille, & se sousmettre au Sereniβime Don Ian quatrieme du nom…* (1641), 4. HAB: A: 57.10 Pol. (4).

[77] L. Bély, *La Société des Princes. XVIe–XVIIIe siècle* (2000).

[78] Tischer, above note 4, at 151–65.

[79] *A declaration of the cavses mooving the Queene of England to giue aide to the Defence of the People afflicted and oppressed in the lowe Countries* (1585).

[80] *A Declaration of His Highness* (1655), above note 25, at 463; *His Majesties Declaration against the States Generall of the United Provinces of the Low-Countreys. Published by the Advice of His Privy Council* (1671/2), BNF: FOL-NC-1082 (30); J. DuMont, *Corps Universel Diplomatique du Droit des Gens*, vol. 7, 2 (1731), at 230.

Remarkably, the concept that was used most and continuously in arguments was liberty or freedom.[81] The terms had many different facets in the early modern period. Basically, they referred to a traditional order, regulated by rights and privileges, or to a state of being independent from any other secular authority. The political use of the terms liberty or freedom still differed from the enlightened idea of an individual or natural freedom, but when during the French Revolution the notion of 'Liberty' became a key argument, the term was already well established on the international stage. War justifications from the sixteenth to the eighteenth centuries argued for the freedom of Italy, the freedom of the whole of Christianity, the Freedom of Europe, with national concepts of freedom like the German or Polish Liberty, with the freedom of the Sea etc. The terms were used in the sense that no prince or other power should be allowed either to dominate the whole community or another sovereign, but they were also used to support the political participation of estates against their ruler. Sometimes different concepts of liberty and freedom stood in opposition, and the pretension of liberty could also be rejected as self-exaltation.

A few powers, in particular Venice, argued for a fundamental idea of freedom as a value in itself. In 1617, a manifesto stated that from its very beginning Venice had pursued and protected liberty.[82] Its enemy, the future Emperor Ferdinand II, was obviously scandalized that a republic challenged him with the argument of freedom or liberty and suggested that the freedom of Venice was mainly an effect of pretension and injustice.[83] By the way, this quarrel demonstrates very well how easily an argumentation shifted away from the actual reason for war. Whereas the two counterparts discussed liberty and freedom, the war itself was about the piracy of the Uskoks in the Adriatic. Venice wanted the archduke, as an overlord of the Uskoks, to take action against them, and Ferdinand in his counter-manifesto admitted explicitly that the Uskoks were pirates. The just cause was thus not even disputed.

4. Conclusion: War Justifications in the Early Modern International Community

The official justifications of war of the early modern European princes show that there was a corporate identity of an international community, a community constituted by communication. It counter-balanced the idea of sovereignty which emerged at the same time. Even if the norms of the international community were often not followed, they were nevertheless affirmed by the numerous public justifications of war. We should also not underestimate their effects, because up to the eighteenth century no prince actually completely ignored the rules. The justifications were theoretically based on an open global concept of a law of nations, but they argued often for specific

[81] Tischer, above note 4, at 179–88.

[82] *Ragioni della Repvblica Venetiana Contro Vscochi. Stampata in Dalmazagho per Antonio Boron 1617.* Herzogin Anna Amalia Bibliothek Weimar: 22, 3: 13 [m].

[83] *Verantwortung/So zu behauptung der/Vrsachen/derentwegen deß Ertzhertzogs/Ferdinandi Durchleuchtigkeit/etc. zu gegenwärtigem Krieg bewogen worden/fürnemblich gestellt/Vnd der Herrschaft zu Venedig publicirten/Schrifft IL MANIFESTO intitulirt, entgegengesetzt wird./Jetzo erst auß der Italiänischen in Teutsche/Sprach gebracht/*... (1618), VD 17: 23:237271R.

Christian and European values of the early modern period and in fact were directed to the society where those values came from. When we suppose that the current international law is rooted in that of the early modern period, we should ask how far specific Christian, inner-European, and early modern values still shape the global international community. On the other hand, the common values became obviously less attractive in the eighteenth century. They were violated seriously by three Great Powers, Prussia, Austria, and Russia, with the partition of Poland, and the community broke in the end with the French Revolution. There are, however, strong reasons to argue that it was not a definite break. Maybe the consequences, in particular the dramatic conversion of Europe by Napoleon, came as a shock that underlined the necessity of common rules. The Congress of Vienna was a new trial to establish such rules. The chapter by Hendrik Simon will show how the idea of justifications of sovereigns lived on in a changed world in the nineteenth century.

5
The Legal Mechanics of Spanish Conquest: War and Peace in Early Colonial Peru

Arnulf Becker Lorca

1. Introduction: Echoes of the Death of Atahualpa in Spain

In 1533, Francisco Pizarro, the captain of the Spanish band that had entered Peru almost two decades earlier, ordered the execution of Atahualpa, the first Inca ruler who met the Christians in Cajamarca. Atahualpa was son of Huayna Capac, the last Inca emperor before the arrival of the Spaniards. When Pizarro arrived, Huayna Capac had died, and two of his sons were fighting over the Empire.[1] The famous first encounter between the Spanish band and an Inca ruler had happened the year before in Cajamarca, an encounter that ended with Atahualpa captured. After Pizarro explained to Atahualpa that they were seeking nothing more than gold for themselves and their emperor, the Inca extended his legendary offer: a room as big as the one where he was held captive would be filled with objects of gold as ransom for his freedom.[2] Instead, though the ransom had been paid, accused of betrayal for plotting a rescue by his forces, tied to a stick in the public square, Atahualpa was garrotted.[3]

As incredible amounts of gold and silver arrived in Spain, the events of Cajamarca and the fate of Atahualpa attracted much attention as well as criticism. As early as 1534, first-hand accounts by *conquistadores* were published in Seville, the same year that the Crown began to respond to communications from Pizarro and to issue instructions to Francisco's brother, Hernando Pizarro, who had travelled to Spain to present the treasure and secure privileges from the Crown. In the same year, priests and theologians discussed the events of Cajamarca, since returning *conquistadores* had been asking for advice and opinion on the legitimacy of *composición*, that is on compounding their rights to the booty with a payment to the Crown that would clear their titles from reproach.

None other than Francisco de Vitoria, the most prominent theologian of the School of Salamanca, in a letter to his superior who had been asked for an opinion

[1] See M. Rostworowski, *Historia del Tahuantinsuyu* (2017), Chapter 5.

[2] C. Mena, 'La conquista del Perú llamada la nueva Castilla', in R. Porras Barrenechea (ed.), *Las relaciones primitvas de la conquista del Perú* (1967), at 88. According to Porras, other early *cronistas* such as Francisco de Xerez and Hernando Pizarro did not recount the offer of a rescue in exchange of freedom, story that became widespread later. Ibid., at 89.

[3] F. Xerez, *Verdadera relación de la conquista del Perú* (1985) [1534], at 155.

Arnulf Becker Lorca, *The Legal Mechanics of Spanish Conquest: War and Peace in Early Colonial Peru* In: *The Justification of War and International Order.* Edited by: Lothar Brock and Hendrik Simon, Oxford University Press (2021).
DOI: 10.1093/oso/9780198865308.003.0005

on *composición*, famously noted that the mere mentioning of the affairs of the Indies 'freezes the blood in my veins'.[4] Without disputing the 'emperor's right to conquer the Indies', Vitoria questioned the justice of the war that led to the death of Atahualpa: 'Neither he or any of his people had ever done the slightest injury to the Christians, nor given them the least ground for making war on them.'[5] Granting that 'all the battles and conquests were good and wholly', Vitoria, however, observed that this war was 'not against strangers', but by admission of the *conquistadores* it was 'against true vassals of the emperor', who even if the emperor had just titles to conquest, did not know this. Indians—Vitoria concludes: 'are most certainly innocents in this war'.[6]

The Crown, on the other hand, also questioned the fate of Atahualpa. Responding to information sent by Pizarro, King Carlos V replied: 'I saw what you tell me about the Justice you made of the chief Atabaliba [Atahualpa] ... the death of Atabaliba being a *señor* has displeased me.'[7] The king's reprimand, however, was barely a slap on the wrist. In the same communication, Pizarro is thanked for the services rendered to the king, for the gold and silver obtained, and is reassured that he will be honoured and granted privileges. Moreover, recognizing that Pizarro had been informed that Atahualpa ordered 'war people to come against you and the Christians', the king concluded that since this course of action 'seemed to you appropriate it is fine for now' until further instructions.[8]

But there was another aspect regarding the execution of Atahualpa that was of concern to the Crown and that likewise deserved response and a mild reprimand. The ransom of Cajamarca had been divided by Pizarro setting aside the royal fifth, the tax owed to the Crown. The king, considering the exceptional nature of the events of Cajamarca claimed rights over the entire treasure rescued by Pizarro. 'In conformity to the law and the rules of our kingdom, when our captains and subjects that go with our forces take a principal lord or commander in chief of the lands they will conquest, the ransom and treasures ... are ours.'[9] Although the Crown notes that 'strictly speaking' it could demand restitution of the entire ransom and treasures, considering the work and services afforded to the king, it decided to let Pizarro and his men to keep and enjoy the whole booty.[10]

In the same year 1534, the first accounts by men who participated in the events of Cajamarca were seen by the Spanish public in print. An anonymous account, later attributed to Cristóbal de Mena, was published in April.[11] Later in July, Francisco de Xerez published *A True Account of the Conquest of Peru*.[12] Both chronicles recount the events surrounding the execution of Atahualpa from the perspective of the

[4] Vitoria, 'Letter to Miguel Arcos', in A. Pagden and J. Lawrance (eds), *Francisco Vitoria, Vitoria: Political Writings* (1991), at 331.
[5] Ibid., at 332.
[6] Ibid., at 332. 'If the Indians are not men but monkeys, they are incapable of injury. But if they are men ... I cannot see how to excuse these conquistadores of utter impiety and tyranny.' Ibid., at 333.
[7] R. Porras Barrenechea, *Cedulario del Perú, siglos XVI, XVII y XVIII*, (1944), vol. 1, at 91.
[8] Ibid., at 191.
[9] Ibid., at 207.
[10] Ibid.
[11] Mena, above note 2.
[12] Xerez, above note 3.

conquistadores. Mena recounts Atahualpa offering to pay a ransom in gold and silver to regain his freedom. 'Governor [Pizarro] promised him that as long as he did not betray him.'[13] In both chronicles it is Atahualpa who breaks the agreement.

While Hernando Pizarro left Cajamarca to collect the ransom, Mena reports that everyone who stayed had to work hard every day because 'that treacherous Atabalipa' (Atahualpa) continuously sent people to attack.[14] Likewise, other plots where revealed when Calicuchima, one of Atahualpa's captains, tortured by Hernando de Soto, confessed not only where to find more gold, but also, to have organized powerful forces to come against the Christians, retreating only because of orders given by Atahualpa who feared being killed by the Christians.[15] It was then, right after the division of the booty, that according to Xerez, a *cacique* informed Pizarro that Atahualpa had ordered his forces, including thousands of '*caribes* that eat human flesh', 'to come against you and your people and kill you all'.[16]

Confronted by Pizarro, Atahualpa, 'hiding his wickedness', laughed off the accusations.[17] After consultation with 'Crown officials, captains and people with experience', the Governor 'sentenced Atabaliba to death, and ordered to burn him for the betrayal he had committed unless converted Christian'. Having converted, Atahualpa was garrotted.[18] Xerez explains the rationale behind Pizarro's decision: 'for the safety of Christians and for the good of all the land and its conquest and pacification; because with Atahualpa dead, all that people will be defeated'.[19] 'For his sins he died,' paying for 'the great evils and cruelties he had done to his vassals.' For it was—Xerez concludes—'under tyranny that he had subjected all that land'.[20]

2. Spanish Colonial Law to Justify and Regulate Conquest

The echoes of Atahualpa's execution could be heard in Spain in different registers as members of the Crown, Church, and *conquistadors* gave respectively pragmatic, critical, and apologetic readings of the events of Cajamarca. However, this brief overview shows also a common language to understand and judge events. Entry to Peru, capture of Atahualpa, and the division of the booty, among other aspects of the conquest, were, as the Crown put it, regulated by the law and regulations—by '*derecho e leys*'.[21]

This was no modern law. The same medieval law and institutions that shaped the Christian kingdoms' expansion over the Islamic Emirates in the Hispanic peninsula, during the centuries-long process commonly known as *reconquista*, proved flexible enough to be transplanted to the Indies. As pre-modern law, this law cannot be easily disentangled from morality. It served to regulate and justify the use of force.

[13] Mena, above note 2, at 89.
[14] Ibid., at 91.
[15] Ibid., at 95.
[16] Xerez, above note 3, at 153.
[17] Ibid., at 154.
[18] Ibid., at 155.
[19] Ibid.
[20] Ibid., at 156.
[21] Porras Barrenechea, above note 7, at 207.

Compounding, for example, blending legal and moral reasoning, cleared both Spanish *conquistadores*' titles and souls. Enslavement is described by the Crown as 'tolerated as something that by law and regulations ... could be done without charge to our conscience'.[22] The known passion that the Spanish showed for legalism, the *requerimiento*—the document that upon encounter should be read to natives—as one of the most famous examples, is in part explained by this double function. Though difficult to disentangle, I will explore the interplay between the law as discourse of justification and as regulation.

Spanish colonial law had not one, but multiple sources: custom, compilations of written law, royal ordinances, agreements, and capitulations between the Crown or colonial officials and bands, instructions, as well as licences issued by the Crown and colonial officials and scholarly arguments about the law (backed by the authority of the scholar making the argument and the prestige of authorities invoked in support). This law formed a vast tangle of conflicting laws, a legal tangle where not one but multiple and conflicting projects of ordering coexisted.[23]

I use the term *legal mechanics* as a term that, recognizing the feudal character of the Spanish colonial law, grasps the ways in which the law shaped conquest. Law and legal arguments were used by the Crown, Church, *conquistadores*, but also Incas as well as other indigenous groups, to pursue and justify their own and divergent interests. In the echoes of Atahualpa's execution and in the rest of this chapter, we see legal argumentation portraying Christians and Incas as *señores* or tyrants, amicable or hostile; Incas as *caribe*, *amigos*, or *súbditos*, namely, enemies, friends, or subjects of the king. And for some it was Pizarro, for others Atahualpa, who broke the agreement, etcetera. Most legal scholars and practitioners would find this completely unsurprising.

Lawyers are accustomed to the messiness of legal argumentation. Legal scholars have identified this malleable and messy character as constitutive of law and explored the intimate relation between law and war.[24] More specifically to Spanish colonial law, David Kennedy has shown not only the malleability of Vitoria's just war, but also how this doctrine constitutes public authority.[25] These insights have travelled from the law of war to other areas of international law and then to international relations and the history of political ideas. These insights, however, have not really been taken up in the history of colonial and international law.

The history of the Spanish conquest of the Indies has been fertile ground for both scholars of *derecho indiano*, that is, Spanish colonial law, and for scholars of

[22] 'Provisión que manda que no se pueda captivar, ni hazer esclauo a ningun Indio, Madrid, 1530', in *Colección de documentos inéditos relativos al descubrimiento, conquista y organización de las antiguas posesiones españolas de Ultramar* (*CDIU*) vol. 10 (1897), at 39.

[23] See S. Stern, *Peru's Indian Peoples and the Challenge of Spanish Conquest: Huamanga to 1640* (1993).

[24] Nathaniel Berman has exposed the instability of the boundaries dividing lawful from unlawful violence, boundaries shifting in legal argumentation to place some wars and warriors within and others beyond what is legally permissible. Berman, 'Privileging Combat? Contemporary Conflict and the Legal Construction of War', 43 *Columbia Journal of Transnational Law* (2004) 1. When the law prohibits it also regulates war, not only empowering states to deploy legalized violence, but also enabling combatants to deflect to the law responsibility for their violence. See D. Kennedy, *Of War and Law* (2009). On the enabling and the restraining functions of international law see the contribution by Simon and Brock to this volume.

[25] Kennedy, 'Primitive Legal Scholarship', 27 *Harvard International Law Journal* (1986) 1.

international law.[26] Many focus on the defence of the Indians by the Church against the abuses committed by *conquistadores*. This defence turned into an interrogation of the titles the Crown could invoke over the New World, and to identifying just titles that entailed the recognition of indigenous peoples as legal subjects and finally to convincing the Crown to slowly change course, enacting new laws and policies that in the words of famed twentieth-century *derecho indiano* scholar Luciano Pereña, amounted to a reconfiguration of the colonial enterprise—the *reconversión colonial*.[27]

Early colonial Peru is central to this narrative as the impact of the execution of Atahualpa in the Spanish public triggered important legal and policy changes.[28] Changes in the law included, for example, the 1556 Prohibition of new wars of conquests, a Prohibition that was reaffirmed in the new laws of populations and discovery of 1573. Changes in colonial policy were reflected in new instructions directing colonial officials to peacefully negotiate with rebellious Indians. In 1543, for example, to secure pacification of the Incas, Carlos V instructed the viceroy Blasco Núñez Vela to pardon *indios* for the commission of crimes related to rebellion against Spaniards.[29] In 1563, as we will see, the governor of Peru, Lope de Castro, is authorized to negotiate with the Incas, and to excuse some of the harms they have caused.[30]

However, almost four decades after the killing of Atahualpa, at a time when the '*reconversión colonial*' was well underway, we will see Atahualpa's fate repeating itself in Tupac Amaru. In 1571, the last independent Inca ruler was executed. While *derecho indiano* scholars would probably explain this away as an exception or as justified, scholars with postcolonial affinities would see the execution as further evidence of the perils Indians faced when included as legal subjects. Anthony Anghie has forcefully demonstrated that the recognition of the Indian as a legal subject by theologians like Vitoria meant not only to have rights under the law of nations, but also obligations.[31] Avenging the violation of an essential duty imposed by the law of nations was a just cause to wage war.

[26] Martti Koskenniemi has challenged the conventional idea about the Spanish origins of international law rooted in the Salamanca School teachings on just titles, pointing at a different legacy. Koskenniemi credits Salamanca for the 'the development of a vocabulary of private rights (of dominium) that enabled the universal ordering of international relations by recourse to private property, contract, and exchange'. 'Spain was followed by the Netherlands and England ... their imperialism was the imperialism of free trade, carried out by private companies through private transactions and private war'. Koskenniemi, 'Empire and International Law: The Real Spanish Contribution', 61 *University of Toronto Law Journal* (2011) 1, at 1 and 36. In this chapter we see a similar challenge to the conventional view, but pointing backwards to an 'imperialist' framework of feudal character, of private/public war, private transactions and companies, but of monopolies, tribute, and grants.

[27] L. Pereña, *La idea de justicia en la conquista de América* (1992), at 135, 153–60. The reconversion is understood as projected by the school of Salamanca. See L. Pereña, *Carta Magna de los indios: fuentes constitucionales, 1534–1609* (1988), at 2. For similar ideas popularized in the Anglo-American context see L. Hanke, *The Spanish Struggle for Justice in the Conquest of America* (1949).

[28] The last two works of Bartolome de las Casas dealt with the Peruvian context, *De Thesauris* (On the Tomb Treasures of Peru) (1562) and in the *Tratado de las doce dudas* (The Treatise of Twelve Doubts) (1564).

[29] 'Provisión que dispone se procure traer de paz los Indios que anduvieren alzados y estuvieren en los montes y tierras, 1543', in *Cedulario de Encinas: Estudio e índices de Alfonso García-Gallo* (1945–46), vol. 4, at 278f.

[30] Guillen, 'Documentos inéditos para la historia de los incas de Vilcabamba: la capitulación del gobierno español con Titu Cusi Yupanqui', 10 *Historia y Cultura* (1976–1978) 47, at 62f.

[31] A. Anghie, *Imperialism, Sovereignty and the Making of International Law* (2005), at Chapter 1.

Unlike Atahualpa, who might have been innocent when resisting the Christians—because, as Vitoria suggested, there was no reason for him to know what the Spaniards intended when showing up in his land—Tupac Amaru, four decades after the first encounter and having killed a Spanish envoy, was guilty. In Vitoria's words, 'if the barbarians nevertheless persist in their wickedness and strive to destroy the Spaniards, they may then threat them no longer as innocent enemies, but as treacherous foes against whom all rights of war can be exercised, enslavement, deposition of their masters, and the institution of new ones'.[32] That Tupac Amaru ended up beheaded, his head on a pike exposed in the public square of Cuzco until it was removed because it became a subject of veneration, would come as no surprise to the postcolonial scholar.

However, the exploration of Spanish colonial law in this chapter with special reference to the continuities identified between 1533 and 1571, between the killings of Atahualpa and Tupac Amaru, vindicates neither the conventional nor the postcolonial scholar. The conventional view is correct to identify changes in attitudes, in colonial policy, and in the law that resulted from the denunciations by the Church of abuses against indigenous people committed by *conquistadores*, as well as from the activism by clergy, most prominently by Bartolome de Las Casas.[33] But the conventional view mistakes changes in the law for changes in the life of the law and legal argumentation. Therefore, it is not so much that changes of law failed at implementation, but rather that even after the reconfiguration of titles, after the *reconversión*, Spanish colonial law contained not only one but multiple and conflicting projects of ordering. *Reconversión*, we will see, did not change significantly the regulation of war.

On the other hand, postcolonial approaches, correctly see underlying religious and civilizational differences between Spaniards and indigenous as a basis for domination of the latter, even if recognized as legal subjects. However, the *modus vivendi* between Spaniards and Incas that emerged between 1533 and 1571 was much more fluid than postcolonial reasoning would have it, including, as we will see, not only diplomatic negotiations between Spanish and Inca (culminating, e.g. in the signing of the Treaty of Acobamba), but also alliances between factions of Spaniards, Incas, and other ethnics groups.

Both conventional and postcolonial perspectives foreground the dynamic of inclusion and exclusion, the former emphasizing inclusion, the latter exclusion, of the indigenous from the realm of the law as the decisive question determining the fate of the indigenous. In this context, both perspectives conceive the regulation of war and peace as depending on the inclusion and exclusion of indigenous. Both perspectives fail to grasp the contradictory and fragmented nature of a Spanish colonial law firmly rooted in a feudal framework. This oversight leads to another misunderstanding regarding the place and nature of war.

War, as a legal institution, as just war, was the most important title for Spanish presence in the Indies, a legitimate title for Spanish conquest, when natives failed to

[32] Vitoria, 'On the American Indians', in above note 4, 231, at 283; see also Vitoria, 'On the Law of war', ibid., at 293, on just war as punishment: 'it is lawful to avenge the injury done to the enemy, and to teach the enemy a lesson by punishing them for the damage they have done', at 305; on plunder, at 317, and on enslavement at 318.

[33] See L. Hanke, *All Mankind is One: A Study of the Disputation between Bartolomé de Las Casas and Juan Ginés de Sepúlveda in 1550 on the Intellectual and Religious Capacity of the American Indians* (1974).

comply with their obligations under the law of nations. As legal institution, war was also central to the legal mechanics of conquest, the actual colonial expansion, and its regulation. In early colonial Peru and specifically in the killing of Atahualpa, for example, the law regulated the constitution of Pizarro's band; capitulations as well as royal ordinances and customary rules governed relations between Crown and the band, and between its members; and most importantly the steps of conquest were also regulated institutions: entry—*entrada*—of the band into Peru to ransom—*rescate*—and then divide in the *reparto*, the booty and other privileges, like land grants and *encomiendas*—that is, royal grants over a group of natives, conferring rights, to extract tribute and labour, and imposing obligations, to protect and evangelize. The two sections that follow examine the double dimension of colonial law, the mechanics of justification, and conquest.

3. The Legal Mechanics of Justification

Historians have long recognized medieval precedents in the Spanish conquest of the New World. The crusading mind-set that emerged from a centuries-long struggle against Muslim kingdoms in the Iberian peninsula and from the myth of the Spanish *reconquista* that resulted from that struggle, explains not only the Spanish drive to expand overseas, but also the use of familiar frameworks to understand and organize the expansion.[34] We see in this chapter expressions of this mind-set in Pizarro's reference to the Incas as infidels, whose salvation would be brought about by Spanish evangelization; in the battle-cry 'Santiago' (patron saint of Spain who was invoked against the Moors during *reconquista*) now being used in Cajamarca to announce the attack against Atahualpa; or in references to mosques for the Inca temples where Pizarro's band rescued gold and to the devil for their deities.[35] The medieval precedents have also more specific legal significance.

Historians who recognize medieval precedents see them operating as frameworks providing a mechanics of expansion.[36] Arguably, legal discourse was one of the frameworks offering such mechanics. Historians like Felipe Fernández-Armesto understand these precedents as 'conceptual continuities' between the ways Spaniards thought about the conquests of Granada, Canaries, and the New World, and the ways they organized these conquests, institutionally, economically, and socially.[37] Exploring these conceptual continuities, Fernández-Armesto notes a special role for legal discourse: 'The juridical passion of Castilian monarchs and their advisers, the obsessive desire to justify in theory—even more, perhaps, than in practice—all their doings and especially all their wars, means that their conceptions were often made explicit.'[38]

[34] As Mario Góngora puts it: 'The legal formulae and the techniques employed in the enterprises of Africa and the Canaries were transferred to the Indies, at first without significant modifications.' M. Góngora, *Studies in the Colonial History of Spanish America* (1975), at 2. See in general, L. Weckmann, *La herencia medieval de México* (1984).

[35] See e.g. Xerez, above note 3, on Santiago, at 112; on idols and devils, at 124; and on mosques, at 127.

[36] R. Romano, *I conquistadores: meccanismi di una conquista coloniale* (1974).

[37] F. Fernández-Armesto, *Before Columbus: Exploration and Colonization from the Mediterranean to the Atlantic, 1229–1492* (1987), at 212–17.

[38] Ibid., at 212.

The juridical passion that results from a passion for moral correctness has marvelled but also daunted legal scholars who have to make sense of vast amounts of legal material. Legal historians like Silvio Zavala have produced massive works describing the colonial legal institutions that the Crown established in the New World. Zavala and others have identified and studied the medieval origins of these colonial institutions, including the law of war.[39] We will explore now the regulation of war as part of the discourse of justification. We will briefly explore the very important but probably over-studied debate on the legal status of the indigenous and the titles of the Crown to the New World, in order to understand the interplay between this law as justification with the law providing a framework for action and conquest.

3.1 Finding Titles and Defining *Indios*

In 1512, the Spanish king signed the *Leyes de Burgos* on the protection of *indios*, the first comprehensive law regulating Spanish conduct regarding the inhabitants of the New World. Conventional *derecho indiano* scholarship sees a strong link between the first denunciations by members of the Church of the abuses committed by *conquistadores* and the Crown's decision to examine and ultimately revise the titles under which the conquest of the Indies would be justified. Conventional scholarship also establishes a strong link between the reconsideration of titles by the Crown and the enactment of laws that recognized *indios* as free subjects and protected them form abuse. What started in 1512 as legislation protecting natives, evolved into a complete legal order that culminated in 1556 with the prohibition of new wars of conquest.

Conventional legal histories point at a linear progression in the protection of natives by showing jurists, theologians, and the Crown refining and reformulating their opinions regarding two interrelated questions: by what right may the Spanish enter the Indies and whether their inhabitants were as human as Spaniards were.[40] These were the same questions that lawyers, canonists, and theologians had asked in relation to infidels for centuries.[41] Answers, though applied to a new context, were deeply rooted in that earlier tradition. In relation to the first question, the Bulls issued by Pope Alexander VI restated the traditional position of the Church. The Pope bestows rights to occupy lands to be discovered and to enjoy monopoly over trade to the Spanish Crown. The Crown assumes the obligation to evangelize and finance the establishment of archdioceses.

The answer to the second question on the legal status of *indios* was full of ambiguities. Infidels, in this case *indios*, depending on circumstances, could be converted and become free subjects, could retain *dominium* without conversion, or could be enslaved. When enslavement of natives by the first explorers clashed with instructions

[39] S. Zavala, *Las instituciones jurídicas en la conquista de América* (1971), at 255. See also Mena García, 'Medieval Factors in the Conquest of America: Organisation and War Practices in the Incursions into Darién', in T. Glick, A. Malpica, F. Retamero, and J. Torró (eds), *From Al-Andalus to the Americas (13th–17th Centuries): Destruction and Construction of Societies* (2018) 130.

[40] Among many others, see A. Dougnac, *Manual de historia del derecho indiano* (1994), at Chapter 2.

[41] J. Muldoon, *Popes, Lawyers, and Infidels: The Church and the non-Christian World 1250–1550* (1979), at 136, Chapter 7, and *passim*.

issued by the Queen to regard *indios* as free subjects, the answer to the question required further clarification. Friar Montecinos' denunciations of abuses of *indios* in 1511 put additional pressure on the need to clarify the status and treatment of *indios*. Thus in 1512, the Crown convened a meeting of theologians and jurists, the Junta de Burgos, which resulted in the signing of the laws of Burgos and then the drafting of the most famous version of the *requerimiento*.

The Laws of Burgos and the *requerimiento*—in the eyes of conventional scholarship—confirmed the theory of Papal donation as title over discovered lands and answered the controversy about the nature of natives with compelling ethical force by defining them as free subjects rather than natural slaves. This ethical impulse that conventional histories identify is then seen as completed by the School of Salamanca's rejection of Papal donation and substitution of this title with others like the rights of Spaniards to travel, trade, and communicate with natives, which strengthened the recognition of *indios* as free subjects capable of possessing *dominium*—that is sovereignty.[42]

There are many reasons to question a linear progression in the law sanctioning the protection of *indios*. The answers formulated in a succession of *Juntas* did not resolve, once and for all, old disputes that continued for at least another century. For example, turning to the Peruvian context, as late as 1648, in his famous *Política Indiana*, Juan de Solórzano Pereira, an influential Spanish jurist who taught at Salamanca and then became judge in Peru—*oidor* at the *Audiencia* of Lima—defended the Bulls as legitimate ground for conquest.[43] Thus, lawyers and royal officials continued invoking Papal donation and its corollary evangelization as legitimate titles well after the laws of Burgos and after the critique of this title by Salamanca.

Similarly, in relation to the second question, even after Burgos and after the activism of clerics, the legal status of *indios* remained unsettled. Again, Solórzano Pereira, for example, defended natives' natural slavery, for 'it cannot be denied that all *indios* when we discovered them ... were slaves—*bozales*—... totally wild ... like beasts, without trace or form of society ... and they ate each other, as even today do in many islands those we call *caribes* and *caníbales*'.[44]

Burgos did not clarify the legal status of *indios*, but rather, the laws contained protections that left the question of legal status open.[45] The ambiguity was narrowed throughout a series of pronouncements by the Crown on the treatment of *indios*, since the beginning of explorations and continuing well after 1512. The Crown pronouncements that followed reflected the trajectory of the controversy, among royal officials, jurists, and theologians. Rather than natural slaves, *indios* became gradually free subjects. However, when translated into legal regulation, a general determination on the status of *indios* did not result in one, but at least in three legal categories. Natives could be: *súbditos*, *amigos*, and *caribes*. And as subjects, friends or enemies, *indios*

[42] See Pereña, above note 27; and Hanke, above note 33.

[43] J. Solórzano Pereira, *Política indiana* (1736) [1648], vol. 1, Book 2, Chapter 9, at 43–48.

[44] Ibid., at 37.

[45] On the mixed record of the Laws of Burgos, see D. Abulafia, *The Discovery of Mankind: Atlantic Encounters in the Age of Columbus* (2008), at 293f.

were under very different regulations. Let us explore now the counterpoint between subjects and enemies and leave *indios amigos* for the end of the chapter.

Converted and paying tribute in exchange for protection, *indios* were subjects, as we saw Vitoria declaring, they were 'true vassals' of the Crown.[46] Between 1495 and 1500, when *indios* began to arrive in Spain to be sold as slaves, the Crown first sanctioned the trade, then suspended the taking of profits from sales while inquiring on the causes of their capture, and finally ordered the liberation and return of some natives.[47] In 1501, in an instruction to Governor Nicolas de Ovando, we see an example of both the change in policy with an explicit recognition of *indios* as subjects, and the translation of that change into regulations.[48] Ovando is instructed to let *caciques* and other principals know that *indios* would be 'well treated as our subjects and vassals'. The governor is also instructed to ensure that clerics convert *indios* gently, to ensure that *indios* are well protected from violence and robbery, and that they be restituted from what has been unjustly taken from them. At the same time, as subjects, the Crown orders *indios* to pay taxes. Moreover, to collect gold and do other tasks, the Crown explains that 'it will be necessary to take advantage of the service of the *indios*'. Ovando is therefore instructed to 'compel them to work ... paying each one the salary that seems just to you'.[49]

Conversion by good will, well-treatment, securing obedience, taxation, and extraction of gold were objectives that the Crown had to carefully balance. And this was not an easy balance: better to negotiate than to enslave, while securing the rescue of gold, affirmed the Crown in a 1511 instruction to Alonso de Ojeda.[50] However, it was clear that stern consequences would follow if any of these objectives were to be resisted by natives. As early as 1503, the consequences for disobeying royal commands were well defined, turning *indios* into enemies worth to be called by their names: 'the said cannibals resist and do not want to receive and welcome in their lands the captains and people that by my command have been sent ... to be indoctrinated in the things of our Catholic faith and to be at my service and obedience'. The same Queen that had been so adamant about the protection of *indios* now gives 'license and power to take them captive ... sell them and take advantage of them'—but not without paying the tax belonging to the Crown.[51] Enslaving disobeying cannibals 'without incurring any penalty'—the Queen clarifies—is not only justified for the wickedness of eating humans, but also, in a usual argument used to authorize slave raids: it is better for slaves to be taken from infidels for their chances of salvations increase if brought under Christian masters.[52]

46 Vitoria, above note 4, at 332.

47 R. Konetzke, *Colección de documentos para la historia de la formación social de Hispanoamérica, 1493–1810* (1953), vol. 1, at 1–4.

48 'Instrucción al comendador frey Nicolás de Ovando, gobernador de las islas y tierra firme del Mar Océano', ibid., at 4–6.

49 Ibid., at 4–6.

50 'of the Indians who resisted our commands, if it would be good if they were taken as slaves, it would be a great inconvenience, because that has to be handled with great skill and some gentleness, to reduce them in good will ... the best negotiation that can be to do ... is to appease and give the Indians the best treatment possible ... and also to try by means of rescue, to extract all the gold that could be extracted', 'Real Cédula Alonso de Ojeda', in J. Toribio Medina, *El descubrimiento del Océano pacífico* (1913), vol. 1, at 18f.

51 'Provisión para poder cautivar a los caníbales rebeldes, 1503', Konetzke, above note 47, 14, at 15.

52 Ibid.

The line separating subjects from enemies, in consequence, was not only subtle, but also depended as much on the law as discourse of justification as on practical considerations of the law as regulation. Cannibalism, an ultimate breach of natural law, coexisted with arguments about *indios* disobeying or attacking Spaniards, and practical arguments about raiding slaves and paying taxes. In the Spanish conquest, these arguments were combined together in a legal document, the *requerimiento*.

3.2 Just War: Where *Requerimiento* and Salamanca Converge

In 1511, the king authorized the enslavement of *caribes*, another ethnic group we know by name. 'We sent some of our captains and clerics', the king explains: 'to preach and indoctrinate them in the things of our Catholic faith and to request—*requerir*—them to be at our service'.[53] To summon infidels, informing them about the reasons why Spaniards had entered their land, about what Spaniards demanded from them, and the consequences of failing to meet these demands, was an old practice originating in the Christian expansion against Moors.[54] In 1513, after the *Junta* of Burgos, that practice was put into writing in the famous *requerimiento*, a document drafted by a Salamanca jurist, Juan López de Palacios Rubios. This document was important, but not for its novelty. The practice was old. Bringing the basic arguments on titles and on the status of natives together in one document, the content of the 1513 *requerimiento* was also old. This version of the *requerimiento* became famous because of being written, and important because it marked the increasing reliance on just war as the central title to justify Spanish conquest.[55]

The purpose of the *requerimiento* was less to inform and convince natives, than to create a legitimate title for the Crown to oppose against other Christian rulers. It was also a title that could be easily invoked without causing controversy with the Pope. James Muldoon has argued that fearing the charge of heresy that a justification based on the rejection of the right of infidels to have *dominium* could spark in the Papacy, Crown officials preferred a justification based on a *requerimiento* relying on just war.[56] For example, the abovementioned 1511 authorization by the king accuses *caribes* of never wanting to hear, nor to welcome captains and clerics, rather they 'defended themselves with their weapons and resisted so that we could not *enter* ... and they killed some Christians, and in this harshness they have persevered ... waging war against the *indios* who are at our service'.[57]

The older legal tradition that Palacios Rubios embodied and the Salamanca school were at odds in relation to the latter's strong rejection of the Pope and emperor's

[53] 'Provisión que los indios caribes se puedan tomar por esclavos, 1511', ibid., 31, at 32.

[54] Patricia Seed traces the origins further back to the practice followed by Islamic rulers on the Iberian Peninsula, to summon, that is, invite and demand conversion to Islam. P. Seed, *Ceremonies of Possession in Europe's Conquest of the New World, 1492–1640* (1995), at 75–79 and in general Chapter 3.

[55] Ibid., at 70. Recognizing that infidels had dominium, the *requerimiento* based Spanish just cause to invade *indios*' lands on the refusal to admit peacefully missionaries. Muldoon, above note 41, at 141f.

[56] Muldoon, above note 41, at 142. Affirming that infidels did not possess *dominium* came close to reformist heretics' belief ideas that *dominium* depended on grace.

[57] Above note 53, at 32.

universal jurisdiction. In relation to just war, however, there was agreement. Christians could wage just war against *indios*, if the barbarians, in Vitoria's words, breached obligations they had in relation to Spaniards. Anthony Pagden has argued that Salamanca scholars, opposing prerogatives to *dominium* over infidels by both the Pope and emperor, considered consent and just war, as the only titles that could give the Crown a claim to *dominium* both in the sense of jurisdictional power and as property rights over land.[58] The Crown was interested in both having jurisdiction over subjects, especially *conquistadores* to tax, and having property, and thus authority to divide land, between *conquistadores* and royal officials, after conquest.

There is another reason for convergence between *requerimiento* and Salamanca. David Kennedy, examining the differences between pre-modern and modern international legal scholarship, has convincingly warned modern readers of Vitoria against seeing his recognition of Indian sovereignty as enabling the resolution, through the language of the law of nations, of conflicts between Spaniards and native rulers. Kennedy shows Vitoria's doctrine circumventing rather than offering resolution to normative conflicts between polities. Vitoria assumes Spanish and Indians to be bound by the same divine and natural order, and ascribes to that order ultimate authority. As there is no question that natural law and specifically that the law of nations is binding, there is no room for disagreeing about the terms under which relations between Spanish and *indios* should be governed.[59]

In this sense, both *requerimiento* as well as Vitoria's *jus gentium*, do not regulate relations between sovereigns, they rather establish legitimate public authority for one side, the one that reads the document, and the one identifying a just title against the other side who lacks titles. In the case of Vitoria, this is clear in his treatment of just war doctrine. Regulation of war, rather than assuming disagreement between sovereigns, becomes a mechanism to establish authority by transferring titles from a guilty party who is no longer sovereign, to an innocent party. The latter, becoming judge of his enemy, inflicts legal punishment according to the scale of his wrongdoing.[60]

The School of Salamanca and the medieval legal tradition expressed in the *requerimiento*, in consequence, converge around the centrality conferred to just war. There are differences between the legal thought behind *requerimiento* and Salamanca, though in relation to the Spanish rights to wage war against *indios*, these are matters of degree and in interpretation. Under the *requerimiento* there is a direct plea to subject and convert. For Vitoria, Spaniards' right to enter the Indies follows their 'right to travel and dwell in those countries', and 'trade among the barbarians', without doing harm to them.[61] If Spaniards can travel and trade, Vitoria notes, they can also 'preach and announce the Gospel in the lands of the barbarians'.[62] If there is no harm to the barbarians, they have 'no just war against the Spaniards', it is thus 'not lawfully for them to bar them from their homeland'.[63]

[58] A. Pagden, *Lords of all the World: Ideologies of Empire in Spain, Britain and France c. 1500–c. 1800* (1995), at 51f.

[59] Kennedy, above note 25, at 11f., 26, and 37–39.

[60] Ibid., at 31–35.

[61] Vitoria, above note 32, at 278.

[62] Ibid., at 284.

[63] Ibid., at 278.

In the abovementioned 1511 royal decree and in Vitoria's teachings there is a difference on what constitutes a just cause, respectively, refusing to convert, versus refusing Spaniards from travel and preaching, but there is agreement on the consequences of barbarians' refusal. As punishment for the crimes *caribes* committed against his subjects, the king gives 'license and authority' to make war, take captive, sell, and take advantage of *caribes*.[64] Vitoria agrees: 'if the Indians ... prevent Spaniards from freely preaching the Gospel ... Indians would be doing injury to Spaniards ... and these would have a just cause of war'.[65]

Legal scholars tend to stress the importance of the question about titles to conquest and about the status of *indios* as a new question prompted by the Spanish expansion in the Indies. Here we have seen that these questions as well as the answers provided for the new context of the Indies and the *indios*, were rooted in the Spanish medieval legal tradition. This continuity is relevant because if one needs to look back at the medieval laws of Spanish expansion, one finds not only the law as justification, but also the rules, doctrines, and institutions that provided a framework for action in war and conquest, that is law as regulation.

4. The Legal Mechanics of Conquest

As is well known, there were no formal military forces of the Spanish Crown conquering the Indies.[66] The *conquistadores*, the band led by Pizarro that entered Peru, for example, was not part of the Spanish Crown, it was rather a private enterprise. There was both written and customary law as well as contractual relations governing private enterprises like Pizarro's band. For example, in a capitulation signed with Pizarro and his partners—*compañeros*—the Crown sanctioned 'discovery, settlement and pacification' of the 'coast of southern sea towards east ... the lands and provinces of Peru'.[67] Licensed by the Crown to pacify Peru, Pizarro was ready to lawfully wage war against the Incas, and order the execution of Atahualpa.

In a typically feudal model of distribution of power and authority, Pizarro's band was both—*compañía*—private company and—*hueste*—host, or more precisely, *cabalgada*, that is, a smaller military group engaged in incursions.[68] Both dimensions were deeply intertwined, claiming for the Crown rights to discover and settle, the host waged war and plundered; as company, the band was formed by the contributions of its members who organized expeditions under their own risk, and distributed profits when a booty was obtained and divided. The combination of legal regulations of bands as private and public enterprises shaped the steps followed from discovery to pacification,

[64] Above note 53, at 33.

[65] Vitoria, above note 32, at 285.

[66] The conventional triumphalist account of a quick military conquest has been challenged by scholarship known as 'New Conquest History'. See Restall, 'The New Conquest History', 10 *History Compass* 10 (2012) 151. See in general M. Restall, *Seven Myths of the Spanish Conquest* (2004).

[67] 'Capitulación y asiento celebrados con Francisco Pizarro para la conquista y población de las provincias del Perú, donde se contienen varias disposiciones para el buen gobierno de ellas', *CDIU* vol. 9, 407, at 407f.

[68] M. Góngora, *Los grupos de conquistadores en Tierra Firme, 1509–1530* (1962), at 10.

shaped the organization and the structure of incentives for Crown, Church, *conquistadores*, and indigenous peoples. In short, the law offered a mechanics of conquest.

4.1 Pizarro's Band as *Hueste*

Before embarking on his third and finally successful expedition to Peru, Pizarro returns to Spain to negotiate with the Crown a capitulation authorizing conquest. In 1529, the capitulation with the queen gives 'license and faculty ... so that for us in our name and the Crown ... you can continue the said discovery, conquest and settlement ... of Peru'.[69] Towards the end of the capitulation a condition is included in relation to the conquest and treatment of *indios*: 'you are held and obliged to keep in everything and for everything contained in the ordinances and instructions that we have done and were made for this ... and in the provisions that we will send you for *encomienda* of these *indios*'.[70]

The Crown refers here not only to specific instructions that will be issued to Pizarro, but also to past ordinances on matters of conquest, arguably, to the entire legal corpus that regulates a conquest. A number of specific military practices, like entry, ransom, cavalcade—*entrada*, *rescate*, *cabalgada*—that are mentioned in instructions and capitulations, practices that taken together constituted conquest, were regulated by customary as well as written law, in compilation of laws going far back to medieval Spain, like the *Siete Partidas*.

For example, the abovementioned 1511 decree on enslavement of *indios caribes*, describes *caribes* as preventing Spaniards from *entering* the islands where they are.[71] And the capitulation of 1529, instructs Pizarro to pay taxes over gold and over other things obtained in *ransom* or *cavalcade*.[72] A band entering foreign land to plunder is a *cabalgada*. But as we see in the capitulation, the term *cabalgada* refers not only to the military band, but also, in the sense of cavalcade, to the raid made on horseback, and as such, it refers to the booty obtained from the raid for which royal taxes have to be paid.

Cabalgada, the practice of crossing a frontier, and *entry*, crossing into enemy land, were old practices rooted in the Spanish kingdoms' military engagement during *reconquista*.[73] Crossing into enemy territory to rescue Christian captives under infidel rulers, *cabalgadas* could also rescue infidel slaves who became part of the booty and could be sold as slaves. Claiming rights over the booty meant that the host, exercising public authority, should provide legal basis to justify the *raid*. In *cabalgadas*, the discourse of justification examined above, intersects with the practice of conquest, for a just cause under just war theory was necessary to claim title over the booty.

[69] Above note 67, at 409.
[70] Ibid., at 419.
[71] Above note 53, at 32.
[72] Above note 67, at 413.
[73] C. Mena García, *El oro del Darién, entradas y cabalgadas en la conquista de la Tierra Firme (1509–1526)* (2011), at 217–24.

The military ordinances issued in 1520 by another famous *conquistador*, Hernando Cortés, offers an example of this intersection between just war and military practice.[74] The term host, *hueste*, it has been pointed out, while used in medieval law, is rarely used in legal materials relating to the Indies.[75] Instead, the Crown describes bands as Spaniards, Christians, *conquistadores*, or *compañeros*. Independently of the label, the status of the band and its leader was clear. The leader of a recognized band, like Cortés and Pizarro, is a commander in chief—*capitán general*—with military powers and duties.[76]

The ordinance issued by Cortés begins with an explanation of the motives behind the wars of conquest. Naturals, as infidels, should be evangelized.[77] War, in consequence, should only be waged to reduce natives to the 'knowledge of our holly faith'.[78] Otherwise, war would be unwise and everything obtained in it should logically be restituted.[79] With *right intentions*, Cortés believes that the ultimate goal can be achieved, namely, claiming lands that legally belong to the Crown.[80]

Proclaiming right motives to 'use war', provides Cortés with legitimacy to enter Mexico and preventing restitution, to keep the booty. The practice of warfare, on the other hand, did not depend on just war theory, but on customary and written rules, among others, giving authority to the captain to organize combat. For example, Pizarro, the captain of the band of *conquistadores* that entered Peru, learning from the experience accumulated by Cortés in Mexico, followed the same strategy based on playing native ethnic groups against each other and on capturing the head of the empire to control the indigenous population. Similarly, both Cortés and Pizarro followed older practices, like rape, mass killing of women and children, cutting the right hand of captured combatants, torturing by fire to obtain information, the use of dogs to harass, torture, and—*aperrear*—kill *indios*, and in general the practice of domination through terror.[81]

[74] 'Hordenanzas militares dadas por Hernando Cortés en Texcallete, 1520', in *Colección de Documentos Inéditos relativos al Descubrimiento, conquista y organización de las antiguas posesiones españolas de América y Oceanía, sacados de los Archivos del Reino, y muy especialmente del de Indias, CDI* (1864–1884), vol. 26, 19.

[75] Torró, 'Partners-in-arms. Medieval Military Associations: From the Iberian cabalgada to the American entrada', in above note 39, at 19f.

[76] Pizarro, in the capitulation is referred as captain, and he is promised to be appointed governor and commander in chief of Peru. Above note 67, at 409.

[77] 'The natives ... have a culture and veneration of their idols, that our God is rendered great disservice and the devil because of the blindness and deception that brings, is very revered by them, and turning them away from so much error and idolatry and reduction to the knowledge of Our Holy Catholic Faith, our Lord will be very served ...' Above note 74, at 21.

[78] Ibid., at 22. 'I exhort and pray to all the Spaniards who in my company went to this war ... that in their main motive and intention, is to separate and uproot all the said idolatries to all the naturals of these parts, reduce them, or at least wish their salvation and that they be reduced to the knowledge of God'. Ibid., at 21.

[79] '... if the war were to be carried out with another intention ... everything obtained form it ... should be obliged to restitution'. Ibid., at 22.

[80] '... my main motive is to wage this war ... to reduce ... the said naturals to the said knowledge ... and then to subdue them, subordinate under the yoke and royal imperial dominion of your ... Majesty, to whom the lordship of all these parts legally belongs'. Ibid., at 22.

[81] On the resort to terror, cruelty, and rape, see E. Mira Caballos, *Conquista y destrucción de las Indias, 1492–1573* (2009), at 220–37. On massacres, including children, see A. Espino López, 'On the Use of Terror, Cruelty and Violence in the Spanish Conquest of the Americas: Some Thoughts', above note 39, 163, at 171, and *passim*. On the use of dogs see Mena García, above note 73, at 341–48. In Peru, killing—*rabonas*—women who assisted their husbands in combat, to deter men, or taking prisoners to mutilate and release

4.2 *Rescate*, Ransoming Slaves

War and economics were interlocked. The *rescate* of gold and silver, but most important of all, the capture of slaves, were the driving forces behind Spanish conquest.[82] We can understand ransom through the lenses of the regulation of slavery. In 1530, the Crown banned the enslavement of *indios*, only to reverse its decision in 1534. Determining the legal limits of native slavery was at the same time moral and legal as well as an economic issue for both *conquistadores* and royal coffers. We may grasp this double dimension by reading these decrees on enslavement. The decree of 1530 recounts the orders we explored above that limited enslavement to *caníbales* and *caribes*. Preventing armed attacks against preachers, just war was waged against *indios*, those taken prisoners could be enslaved.[83]

The relevant question was not the status of *indios* as free subjects or natural slaves, but the distinction between *indios* as subjects (who should not be enslaved), from *indios* who are enemies (and thus lawfully enslaved) and *indios* with whom there is less contact and who have neither become subjects, nor enemies and therefore are objects of *rescate* (who could be lawfully enslaved).

The Crown makes clear in the 1530 decree that war can be waged against enemies, killing, but not enslaving *indios*. Yet, less clear is the fate of *caníbales* and *caribes*, who became a general category to describe any other indigenous group turned enemy, a useful category that with the opprobrium of eating human flesh, a flagrant violation of natural law, offered a potent additional *causa bellis*. For example, in the abovementioned chronicle by Xerez, Atahualpa's forces are described as *caribe*.[84] Moreover, the *requerimiento* drafted for Pizarro reads: '… the summon that in the name of his Majesty should be made to the *Indios Caribes*, insurrected of the province of Peru'.[85] This was a general and abstract category to justify war against indigenous groups across the continent, rather than a claim to punish actual crimes, since if it is for cannibalism, there never were just war arguments against Spaniards, who when starving showed appetite not only for *indios*, but also for their fellow Christians.[86]

Enslavement in *rescate* followed a different rationale. When banning slavery and then when reversing its decision in 1534, the Crown distinguishes between the captivity of *indios* taken in just war, from those taken as ransom—*a título de rescate*.[87] The Crown gave licence to Christians to ransom—*rescatar*—*caciques* who possessed

them, Spaniards resorted to terror to instill fear among the population. See G. Lamana, *Domination without Dominance: Inca-Spanish Encounters in Early Colonial Peru* (2008), at 140f.

[82] See V. Huber, *Beute und Conquista: Die politische Ökonomie der Eroberung Neuspaniens* (2018).

[83] 'Provisión que manda que no se pueda captivar, ni hazer esclavo a ningún Indio, Madrid, 1530', *CDIU*, vol. 10, 38, at 39.

[84] Xerez, above 3, at 153.

[85] 'La forma y orden que se ha de tener en el requerimiento que de parte de su Majestad se ha de hazer a los Indios Caribes alçados de la provincia del Peru', in above note 29, at 226.

[86] Mira Caballos, above note 81, at 11f.

[87] 'Real Provisión donde se declara la forma y orden que se ha de guarder en hacer esclavos en la Guerra y con rescates, 1538, Toledo', *CDIU* vol. 10, 192, at 193. Cortés' ordinance distinguishes between things taken in and outside war. Above note 74, at 28.

slaves, natives slaves that had been taken in their own wars, or according to their laws and customs.[88] These licences were revoked by the Crown in 1530, because *conquistadores*, in their 'rampant greed', made wars and took captives *indios* that should not have been enslaved.[89] When reversing course in 1534, noting that without slavery, *indios* had become more audacious fighting Christians, the Crown appealed to the old medieval argument about slaves being better off under Christian masters to justify *rescate*.[90] But the Crown appealed also to reasons of economic expediency: without *rescate* the settlement of new lands would be impossible.[91] A more practical outlook guided also the regulation of slavery once reauthorized: women and children should not be enslaved but serve as *naborias*—domestic servants; a registry of slaves should be established, after being recorded, and branded on the body with a scorching royal iron (iron that should be kept under the custody of royal officials or priests), enslaved *indios* were ready to be traded.[92]

4.3 The Band as *Compaña* and the Division of the Booty

The reasons for the Crown's practical outlook are evident if the economic interests and incentives of the Crown and *conquistadores* are considered. The controversies on the status of *indios* and the loud assertions by royal officials and jurists about natives being free subjects under the protection of the Crown as vassals, responded to moral and legal considerations as well as the need to counter the excessive economic power that bands of *conquistadores* had accumulated.

The economic benefits of military tactics were reaped by *conquistadores* from selling goods obtained in the booty. After each *rescate*, spoils where divided between members of the band according to their contribution to the expedition, after having deducted the *quinto real*, the tax owed to the Crown. For these incredible opportunities to materialize, bands had first to organize expeditions under their own risk and discover new lands to conquer. As mostly self-constituted and self-regulated enterprises, bands were companies.[93] Bands formed a *compaña*, the medieval institution regulating the relations between members of a military band outfitting themselves and receiving, rather than wages, a share of the benefits obtained from an expedition.[94] Medieval law regulated the division of the booty. How these rules were applied

[88] Above note 22, at 39.

[89] Ibid., at 39–40.

[90] '... under the natives, *indios* would remain in their idolatry and other vices and abominable customs ... keeping them as slaves under the power of our Christian subjects they would more easily be instructed in our holy Catholic faith and would stop committing these vices and sins'. Above note 87, at 194.

[91] With *rescate* 'trade and commerce of our subjects, both Spanish Christian and *indios* would increase, without which the land could not be sustained and settled'. Ibid., at 194f.

[92] Ibid., at 197–200.

[93] Mellafe, 'Descubrimiento del Perú', in R. Mellafe and S. Villalobos, *Diego de Almagro* (1954) 5, at 17.

[94] As the first Caribbean phase of the conquest gave way to the conquest of the mainland, the initial Italian commercial company model with employer-investors and employees, was progressively replaced with the revival of the old *reconquista* war-like company. Góngora, above note 68, at 39. J. Lockhart, *The Men of Cajamarca: A Social and Biographical Study of the First Conquerors of Peru* (1972), at 66. R. Varón Gabai, *La ilusión del poder: Apogeo y descandencia de los Pizarro en la conquista del Perú* (1996), at 18.

depended on the power dynamics of each division.[95] However, laws regulating the *compaña* were important, especially when there were no rules regarding *reparto* included in capitulations.[96]

The *compaña*, representing the whole band in matters of common interest, coexisted with separate partnerships between investors and captains or between members of a band. These partnerships, concluded in contracts or capitulations, constituted a *compañía*—a company. In Pizarro's band, there were many *compañías*, partnerships between members, sharing, for example, horses in common. But the most important *compañía* was the partnership between Pizarro and Almagro, the organizers of the Peruvian expeditions, who reinvested profits obtained in past enterprises and subscribed debt against future earnings to finance their enterprise. The general partnership between the two captains had a long history dating back to the conquest of Panama, where both had large *encomiendas*. Having embarked together on various explorations, Pizarro and Almagro as well as Fernando de Luque a priest and investor, constituted the *Compañía de Levante* to organize the explorations that lead to Peru.[97]

When leaders of smaller explorations got indicia of a large kingdom and thus untapped treasures, as in the case of Cortés in México and Pizarro in Peru, they returned to Spain to obtain from the Crown an exclusive licence to discover and conquer. This was a crucial step, not only to fence off contending bands of *conquistadores*, but also to raise money necessary for a larger expedition, and to convince members to join the expedition, who in turn, trusting the licence, risked their own capital or borrowed money to participate. Organizers also negotiated with the Crown titles and privileges. To support these expeditions, the Crown offered also tax incentives, subsidized weapons and horses, and licences to take slaves from Spain.[98]

5. A Fragmented Legal Mechanics: What Room for the Inca?

We have seen band leaders like Pizarro negotiating with the Crown capitulations specifying the reciprocal rights and obligations that tied them within a feudal legal structure. There was a similar space open for native rulers and their descendants to negotiate with the Crown the meaning of their place as vassals of the king, that is, the reciprocal rights and obligations binding native lords and the Crown together. This was possible within a medieval legal framework that pushed conquest without centralizing all power on a band's commander-in-chief turned governor as Pizarro in Peru, nor on

[95] The military ordinance issued by Cortés, e.g. restates the general customary principle that the spoils 'must be distributed according to what each one serves and deserves'. Then the ordinance orders *compañeros* to immediately take goods obtained in *rescate* to Cortés or to another appointed member, disobedience being punished by death and loss of all property. Above note 74, at 28.

[96] The treasure, as Lockhart has shown, was only distributed between members who participated in the capture of the Inca in Cajamarca, though some absent groups got token payments to appease tensions. Shares of captains and mainly the Pizarro brothers, were much higher than those of horsemen. Footmen obtained shares according to their performance in battle, while others, as retainers of the Pizarros, got favours but no shares in the treasure. Lockhart, above note 94, at 78–82.

[97] Varón Gabai, above note 94, at 21.

[98] See e.g. the capitulation concluded between the Crown and Pizarro, above note 67.

the royal officials sent by the Crown to the Indies. The model that emerged in the early conquest left power diffused not only among newcomers—Crown, *conquistadores* and the many orders of the Church present in the Indies—but also between them and *indios amigos*, specifically, *señores naturales*—natural lords—*caciques*, or in the Inca context, *kurakas*—indigenous chiefs—who had not become enemies.

The feudal model where everyone is a subject (but for enemies and at times also infidels), left everyone, both Spaniards and *indios*, hierarchically placed within the law. *Conquistadores* who, participating in the division of a treasure and privileges, became *encomenderos*, as prominent neighbours of the towns and cities founded by the Spaniards, participated in the *cabildo*—the town council—shared power with royal officials. Unlucky *conquistadores* remained poor and protected by medieval institutions as *miserables*.

Similarly, *indios amigos*, chiefs or natural lords, especially those who had allied and contributed to the success of *conquistadores*' enterprise, were rewarded with land grants and *encomiendas* of *indios*. In very exceptional cases, other *indios*, neither chiefs nor lords, helpers—*indios auxiliaries*—who served Spaniards in battle and as translators (but also African slaves who contributed in battle), could participate in the division of the treasure and be rewarded with *econmiendas*.[99] However, outside these exceptional cases, commoners, 'plebeian' *indios*—*indios del común*—would become subjects—*súbditos*—vassals of the King, *indios* who would simply transfer their fidelities and duties from native lords to the Spanish Crown.[100]

Tributary relations between 'plebeian' *indios* and their native lords would be replaced, by consent or force, with a new type of relation established by *encomienda*. *Indios* would be subject now to labour obligations for their *encomenderos* and to the obligation to remain within the land, or in some cases to be transferred to another territory and *encomienda*. On the other hand, *indios* should be protected by their *encomendero*, in the feudal sense of protection from external harms. Protection from abuses by *encomenderos* should be afforded by clerics who were paid by the *encomendero* to fulfil his obligation to evangelize *indios*. Protection should also come from royal authorities appointed as protectors of *indios*, in the regime of protection of *indios* as *miserables*. And overall protection, as vassals, should come from the Crown.[101]

Although in a fragmented framework where power was diffused, harms suffered by indigenous peoples were enormous, from the devastation caused by war and by the collapse of whole empires, like the Inca Empire, to the resulting population transfers, and after the initial encounter, from meeting compulsive labour and religious obligations, to the violence from *encomenderos* and clerics policing the discharge of

[99] For a reward to Pizarro's helper, Martín see J. Hemming, *Conquest of the Incas* (2003) [1970], at 271f. For *encomiendas* granted to blacks and mulattos like Juan Beltrán, see Restall, *Seven Myths*, above note 66, at 62f.

[100] Góngora, above note 34, at 20.

[101] See Góngora ibid. Protection of vassals was an obligation shared between the king and neighbours, especially *encomenderos* who had the duty to assist the king in the protection of the kingdom. For example, *encomenderos* and members of the *cabildo* of Cuzco participated in the final campaign against Vilcabamba, see below at note 133 and accompanying text.

these obligations.[102] Then again, resisting Spanish presence by turning back to pre-Columbian times was the goal of innumerable indigenous rebellions. However, there was also room for resistance within the feudal framework, resisting demands from *encomenderos* and royal officials, by appealing to the Church and the Crown, from the *protector de indios*, to royal courts, to the *Consejo de Indias*, were common indigenous strategies of resistance.[103] Moreover, in the pre-modern model of governance of the early colonial period, there was room for flouting obligations, and for keeping pre-Columbian languages as well as cultural and religious practices alive.[104]

Historians have shown that native legal claims began as soon as Europeans arrived to conquer their lands.[105] Petitioning the Crown, appealing to courts, and in general channelling legal discourse to protect their interests, suggest that law and legal discourse played an important role in enabling and shaping indigenous agency in the Indies.[106] These avenues for agency were open to all *indios* as vassals, but were particularly relevant for *indios* as part of the native 'nobility', *caciques* and *señores naturales*, or their descendants, who sought in front of royal officials, if not in front of the king himself, recognition of tax privileges, land grants, *encomiendas*, arranged marriages to consolidate land and privileges under one line of descendants, coats of arms, etcetera. The historian José Carlos de la Puente Luna, for example, has shown the strong interconnections that native claim-making created between Spain and Peru, including, among others, the journey of Melchor Carlos Inca, a grandson of one of Atahualpa's sons, Paullu Inca, who travelled to Spain to request privileges.[107]

It would, on the one hand, be anachronistic to point to native claimants as example of relations between *indios* and Spaniards being governed by an emerging international law. If international law regulates relations between sovereigns, then there is no international law here. This would, furthermore, be confirmed by the earlier exploration of the debate about titles and about the legal status of *indios*, which justified Spanish authority, rather than outlined rules for inter-polity relations.[108]

On the other hand, when political and military circumstances on the ground forced Spaniards to confront native groups and their rulers as potential allies, or as enemies who could not be pacified, then negotiations and diplomacy (as basic practice of sending envoys who enjoyed immunity) followed. These diplomatic negotiations were

[102] On cruelty and violence as methods of Christian instruction, see Stern, above note 23, at xxxii and sources quoted.

[103] See sources below note 105, 106 and 107.

[104] For example, at the end of the 1550s, Incas in Cuzco continued venerating mummies of dead Incas they had concealed. See Hemming, above note 99,at 287.

[105] Belmessous, 'Introduction: The Problem of Indigenous Claim Making Colonial History', in S. Belmessous (ed.), *Native Claims: Indigenous Law against Empire, 1500–1920* (2011) 3.

[106] See among many others, Yannakakis, 'Indigenous People and Legal Culture in Spanish America', 11 *History Compass* (2013) 931; B. Owensby, *Empire of Law and Indian Justice in Colonial Mexico* (2008); Herzog, 'Colonial Law and "Native Customs": Indigenous Land Rights in Colonial Spanish America', 69 *The Americas* (2013) 3.

[107] J.C. de la Puente Luna, *Andean Cosmopolitans: Seeking Justice and Reward at the Spanish Royal Court* (2018), at 2–7. See also N. Van Deusen, *Global Indios: The Indigenous Struggle for Justice in Sixteenth-century Spain* (2015).

[108] On the use of inter-polity law as a wider notion than international law, comprehending relations between European and non-European polities, see Benton, 'Possessing Empire: Iberian Claims and Interpolity Law', in above note 104, 19, at 19 and footnote 7 at 37; also see Benton's chapter in this volume.

in part framed by legal discourse. Paradoxically, the law that was relevant here was not exactly *jus gentium*, the law of peoples that the Salamanca school proposed as an order under which both Spaniards and 'barbarians' should find themselves with rights and duties, the law of peoples that is conventionally invoked as originating international law. The feudal law that has been described in this chapter as offering a legal mechanics of conquest proved to be more significant. When these legal mechanics shaped a negotiated pacification of native rulers, did the law enable dialogue across the deep cultural and religious divide separating natives from conquerors?

There is no easy answer to this question. Postcolonial legal histories and postcolonial scholarship in general, reads the early colonial encounter as marking a deep, almost ontological divide between Christians-Europeans and natives.[109] Conventional legal history, on the contrary, reads the Spanish quest for just titles and the recognition of *indios* as free subjects as demonstration that *jus gentium* and *derecho indiano* can establish a common and just base for relations between Spaniards and indigenous peoples. This last section briefly explores the law behind the process of Spanish–Inca negotiation, showing inter-polity arrangements that were more fluid than the fixed European/native divide proposed by postcolonial readings, arrangements that, however, were much less benevolent than the protective *derecho indiano* of conventional readings (see also the contribution by Lauren Benton).

5.1 The Law behind Alliances and Negotiations: Paullu Inca, Sayre Tupa, and Titu Cusi

After Atahualpa's execution, and the death of his successor, another half-brother of Atahualpa, Manco Inca, became the new emperor. Pizarro and Manco Inca entered together Cuzco in 1533, inaugurating a period of coexistence in which both rallied to defeat the remaining forces of Atahualpa's generals, while throughout the provinces of the empire, Spaniards forged post-Inca alliances with ethnic groups that had been under Inca domination.[110] In Cuzco, coexistence between Pizarro and Manco Inca ended in 1536. After a failed year-long siege of Cuzco and Lima, the city founded by the Spaniards, Manco Inca retreated to Vilcabamba, a remote region of the empire, where a neo-Inca state subsisted for another four decades. Cuzco was retaken in 1537 by the forces of a Spanish–Inca alliance between Paullu Inca, brother of Manco, and Diego de Almagro, Pizarro's partner and captain of the band.

Diego de Almagro's alliance with Paullu followed increasing dissatisfaction within the Spanish band over Pizarro's division of the booty and distribution of privileges, among others. Conflict between Almagro's faction and the faction lead by Francisco Pizarro and his brothers, escalated into internecine fighting that left Almagro defeated and executed by Hernando Pizarro in 1538, Francisco Pizarro murdered by Almagro's son, and Gonzalo Pizarro beheaded, after failing in his rebellion against the viceroy Blasco Núñez Vela, who tried to enforce the New Laws of 1542—laws that continued

[109] See respectively, Anghie, above note 31 and Quijano, 'Coloniality of Power and Eurocentrism in Latin America', 15 *International Sociology* (2000) 215.

[110] Stern, above note 23, at 27–49.

the trend set by the laws of Burgos, of counter-balancing the power of *encomenderos* with measures to protect *indios*. Paullu, after retaking Cuzco with Almagro and being crowned in 1537, allied himself with the Pizarros, becoming the Inca sitting in Cuzco, while his brother Manco, continued ruling the neo-Inca state from Vilcabamba.[111]

After his death in 1544, Manco was succeeded by his son, Sayre Tupa, who negotiated his peaceful withdrawal from Vilcabamba to take residence in Cuzco and the Yucay Valley.[112] Sayre Tupa joined the Inca nobility that had stayed in Cuzco with Paullu, but died soon after in 1561. Titu Cusi, another son of Manco Inca who had remained in Vilcabamba, became the new ruler. Titu Cusi continued negotiating from Vilcabamba and signed in 1566 a treaty, the Capitulation of Acobamba, which was ratified by Felipe II in 1569.[113] Although the capitulation sanctioned Titu Cusi's peaceful withdrawal, he never left Vilcabamba where he died in 1571. The same year, his brother and successor, Tupac Amaru, was finally captured and executed in Cuzco.

The contrast between Paullu Inca and Manco Inca, as Gonzalo Lamana has argued, is not as sharp as had been previously suggested. Rather than following opposite strategies, Paullu collaborating and establishing alliances with Spaniards in Cuzco, versus Manco's fighting Spaniards from his retreat in Vilcabamba, both Paullu and Manco, but also Spaniards—Pizarro and Almagro—shifted between strategies and alliances, negotiating their place in front of the emerging power structures.[114] In this uncertain environment, the legal mechanics of conquest offered a framework, if not to regulate relations between polities horizontally organized, to stabilize relations within a common, simple, and malleable structure of rights and obligations organized hierarchically.

Inca rulers and the Inca 'nobility' were to place themselves at the service of the king. Recognizing vassalage entailed acceptance of various obligations. Allowing preaching in their lands, frequently converting to show goodwill, but many times keeping their own beliefs, or practising religious syncretism, Incas had to accept that most of their subjects would be placed under Spanish *encomenderos* while some were to remain under their rule and *encomienda*. The rights Inca rulers enjoyed included the grants and privileges already mentioned, as well as royal pardons for having battled against Spaniards.

This feudal based arrangement worked as a framework for Spanish–Inca cohabitation, in cases like Paullu, who settled in Cuzco, and was at the centre of an emerging syncretic culture.[115] When alliances broke, this arrangement also offered a model

[111] Kubler, 'The Neo-Inca State (1537–1572)', 27 *The Hispanic American Historical Review* (1947) 189.

[112] Hemming, above note 99, at 290 and on the privileges granted by the king to Sayre, ibid., at 365f. Merluzzi, 'La monarquía española y los últimos incas ¿una frontera interior?', 32 *Manuscrits: revista d'història moderna* (2014) 61, at 69–72.

[113] For a summary and recording of the capitulation see *CDIU* vol. 15 II, 270–76. For the text, addendum, certification of the document, and ratification by Titu Cusi, see Guillen, above note 30, at 63–71.

[114] 'Conquerors and Incas copied each other's ways of waging war, dressing, and practicing politics; Spanish clergymen tried to appropriate Inca religious forms and cosmology, while did likewise with Christian ones.' Lamana, above note 81, at 3. See also ibid., at 18 and Chapter 4. For the view that Paullu and Manco were in contact and collaboration, see Medinaceli, 'Paullu y Manco ¿una diarquía inca en tiempos de conquista?', 36 *Bulletin de l'Institut français d'études andines* (2007) 241.

[115] See MacCormack, '¿Inca o español? Las identidades de Paullu Topa Inca', 8 *Boletín de Arqueología PUCP* (2004) 99.

for re-establishing Spanish–Inca relations, as in the case of Manco and his successors at Vilcabamba. This is not to say that these arrangements were successful and long-lasting. If they helped to arbitrate conflicts, it was mediation in a world turned 'upside down'.[116] If rules were malleable, they were not only bent to support cohabitation, but also bent to shift the colonial enterprise towards a more 'vertical empire'.[117] Having been considered *señores naturales*, in the wake of the final campaign against Vilcabamba, Incas were redefined as usurpers and tyrants. At the same time, however, someone like Titu Cusi Yupanqui, during discussions with Spanish envoys to materialize the peaceful withdrawal negotiated in the capitulations, did not passively accept the terms offered within the Spanish legal framework, but actively appropriated legal arguments. In his famous 1570 *instruction to be presented to the king*, Titu Cusi encloses a history of 'how the Spaniard arrived in Peru', which not only defends a negotiated withdrawal within the feudal framework, but also decries Spanish presence as illegitimate.[118]

A failed armed campaign against Vilcabamba, then internecine fights between Spaniards, and then a shift in Crown policy with the enactment of the New Laws, left a peaceful withdrawal of Manco Inca as the best alternative to pacify Vilcabamba Incas. In 1543, after the New Laws were enacted, the incoming viceroy Blasco Núñez Vela was instructed by Emperor Carlos V to bring back *indios* who because of the maltreatment of Spaniards had escaped to the wilderness. The old goal of making '*indios* come in peace, and in the knowledge of our Catholic faith', was restated, but now cautioning that teaching Christian doctrine should be done with care, without war, death, or robbery to *indios*.[119] To achieve this goal and 'bring down *indios* to the service of the King', Núñez Vela was to supervise clergymen, and was moreover authorized to grant pardons for crimes committed during rebellion and confer exemptions from tribute.[120] General instructions like this translated then into more concrete proposals.

In 1552, then Prince Felipe II, sent a letter to Sayre Tupa. Recognizing that Manco Inca, his father, had rebelled because of 'some mistreatments that Spaniards had done' to him, and having been told that Sayre Tupa was willing to become Christian and come to the 'service and obedience' of his majesty, Felipe II informs the Inca that he has given a decree to Antonio de Mendoza, the viceroy, to forgive him for past wrongs and restore his father's property.[121] After long negotiations, in 1557 the viceroy issued a full pardon, Sayre Tupa then left Vilcabamba and travelled to Lima where Mendoza granted him estates that had been confiscated from the Pizarros. Settled in Cuzco and his estates in the Yucay valley, obtaining a special dispensation from the Pope, Sayre

[116] S. Ramírez, *The World Upside Down: Cross-cultural Contact and Conflict in Sixteenth-century Peru* (1998).

[117] 'Spaniards in Peru came to believe that the challenges of the Andean environment and its people necessitated a centralized, authoritarian government, and forms of social engineering unknown in Europe.' J. Mumford, *Vertical Empire: The General Resettlement of Indians in the Colonial Andes* (2012), at 9.

[118] T. Cusi Yupanqui, *History of how the Spaniards Arrived in Peru* [1570], (2006) (Catherine Julien trans.) For a more detailed analysis of Titu Cusi's account as an intervention in legal discourse see Becker Lorca, 'The *Tawantisuyo* Invaded by Demons: Inca Interventions in Spanish Colonial Law' (forthcoming).

[119] Above note 29, at 278f.

[120] Ibid., at 278.

[121] Letter included in judicial proceedings initiated by Ana Maria de Loyola Coya, grandchild of Sayre Tupa, see 3–4 *Historia y Cultura* (1969–1970) 165, at 166–67.

Tupa married his sister, the couple were then baptized in 1558.[122] In spite of Sayre Tupa's peaceful withdrawal, Vilcabamba Incas did not abandon their retreat. When Sayre Tupa died in 1561 pacifying these Incas and negotiating with his successor became imperative.

The incoming president of the *Real Audiencia* of Lima and acting governor, Lope Garçía de Castro, another lawyer who studied and taught in Salamanca, was instructed by the Crown to renew efforts to obtain from the Incas peace and withdrawal from Vilcabamba. Thus, in 1565, Castro sent another envoy, García de Melo, to restart negotiations.[123] Pressure to obtain a negotiated solution increased when Spaniards thought that Vilcabamba Incas were allying with other ethnic groups to mount a general rebellion and when an autochthonous religious movement, the *Taqui Onqoy*, advocating a return to pre-Columbian traditions to expel the Spaniards, gathered momentum among indigenous of the Andes. Facing these new dangers, the *corregidor de indios* of Cuzco, Diego Rodríguez de Figueroa, and the prominent jurist and *Oidor* Juan de Matienzo set out to Vilcabamba to join the negotiations. After a famous encounter and exchange of views between Matienzo and Titu Cusi, and after subsequent negotiations by García and Rodiríguez, two capitulations were concluded in 1566 and 1567, known as the capitulations of Acobamba.[124]

The second capitulation was ratified by Felipe II in 1569, sanctioning Inca autonomy and coexistence under vassalage.[125] 'The said Inca pledges and promises, as vassal of the kings of Castile that he is, that neither he, nor his captains, nor his people, will harm the towns or *indios* ... of this kingdom ... but will have perpetual peace.'[126] The capitulation also established that 'if the Inca and his captains from now on did any damage, war can be made against them, without further warning—*sin más* apercibirle'.[127] In exchange, Titu Cusi would get *encomienda* over the *indios* he has, and marriage of his son with Sayre Tupa's daughter, including the tribute of *indios* of the lands that had been granted to Sayre Tupa. Titu Cusi, however, did not leave Vilcabamba.

6. Conclusion: The Execution of Tupac Amaru

It was for the new viceroy, Francisco de Toledo, to implement the treaty negotiated by Castro, as well as to exert greater royal control over the kingdom. Toledo continued the effort to convince the Incas to leave Vilcabamba peacefully. Toledo, however, not really convinced by this strategy, challenged a basic assumption underlying diplomatic negotiations, the idea that the Incas were natural lords of their land. When he arrived in Peru, Toledo organized field inquiries throughout the provinces interviewing locals about the origins of Inca rule, and charged Pedro Sarmiento de Gamboa, a well-educated sea captain and cosmographer of the viceroyalty, with the writing of a history of the Incas based on first-hand oral accounts. The *informaciones*, the reports of the

[122] Hemming, above note 99, at 282–86.
[123] Guillen, above note 30, at 62f.
[124] *CDIU* vol. 15 II, at 270–76. Guillen, above note 30, at 63–68.
[125] *CDIU* vol. 15 II, at 276.
[126] Guillen, above note 30, at 65.
[127] Ibid.

inquiries, as well as the *Historia de los Incas*, Gamboa's manuscript, described Inca rule as resulting from a recent conquest. Rather than legitimate bearers of rights over their lands, Incas were usurpers; and rather than natural lords presiding over a well-organized empire, human sacrifices, indolence, large tributes on their subjects, among others, rendered Inca tyrants.

This was not a new idea, for we saw Xerez invoking the tyranny of Atahualpa to justify his execution.[128] But almost four decades later, the tyranny of the Incas had become a full-fledged argument about the legitimacy of the Spanish conquest of Peru. The argument was presented in a 1567 juridical tract by Juan de Matienzo. In the first chapter, entitled 'On the government and tyranny of the Incas, and on how they were not natural lords of these kingdoms of Peru', Matienzo lists the reasons why Incas were tyrants. Although Incas possessed their kingdom, they had acquired it by force, fraud, and treason. And once in power, according to Matienzo, Incas have sought their own benefit, rather than the 'public good' and interests of their vassals.[129]

The consequences of the argument were clear for Matienzo. Mentioning the whole array of Spanish titles, Matienzo clarifies that Inca's tyranny would be enough to establish a just title for his majesty.[130] Gamboa strengthened the historical basis for this argument. Toledo's reports turned the argument into policy to justify war against Vilcabamba. And the tyranny of the Incas as title for conquest continued well passed Toledo's war.[131]

In 1571, Toledo wrote a letter appealing to the Inca to leave Vilcabamba in peace.[132] By then Titu Cusi had died, and since the Incas thought he had been poisoned, Tupac Amaru was not willing to continue negotiations with the Spaniards. The envoy in charge of bringing the letter was killed, giving an additional reason to wage war against Vilcabamba. But the decision was already taken and a military band formed by neighbours and *encomenderos* of Cuzco and some royal combatants, had already been dispatched.[133] According to Toledo's reports, with the agreement of the wisest people of Cuzco, the *cabildo* resolved 'to remove that hideout and bugbear—*ladronera y espantajo*—and make war on the Inca, as apostate, prevaricator and murderer, rebel and tyrant'.[134]

Almost four decades after the execution of the first Inca ruler meeting the Spaniards, Atahualpa, the last Inca ruler, Tupac Amaru, was captured in the jungle and taken

[128] Xerez, above note 3, at 156. Also Vitoria, in spite of arguing that compelling barbarians to give up their sins against the law of nature was an unjust title, he accepts the defence of the innocent against tyranny as a just title. There is no great inconsistency, for it is the tyranny of the barbarians' masters against their subjects that entitles Spaniards to defend the innocent from unjust death, such as human sacrifice and cannibalism. Vitoria, above note 32, at 273, 288f.

[129] J. Matienzo, *Gobierno del Perú*, G. Lohmann Villena ed. [1567] (1967), at 7–10.

[130] The Incas, having oppressed the natural kings of the *indios* ... and having taken their lands by force ... Pizarro ... [liberated] the *indios* ... of the tyranny and oppression in which they were ... because God commanded to anyone to liberate his neighbour from oppression ... the Spaniards could very well and justly make war against the Incas.' Ibid., at 11. Note the similarities to Vitoria: 'barbarians are all our neighbors, therefore anyone, and especially princes, may defend them from such tyranny and oppression. (...) It makes no difference that (...) they refuse to accept Spaniards as their liberators in this matter'. Vitoria, above note 32, at 288.

[131] See J. Acosta, *De Procuranda Indorum Salute* [1588] (1984), at 4–8; Solórzano Pereira, above note 43, at 37.

[132] *CDI* vol. 8, at 266f.

[133] Julien, 'Francisco de Toledo and his Campaign against the Incas', 16 *Colonial Latin American Review* (2007) 243, at 245.

[134] *CDI*, vol. 8, at 271.

with a golden chain around his neck to Cuzco, where he was baptized, beheaded, and his head displayed on a pike in the *plaza mayor*.[135] Opening and closing this chapter with these two executions is intended to correct a common bias in conventional legal history. In the celebrated formulation by the Italian legal historian Luigi Nuzzo, *derecho indiano* produced an image of colonial law: 'without Indians, without the Indies, a legal history of conquest without conqueror and conquered'.[136] This chapter has placed war and the legal mechanics of conquest at the centre.

Both Incas were executed accused of tyranny and of committing specific offences. These similarities point to the importance of just war doctrine in the Spanish conquest. The conventional narrative about the civilizing influence of Salamanca's *jus gentium* is difficult to reconcile with the convergence between *requerimiento* and Salamanca around just war suggested in this chapter. Offering a just war justification in both executions, we see the law being used less as inter-polity regulation, than as legal support for the bands' protection of the booty from royal confiscation.[137]

That both executions followed the capture of a booty highlights the interplay between the two dimensions of colonial law explored in this chapter, the mechanics of justification and of conquest. It was under the latter—not a legacy of Salamanca but of the decentralized legal framework brought from feudal Spain—that we have seen spaces for native agency and inter-polity law. In both executions, we see Inca negotiation and resistance, and Spanish fighting alongside native forces, in ways that do not fit well with postcolonial accounts. But the killing of Tupac Amaru marks a turning point, the era where the legal mechanics which shaped the conquest of Peru came to an end, replaced by a more centralized governmental project.[138] Paradoxically, with the end of the legal mechanics of conquest came also the end of the feudal, decentralized framework of coexistence.

Acknowledgements

I thank Martti Koskenniemi, David Kennedy and Manuel Núñez, for support while doing research for this chapter, at respectively the Erik Castrén Institute at the University of Helsinki, the IGLP at Harvard Law School and Pontificia Universidad Católica de Valparaíso. I presented earlier versions at the Minnesota International Relations Colloquium, University of Wisconsin Law School, University of Colorado Law School, Harvard International Law Workshop, Conversaciones Porteñas sobre Historia del Derecho and Harvard Legal History Workshop. I thank participants and especially Bill Alford, Gabby Blum, Kristen Carpenter, Charlie Donahue, Tamar Herzog, Alexandra Huneeus, Helen Kinsella, Fernando Pérez, Carlos Salinas and Justin Desautels-Stein. I thank the editors Lothar Brock and Hendrik Simon for their feedback and for comments and much more, I thank Anna di Robilant.

135 *CDI*, vol. 8, at 279–82; Hemming, above note 99 , at 437–40.
136 Nuzzo, 'Dall'Italia alle Indie: un viaggio del diritto comune', 12 *Rechtsgeschichte* (2008) 102, at 105.
137 On the booty of this expedition see Hemming, above note 99, at 437.
138 For this turning point see Mumford above note 117 and M. Merluzzi, *La pacificazione del regno: negoziazione e creazione del consenso in Perù, 1533–1581* (2010).

6

Capitalism, British Grand Strategy, and the Peace Treaty of Utrecht: Towards a Historical Sociology of War- and Peace-making in the Construction of International Order

Benno Teschke

> England became [after the Treaty of Utrecht (1713)] the representative of the universal maritime sphere of a Eurocentric global order, the guardian of the other side of the *ius publicum Europaeum*, the sovereign of the balance of land and sea—of an equilibrium comprising the spatially ordered thinking of this international law. The English island remained a part or rather the center of this European planetary order, but simultaneously distanced herself from the European continent and assumed the world-historical, intermediary position that for more than three centuries made her 'of Europe, but not in Europe.[1]
>
> Throughout the world of capitalism, and specifically among the elements formed by capitalism in modern social life, there has arisen a fundamental opposition to war, expansion, cabinet diplomacy, armaments, and socially entrenched professional armies. This opposition had its origins in the country that first turned capitalist—England—and arose coincidentally with that country's capitalist development.[2]

1. Introduction: After 'Westphalia'

The Treaty of Utrecht (1713) brought an end to almost fifty years of French expansionism under Louis XIV, marked the beginning of British ascendance in European great power politics, and formally enshrined the principle of the balance of power in international politics. This makes it one of the most important events in international

[1] C. Schmitt, *The Nomos of the Earth in the International Law of the Ius Publicum Europaeum* (2003), at 173.

[2] Schumpeter, 'The Sociology of Imperialism', in J. Schumpeter, *Social Classes and Imperialism* (1955) 70.

Benno Teschke, *Capitalism, British Grand Strategy, and the Peace Treaty of Utrecht: Towards a Historical Sociology of War- and Peace-making in the Construction of International Order* In: *The Justification of War and International Order.* Edited by: Lothar Brock and Hendrik Simon, Oxford University Press (2021). © the many contributors. DOI: 10.1093/oso/9780198865308.003.0006

history. While the tercentenary of the Peace Accords was widely commemorated in the fields of History and International Law, no fresh re-examination of the standing of the Peace Congress and the meaning of the balance of power has been forthcoming in the contemporary International Relations (IR) literature.[3] This neglect is however not restricted to the more recent IR publications on early modern international orders.[4] Serious treatments of the significance of Utrecht in the wider development of the inter-state system are rare in the older IR literature as well, de-emphasizing rather than sharing Carl Schmitt's and Joseph Schumpeter's dramatic assessments of Utrecht's world-historical significance.[5] In fact, the theoretically informed IR literature—Realist, Neo-Realist, English School, Constructivist, and World-Systems/Neo-Gramscian alike—misinterprets the Peace Accords and the post-Utrecht transformation of international ordering. The IR literature thereby fails to identify and explain the most significant moment in the transition from dynastic-absolutist Westphalian towards British-managed early modern international relations during the 'long eighteenth century' (1688–1815) and beyond.[6]

This chapter argues that the Treaty of Utrecht was the single most important event in the history of early modern international re-ordering, constituting the first international expression of Britain's rise to European and, later, global pre-eminence. Within the wider Settlement, the balance of power was neither the automatic manifestation of the systemic imperatives of anarchy, threatened by a French bid for 'hegemony', nor a rational choice for offshore balancing by a rational state, determined by Britain's insular geography and the stopping power of water, nor the result of the consensual construction of a nascent norm-based 'international society', nor an aspect of British hegemonic leadership designed to generalize its specific capitalist state/society complex. It was a new, decisive, and uniquely British foreign policy technique, enabled by domestic socio-political

[3] M. Espenhorst and H. Duchhardt (eds), *Utrecht—Rastatt—Baden, 1712–1714: Ein Europäisches Friedenswerk am Ende des Zeitalters Ludwigs XIV* (2013); B. Fassbender and A. Peters (eds), *The Oxford Handbook of the History of International Law* (2014); F. Dhondt, *Balance of Power and Norm Hierarchy: Franco-British Diplomacy after the Peace of Utrecht* (2015); R. De Bruin, C. van der Haven, L. Jensen, and D. Onnekink (eds), *Performances of Peace: Utrecht 1713* (2015).

[4] D. Philpot, *Revolutions in Sovereignty: How Ideas Shaped Modern International Relations* (2001); D. Nexon, *The Struggle for Power in Early Modern Europe: Religious Conflict, Dynastic Empires and International Change* (2009); J. Quirk, S. Suzuki, and Y. Zang (eds), *International Orders in the Early Modern World: Before the Rise of the West* (2013); J. Branch, *The Cartographic State: Maps, Territory and the Origins of Sovereignty* (2013); B. Allan, *Scientific Cosmology and International Orders* (2018).

[5] K. Holsti, *Peace and War: Armed Conflicts and International Order, 1648–1989* (1991); Layne, 'The Unipolar Illusion: Why New Great Powers will Rise', 17 *International Security* (1993) 4; Schroeder, 'Historical Reality versus Neo-Realist Theory' 19 *International Security* (1994) 1; Schroeder, 'Not Even for the Seventeenth and Eighteenth Centuries: Power and Order in the Early Modern Era', in E. May, R. Rosecrance, and Z. Steiner (eds), *History and Neorealism* (2010) 78; G. Arrighi, *The Long Twentieth Century: Money, Power, and the Origins of Our Times* (1994); A. Osiander, *The States System of Europe, 1640–1990: Peacemaking and the Conditions of International Stability* (1994); C. Reus-Smit, *The Moral Purpose of the State: Culture, Social Identity, and Institutional Rationality in International Relations* (1999); Sofka, 'The Eighteenth Century International System: Parity or Primacy?', 27 *Review of International Studies* (2001) 5; J. Mearsheimer, *The Tragedy of Great Power Politics* (2001); P. Bobbitt, *The Shield of Achilles: War, Peace and the Course of History* (2002); I. Clark, *Legitimacy in International Society* (2005).

[6] Westphalian international relations are here defined in terms of dynastic sovereignty, royal marriage strategies that forged composite dynastic unions, wars of succession and mercantilist trade wars, plus the idea of equilibrium through the predatory *convenance* system in pre-capitalist Europe; see also B. Teschke, *The Myth of 1648: Class, Geopolitics and the Making of Modern International Relations* (2003).

and institutional changes after the Glorious Revolution of 1688, and unilaterally imposed at Utrecht on its former friends and foes alike through successful coercive diplomacy. Yet, the British-sponsored balance of power formed only one dimension, if a central one, of a much wider British grand strategy—the blue-water policy—for the geopolitical management of continental inter-dynastic relations.[7] Utrecht, I suggest, did not simply register a shift in the identity of the key great power—from France to Britain, both subscribing to the management of an otherwise static international anarchy—but expressed the first institutionalization of a dynamic grand strategy for the reorganization of European and colonial order executed by the first capitalist-parliamentarian state, post-1688/1707 Britain.[8] British blue-water policy cleaved into two geographically distinct but joined-up aspects: defensive and concerned with security interests versus the Continent, prosecuted through power-balancing, and offensive and concerned with commercial interests overseas, prosecuted through an aggressive navalism.[9] The British peace plan, elaborated, pursued, and successfully implemented at Utrecht, constitutes a *sui generis* phenomenon—creating a new system of British national security and international ordering—different from generic IR concepts, including automatic power-balancing, off-shore balancing, uni-polarity, hegemony, international society, anarchy, imperialism, or collective security.

This chapter suggests furthermore that Utrecht's world-historical significance cannot be captured through extant IR theories, but requires an Anglo-centred (but not Anglo-centric) historical sociology of international politics that brings grand strategy and international diplomacy 'back in'.[10] It seeks to address the dual theoretical challenge of moving beyond structuralist conceptions of International Historical Sociology by demonstrating the efficacy of foreign policy and international diplomacy, without subscribing to arguments about the autonomy of international history as diplomatic history, divorced from specific historical-sociological contexts.[11]

[7] The concept of 'blue-water policy' was developed by Daniel Baugh. While it is widely used in the historiographical literature, it is absent in the IR canon. Baugh notes that contrary to its connotations, blue-water policy is not synonymous with a purely maritime strategy, but captures maritime and continental objectives in a joined-up strategy where continental strategy was an extension of maritime strategy; Baugh, 'Great Britain's "Blue-Water" Policy, 1689–1815', 10 *International History Review* (1988) 1; Baugh 'Maritime Strength and Atlantic Commerce: The Uses of a "Grand Marine Empire"', in L. Stone (ed.), *An Imperial State at War: Britain from 1689–1815* (1994).

[8] I refer to England for the period up to 1707 (Act of Union with Scotland) and to Britain thereafter.

[9] J. Brewer, *The Sinews of Power: War, Money and the English State, 1688–1783* (1989); J. Black, *A System of Ambition? British Foreign Policy, 1660–1793* (1991); J. Black, *Parliament and Foreign Policy in the Eighteenth Century* (2004); P.J. Cain and A.G. Hopkins, *British Imperialism: Innovation and Expansion, 1688–1914* (1993); Maltby, 'The Origins of a Global Strategy: England from 1558 to 1713', in W. Murray, M. Knox, and A. Bernstein (eds), *The Making of Strategy: Rulers, States, and War* (1994) 151; Simms, '"Ministers of Europe": British Strategic Culture, 1714–1760', in H. Scott and B. Simms (eds), *Cultures of Power in Europe during the Long Eighteenth Century* (2007) 353; S. Pincus, *1688: The First Modern Revolution* (2009); Ostwald, 'Creating the British Way of War: English Strategy in the Spanish War of Succession', in W. Murray and R. Sinnreich (eds), *Successful Strategies: Triumphing in War and Peace from Antiquity to the Present* (2014) 100.

[10] The lament over the decline of international history *qua* diplomatic history, often associated with a restrictive Rankean focus on the interactions between foreign policy elites, is long-standing. See also Haber, Kennedy, and Krasner, 'Brothers under the Skin: Diplomatic History and International Relations', 22 *International Security* (1997) 1.

[11] It also diverges from the 'war-justifications paradigm' that informs much of this volume. Simon and Brock, 'Introduction: The Justification of War and International Order: From Past to Present'; Tischer,

Both Neo-Weberian and Neo-Marxist International Historical Sociologies routinely encounter the former problem.[12] The latter problem is the preserve of the genre of international history *qua* classic and 'new' diplomatic history.[13] Both IR traditions have so far failed to engage with each other. Overcoming this divide suggests a move from structural conceptions of International Historical Sociology and agentic conceptions of Diplomatic History as 'practice' to a new Historical Sociology of International Politics. The ultimate pay-off of this historicist method consists in identifying and validating the difference that foreign policy encounters and diplomacy makes to structurally determined expectations. This also implies the rejection of grand theory building in IR that pursues the ideal of generalizing explanations. A Historical Sociology of International Politics privileges the explanation of specific outcomes, rather than their subsumption under one general logic, while its method of analysis can be transported to different historical contexts.

Accounting for British success requires a qualitative—historical-sociological—analysis rather than a quantitative account of pre-war Anglo-French power differentials, which altered the institutional context for British foreign policy-making—the post-1688 parliamentarization of foreign policy. This enabled the domestically contested and deliberate making of a new grand strategy that was developed and militarily tested during the War of the Spanish Succession and imposed through secret diplomacy at Utrecht on Britain's former allies (the Dutch and the Austrians).

I argue that the blue-water policy, in striking contrast to foreign policy-making in Absolutist France, rested on a revolution in foreign affairs after 1688 and the prior reorganization of British political and social relations resulting from the rise of agrarian capitalism in the countryside. Britain's new grand strategy was formed after the transition from dynasticism and the conception of foreign policy as part and parcel of royal executive powers to the co-articulation of foreign policy by Parliament. The new institutional basis of policy-making meant that party politics and social interests, unlike in continental Old Regimes, came to influence public policy, including grand strategy formation. This led to a process in which the conflicting foreign policy orientations between the Whigs (military continental commitment) and the Tories (oceanic

'Princes' Justifications of War in Early Modern Europe: The Constitution of an International Community by Communication'; A. Tischer, *Offizielle Kriegslegitimationen in der Frühen Neuzeit: Herrscherkommunikation in Europa zwischen Souveränität und korporativem Selbstverständnis* (2012). Reconstructing the history of war and international order by granting analytical priority to the self-justifications of war by practitioners who constitute an international communication community, framing the normative terms of politics, externalizes the prior question of the fundamental presence of power politics in early modern Europe.

[12] T. Skocpol, *States and Social Revolutions: A Comparative Analysis of France, Russia, and China* (1979); C. Tilly, *Coercion, Capital and European States, A.D. 990–1990* (1992); S. Hobden and J. Hobson (eds), *Historical Sociology of International Relations* (2001); Rosenberg, 'Why Is There No International Historical Sociology?', 12 *European Journal of International Relations* (2006) 3. The genre of 'Global Historical Sociology' equally brackets foreign policy and international politics. See also J. Go and G. Lawson (eds), *Global Historical Sociology* (2017).

[13] H. Kissinger, *Diplomacy* (1994); P. Schroeder, *The Transformation of European Politics, 1763–1848* (1996); C. Jönsson and M. Hall, *Essence of Diplomacy* (2005); P. Sharp, *Diplomatic Theory of International Relations* (2009); B. Rathbun, *Diplomacy's Value: Creating Security in 1920s Europe and the Contemporary Middle East* (2014); O. Sending, V. Pouliot, and I. Neumann (eds), *Diplomacy and the Making of World Politics* (2015).

navalism) merged during the prosecution of the War of the Spanish Succession into a 'new way of war', a specific peace time strategy and a distinct pro-British way of international ordering. It envisaged the 'rationalization', that is the de-ideologization, the de-dynastification, the de-confessionalization, and de-territorialization of British ambitions on the Continent, the recalibration of continental political geography more conducive to power-balancing, the invention and active manipulation of the balance of power, and the pursuit of oceanic commercial primacy, premised on the new principle of the 'national interest'. These new principles rested on a sober calculus of the secular interests of the 'political nation', as opposed to the whims of dynastic interests. Blue-water policy represented the first conscious attempt to develop a secular, advantageous, and comprehensive strategy for rationalizing geopolitical and geo-economic space in line with the British 'national interest', even if Britain never intended nor was capable to conclusively 'settle' or fully transform the 'Westphalian' international affairs of Europe and its overseas colonies.

The argument proceeds in four steps. I start with a brief summary of the historical Utrecht context and discuss the problems of generalizing, logico-deductive and nomological IR theories, including those deriving from the sub-field of International Historical Sociology, for capturing historical specificity. To rectify these shortcomings, the section sets out the basic parameters of a radically historicist sociology of international politics that brings statecraft, foreign policy, grand strategy, and international politics 'back in' by privileging contextualized diplomatic agency and counter-agency. Section 3 provides a comparative perspective on the diverging Anglo-French socio-political and institutional contexts for late seventeenth-century foreign policy-making. The differential Anglo-French social relations of sovereignty explain the power differentials between both polities, and their dissimilar fiscal, financial, and foreign policy institutional arrangements. These different contexts facilitated incommensurable war-time military strategies—parliamentarized British grand strategy versus royal-executive French strategy. Section 4 reconstructs how the prosecution of the War of the Spanish Succession shaped blue-water thinking amongst British policy-makers, and how two contextual events—the victory of the Tories in the 1710 general election and the enthronement of the new Austrian Emperor Charles VI in 1711—changed the direction of anti-French British grand strategy, embarking on a course of pre-Utrecht secret diplomacy with France. The splitting of the opposition informed the articulation of a distinct peace plan—the Bolingbroke Plan—which was skilfully imposed at Utrecht as a *Diktat* on Britain's erstwhile war-time allies, the Dutch and the Austrian Habsburgs. Section 5 shows how British blue-water policy was converted into the terms of the Utrecht Settlement. The Treaty provisions secured the post-revolutionary domestic constitutional arrangements (the Protestant Succession and the primacy of Parliament) and enshrined a geopolitical vision that reconfigured the political geography of the European Continent conducive to the operation of British power-balancing. It simultaneously enabled British colonial-commercial expansion overseas. The section also demonstrates the limited understanding of British concerns for 'European liberties' and peace, restricted primarily to the external maintenance of a sufficient number of power-politically near-parity actors that could be balanced against each other. The conclusion sets out how British blue-water policy innovated a dualistic long-term system of national security and international ordering, which defensively maintained and managed Westphalian order on the Continent in pro-British ways and offensively established colonial expansion throughout the 'long eighteenth century'—a strategy that fortified British global primacy for another near-century after the Vienna Settlement.

2. The Utrecht Settlement and International Historical Sociology: Bringing Grand Strategy, Diplomacy, and International Politics 'Back In'

The Treaty of Utrecht ended the era of French expansion, including the War of Devolution (1667–68), the Dutch War (1672–78), the Nine Years' War (1688–97), and the War of the Spanish Succession (1701–13/14).[14] The alliance patterns during the Nine Years' War arrayed the League of Augsburg (1686), comprising Austria, Spain, Sweden, Brandenburg, Bavaria, and various smaller German principalities, against France. The League was expanded to form the Grand Alliance in 1689, when England and the Dutch Republic, ruled in personal union by William III, joined the anti-French coalition. The *casus belli* for England's entry into the War of the Spanish Succession—a classic example of Old Regime Wars occasioned by the death in 1700 of the childless King Charles II of Spain—derived from the complex inheritance claims to the Spanish dominions that were contested by Louis XIV of France and the Austrian Emperor Leopold I. The takeover of Spain by either the French Bourbons or the Austrian Habsburgs augured the possibility of 'universal monarchy'. As the French invaded the Spanish Netherlands, the Duchy of Milan, and the West Indies in 1701, England concluded in September 1701 the Treaty of the Grand Alliance in The Hague with the Dutch Republic and the Austrian Emperor 'for the Preservation of the Liberties of Europe, the Property and Peace of England, and for reducing the Exorbitant Power of France'.[15] War was declared by the Grand Alliance in May 1702. The Alliance was strengthened during the war through the accession of the Holy Roman Empire of the German Nation, Prussia, Portugal, and Savoy, while Wittelsbach Bavaria and the Spanish Bourbons defected to the French side. As the fortunes of war swung after Marlborough's victorious, but costly, campaigns in Flanders in the direction of the anti-French Grand Alliance and as the French repeatedly sued for peace at The Hague (1709) and, again, at Geertruidenberg (1710), the British defected from the Grand Alliance and entered a period of secret diplomacy with France, leading to the Anglo-French London Preliminaries (1711). This formed the blueprint for the Utrecht Settlement, which was ultimately accepted under protest by Britain's former allies, the Dutch and the Austrians.

Technically, the term 'Utrecht' refers to a cluster of eleven bilateral treaties that were concluded between 1713 and 1715, comprising also the Treaties of Rastatt (1714),

[14] Pitt, 'The Pacification of Utrecht', in J. Bromley (ed.), *The New Cambridge Modern History. Vol. 6: The Rise of Great Britain and Russia, 1688–1715/25* (1970) 446; Hill, 'Oxford, Bolingbroke, and the Peace of Utrecht', 16 *The Historical Journal* (1973) 2; K. Kluxen, *Geschichte Englands: Von den Anfängen bis zur Gegenwart* (1985).

[15] Hattendorf, 'Alliance, Encirclement, and Attrition: British Grand Strategy in the War of the Spanish Succession, 1702–1713', in P. Kennedy (ed.), *Grand Strategies in War and Peace* (1991) 11, at 16.

Baden (1714), and Madrid (1715), stretching possibly to the Treaty of The Hague (1720). This geographical and chronological pattern indicated the fragmented nature of the peace process and the often uncoordinated, mainly bilateral, negotiations, which were also occasionally interrupted by the resumption of hostilities. Ending a long period of European-wide conflict, stretching into the overseas colonies in the Americas, the Utrecht Settlement fundamentally altered the political geography of Europe and its colonies, enshrined for the first time the principle of the balance of power into an international peace accord, but retained the dynastic principle on the Continent. The peace provisions, mainly territorial in nature, trimmed the size of France back to its old 1688 borders (ceding territories east of the Rhine, but not west of the river), rather than pressing for its partition, blocked the personal unions between Spain and France and between Spain and Austria, divided the Spanish Empire into peninsular Spain and the West Indies under Philip V, and added formerly Spanish possessions (Milan, Sardinia, Parma, Modena, and Naples) to the Habsburg Empire. They established a buffer zone in the newly created Austrian Netherlands against French expansion, kept the Dutch Republic out of Flanders to avoid a common border with France while allowing it to erect, garrison, and control a barrier system that stretched from the North Sea to the Moselle River, and raised the status of Savoy from a duchy to a kingdom and enhanced its territory in Northern Italy to act as a buffer—analogous to the Austrian Netherlands—between France and Habsburg Italy. They also safeguarded the independence of middle powers like Portugal, Prussia, the Dutch Republic, Savoy, and various smaller German and Italian principalities. The British gains looked *prima facie* innocuous, reduced to securing political control over strategically important naval bases in the Mediterranean (Gibraltar and Minorca), the destruction of the French naval base at Dunkirk, the acquisition of French possessions in Canada (Hudson Bay, Nova Scotia, and Newfoundland) and in the West Indies, and the transfer of the *asiento* from Spain—the monopoly of the slave trade with the Spanish colonies—to the British South Sea Company. Utrecht rearranged the political geography of Europe and was an international event comparable in its significance to the Peace of Westphalia (1648) and the Peace of Vienna (1815).

How should International Historical Sociology be reconceived to capture this event? Macro-sociological attempts—whether Neo-Marxist or Neo-Weberian—to theorize international relations through the recovery of the differential domestic composition of social and authority relations of polities that constitute international orders ultimately tend to relapse, like Neo-realism, into functional explanations of international politics. By either erasing foreign policy or by claiming that policies are a function of deeper structures, they suppress the open-ended process of foreign policy-making, the efficacy of military and diplomatic agency, and the further effects of international politics.[16] Structural, generalizing, and nomological theories of IR or International Historical Sociology externalize foreign policy from the remit of explanation.[17] This

[16] A. Anievas and K. Nisancioglu, *How the West Came to Rule: The Geopolitical Origins of Capitalism* (2015); see also Skocpol, Tilly, and Rosenberg, above note 12. For a critical engagement with nomothetic theories of International Historical Sociology see Teschke, 'IR Theory, Historical Materialism, and the False Promise of International Historical Sociology', 6 *Spectrum: Journal of Global Studies* (2014) 1.

[17] On the nomothetic/ideographic distinction see also Levy, 'Explaining Events and Developing Theories: History, Political Science, and the Analysis of International Relations', in C. Elman and M. Elman

raises an acute analytical question for IR theory. What theoretical revisions are necessary to capture and validate the efficacy of foreign policy-making, diplomacy, and international politics? How should International Historical Sociology be readjusted to escape the nomological and functionalist-structuralist trap for purposes of framing our object of study—a specific event?

This chapter adopts a radical historicist approach to capture the efficacy of foreign policy, grand strategy, diplomacy, and international politics in the construction of the early modern international order. I define diplomacy as the coercive and non-coercive pursuit of foreign policy interests, grand strategy (as part of foreign policy) as the articulation of essential long-term security interests and objectives that are pursued through the domestic coordination of multiple policy areas, and international politics as the contested encounter of multiple foreign policies, leading at times to major systemic wars and comprehensive peace settlements, which reshape the nature of international order. Rather than deducing strategies, decisions, and outcomes from a series of antecedent structural causes, whether international (the distribution of power/international competition), domestic (the rise of capitalism), or both (uneven and combined development resulting from the interactivity between advanced and backward polities), this shifts the explanatory burden away from external and internal imperatives to the creative responses developed by situated agents to such contextual pressures, the inter-subjectively (both domestically and internationally) contested and negotiated resolution of these multiple agentic strategies, and their intended and unintended consequences. This brings international politics 'back in'. Deploying a historicist framework of analysis that centres on contextualized agency requires a trade-off between generalizing theories that operate at levels of abstraction that abandon the concreteness of international politics while subsuming history under preconceived laws and concepts, and retrieving and activating the situated consciousness, actions, and decisions of contextualized agents. Failing that, decisions are not taken, but reduced to outcomes of antecedent pressures. The explanatory valued-added consists in the ability to account for historical specificity.

This procedure also requires that we abandon, against Karl Marx and Joseph Schumpeter, a generic understanding of capitalism—or any other macro-sociological category—as a purely theoretical concept from which foreign policy can be logically and abstractly deduced, and reconceive it as a historical process category: as a contested social relation, whose international 'effects' cannot be logically or organically extrapolated, but require radical contextualization, historicization, and politicization.[18] Capitalist pressures are never translated one-to-one or functionally into a determinate foreign policy. Social interests, rather than organic tendencies, become domestically articulated and inter-subjectively constructed as international relations in wider geopolitical contexts. Diplomatic history can neither be detached from, nor causally

(eds), *Bridges and Boundaries: Historians, Political Scientists, and the Study of International Relations* (2001) 39.

[18] On the controversial history of the concept of 'capitalism', see Teschke and Wenten, 'Capitalism', in F. Berenskoetter (ed.), *Concepts in World Politics* (2016) 157; on the problem of foreign policy in Marxism, see Teschke and Wyn-Jones, 'Marxism in Foreign Policy', in C. Thies (ed.), *Oxford Research Encyclopedia of Politics* (2017).

reduced to, the specific socio-political trajectories and divergent constitutional forms of the Settlement's constitutive powers. Ultimately, it is the efficacy of specific agents that explains specific outcomes. While these outcomes are retrospectively intelligible, other courses of action—a full-scale British Whig-led 'European commitment' or a Tory-led 'peripheral navalism', rather than the blue-water policy—remained distinct possibilities. The emphasis on contextualized and relational agency rejects thus deterministic modes of reasoning.

Yet, this is not a return to the autonomy of classic diplomatic history. The requirement of agentic contextualization raises three demands. First, to account for differential power increases and decreases amongst polities by tracking nationally divergent social property relations that generated differential economic growth rates and modes of public revenue procurement. Second, to explain the historical formation of different institutional contexts for foreign policy-making, and to show how the open-ended question to what purpose power is being used and deployed is being resolved, that is to account for foreign policy formation, given that states are never unitary and rational actors endowed with pre-fixed and ready-made national interests. Third, to show how foreign policy becomes internationally executed, modified, and enacted through dynamic interaction between multiple polities in the sphere of international politics. This move from structural macro-determinations to the micro-politics of international statecraft remains central to account for the specific outcomes of the Utrecht Settlement. At the centre of this Historical Sociology of International Politics stands the encounter between two divergent foreign policy projects—English grand strategy and French strategy.

I further define grand strategy by leaning on the wider political and hyper-Clausewitzian conceptualization in the literature that encompasses the political objectives that subordinate military strategy to the superior considerations of political purpose.[19] This includes the question of how to win the peace after victory in war. Yet, strategic performance—success or defeat—cannot be easily read back into strategy formation. I suggest therefore to draw on Williamson Murray and Mark Grimsley's more dynamic and interactive reformulation of grand strategy that assumes that 'strategy is a process, a constant adaptation to shifting conditions and circumstances in a world where chance, uncertainty, and ambiguity dominate'.[20] Grand strategy, like all human action, is a practical and inter-subjective art, rather than a science. This interactive conception—vertical/domestic and horizontal/international—incorporates domestic and international politics. The following interpretation tries to evade the fallacy of definitional over-coherence by tracking the process of grand strategy-making and execution by placing domestic and international agency and counter-agency at the centre of analysis. For the English blue-water policy emerged in rapidly changing domestic and international contexts, as key actors created, modified, and enacted it in

[19] Kennedy 'Grand Strategy in War and Peace: Toward a Broader Definition', in P. Kennedy (ed.), *Grand Strategies in War and Peace* (1991); Rosecrance and Stein, 'Beyond Realism: The Study of Grand Strategy', in R. Rosecrance and A. Stein (eds), *The Domestic Bases of Grand Strategy* (1993) 3; Murray and Grimsley, 'Introduction: On Strategy', in W. Murray, M. Knox, and A. Bernstein (eds), *The Making of Strategy: Rulers, States, and War* (1994) 1; C. Gray, *Modern Strategy* (1999).

[20] Murray and Grimsley, above note 19, at 1.

a process of 'trial and error'. Still, while Britain did not get all she wanted at Utrecht, she got most of it.

3. The Divergent Power Resources and Institutional Contexts of Anglo-French Foreign Policy-Making

What accounts for Britain's success to imprint its designs on the Settlement? Britain's growing power—commercial-maritime and financial-military—played a dominant role, as did the institutional reconfiguration of the British state in the late seventeenth and early eighteenth centuries. This is widely recognized in the literature with reference to the concept of the rise of the 'fiscal-military' state.[21] While this concept has been usefully deployed in the Neo-Weberian literature on state-formation, it suffers from the assumption of a quasi-isomorphic interpretative template generalized across very diverse processes of state-formation. It obscures the qualitatively different social sources and outcomes of state-formation in early modern Europe.[22] For post-revolutionary Britain came to express a new set of social relations generated by England's revolutionary seventeenth century, rooted in the rise of agrarian capitalism.[23] This led to a class-constellation in which an entrepreneurial aristocracy, supported by the new 'interloping merchants', entered into a period of intense conflict with the monarchy, the old colonial merchant class, and surviving feudal magnates over the form and control of the English state.[24] While the Stuarts tried to establish absolute authority, the capitalist aristocracy sought to construct a state that was responsive to the needs of private property protection, limited taxation, and capital accumulation, encapsulated in the programmatic call for 'political liberties'. This conflict between 'court' and 'country' culminated in the Glorious Revolution and the new notion of the 'King-in-Parliament'—a formula that essentially codified the Crown's concession of crucial powers to Parliament, which became the locus of British sovereignty. Between 1688 and 1701, the parliamentary classes consolidated their power by passing a series of fundamental constitutional acts, increasingly accepted as statutory law. This included the 1689 Bill of Rights, which established the rights of regular parliaments, free elections, and freedom of speech in Parliament, and prohibited the existence of a standing army during peacetime without the consent of Parliament; the 1689 Mutiny Act, which made the army and the defence budget responsible to Parliament; the 1694 Triennial Act, which empowered Parliament to meet annually

21 Brewer, above note 9.

22 C. Storrs (ed.), *The Fiscal-Military State in the Eighteenth Century: Essays in Honour of P.G.M. Dickson* (2009)

23 Brenner, 'The Agrarian Roots of European Capitalism', in T.H. Aston and C.H.E. Philpin (eds), *The Brenner Debate: Agrarian Class Structure and Economic Development in Pre-Industrial Europe* (1985); Comninel, 'English Feudalism and the Origins of Capitalism', 27 *Journal of Peasant Studies* (2000) 4; S. Dimmock, *The Origin of Capitalism in England, 1400–1600* (2014); O'Brien, 'The Formation of States and Transitions to Modern Economies: England, Europe, and Asia Compared', in L. Neal and J. Williamson (eds), *The Cambridge History of Capitalism. Vol. 1: The Rise of Capitalism: from Ancient Origins to 1848* (2014) 357.

24 R. Brenner, *Merchants and Revolution: Commercial Change, Political Conflict, and London's Overseas Traders, 1550–1653* (1993).

and to hold general elections every three years; and the 1701 Settlement Act, which disqualified any Roman Catholic from inheriting the throne by reserving the succession to the Protestant Hanoverian line. Private and public, society and state came to be increasingly differentiated, yet mediated through Parliament's translation of contested private and social interests into public policy.[25]

This new form of sovereignty, no longer personal-dynastic, but abstract-national sovereignty, drove the concomitant revolution in public administration—the Fiscal Revolution, the Financial Revolution, and the Revolution in Foreign Affairs.[26] By 1688, sovereignty lay no longer with the king, but with Parliament, articulating the interests of the propertied classes.[27] While the Crown continued to enjoy certain prerogatives, 'the authority of ministers rested on their ability to maintain the support of the majority in the House of Commons'.[28] Foreign policy was made accountable to a permanent elected assembly.[29] With the Glorious Revolution control over the budget also passed to Parliament—a shift from the extraordinary taxation levied by the king on the people, especially during times of war, to taxation as a national affair, subject to deliberation and co-determination in Parliament, particularly with regard to military expenses and the setting of revenue levels. This amounted to the self-taxation of the 'political nation', now in form of either the land tax or the older expedients of customs and excise.[30]

'The results were astonishing. Under Charles II, annual Crown revenues normally amounted to between 1.3 and 1.4 million pounds. In contrast, revenues during the Nine Years' War totalled 32.7 million pounds, and during the War of the Spanish Succession 64.2 million pounds. The largest single source of this increase was the land tax, introduced in 1692'.[31]

While war-driven fiscal revenues increased dramatically, taxation and war-related expenditures became subject to a rational and secular calculus and accountable to public debate and scrutiny. This was most clearly evidenced in the creation of the Treasury Board, which made all government departments' expenses and receipts accountable to its oversight, and the Public Accounts commissions of the House of Commons, set up to eradicate waste and corruption during wartime spending. The British experience of parliamentary control over the budget—indeed, the use of the technical term 'budget' to record and quantify public revenues and expenditures—contrasted sharply with contemporaneous French practices. Here, in spite of repeated requests by finance ministers like Colbert to the king to subject war-declarations to a

25 E. Wood, *The Pristine Culture of Capitalism: A Historical Essay on Old Regimes and Modern States* (1991).

26 Brewer, above note 9, at 58.

27 Black and Pincus, above note 9; Brenner, above note 24, at 713–16.

28 P. Harling, *The Modern British State: An Historical Introduction* (2001), at 21f.

29 Maltby, above note 9, at 162f.

30 O'Brien and Hunt, 'England, 1485–1815', in R. Bonney (ed.), *The Rise of the Fiscal State in Europe, c. 1200–1815* (1999) 53; Bonney, 'Revenues', in R. Bonney (ed.), *Economic Systems and State Finance* (1995) 423, at 481.

31 Maltby, above note 9, at 158.

prior economic calculus on their financial sustainability, 'budgets ... were not yet established as a part of domestic government' until the late eighteenth century.[32]

Additionally, public finance was put on a new and secure footing through the creation of the National Bank of England in 1694.[33] This dramatically increased the government's ability to raise funds. The public debt was no longer the king's debt, liable to fraud and defaults, but precisely the National Debt, generating lower but more secure interest rates, and uniting the lending classes behind the war efforts. This enabled 'the state to borrow on comparatively favourable terms by providing a parliamentary guarantee for the swift repayment of interests'.[34] The revolution in foreign affairs was thus supplemented by a revolution in fiscal and financial affairs. This underpinned Britain's rise to Europe's pre-eminent power during 'the long eighteenth century'.

But the parliamentarization of foreign policy, taxation, and defence spending had even deeper ramifications that reverberated throughout British society at large. For as the Tory–Whig rivalry over the direction of public policy, including foreign policy, came to inform public debate amongst an ever-expanding electorate (the franchise was qualified by relatively low property requirements), the three elections of 1710, 1715, and 1722 drew large numbers of voters into the deliberation and, ultimately, indirect co-determination of foreign policy formation.[35] Thus, in sharp contrast to the prevailing conception of foreign policy in continental absolutist states as an exclusive executive prerogative of the ruler's *arcana imperii*, foreign policy formation in Britain became not only parliamentarized, but also 'nationalized' as a subject of interest, debate, and engagement in an expanding wider public sphere. The combination of revolutionary institutional innovations—Britain's naval superiority and exceptional fiscal-financial responsiveness in the face of external military pressure on the basis of a self-sustaining capitalist economy—gave Britain the decisive *comparative economic, fiscal, financial, military, and administrative, though not necessarily strategic, advantage* over its continental competitors. This consolidated British exceptionalism. There was, in short, a new 'reason of state', different from the 'reason of princes'.

British constitutional developments stood in striking contrast to absolutist France. By the mid-seventeenth century, the demise of independent feudal centres of power finally meant that French sovereignty became internally more consolidated, since the Bourbon polity was no longer a fragmented ensemble of lordships that defined the parcellized sovereignty of the medieval polity, but a kingdom in which the Crown claimed sovereignty. It would be a fundamental mistake, however, to confound 'absolutist' with modern sovereignty, since the king regarded the realm as his patrimonial property. *Raison d'État* meant *raison de roi*, embodied in the personalized sovereignty of the ruling dynasty. 'Divine kingship' became the dominant mode of legitimation rather than a secularized discourse and praxis of 'parliamentary sovereignty' or 'the national interest'. But 'absolutism', as the revisionist literature has confirmed, never

[32] Scott, 'The Fiscal-Military State and International Rivalry during the Long Eighteenth Century', in C. Storrs (ed.), *The Fiscal-Military State in the Eighteenth Century: Essays in Honour of P.G.M. Dickson* (2009) 23, at 25.

[33] P.G.M. Dickson, *The Financial Revolution in England: A Study in the Development of Public Credit, 1688–1756*; Cain and Hopkins, above note 9, at 60–71.

[34] Harling, above note 28, at 3.

[35] Ostwald, above note 9, at 109; Maltby, above note 9, at 162f.; Harling, above note 28, at 24f.

implied unlimited or unchecked executive royal power, but rather institutionalized a new and ultimately unstable *modus vivendi* between king and privileged groups, most notably the sword and office nobility and the higher clergy.[36] The relations between the Crown and the nobility and between them and the peasantry remained governed throughout the *ancien régime* by political conflicts over access to and distribution of the peasant produce. Consequently, taxation became the key arena of domestic conflict. To secure noble allegiance, the nobility was, as a rule, exempted from taxation and therefore not represented in a national forum. The Estates General had met for the last time before the French Revolution in 1614.[37]

Since in France the 'nation' was not represented in a national assembly, absolutism implied that foreign policy-making remained the executive prerogative of the Crown. 'Louis XIV debated the issues, made the decisions, and monitored their execution. Freed from the influence of traditional rivals, he exercised complete authority for the formulation of foreign policy and in the drafting of strategy.'[38] Foreign policy was decided in the *Conseil d'en haut* (Council of State), convened, constituted, appointed, and chaired by the king, and normally composed of the secretaries of state for foreign affairs, war, and navy, and the controller-general of finances. Decisions were subject to his consent. Louis XIV's personal regime meant that military operation and strategy was directly decided in the Council of State, leading to the so-called *guerre de cabinet*.[39] French strategy—not grand strategy—remained tied to the classical objectives of dynastic states, defined in terms of permanent bellicosity and territorial expansion, expressed in the pursuit of *gloire*, rather than the national interest. 'The realization that Louis did not evaluate his actions in terms of their benefit for the French state or people, but in terms of their effects on his *gloire*, explains his tendency to pursue his wars to excess and his willingness to initiate them when he should have looked for peace.'[40] The precarious financing of wars also lent a self-propelling dynamic to the French way of war, since French troops had to live off the country so that 'war fed off itself'. Absolutist wars were wars of attrition and peace accords only temporary respites before the quest for territory resumed.

In short, the 'permanent war-state' drew on a limited resource base, circumscribed by the taxation of a pre-modern agrarian economy, the exemption of the nobility from taxation, the non-rationalized provision of lines of credit, and inefficient and socially divisive fiscal and financial systems, while being conducted by the dynastic-territorial ambitions of its absolutist ruler. While France was more populous and commanded a greater territory than seventeenth-century Britain, its agrarian economy remained mired in Malthusian cycles and pre-capitalist patterns of growth, so that the total tax

[36] W. Beik, *Absolutism and Society in Seventeenth-Century France: State Power and Provincial Aristocracy in Languedoc* (1985); Beik, 'The Absolutism of Louis XIV as Social Collaboration', *Past & Present* (2005) 195; D. Parker, *Class and State in Ancien Regime France: The Road to Modernity?* (1996); H. Gerstenberger, *Impersonal Power: History and Theory of the Bourgeois State* (2007).

[37] Bonney, 'The Eighteenth Century II. The Struggle for Great Power Status and the End of the Old Fiscal regime', in R. Bonney (ed.), *Economic Systems and State Finance* (1995), at 325.

[38] Lynn, 'A Quest for Glory: The Formation of Strategy under Louis XIV, 1661–1715', in Murray, Knox, and Bernstein, above note 9, 178, at 179; see also K. Malettke, *Frankreich, Deutschland und Europa im 17. und 18. Jahrhundert* (1994), at 303–331.

[39] Lynn, above note 38, at 183.

[40] Lynn, above note 38, at 187.

base generated a lower return per capita than its smaller neighbour on the other side of the Channel.[41] The Crown failed to establish a central bank or secure lines of credit, while being forced to borrow on short-term loans at high interest rates from a class of wealthy financiers, who were themselves often tax-farmers. The desperate reform attempts by Colbert to establish a rational and national fiscal and financial system failed consistently 'so long as France's strict social hierarchy conceded great privileges, including tax exemptions, to those at the top'.[42]

The social dynamics and geopolitics of the *ancien régime* could not and did not lead to a rationalized, efficient, and 'modern' bureaucratic state. In fact, military rivalry reinforced and intensified rather than resolved the pre-modern social tensions that structured Old Regime France. It was this pre-capitalist complexion that gave the early modern continental system of 'states' its over-militarized, bellicose, and crisis-ridden character, both in its quantitative sense regarding the frequency of war and its qualitative sense regarding the catastrophic ratio between war expenditures and fiscal income, that is the economic sustainability of war.[43] The continental dynastic states system—the Westphalian system—remained thus defined by very specific foreign policy practices: the war-driven accumulation of territories; the predatory and compensatory logic of dynastic equilibrium (*convenance* rather than power-balancing); control over exclusive and monopolistic trading routes secured by politico-military means; the elaborate dynastic strategies of territorial aggrandizement through marital policies; the resulting dynastic unions and composite monarchies and their flipside, the endemic wars of succession, including a general drive towards territorial empire-building.

4. The English Prosecution of the War of the Spanish Succession: The Making of Blue-Water Policy

What difference did British party politics, military strategy, and diplomacy make for converting structural conditions—the rise in power resources, the new post-1688 constitutional arrangements, and the parliamentary foreign policy prerogative—into specific outcomes? For neither power per se, as we established earlier, even if grounded in 'differential growth rates', nor 'capitalism', nor even 'institutions' tell us to which uses power was deployed. To answer this, we have to turn to the actual making of grand strategy—bearing in mind the distinction between strategic planning and strategic performance—and thus ultimately to international politics. We need to show additionally how the conduct and conclusion of the War of the Spanish Succession tested and developed blue-water policy and how British international diplomacy succeeded in dominating the peace agenda at the Congress of Utrecht, whose post-war parties comprised, *inter alia*, capitalist-parliamentarian Britain, the mercantile Dutch Republic, and the Austrian Old Regime on the winning side, and the French and

[41] Bonney, 'France, 1494–1815', in R. Bonney (ed.), *The Rise of the Fiscal State in Europe, c. 1200–1815* (1999), at 144–47.
[42] Lynn, above note 38, at 189.
[43] Bonney, above note 41.

Spanish Old Regimes on the losing side. Yet, while the Utrecht Accords were formally the result of inter-polity negotiations, the exclusive focus on official summit diplomacy overplays the power symmetries between the peace parties, neglects the decisive pre-Utrecht Anglo-French politics of secret diplomacy, which diminished the significance of the negotiations *in situ*, and downgrades the overriding role played by Britain in the Settlement's design and implementation. It is widely noticed in the literature that only the British had, in effect, an actual peace plan, masterminded by Foreign Minister Viscount Bolingbroke and assisted by Chief Minister Robert Harley.[44] This formed the blue-print for a wider settlement that went beyond the short-term accommodation of conflicting, narrow, and ad hoc individual interests—'old diplomacy'—of the diverse peace parties by providing a draft nominally designed in the 'public interest' of Europe as a whole. Nevertheless, even though the Bolingbroke Plan was wider in scope and ambition than the peace aims of the co-signatories, its provisions never amounted to a comprehensive multilateral settlement, a peace system, or a system of collective security, or even British hegemony.

The following interpretation sides thus, in a first step, with 'second-image' IR theories that emphasize the 'domestic bases of grand strategy' by showing how the British devised and constructed a specific way of war, a new grand strategy, and distinct peace plan to turn power into policy.[45] Yet, it transcends this 'inside-out' perspective through an historical sociology of international politics by tracing additionally how British grand strategy had to be constantly and interactively modified, leading to pro-French concessions, due to supervening domestic and international contextual changes. Domestic policy objectives did not translate one-to-one into international outcomes, precisely because of the effects of international relations. This made a difference to the peace provisions.

At the end of the seventeenth century, Whigs and Tories represented different constituencies, roughly speaking the division between financial-commercial and landed-commercial wealth. This also informed their differences in foreign policy orientation, but this time based on interests and pragmatics, rather than ideology, religion, or dynasticism.[46] As the 'national interest' is never something generically given, but defined and redefined through domestic politics, the question of the shape of global strategy, suspended between a colonial peripheral navalism and a full-scale commitment to continental warfare, and the consequent tax burdens became a key issue in the controversy between Whigs and Tories. As a rule and under the mounting costs of the Nine Years' War and the War of the Spanish Succession, the post-revolutionary foreign policy of the Tories was committed to the containment—rather than defeat—of France, and content to protect the Constitutional Settlement of the British Isles, preventing any continental hegemony, and expanding overseas. Geopolitics and geo-economics trumped ideology and religion. The Whigs, in turn, while first favouring a

[44] Maclachlan, 'The Road to Peace, 1710–1713', in G. Holmes (ed.), *Britain after the Glorious Revolution, 1689–1714* (1969) 197; Hill, above note 14; H. Duchhardt, Gleichgewicht der *Kräfte, Convenance, Europäisches Konzert: Friedenskongresse und Friedensschlüsse vom Zeitalter Ludwig XIV. Bis zum Wiener Kongress* (1976), at 41–48; H. Duchhardt, *Balance of Power und Pentarchie: Internationale Beziehungen, 1700–1785* (1997); D. McKay and H. M. Scott, *The Rise of the Great Powers, 1648–1815* (1983), at 63–66.

[45] Rosecrance and Stein, above note 19.

[46] Pincus, above note 9, at 350–65.

full-scale invasion of France and a pro-European policy, driven by their commitments to anti-absolutism and anti-Catholicism, scaled back over time their war aims and were content to assemble a multi-confessional and multi-ideological alliance against the French.

England's entry into the war in 1702 led to the 'principal-auxiliary debate' amongst Whigs and Tories about whether England should adopt a leading or a subsidiary position in the allied war effort.[47] Although nominally headed by a moderate Tory, the Lord High Treasurer Sydney Godolphin, growing Whig influence in the government carried the day. England committed to adopt the principal role in the coordination of allied campaigns, even if the ultimate post-war objectives diverged among the members of the Grand Alliance. From the start, English strategy was not predicated on the decentralized operational decisions of its leading commanders—notably Marlborough—in the field and at sea, but politically orchestrated as grand strategy back home by the government in Westminster, ultimately answerable to Parliament.[48]

There was considerable unity amongst Whigs and Tories as to the English core strategic interests: control of the Channel, which meant keeping the French out of the Spanish Netherlands, protecting the right of Parliament to regulate the succession (threatened by Louis XIV's proclamation of James III as King of England in 1701), preventing the union between France and Spain under the Bourbons, and protecting overseas trade and colonies by keeping the French, Spanish, and Dutch out. Still, there was considerable disunity as to their prioritization and best realization. As a rule, the Tories favoured a strategy of peripheral navalism in Europe, occasional amphibious 'descents' on the French coastline, and oceanic naval supremacy. The Whigs were committed to substantial land warfare, notably in the Low Countries. During the Whig Administration under Godolphin, these diverging foreign policy orientations gradually combined to form a specific and new 'British Way of War', which included alliance-formation, naval encirclement, continental commitment, subsidization of smaller powers, and overseas navalism, leading to a war of attrition.[49] This implied a division of labour between Britain and her allies, committing Britain primarily to the deployment of her superior naval forces in the Channel, the Mediterranean, and overseas to encircle France, while the superior armies of the Dutch and the Austrian Habsburgs were deployed on land, notably in the Low Countries, Western Germany, and Northern Italy. Encirclement implied also enhancing the Alliance through coercive diplomacy by manipulating Portugal, confirmed in the Methuen Treaty of 1703, and the Duchy of Savoy in the same year to switch sides.

As the French repeatedly sued for peace in 1709 and 1710, two events—elections and the enthronement of Emperor Charles VI—decisively changed the domestic and international contexts for the prosecution and conclusion of the war. This led to the dynamic adaptation of British grand strategy and the redefinition of war aims through secret diplomacy with France. Domestically, the 1710 British parliamentary elections brought the Tories, headed by Robert Harley (Earl of Oxford), to power on a disengagement and peace platform. The diverging foreign policy orientations of Tories and

[47] Ostwald, above note 9, at 107.
[48] Maltby, above note 9, at 162f.; Hattendorf, above note 15, at 12; Ostwald, above note 9, at 107–09.
[49] Ostwald, above note 9, at 101; Hattendorf, above note 15, at 19–21.

Whigs revealed a deep socio-political split within Britain's political class. As a rule, the Tories represented 'the country', which meant the landed classes that defended the hereditary succession (even if Catholic) and resented the growth of the 'fiscal-military' state and its growing war-time expenditures. War-induced taxes fell disproportionally through the land tax on agrarian landlords. The Whigs, in turn, represented 'the court' in the sense of supporting the anti-Catholic (and thus anti-French and anti-absolutist) Protestant Succession, subject to Parliament, through the enthronement of the Oranian Prince William III. This translated into Whig support for a full-scale 'European commitment'. War expenditures enhanced the growth of a politically dependent 'moneyed interest', represented by Whig city-based financiers, who had an interest in prolonging the war by profiting from higher returns (compared to the returns on ownership from land) on the buying and selling of government securities in the exponentially growing London stock market.[50] The incoming Tory government recalibrated the British, but not the Alliance's, war aims by switching from the more offensive objective of French defeat and possible partition to the more prudent and defensive idea of preventing the rise of any continental hegemon.

One year later, the spectre of a renewed reunion of the Habsburg lands under the new Emperor Charles VI (enthroned in 1711), combining Austria and Spain, came to replace the spectre of the union of the Bourbon lands, combining Spain and France. In this altered context, Tory-ruled Britain defected from the anti-French alliance and entered into secret diplomacy with the French.[51] Without the consultation of Britain's war-time allies, a plan was hammered out bilaterally between Bolingbroke and the French foreign minister Torcy in the London Preliminaries (1711). This joint Anglo-French preliminary peace plan was presented at Utrecht in 1712. Here,

> the British ministers planned to repeat the tactics of the preliminary negotiations by conducting the essential business with France behind the backs of the other participants until the time was ripe to force the latter either to accept the terms thus agreed or to lose British military and diplomatic support.[52]

The overriding rationale of the plan was to ensure that France, much to the dismay of the Dutch and the Habsburgs, would be re-admitted as a full, if diminished, member into the Settlement. The exclusion of the Bourbon monarchy from the field of international politics would have generated a power vacuum that Austria would have amply filled. Militarily weakened, as Britain withdrew from military action during the course of 1712, strategically out-flanked and diplomatically out-manoeuvred, the Dutch, the Austrians, and the Spanish had to accede as junior partners to the table. The Austrian emperor, incandescent because of British deceitfulness and the terms of the Anglo-French preliminaries, withdrew from the negotiations, and continued to fight against Louis XIV. Ultimately, he signed the separate Peace of Rastatt (1714) without changing the core provisions of the Utrecht Settlement. A militarily and financially exhausted

50 MacLachlan, above note 44, at 200f; Harling, above note 28, at 22–27; Pincus, above note 9, at 305–65; Ostwald, above note 9, at 105.

51 MacLachlan, above note 44; Pitt, above note 14, at 456–61; Hill, above note 14.

52 Hill, above note 14, at 252f.

France accepted the more lenient terms of the British over the more Carthaginian ambitions of the Dutch and the Habsburgs. The Peace Treaty of Utrecht largely ratified the essentials of Bolingbroke's preliminary peace plan. Utrecht's 'material results were already agreed upon by secret diplomacy before the congress'.[53] Utrecht was a unilaterally imposed *Diktat*—not imposed on the common enemy, but rather on Britain's alliance partners.[54]

While the Tories masterminded the peace plan, Utrecht partly accommodated Whiggish interests by maintaining a territorially much-reduced absolutist France on the Continent, curbing its possessions overseas, and increasing the number of independent polities—absolutist and Republican alike—on the Continent as fodder for the balance of power. The Whig commitment to the liberty of Europe was therefore not a call for the substantive constitutional liberty of European polities—the internationalization of the English Revolution—but reduced to support for their numerical plurality. The dualistic outcome of these opposing but increasingly narrowing partisan commitments was formalized in the blue-water policy—a new system of national security, which included minimal continental commitment based on power-balancing in the European theatre, plus unlimited overseas expansion based on naval power and maritime supremacy. 'It became apparent that "Whig" and "Tory" strategies were not necessarily incompatible and that England—Britain after 1707—was strong enough to afford both.'[55]

5. The British Peace Plan: Institutionalizing Blue-Water Policy

How was British blue-water policy converted into the terms of the Treaty? Overall, blue-water policy divided into a geopolitical and a geo-economic aspect, the first geared towards the Continent, the second geared towards overseas. Geopolitically, it implied a policy of 'rationalizing' political geography, as outlined in section 3, through the redistribution of power in terms of territoriality, the prevailing currency of power in pre-capitalist polities. This redistribution of space established the territorial presuppositions for British power-balancing, designed to guarantee the post-Utrecht political geography of Europe. The balance of power was thus for the first time in the history of European peace congresses codified in the Settlement. This is widely noticed in the IR literature.[56] Yet, the balance of power was not conceived as a timeless and automatic self-equilibrating and self-adjusting mechanism, enforcing a subjectless and systemic gravitational pull across the spectrum to prevent uni-polarity or hegemony, as much of Neo-Realist IR theory suggests. Nor was it a general maxim universally subscribed to by the members of 'international society'. It was rather contrived as a conscious

[53] Duchhardt, 'Peace Treaties from Westphalia to the Revolutionary Era', in R. Lesaffer (ed.), *Peace Treaties and International Law in European History: From the Late Middle Ages to World War One* (2004) 45, at 52.

[54] According to the French diplomat Melchior de Poligniac, addressed to the Dutch, negotiations were held '*de vous, chez vous, sans vous*'.

[55] Maltby, above note 9, at 165; see also Brewer, above note 9, at 168f.

[56] Schroeder, 'Historical Reality versus Neo- Realist Theory' above note 5, at 92.

strategy by Britain to hold the balance in its hands.[57] It became the key policy for the achievement, maintenance, and enhancement of British primacy and security, but not hegemony. The balance of power was a political construct—a foreign policy practice—routinely invoked and reaffirmed in the annual Mutiny Acts of the British Parliament, which fixed the essential strategic aims of the British army and navy, and was already successfully tested in the War of the Spanish Succession itself. At Utrecht, it was grafted onto an altered territorial floor plan, more conducive to its actual manipulation.

Substantively, it meant subsidizing smaller powers, primarily financially but also through the hiring of mercenaries, orchestrating alliance-formation against any potential would-be hegemon, dispatching smaller British expeditionary forces, and considering intervention and war as the *ultima ratio* if everything else failed: buck-passing if possible and balancing (intervention) if necessary! This went hand-in-hand with a clear commitment to the de-territorialization and de-dynastification of British claims to continental possessions—the Hanoverian stemlands being a constant source of irritation for Parliament[58]—while jealously guarding control over strategic naval bases. The overriding issue here was security interests, which trumped narrower economic interests, which Britain may have entertained vis-à-vis the Continent. This was essentially a defensive policy, controlling continental territorial might 'on the cheap' and by 'remote control' through the active manipulation of the balance of power and its associated sub-policies.[59] It was a strategy of containing the belligerent aggressions of primarily absolutist powers by making wars militarily and financially risky and unattractive—a policy of deterrence, flexible alliance-formation, subsidization, and encirclement, plus financial attrition, plus, if needed, direct intervention.

But this innovation could not and did not lead to the general acceptance of the balance of power as a new international principle on a system-wide basis to stabilize the new territorial order, as the British overriding strategic aim was not to freeze a specific territorial *status quo* 'in perpetuity'. Rather, the objective was to flexibly maintain a sufficient number of polities of roughly equal size, while preventing any anti-British coalition amongst them or the formation of a continental hegemon that could single-handedly challenge British supremacy. Power-balancing became an unprincipled principle. Europe was again at war fifteen years after Utrecht. British power-balancing led thus to a series of territorial swaps, adjustments, and personal unions, which yoked very disparate regions under the umbrella of one nominal ruler, while supporting the recognition and independence of middling powers like the Dutch Republic, Prussia, Portugal, and Savoy. The overriding aim of power-balancing for the geopolitical management of space was thus never the protection of the independence and freedom of each and every political community in the 'society of states'. It was never intended to guarantee peace or secure the *status quo*.[60] Its core rationale consisted in the pragmatic

[57] This is noted by classical realists, but largely explained in terms of Britain's geographical position plus William of Orange's individual genius. Kissinger, above note 13, at 70f.; H. Morgenthau, *Politics Among Nations: The Struggle for Power and Peace* (1985), at 199.

[58] Harding, 'British Maritime Strategy and Hanover: 1714–1763', in B. Simms and T. Riotte (eds), *The Hanoverian Dimension in British History, 1714–1837* (2007) 252; Gibbs, 'The Revolution in Foreign Policy', in G. Holmes (ed.), *Britain after the Glorious Revolution, 1689–1714* (1969), at 67.

[59] Teschke, above note 6, at 256–64.

[60] Duchhardt, Gleichgewicht der *Kräfte*, above note 44, at 17.

handling of a numerically ever-dwindling group of states in the interest of British security through the flexible formation of rapidly changing alliances in which individual smaller polities were promoted or abandoned as pawns of the game. 'Albion' became 'perfidious'.[61]

6. Conclusion: Towards a Historical Sociology of War- and Peace-making in the Construction of International Order

By reducing the territorial and commercial power resources of continental polities to never perfect relations of near-parity, by diverting their navalist energies and strategies into terrestrial conflicts by tying them to the European theatre, and by retaining a commitment to military intervention on the side of weaker powers as a last resort, Britain devised a grand strategy that had the triple advantage of being geopolitically effective, economically affordable, and ideologically ostensibly neutral and in the 'public interest' of the 'liberties of Europe' as a whole. This preference for pragmatic adjustments to the realities of power shifts in a dynamic inter-state order, rhetorically pursued in the name of equity and harmony, came to form a long-term strategy of the British state into the early twentieth century: *divide et impera*! 'Utrecht therefore marked a "strategic moment", though it is doubtful that anyone fully appreciated it at the time. Its terms—a product of compromise and consensus even within Britain—contained many of the elements of later strategic concern.'[62]

Power-balancing, with Britannia holding the scales in her hand, implied the disengagement from the continental dynastic game of territorial geopolitics with its endless wars of succession, political marriages, and dynastic unions. With the Treaty of Utrecht, Britain largely withdrew from direct territorial aspirations on the Continent and started to regulate the states-system by means of rapidly changing alliances, with its monetary subsidies and mercenary expedition corps to smaller powers always ready to counter any emergent continental hegemony. Britain started to 'drop out' of the archaic practices of continental 'Westphalian' geopolitics while steering it from offshore. Simultaneously, it built up its colonial empire overseas and rose to global hegemony by the end of the Seven Years' War (1756–63), a position fortified for another century by Waterloo and sealed by the Vienna Settlement. Whereas Britain upheld a commitment to power-balancing from 1713 onwards as a grand strategy translating into support for *Kleinstaaterei* (mini-state proliferation) on the Continent, it imperiously established itself as the colonial master of the seas, ruling the waves.

Geo-economically, blue-water strategy meant widening the oceanic and Mediterranean sphere of trading monopolies for British commerce against Dutch, Spanish, and French mercantilist competition. It is in this context that Britain's Utrecht gains become significant. For the curtailing and confiscation of French colonies and trading posts in the Americas, the transfer of the potentially lucrative right of the slave contract, the *asiento*, from the Spanish Crown, and the cession of Gibraltar and Minorca turned not only the Mediterranean into a British lake to access the riches of

[61] L. Dehio, *The Precarious Balance: The Politics of Power in Europe, 1494–1945* (1963), at 84.
[62] Maltby, above note 9, at 176.

the Levante trade, but strengthened British pre-eminence in the Atlantic and the wider global world of sea-borne commerce.[63] Blue-water strategy implied lessening the dependence of British trade on the continental mainland, circumventing the recurring threats of Dutch or French trading blockades, redirecting export and import market shares away from Europe to overseas trade, while pursuing an uncompromising and aggressive policy of commercial and imperial expansion across the globe.[64] This policy should therefore not be conceived as a—peaceful and anti-imperialist—capitalist foreign policy per se, measured against a generic *abstractum* of a logically pure and idealized conception of capitalist rationality, as Schumpeter implied, nor as an outcome of Britain's insular geography, as Schmitt suggested.[65] It should rather be seen as a historically specific construction intelligible within a historically specific domestic and geopolitical context. Blue-water strategy did not spread capitalist principles across the Continent, nor did it introduce capitalist social property relations overseas, but served nevertheless dominant ruling class interests—Tory and Whig alike—in safeguarding the domestic 'liberties', that is the parliamentary prerogative over the constitutional settlement which guaranteed domestic private property, the self-organization of the 'political nation', that is the propertied classes in Parliament, and the nascent liberalization of the British home market and, to some degree, intra-imperial British trade. Utrecht established a British 'maritime-imperial' system within a hostile mercantilist world of geopolitically competing pre-capitalist states.[66] While Utrecht protected and promoted the deepening and expansion of British capitalist practices, Europe, throughout the eighteenth century, could only be geopolitically managed and contained, but not constitutionally or economically transformed.

Utrecht constituted the first international institutionalization of Britain's new grand strategy, which translated into a new type of international ordering, with Britain as the two-faced pivot—the regulator of the system. This reveals Utrecht not so much as a peace treaty, but as part of a wider and specifically British global strategic project, pivoting around power-balancing, for the novel management of geopolitical space and international politics—the first in a series, which altered European and colonial political geography and international relations in qualitatively new ways. Yet, none of the specific Utrecht outcomes was 'written into', preordained by, or logically derivable from the growth in British power, the rise of capitalism, the new foreign policy institutions, party politics, or even the blue-water policy itself. While all of these structural conditions mattered, the specific outcomes of Utrecht are ultimately only intelligible as the result of the creative agency of politicians and diplomats, notably Anglo-French secret diplomacy, situated within rapidly changing domestic and international contexts.

[63] Maltby, above note 9, at 168–75.

[64] Cain and Hopkins, above note 9, at 84–101; Dudley, 'Party Politics, Political Economy, and Economic Development in Early Eighteenth Century Britain', 66 *Economic History Review* (2013) 1084.

[65] For a critique of Schmitt's history and theory of early modern European geopolitics, see Teschke, 'Decisions and Indecisions: Political and Intellectual Receptions of Carl Schmitt', 67 New Left Review (2011) 61; Teschke, 'Carl Schmitt's Concepts of War: A Categorical Failure', in J. Meierhenrich and O. Simons (eds), The Oxford Handbook of Carl Schmitt (2016) 367.

[66] Baugh observes that 'the most pervasive cause of historical misunderstanding . . . has been the founding text of modern economic science, Adam Smith's *Wealth of Nations* (1776). The book's whole approach—on which its analytical purity depends— is grounded in an economic idealism that presupposes an eternally peaceful world featuring unfettered and harmonious marketing.' Baugh, 'Maritime Strength and Atlantic Commerce', above note 7, at 187.

7

Kant's Rejection of Just War: International Order between Democratic Constitutionalism and Revolutionary Violence

Oliver Eberl

1. Introduction

The tradition of *just war* offered the standard justification for war in early modernity.[1] In legal and political theory, it had reached its peak in the eighteenth century, when contemporary theories of natural international law had modernized it with ideas of 'regular' or 'formal' war. Kant's primary effort was to refute these theories and ensure that all parties remained committed to peace even in times of war.[2] This explains why the French Republic became a shining example for Kant: it had enshrined the imperative of peace in its constitution and renounced wars of aggression and conquest.[3] Kant must have felt encouraged in his belief that the republic offered the *clou* to containing war. Where kings in absolutist states had the uncontested right to declare war, transferring it to a democratic legislature would see decisions of war being made by those who would have to shoulder its burdens.[4]

Following the revolution of 1789 and the internationalization of the conflict over a republican constitution, Kant published not only *Common Saying* (1793) but in *Perpetual Peace* (1795) and *The Metaphysics of Morals* (1797/98) two works of particular importance for this topic. On the occasion of the Peace of Basel between France and Prussia in 1795, Kant reflected on the hopes it brought in *Perpetual Peace*, and connected these hopes with a general overhaul of international law. Understanding this historical context enables us to understand Kant's critique of *just war*. Kant denies that theories of just war can be of any use in preventing war. He systematically lays out the weaknesses and consequences of the theories of just war and natural international law, and he explicitly rejects them as a foundation for an international law geared towards peace.

At the same time, Kant translates the substance of *Perpetual Peace* into the presentational form of natural law. When he speaks the language of natural international law

[1] See Anuschka Tischer's chapter in this volume.

[2] Ripstein, 'Just War, Regular War, and Perpetual Peace', 107 *Kant-Studien* (2016) 179.

[3] Title VI of the French Constitution of 1791 reads: 'The French nation renounces the undertaking of any war with a view of making conquests, and it will never use its forces against the liberty of any people.'

[4] Kant, 'Toward Perpetual Peace', in *The Cambridge Edition of Works of Immanuel Kant*, edited by P. Guyer and A. W. Wood, volume 'Practical Philosophy', edited and translated by M.J. Gregor (1996) 322 (AA 8:350).

Oliver Eberl, *Kant's Rejection of Just War: International Order between Democratic Constitutionalism and Revolutionary Violence*
In: *The Justification of War and International Order.* Edited by: Lothar Brock and Hendrik Simon, Oxford University Press (2021).
 DOI: 10.1093/oso/9780198865308.003.0007

in *The Metaphysics of Morals*, he does so only to describe the limitations of natural law as law in the state of nature. This is precisely what his thinking is meant to overcome. The function of Kant's reference to natural law and its practical application in inter-state relations is to describe the weaknesses of the law that exists in the state of nature and which can only be overcome under public law. The latter, however, is barely addressed in Kant's legal theory. It is treated in more detail in the earlier *Perpetual Peace*.[5] This point is important when dealing with the allegation that Kant takes a bellicist position in his *The Metaphysics of Morals*. This allegation, if kept up, would have to explain what exactly could have caused such dramatic change in Kant's attitude between 1795 (seeking peace) and 1797 (war-prone). The only demonstrable change of attitude in Kant is related to the French Revolution as a historical event: Kant's thoughts on international law, which he developed in his *Idea for a Universal History with a Cosmopolitan Purpose* (1784), could not yet be supported by a republican practice (simply because it did not exist, at least in Europe). As a consequence he favoured solutions revolving around a world state. Later on, the events of the revolution led him to develop the idea of a league of nations which would have no coercive powers but would institutionalize international law and thus facilitate democracy.[6]

This fundamental contrast in Kant's interpretations of the design for international law is closely related to our situation today: just war arguably still informs contemporary practices of foreign policy.[7] Under the labels of 'humanitarian intervention' and 'war against terror', elements of the unilateral justifications of war persist despite the confirmation of a 'Responsibility to Protect' by the General Assembly of the UN and its General Secretary (Ban Ki-moon).[8] Against this background, the present chapter intends to contextualize Kant's writing in order to prevent appropriating interpretations and to offer arguments which may help to deal with today's issue of international intervention.

To this purpose, I begin by tracing the historical situation in which Kant's discussion takes place, that is the internationalization of the conflict about the republican constitution in France. Then I present the influence of the tradition of peace plans and Hobbes's idea of the state of nature on Kant's thinking. In section 4, I outline Kant's contribution to the refutation of just war. The linking of the state of nature, social contract theory, and the idea of the world republic by Anacharsis Cloots is explained in the fifth section. This leads to the *unjust enemy* in the tradition of natural law and its revival in the revolutionary wars. Kant's dealing with the *unjust enemy* is the topic of the last part. The fact that republican France, which at first seemed to have overcome war, soon threatened to regress into the mind-set of just war through the introduction of the *unjust enemy*, is of particular importance in this context. Kant's actual ambition

[5] See Eberl and Niesen, 'Kommentar', in I. Kant, *Zum Ewigen Frieden* (2011 [1795]), at 125.

[6] Maus, 'From Nation-State to Global State, or the Decline of Democracy', 13 *Constellations* (2006) 465; Kleingeld, *Kant and Cosmopolitanism. The Philosophical Ideal of World Citizenship* (2012), at 44–58; Eberl and Niesen, above note 5, at 232–48.

[7] As Chris Brown's chapter and other contributions in Part VII show.

[8] Brock and Simon, 'Die Selbstbehauptung und Selbstgefährdung des Friedens als Herrschaft des Rechts. Eine endlose Karussellfahrt?', 59 *Politische Vierteljahresschrift (PVS)* (2018) 269. See also Beate Jahn's contribution on 'humanitarian intervention' and Michael Stohl's contribution on the 'war on terror' to this volume.

is to remind the practitioners of revolutionary war of their original principles and to lead them onto appropriate paths at a moment when they were in danger of leaving their principles behind. Kant's work on the *unjust enemy* serves primarily to restrain both the use of the term and the treatment of the enemy as unjust, by showing the wrongness and injustice of such a practice.

2. The Republican Constitution and International Law

Kant's work perhaps reflects the connection between domestic democratization and international order more clearly than any other. After 1784, Kant dedicated much effort to questions of international peace. With the revolution, France had proven that 'a powerful and enlightened people can form itself into a republic'.[9] After this epochal event, Kant doubled his efforts to elaborate on the connection between the project of an international peace order and national constitutions.

After the revolution, constitutional law and international law could no longer be treated separately. In particular, Kant had to grapple with the fact that France's constitution had become the object of international conflict: the republican constitution provoked intervention. Moreover, constitutional change had become a war aim. The theory of natural international law had always ruled out that national constitutional change should serve as a justification for intervention.[10] The *Declaration of Pillnitz*, however, framed the continued existence of the French monarchy as 'a matter of communal interest for all sovereigns of Europe' and thus abandoned the principle of neutrality in domestic issues. Kant reacted to this development with Preliminary Article 5 of his *Perpetual Peace*: 'No state shall forcibly interfere in the constitution and government of another state.'[11] Notably, this article forbids intervention in either direction: intervention is unjustified, whether against or in favour of republican constitutions. The war against France provoked the French Republic into, at first, supporting democratic constitutions in conquered territories, and later, establishing them top-down through intervention or by incorporating the regions concerned into France. This development is reflected in Kant's writings. As the process of democratization led to international conflict, the defence of the French Republic turned into a war of conquest and 'democracy export'. Just as in the theory of just war, the French Republic justified conquest and expropriation, and even 'regime change' or the introduction of constitutions.[12]

Kant was forced to react to this development because it suggested a return of just war into republican theories of international law. Just war theory considered the states as equals and suggested a 'natural freedom' of states to go to war. That is why the

[9] Kant, above note 4, at 327 (AA 8:365).

[10] Steiger, 'Das natürliche Recht der Völker. Die Debatten der Französischen Revolution 1789–1793', in H. Steiger, *Universalität und Partikularität des Völkerrechts in geschichtlicher Perspektive. Aufsätze zur Völkerrechtsgeschichte 2008–2015* (2015), at 135.

[11] Kant, above note 4, at 319 (PP 8:346).

[12] Steiger, ‚Das Völkerrecht und der Wandel der Internationalen Beziehungen um 1800', in Steiger, above note 10, at 490–94; see also the corresponding passages on the French war justifications in Hendrik Simon's contribution to this volume.

occurrence of war could not be ruled out since there was no judge above states and no right to punish the use of force.[13] However, there was always the possibility that a state would violate the rules of the *ius in bello*.[14] It is here that the practice of revolutionary war revived the 'unlawful enemy' of earlier theories of just war. In the revolutionaries' view, the European monarchies had become such 'unjust enemies' because they tried to change the republican order back into an absolutist monarchy. Revolutionary France concluded from this that it had special rights against these unjust enemies, as for example to change their constitution, confiscate goods and property, and incorporate territory into the French state. Kant reacted to this with a discussion of the 'unjust enemy' in § 60 of *The Metaphysics of Morals*, in an attempt to curtail revolutionary incursions in the constitution, territorial order, and civil property of the defeated countries.

Kant's treatment of the unjust enemy in his legal theory appears to contradict his writings in *Perpetual Peace*. In his later *The Metaphysics of Morals* he defines the 'unjust enemy' as a state that renders peace impossible and which, once defeated, 'can be made to adopt a new constitution that by its nature will be unfavourable to the inclination of war'.[15] This passage is frequently read as Kant's approval of fighting wars in order to spread republican constitutions, against an 'unjust enemy'—and thus, as the re-emergence of just war in his own writings. It is worth noting that this interpretation has appeared in two versions, one against and the other one in favour of just war. Notably, both versions present Kant as a proponent of just war: in the one, Kant appears as a respected voice whose support for just war elevates the writer's own argument; in the other, Kant's 'unjust enemy' lends prominence to liberal interventionist theory.[16] In order to counter these readings and demonstrate that Kant was not a theoretician of just war, but instead intended to replace the concept theoretically, it is both helpful and necessary to consider the historical context in which Kant developed his theory of international law.

The interpretation of Kant as a theoretician of just war refers not only to the notion of the 'unjust enemy' but also to those parts of *Metaphysics* that discuss the 'right to go to war'.[17] This terminology, however, follows from Kant's choice of different systemic frameworks for the conceptualization of law in *The Metaphysics of Morals* and in *Perpetual Peace*. The passage from *Metaphysics* refers to law in the state of nature, and Kant leaves no doubt that 'in the state of nature generally—each state is judge in its own case'.[18] In *Perpetual Peace*, Kant had chosen a contractual approach. The point to be made here is that Kant organized the legal theory of *Metaphysics* in *Paragraphs*

[13] Achenwall and Pütter, *Anfangsgründe des Naturrechts*, edited and translated by Jan Schröder (1995) § 920, at 307: *'Unter Völkern gibt es kein Gesetz eines Herrschers, Rechtsprechung, Gericht, Richter, Richterspruch, Strafe.'*

[14] Ibid., § 365, at 321: *'Wer diese Grenzen überschreitet, wird zum unrechtmäßigen Feind.'*

[15] Kant, 'The Metaphysics of Morals', above note 4, 353, at 487 (AA 6:349).

[16] Orend, 'Kant's Just War Theory', 37 *Journal of the History of Philosophy (JHP)* (1999) 323; Müller, 'Evilization in Liberal Discourse: From Kant's "Unjust Enemy" to Today's Rogue State', 51 *International Politics (IntlPol)* (2014) 475; Müller, 'Kants Schurkenstaat: Der "ungerechte Feind" und die Selbstermächtigung zum Kriege', in A. Geis (ed.), *Den Krieg überdenken. Kriegsbegriffe und Kriegstheorien in der Kontroverse* (2006) 229.

[17] Kant, 'The Metaphysics of Morals', above note 4, at 484 (AA 6:346).

[18] Ibid., at 487 (AA 6:349).

according to the model of contemporary theories of natural law, in particular Achenwall's *Elementa iuris naturae*.[19] Unlike the systemic structure of international peace treaties, theories of natural law are not organized into preliminary articles, definitive articles, appendices, and supplements. Instead, they begin from a descriptive account of the state of nature and extract from natural law a right to war (*ius ad bellum*), a right in war (*ius in bello)*, and a right after war (*ius post bellum*). In no way does this approach suggest different, let alone contradictory, positions which would allow us to distinguish between a pacifist and a bellicist (or realist) Kantian theory.[20] Rather, Kant's theoretical discussion addresses different contemporary approaches to international order in a new way. Two of the most important perspectives are the tradition of peace plans and the idea of a state of nature.

3. Peace Plans and the State of Nature

The crucial development in international legal theory in the eighteenth century is the supplement of the theory of just war, the defining principle of the war discourse since the Middle Ages, by a theory of *regular war* that highlights the diplomatic and formal aspects of wartime decisions as developed by Hugo Grotius, Emer de Vattel, and Christian Wolff.[21] These were the negative points of reference in Kant's reflections, while the tradition of peace plans played an important positive role for Kant as a normative ideal. The peace plans were designed to facilitate a politics of peace by appealing to the enlightened self-interest of the rulers. They were intended to institutionalize international law in a league of nations.[22] Apart from the Abbé de Saint-Pierre, the contractualist Rousseau was the most popular proponent of the idea of a league of nations, even though he did not develop his own proposals and only popularized and critically discussed Saint-Pierre's plan.[23] Kant's commitment to this project was as consistent as that to the republic, but he was also aware of its limitations. The appellative character of the peace plans stood in contrast to the teachings of 'just war' and the 'balance of power' in political practice, which were more palatable to the rulers of the time. Kant's rejection of traditional peace proposals is coupled with the ambition to take up practical problems and resolve them within the theoretical rejection of just war.[24]

[19] See H. Williams, *Kant and the End of War. A Critique of Just War Theory* (2012), at 75.

[20] But see Orend, above note 16; B.S. Byrd and J. Hruschka, *Kant's Doctrine of Right: A Commentary* (2010); S.M. Shell, *Kant and the Limits of Autonomy* (2009), at 212–47. For the opposing view, see Williams, above note 19, at 83, 91–112; Eberl and Niesen, above note 5, at 125.

[21] Kalmanovitz, 'Sovereignty, Pluralism, and Regular War: Wolff und Vattel's Enlightenment Critique of Just war', 46 *Political Theory* (2018) 218. For the continuing relevance of 'just war' criteria for European political practice see the contributions by Anthony Lang, Jr, Anuschka Tischer, Arnulf Becker Lorca, and Hendrik Simon to this volume. For a critique of international law from a 'just war' perspective see Chris Brown's chapter.

[22] K. von Raumer, *Ewiger Friede. Friedensrufe und Friedenspläne seit der Renaissance* (1953).

[23] C.I. de Saint-Pierre, *Projet pour render la paix perpétuelle en Europe* (1986); Rousseau, 'Abstract and Judgement of Saint-Pierre's Project for Perpetual Peace', in D. Hoffmann and D. Fidler (eds), *Rousseau on International Relations* (1991) 52.

[24] This is particularly true for the six preliminary articles in *Perpetual Peace*. With Peter Niesen, I understand these as rules of international law, designed to enable and ensure the transition from natural international law to public international law, see Eberl and Niesen, above note 5.

Hobbes's description of the state of nature as a state-less situation lacking a monopoly of violence is of particular importance for Kant's theorizing on international order. The state of nature is characterized by the constant threat of attack and war.[25] Kant accepts this description of the state of nature and, like Hobbes, states that this threat does not require actual violence to occur; it is caused by mere co-existence of people in the absence of law.[26] At the same time, Kant understands the state of nature as a state of unilateral right, that is, legal uncertainty: this characterizes the state of nature as a 'state *devoid of justice* (*status iustitia vacuus*), in which when rights are *in dispute* (*ius controversum*), there would be no judge competent to tender a verdict having rightful force'.[27]

Escaping the state of nature thus means ending legal uncertainty. Only public law that is both legislated by the general will and authorized to coerce[28] can overcome the private law of the state of nature: 'For without this there is no public right, and any right that one may think of outside it (in a state of nature) is instead merely private right.'[29] People leave the state of nature through the act of forming a state which then needs to be reshaped from its absolutist and private-right origins into a republican form characterized by public right. This is the reason Kant welcomes the French Revolution, even though he prefers reform to revolution as the less dangerous option.

According to Kant, the state of nature also persists among states.[30] For many theorists, the transformation of this state of nature into public law can best be accomplished by establishing a world state. Kant, too, had initially followed this path but then changed course and left the analogy between the state of nature among people, and the state of nature among states behind. Instead, he attempted to identify forms of right and justice in a league of nations. The primary reason for this shift is the idea that the national state would facilitate democratization whereas the creation of a world state would lead either to an increasingly despotic 'universal monarchy' or a war that would swallow nations along with their differences in language and culture, an outcome that Rousseau had already warned against.[31] To Kant, the only possible world republic was to be a republic of republics, not a world state. This is why institutional issues regarding the league of nations are so important to Kant.

Whenever Kant discusses the issue of a monopoly of violence, that is, following Hobbes, the issue of state-ness, he also discusses the issue of democracy. He can neither welcome the absolutist state nor the world state as means to escape the state of nature, because neither can ensure the realization of democracy—in Kant's language, the realization of the republic and of public law. In his eyes, the risk that a world state would turn despotic and thus would provoke new (civil) wars is too high. Kant finds both alternative solutions of putting one's faith in freewheeling states or establishing a world state equally unconvincing.

[25] T. Hobbes, *Leviathan. Revised Student Edition*, edited by R. Tuck (1996), ch. 13.
[26] Kant, 'Toward Perpetual Peace', above note 4, at 322 (AA 8:349).
[27] Kant, 'The Metaphysics of Morals', above note 4, at 457 (AA 6:313).
[28] See ibid., at 388 (AA 6:231).
[29] Kant, 'Toward Perpetual Peace', above note 4, at 350 (AA 8:385).
[30] Ibid., at 326 (AA 8:355); Kant, 'The Metaphysics of Morals', above note 4, at 483 (AA 6:344).
[31] Kant, 'Toward Perpetual Peace', above note 4, at 336 (AA 8:367).

4. Regular War: The Just War Theory of the Absolutist Era

In the just war tradition the warring parties judge for themselves whether a war is just or not, as Chris Brown underlines with a view to the present situation.[32] The party which considers going to war has to check whether it has a just reason to do so. War, in shorthand, is considered just when it is found to constitute a reaction to an injury as determined by the injured. The procedure of decision-making is unilateral amounting to sovereign judgement.

The idea of just cause, which had long been at the core of just war theory, becomes revised with the spread of absolutism after the Reformation: the just war discourse is now supplemented or even dominated by reference to regular or formal war. The emergence of 'regular' or 'formal' war was formulated most prominently by Christian Wolff and Emer de Vattel. However, recent research brings up the question whether there actually occurred a transformation from just war to legal war or whether a dualism of both concepts emerged in early modern legal thought.[33] The resulting dispute is central for an assessment of early modern legal reasoning: according to the narrative of a transformation from just war to legal war, the sovereignty of kings was no longer seen as limited by considerations concerning the justice of war-making; to the contrary, according to this narrative, some early modern political and legal theorists argued that war was always justified if seen as necessary.[34] At the same time, however, the monopolization of war by the state led to an increasing formalization of international law and diplomatic conventions.[35] As Anuschka Tischer argues in her chapter, the cause of war had to be presented and explained to the enemy. This highlights the importance of bridging the gap between narratives of war justifications in legal theory on the one side, and political practice on the other. In political practice, the reference to just causes continued. In legal theory, however, the importance of the legitimacy of authority increased.[36] As Kalmanovitz has argued, Vattel held that nations and their rulers had 'a right of judging what [their] duties require'.[37]

[32] See also Chris Brown's contribution to this volume. For a debate with Chris Brown on this point see Brock 'Between Sovereign Judgement and the International Rule of Law', in M. Albert and A.F. Lang, Jr (eds), *The Politics of International Political Theory. Reflections on the Works of Chris Brown* (2019) 87.

[33] For the first opinion see Lutz-Bachmann, '"Peace" and "Justice" in the Tradition of "Just War-Theories"', in G. Hellmann (ed.), *Justice and Peace: Interdisciplinary Perspectives on a Contested Relationship* (2013) 90, at 94; for the second opinion see Lesaffer, 'Too Much History: From War as Sanction to the Sanctioning of War', in M. Weller (ed.), *The Oxford Handbook of the Use of Force in International Law* (2015), 35; and Zurbuchen, 'Emer de Vattel on the Society of Nations and the Political System of Europe', in S. Kadelbach, T. Kleinlein, and D. Roth-Isigkeit (eds), *System, Order, and International Law: The Early History of International Legal Thought from Machiavelli to Hegel* (2018) 263; On this scientific dispute see H. Simon, *Der Mythos vom 'freien Recht zum Krieg'. Zu einer Genealogie der modernen Kriegslegitimation* (forthcoming), Chapter 4.2.

[34] For a critical examination of this early modern narrative of transformation and the resulting 'myth of *liberum ius ad bellum*', see Hendrik Simon's contribution to this volume, and Simon, above note 33.

[35] Brock and Simon, above note 8, at 7–8.

[36] Kalmanovitz, above note 21, at 221. See E. de Vattel, *The Law of Nations* (2008 [1758]); C. Wolff, *The Law of Nations Treated According to a Scientific Method* (1934 [1750]).

[37] Kalmanovitz, above note 21, at 221; de Vattel, above note 36, Preliminaries §16.

Kalmanovitz interprets this shift as a 'reorientation of the just war' paradigm.[38] What is important here is that 'regular war' does not outlaw war as such, it only ties it to the condition that war must be formally declared, but it changes the relationship between the warring parties. Kalmanovitz describes the process by which regular war replaces the notion of righting a wrong from older theories with the notion of 'belligerent equality' based on the assumption of 'sovereign judgment':[39]

> Wolff and Vattel perfectly understood that the absence of centralized adjudication in the law of nations required *sui generis* procedures to establish those boundaries. They substituted the analogy of criminal trials, which makes the (allegedly) injured party become judge of his enemy, by the duty to publicly state the reasons for waging war in an international public forum.[40]

This announcement for the reasons to go to war is no real restriction of the war, it is a formalization of war. Because absolutist sovereignty cannot be restricted at all, it cannot be restricted by another sovereign. Belligerent equality is ultimately a result of the acceptance of absolutism. Natural international law is not intended to abolish war, but only to constrain unilateral judgement through diplomatic procedures.

Arthur Ripstein has highlighted the legitimacy of war that emerges from these new teachings:

> The regular war view conceives war as a procedure for resolving disputes. Sovereigns resort to it because no court or procedure has jurisdiction over them—that is the sense in which they are sovereign. For these writers [Grotius, Pufendorf, Vattel], the central question is whether a war is conducted in accordance with the procedure; questions about just cause are framed as questions of whether the party starting the war has what lawyers call a 'cause of action' that is, whether there is a genuine dispute about the respective rights of the two states.[41]

Whereas previously, an injury to a state needed to be determined in order to justify (just) war, according to Ripstein, it now suffices to determine the conflict. Where older theories of just war demanded that the (unilateral) verdict preceded the war, it is now postponed to the outcome of the war, in what Kant refers to as a 'so-called judgement of God'.[42] In regular war, the judgement is determined through war, whereas 'just war' executed a preceding judgement. The new theory accepts the impossibility of judging in a quarrel between equals in the absence of a higher judge—herein lies the core of belligerent equality—and following from that they accept war as the resolution of conflict. The ideas of just war and of regular war are in agreement, then, insofar as the possibility to wage war is not outlawed in either. No matter whether one

[38] Kalmanovitz, above note 21, at 221; for a different interpretation of Vattel see Lesaffer, above note 33, and Zurbuchen, above note 33.

[39] Kalmanovitz, above note 21, at 225; see also Brown's contribution to this volume.

[40] Kalmanovitz, above note 21, at 230.

[41] Ripstein, above note 1, at 179–95, 183.

[42] Kant, 'Toward Perpetual Peace', above note 4, at 320 (AA 8:347).

assumes a transformation from just war to formal war[43] or a co-existence of both concepts,[44] neither just war nor formal war prohibits war in positive legal terms. To the contrary, as Ripstein points out, the new doctrine of formal war can be seen as 'even more permissive'.[45]

Arguing against the proponents of natural international law, writers such as Rousseau and Kant insist that the absolutist constitution of states was itself a cause of war, and that an international law which combined legislative absolutism with legal unilateralism would establish war as a permanent means of politics. Wolff and Vattel tried to find a way out of this dilemma by introducing diplomatic procedures prior to war. But they did not succeed. This explains Kant's criticism of monarchical practices in his time: a monarch can decide to go to war 'as upon a kind of pleasure party, ... and can with indifference leave the justification of the war, for the sake of propriety, to the diplomatic corps, which is always ready to provide it'.[46] To Kant, diplomatic procedures are not designed as a serious means of seeking the truth, and they merely euphemize rather than contain war. The natural right of regular war does nothing to mitigate this flaw:

> Grotius, Pufendorf, Vattel, and the like (only sorry comforters)—although their code, couched philosophically or diplomatically, has not the slightest lawful force and cannot even have such force (since states as such are not subject to a common external constraint)—are always duly cited in justification of an offensive war, though there is no instance of a state ever having been moved to desist from its plan by arguments armed with the testimony of such important men.[47]

As Kant sees it, regular war can solve the problem of war neither in theory nor in practice. As a theory, it rather obscures the problem by suggesting that diplomatic procedures can chart a course of justice similar to court procedures. In practice, the final decision rests with the outcome of the war. Kant notes his opposition to the notion of regular war when he holds that 'right cannot be decided by war and its favourable outcome, victory'.[48] But his comment only describes the problem rather than solving it. Neither could contemporary peace plans suggest an international legal order beyond the analogy of a coercive intra-state law in the form of a world state.

5. State of Nature and the World Republic: Anacharsis Cloots

This issue returns to the forefront of philosophical argumentation and international politics after the French Revolution. Social contract theory insists that only the people themselves can act as legislator, but only after they have given themselves a

[43] Kalmanovitz, above note 21; Ripstein, above note 1; Lutz-Bachmann, above note 33.
[44] Lesaffer, above note 33; Zurbuchen, above note 33; Simon, above note 33.
[45] Ripstein, above note 1, at 179–95, at 184.
[46] Kant, 'Toward Perpetual Peace', above note 4, at 324 (AA 8:350).
[47] Ibid., at 326 (AA 8:355).
[48] Ibid., at 327, (AA 8:355).

constitution.[49] This demand is formulated not only in Rousseau's *Social Contract* but also by Emmanuel Sieyès, who defined the establishment of a republic as the passing of its constitution.[50] Kant follows this idea and hopes for the emergence of more republics. He discusses the possible shape of a league of nations that would include republics and non-republics alike in *Perpetual Peace*.[51] However, this still did not solve the problem: logically speaking, a world republic remained the less contradictory solution, but Kant opposed it because he declared that the conclusion of the social contract, that is, the escape from the state of nature through the creation of a state, 'cannot hold for states in accordance with the right of nature'.[52]

Some republicans disagreed. The Jacobin Anacharsis Cloots had argued at length that the only correct conclusion from social contract theory can be the 'universal republic'. In the course of the revolutionary wars, Cloots came to see republican France itself as the core of this universal republic, of which he thought not as France but as world republic. From the foundation of the republic in 1792, its foreign policy was informed by the idea that it could bring freedom to the peoples it defeated.[53] When the French convent could not find agreement on the treatment of conquered regions, Cloots intervened with the most radical proposal: they were to be united with France into a universal republic and would ultimately dissolve into a world republic. As Pauline Kleingeld puts it: 'he argued against turning the conquered territories into federated republics, advocating their merger with what he started to term "the country formerly called France" but now properly called "universal republic" '.[54]

The idea of a world republic does indeed follow from the logic of the social contract and presents an important challenge that is still considered 'consistent' and discussed to this day.[55] Nonetheless, Kant's proposal of a league of nations in *Perpetual Peace* rejects not only the theoretical conclusions of escaping the state of nature by establishing a world republic but also the French practice of 'reunions'.[56] On the one hand, Kant notes the republic's potential to 'provide a focal point of federative union for other states, to attach themselves to it and so to secure a condition of freedom of states conformably with the idea of the right of nations; and by further alliances of that kind, it would gradually extend further and further'.[57] But he also insists that states 'are not to be fused into a single state'.[58] The question, then, is *why not?* This takes us back to Cloots's point. 'Until he has answered this question, Kant has not yet fully met Cloots's

[49] Kant, 'The Metaphysics of Morals', above note 4, at 457 (AA 6:313); I. Maus, *Zur Aufklärung der Demokratietheorie. Rechts- und demokratietheoretische Überlegungen im Anschluss an Kant* (1994).

[50] Sieyès, 'What is the Third Estate?', in P.H. Beik (ed.), *The French Revolution. The Documentary History of Western Civilization* (1970) 16.

[51] Kant, 'Toward Perpetual Peace', above note 4, at 326f. (AA 8:355–356).

[52] Ibid., at 326 (8:355).

[53] Steiger, above note 10, at 144–146.

[54] Kleingeld, above note 6, at 41; A. Cloots, *L'orateur du genre humain* (1791), at 147; A. Cloots, *Bases constitutionelles de la république du genre humain* (1793), at 29–30, both cited in Kleingeld, above note 6, at 43.

[55] See also Kleingeld, above note 6, at 43; also Cheneval, 'Der kosmopolitische Republikanismus—erläutert am Beispiel Anacharsis Cloots', 58 *Zeitschrift für philosophische Forschung (ZPhF)* (2004) 373, at 392–94.

[56] Steiger, above note 10, at 159–61.

[57] Kant, 'Toward Perpetual Peace', above note 4, at 327 (AA 8:356).

[58] Ibid., at 326 (AA 8:354).

challenge', states Kleingeld and stresses the importance of this issue.[59] Kleingeld hopes to meet the challenge by turning to Kant's fear of the despotism of a world state, even of a world republic, which would end in 'universal monarchy' and 'soulless despotism'.[60] According to Kleingeld, Kant is primarily interested not in maintaining sovereignty but in preventing the imperialism of a world state.[61] In terms of historical context, however, 'Cloots's challenge' also suggests an evaluation of republican France's foreign policy and of the possibility of its universalization.

As Kleingeld shows, the systematic core of the issue is the difference between two distinct states of nature; that between individuals and that between states. In her interpretation, this difference leads Kant to reject the world state even in the shape of a world republic and to renounce the use of force as a means of integrating states into a league of nations.[62] Although Kleingeld is correct in her analysis of Kant's vision of a voluntary league of nations made up of both absolutist and republican states,[63] a discussion of the systemic function of Kant's use of the social contract requires an even stronger engagement with the practices of the French Republic. When Kant declares that states should not fuse, he is also commenting on the French practice of incorporating a series of conquered territories. While this was initially done at the request of republicans in those regions—Savoyen in 1792, Nice, Monaco, Basel, and Mainz in 1793, as well as multiple times in Belgium after formal deliberations—the practice changed fundamentally over the course of the war.[64]

The first constitution had defined foreign policy as the attempt to combine the new principles with the positive right of *droit public de l'Europe*.[65] In the republic, the new principles of freedom and human rights were now understood as antagonistic to the old order: 'The discussion and practice of foreign policy were informed by the spirit of a missionary task, a crusade in the name of the universal, natural freedom of peoples.'[66] A war that had started out in defence of the republic's own constitution turned into a war over the constitutions of enemy states.

Such charges of 'missionary' warmongering are the mirror image of today's liberal interventionism. Alyssa Bernstein criticizes Kleingeld for neglecting the issue of enemy states' constitutions and thus for overlooking Kant's support for such constitutional wars. Bernstein's argument revolves around the notion of the 'unjust enemy'.[67] Elaborated in Kant's theory of right, this figure had long been neglected before rising to prominence in recent interpretations that seem to prove Kant's interventionism. In fact, however, Kant's idea of an 'unjust enemy' suggests just the opposite, as I already hinted above and shall show in more detail in the next section. The 'unjust enemy' is a logical part of just war theory, not the proof of a final acceptance of republican imperialism and therefore not a starting point of liberal interventionism. In dealing with

[59] Kleingeld, above note 6, at 61.
[60] Kant, 'Toward Perpetual Peace', above note 4, at 336 (AA 8:367).
[61] Kleingeld, above note 6, at 61.
[62] Ibid., at 50–58.
[63] See Eberl and Niesen, above note 5, at 235.
[64] Steiger, above note 10, at 162.
[65] Ibid., at 165.
[66] Ibid., at 185. Translation mine.
[67] Bernstein, 'The Rights of States, the Rule of Law, and Coercion: Reflections on Pauline Kleingeld's *Kant and Cosmopolitanism*', 19 *Kantian Review* (2014) 233.

it, Kant attempts to constrain republican warfare even against an 'unjust enemy', an enemy that has violated natural law. The 'unjust enemy' is the last proof that natural law is self-contradicting and will not help to resolve war.

6. Kant's Rejection of 'Just War'—Even against the 'Unjust Enemy'

The image of the 'unjust enemy' addressed above closely combines constitutional issues and international law. This explains why Kant's treatment of the issue is such an important topic of debate. Kleingeld did not detect any right to coercive constitution-making in Kant. Bernstein takes up Kleingeld's 'statement that brutally oppressed populations should be put in a position to determine the shape of their own political institutions, and then allowed to decide whether to join a federation',[68] a position she supports, but with reservation concerning the 'unjust enemy'. She limits Kant's proscription of coercion to non-aggressive states. This implies that it is generally illegitimate to coerce states into adopting the correct constitution or into joining the league of nations. Coercion is 'almost never' justified and 'nearly always' wrong—with the one exception of the 'unjust enemy'.[69]

Bernstein's argument opens the door to intervention. This is the direction Susan Meld Shell takes: she sees Kant preparing the way for a

> (new) distinction between states that do and states that don't count as full-fledged members of the community of nations. The latter ('failed' and 'rogue') states do not receive the normal prerogatives of sovereignty to which members 'in good standing' of the community of nations ... are generally entitled.[70]

The peace researcher Harald Müller, who is opposed to such liberal interventionism, turns this reading against Kant. Müller considers the identification of the 'non-democratic other' as an example of ' "evilization" as an extreme form of a securitization speech act'.[71] His point is that the liberal mind-set—exemplified by Kant—is unable to *not* label the Other in this way: 'liberal thought is not immune against this practice as even the most rational and close to pacifist version of liberalism embodied in the work of Immanuel Kant subscribes to the notion "unjust enemy" as a deductively necessary template that both enables and constrains the possibilities for evilization in liberal policymaking'.[72] This strong charge comprises two assertions: one, Kant's work supports 'evilization', and two, the 'unjust enemy' is 'deductively necessary'. The latter point refers to a concept of progress that demands the overcoming of obstacles, one of which is the 'unjust enemy'. Müller himself considers this a frame of mind necessary to achieve the transition from the state of nature to international law, a position he illustrates

[68] Ibid., at 241.
[69] Ibid., at 246f.
[70] Shell, above note 20, at 246.
[71] Müller, 'Evilization', above note 16, at 476.
[72] Ibid.

with 'Hitler's Germany' as the example of evil in international relations.[73] This reading supports the idea that Kant's treatment of the 'unjust enemy' fuelled the missionary zeal of liberals and revolutionaries.

But how did Kant determine the 'unjust enemy'?—By asking:

> what is an unjust enemy in terms of the concepts of the right of nations, in which—as is the case in the state of nature generally—each state is the judge in its own case? It is an enemy whose publicly expressed will (whether by word or by deed) reveals a maxim by which, if it were made a universal rule, any condition of peace among nations would be impossible and, instead, a state of nature would be perpetuated.[74]

In light of Kant's plan for a league of nations, it may seem as if he introduced the 'unjust enemy' in order to identify the enemies of republic and international law. In fact, however, Wolff and Vattel had written about the 'unjust enemy' before Kant,[75] and Kant himself referenced Achenwall. Achenwall had discussed legitimate means of war and proposed that any party could turn into an 'unjust enemy' when its war aims went beyond the restitution for damages and putting an end to incurred injury or threat.[76]

The image of the 'unjust enemy' is thus not the expression of a genuine liberal interventionism (although recent liberal interpretations suggest this understanding) but part of just war theory. It is important to note that the unjust enemy in Achenwall and Vattel[77] is a problem that belongs to the category of right *in* war, not right *to* war. Within this logic of war aims and legitimate means of war, the 'unjust enemy' is indeed *necessary*, because the question of transgression automatically comes up in a practice-oriented theory of just war. This is the type of 'unjust enemy' Kant has in mind. The opening of the *Metaphysics*' article defining the rights against an 'unjust enemy' confirms this reading: the legitimate means against such an enemy are the same as those which are legitimate generally.[78]

As demonstrated, Kant neither 'introduced' the 'unjust enemy' and revived the theory of just war, nor did he allow for exceptional means against the 'unjust enemy'. To the contrary, Kant made it clear that the concept of 'unjust enemy' could not actually be used in any meaningful way, because it is 'pleonastic ... to speak of an unjust enemy in a state of nature; for a state of nature is itself a state of injustice. A just enemy would be one that would be doing wrong by resisting; but then he would also not be my enemy.'[79] This quotation shows, again, that Kant defined the 'unjust enemy' as a figure that belonged to the state of nature, and that the assumption of an enemy's

[73] On the use of Nazi Germany as a historical example in the debate on the 'unjust enemy', see Eberl and Niesen, 'Kein Frieden mit dem "ungerechten Feind"? Erzwungene Verfassunggebung im Ausgang aus dem Naturzustand', in O. Eberl (ed.), *Transnationalisierung der Volkssouveränität. Radikale Demokratie diesseits und jenseits des Staates* (2011) 219, at 226. By using this example, however, Müller simultaneously approves of 'evilization' as a historically logical necessity, and it becomes unclear what exactly he seeks to analyse as 'othering'.

[74] Kant, 'The Metaphysics of Morals', above note 4, at 487 (AA 6:349).

[75] Kalmanovitz, above note 21, at 220.

[76] Achenwall and Pütter, above note 13, § 959–64; Eberl and Niesen, above note 73, at 229.

[77] de Vattel, above note 36, Book III, Ch. X, § 174.

[78] Kant, 'The Metaphysics of Morals', above note 4, at 487 (AA 6:349).

[79] Ibid. (AA 6:349f.).

unjustness was indeed a requisite for just war. Regular war does not entirely replace this assumption with belligerent equality, but rather postpones it until the outcome of the war. If a state would not consider itself justified to wage war against another state, there would be no justification for war at all. In the process of war any state can violate the *ius in bello* which would be determined by a warring party that would then judge his opponent an 'unjust enemy' and so lead to a further escalation of warfare on the basis of a unilateral judgement. This is Kant's point, and it is valid for his rejection of both traditional just war theory and modernized regular war theory. Kant rejects these theories not only because of their practical ineffectiveness concerning peace, but because they either ignore the problem of finding justice or equate justice with the outcome of war. An enemy can be unjust only when a judge declares him to be, and this is only possible after the transition from the state of nature to public law. Just war always carries with it the hidden assumption that the enemy is 'unjust'.

Kant makes his position clear in a passage of *Perpetual Peace*:

> war is, after all, only the regrettable expedient for asserting one's right by force in a state of nature (where there is no court that could judge with rightful force); in it neither of the two parties can be declared an unjust enemy (since that already presupposes a judicial decision), but instead the *outcome* of the war (as in a so called judgement of God) decides on whose side the right is; but a punitive war (*bellum punitivum*) between states is not thinkable (since there is no relation of a superior to an inferior between them.[80]

This passage from the 5th preliminary article also insists that the 'unjust enemy' may not be punished. Kant says the same in § 60 of *The Metaphysics of Morals*, when he discusses the case of a state violating public contracts which affect the freedom of all states. Even then, the state's allied enemies are 'not called upon to divide its territory among themselves and to make the state, as it were, disappear from the earth, since that would be an injustice against its people, which cannot lose its original right to unite itself into a commonwealth, though it can be made to adopt a new constitution that by its nature will be unfavourable to the inclination of war'.[81] This sentence seemingly combines the proscription of dissolving a state into separate territories on the one hand, with the legitimation of forcing upon it a new constitution, on the other. It is important to note, however, that the translation here suggests a clarity not conveyed in the German formulation that reads '*eine neue Verfassung annehmen zu lassen*'. David Colclasure's alternative translation as 'let it accept a new constitution' is preferable here because it avoids suggesting that Kant clearly favours enforced constitutional change.[82] 'Let it accept' preserves the original right of a people to give itself a constitution, as opposed to the disappearance of this facet in 'can be made to adopt'.

[80] Kant, 'Toward Perpetual Peace', above note 4, at 320 (AA 8:346–347).
[81] Kant, 'The Metaphysics of Morals', above note 4, at 487 (AA 6:349).
[82] In P. Kleingeld (ed.), *Toward Perpetual Peace and Other Writings on Politics, Peace, and History* (2006), at 145.

The right of the people to give themselves a constitution had not only become lost in the translation of our days but also in the practice of revolutionary wars. The French victors initially wanted to allow the Dutch people to give themselves a constitution, but this intent was soon replaced by a number of annexations and the order to introduce a constitution mirroring that of the French *Directoire*.[83] The Dutch people rejected this constitution as too undemocratic in a referendum in 1797. As a result, 1798 saw a coup under French leadership and the institution of a directorate of five, following the model of the French government at the time.[84] At that point, Kant reminded the warring French that a people retains the right to give itself a constitution even under an assumed unjust enemy. Whom else could he have addressed? His reminder does not justify, but indeed rejects missionary zeal, which essentially is nothing but the bourgeois urge for territorial expansion. All interpretations that ascribe to Kant the legitimation of intervention in the name of constitutional change overlook that he narrowly constrained the rights of victors and did not justify the external imposition of a constitution. This would also hold in the event of a victory by the absolutist powers: they, too, would not be allowed to force on France a constitution that would restore pre-revolutionary forms of government. Even in the still existing state of nature, then, the victorious party has no right to dictate a constitution to others.

7. Conclusion

Kant's reminder, that people have the original right to unite and give themselves a constitution, is directed at those who claim to be justified to act as legitimate authority vis-à-vis an 'unjust enemy'. Kant reminds them that they cannot take on the role of judge and that, in any case, they have no right to decide on a constitution for another country. Only the citizens of that country have the right to do so. The task of international law then is to put all nations in a position allowing them to pursue their own path of constitutional reform. Kant separates constitution and international law. No war party ever attains constitutional powers. This feature will persist even in the future public international law for two reasons. First because forcing a constitution on people from the outside would violate the people's original constituent right, and secondly because the structure of international law is thoroughly unsuited for intra-state constitution-making. The point of escaping the state of nature through international law is to change the resolution of conflicts between states from war to legal procedures. Neither regular nor just war can contribute to this goal. Both concepts rely on the assumption of an 'unjust enemy', because the idea of a 'just enemy' does not make sense (the just enemy would be correct in its demands against me and therefore I could not rightfully resist him). Kant refuted the logic of just war and 'unjust enemy', and committed international law to the task of conflict-solving by legal procedures in a league of nations.

[83] S. Schama, *Patriots and Liberators. Revolution in the Netherlands 1780–1813* (1977), at 265f.
[84] Ibid., at 269.

PART III
THE NINETEENTH CENTURY AS THE BIRTH ERA OF THE MODERN WAR DISCOURSE

8

Anarchy over Law? Towards a Genealogy of Modern War Justifications (1789–1918)

*Hendrik Simon**

1. Introduction: On 'Anarchy' and 'Progress'—Two Modern Narratives of War Justification and International Order

Modern historiographies of the justification of war and international order usually depart from the thesis that in the course of early modernity a fundamental transformation of the European war discourse occurred which came to a completion in the nineteenth century: according to the overwhelming majority of recent literature on the history of international law and international relations, including standard books such as those by Wilhelm G. Grewe,[1] Martti Koskenniemi,[2] Stephen C. Neff,[3] or Barry Buzan and George Lawson,[4] in nineteenth-century legal theory and political practice, the discourse of 'just war' (*bellum iustum*) had been replaced by the sovereign states' 'free right to go to war' (*liberum ius ad bellum*). As one of its main proponents, Grewe summarizes this thesis as follows:

> To the international legal positivism of the nineteenth century the 'justice' of wars was fundamentally, and more decidedly and radically than for many teachers of natural law in preceding centuries, a juridically irrelevant problem of political ethics.[5]

Accordingly, war did not require any international justification until its legal prohibition in the twentieth century. European sovereigns were only guided by their own

* Many thanks to Lothar Brock for helpful comments on this chapter. Further thanks to Lothar Brock, Anuschka Tischer, and Miloš Vec for ongoing discussions related to its subject. This chapter condenses central arguments of my *Der Mythos vom 'freien Recht zum Krieg'. Zu einer Genealogie der modernen Kriegslegitimation* (forthcoming). In sections 3 and 4, it partly draws on material I have presented in Simon, 'Über das "freie Recht zum Krieg" in Politik und Völkerrecht des 19. Jahrhunderts', *Research Paper/ Max Planck Institute for European Legal History* (2012); and Simon, 'The Myth of *Liberum Ius ad Bellum*. Justifying War in 19th-Century Legal Theory and Political Practice', 29 *European Journal of International Law* (2018) 113; all translations mine.

[1] W.G. Grewe, *The Epochs of International Law*, translated and revised by Michael Byers (2000), at 531.

[2] M. Koskenniemi, *The Gentle Civilizer of Nations. The Rise and Fall of International Law 1870–1960* (2002), at 86; see also Koskenniemi, 'The Politics of International Law', 1 *European Journal of International Law* (1990) 4, at 6.

[3] S.C. Neff, *War and the Law of Nations: A General History* (2005), at 197.

[4] B. Buzan and G. Lawson, *The Global Transformation: History, Modernity and the Making of International Relations* (2015), at 86.

[5] Ibid.

Hendrik Simon, *Anarchy over Law? Towards a Genealogy of Modern War Justifications (1789–1918)* In: *The Justification of War and International Order.* Edited by: Lothar Brock and Hendrik Simon, Oxford University Press (2021). © the many contributors.
DOI: 10.1093/oso/9780198865308.003.0008

conscience and the *raison d'État*, as Reinhart Koselleck has claimed.[6] In the words of Stephen C. Neff: 'The idea of law governing the resort to war—the *ius ad bellum* of lawyers—shrivelled into virtual nothingness in the face of the positivist challenge. The decision to resort to war ... was the prerogative of policy, not of law.'[7] With this assumed triumph of state-preferences, the international community's authority to restrict the *ius ad bellum* was allegedly abandoned: 'If states wanted to go to war, they could', argue Barry Buzan and George Lawson with reference to Neff.[8] In this sense, Anne Peters observes that '[s]overeignty was conceived as absolute. The sovereign's right to wage war, the *liberum ius ad bellum*, became the epitome of sovereignty ...'[9]

In view of this omnipresence of the assumption of *liberum ius ad bellum*, it may come as no surprise that the nineteenth century is remembered as a 'laissez-faire era for war-makers'[10]—as a pre-modern era of anarchy, as famous German lawyer and legal pacifist Hans Wehberg once called it.[11] Interestingly enough (and yet unrecognized by research so far), the two central narratives structuring the modern historiography of war justifications and international order—contradictory as they may be with regard to each other's assessment of the role of law in ordering war—*both* proceed from the thesis of *liberum ius ad bellum*.

The first narrative, here referred to as the *narrative of anarchy*, is that of political *Realists*. For *Realists*, law is an expression of the factual power of the sovereign state. Its authority does not originate in the authority of the international order, but from state sovereignty as a *Rechtsvoraussetzungsbegriff* (Carl Schmitt).[12] Accordingly, the 'free right to go to war' tells us basically everything we need to know about the interrelationship between law and politics when it comes to the justification of war: in an anarchical state of nature (*status naturalis*) sovereign states are not obliged to follow the rules of an—anyway 'ineffective'[13]—international legal order: 'War is just, if it is necessary' (Machiavelli).[14] Much in this vein, Carl Schmitt, inspired by early modern authors like Alberico Gentili and Richard Zouch ('the true founders of European international law'),[15] assumed that by the nineteenth century, 'war was transformed

[6] R. Koselleck, *Kritik und Krise* (13th ed., 2017), at 33f.

[7] Neff, above note 3, at 161f.

[8] Buzan and Lawson, above note 4, at 50.

[9] Peters, 'Membership in the Global Constitutional Community', in J. Klabbers, A. Peters, and G. Ulfstein (eds), *The Constitutionalization of International Law* (2009) 153, at 183.

[10] Neff, above note 3, at 161, 168.

[11] Wehberg, 'Universales oder Europäisches Völkerrecht? Eine Auseinandersetzung mit Professor Carl Schmitt', 41 *Die Friedens-Warte* (1941) 157, at 162.

[12] See also M. Koskenniemi, *From Apology to Utopia: The Structure of International Legal Argument* (2005), at 227; and Lothar Brock's and my introductory chapter.

[13] In his stimulating and thought provoking chapter, Chris Brown argues that international law has been unable to effectively order the use of force in international affairs. At first sight, this seems to be a straightforward *realist* perspective on international law. However, Chris Brown differs from *Realists* in an important point: He assumes that instead of international law, 'just war' as a practice of political judgement is suitable to order the justification of war. Why, however, the unilateral vocabulary of political judgement should be able to develop greater binding force than legal conventions drawn up multilaterally in a broader communication community, remains open to debate. For this debate, see Brown in this volume, and the concluding chapter.

[14] See also I. Brownlie, *International Law and the Use of Force by States* (1963), at 5.

[15] C. Schmitt, *The Nomos of the Earth in the International Law of the Ius Publicum Europeaum*, translated by G.L. Ulmen (2004), at 309.

into a relation between mutually equal and sovereign states'.[16] *Liberum ius ad bellum* seemingly was in line with Clausewitz. This may explain its particular popularity in Germany.

Surprisingly, however, the proposition of the existence of a nineteenth-century 'free right to go to war' is still shared even by those who have little in common with a *Hobbesian* or *Clausewitzian* agenda for dealing with war:[17] for instance, Jürgen Habermas (as a critical Frankfurt School thinker in the liberal tradition of Kant)[18] assumes that *liberum ius ad bellum* constituted the structural core of classic international law.[19] And even Michael Walzer (as one of the most famous current 'just-war' theorists) concludes that from early modernity up to the nineteenth century justice was overcome by political prudence in the sense of Niccolò Machiavelli.[20] Similarly, despite their rich research on the justification of war between 1492 and 1945,[21] Yale lawyers Oona A. Hathaway und Scott J. Shapiro adhere to the thesis that in an 'old world order' before 1928/45, war was 'simply a continuation of politics by other means'—'might was right'.[22] Yet, their argument is blurred: on the one hand, Hathaway and Shapiro argue (correctly, as will be shown) that 'war was an instrument of justice', a 'legitimate means of righting wrongs', and that 'sovereigns rarely, if ever, went to war unless they could assert that their cause was in some way just'.[23] This is an assumption that coincides both with 'just-war' theory[24] and this book's basic thesis of a historically consistent practice of war justifications (see also our introductory and concluding chapters). On the other hand, however, Hathaway and Shapiro argue in the sense of *liberum ius ad bellum* that war was an instrument of state policy, and that the 'old world order' not only sanctioned war, 'but that it relied on and rewarded it' (for instance, in the 'right of conquest').[25] This second argument seems problematic to me, and it will be contested in this chapter.

What distinguishes liberal approaches like Habermas' or Hathaway's and Shapiro's from that of *Realists*, however, is the role *Liberals* attribute to *liberum ius ad bellum* in their understanding of the history of modern international law. By constructing

[16] Ibid.; See also Schröder, 'Carl Schmitt's Appropriation of the Early Modern European Tradition of Political Thought on the State and Interstate Relations', 33 *History of Political Thought* 2 (2012) 348.

[17] See also the references above; see also Chris Brown in this volume, and M. Abbenhuis, *An Age of Neutrals: Great Power Politics, 1815–1914* (2014), at 193; O.A. Hathaway and S.J. Shapiro, *The Internationalists. How a Radical Plan to Outlaw War Remade the World* (2017), at xv; M.M. Payk, *Frieden durch Recht? Der Aufstieg des modernen Völkerrechts und der Friedensschluss nach dem Ersten Weltkrieg* (2018), at 2, 498. For a for a more comprehensive list of references to the thesis of *liberum ius ad bellum* see Simon, *Der Mythos*, above note *.

[18] However, Habermas himself has criticized a strict dichotomy between *Realism* and *Liberalism* in the case of Kant with good reasons: Because of Kant's engagement with Hobbes's political theory, Habermas has called Kant a 'realist Liberal', see also J. Habermas, *The Postnational Constellation: Political Essays* (2001).

[19] Ibid., at 118.

[20] Walzer, 'The Triumph of Just War Theory (and the Dangers of Success)', 69 *Social Research* 4 (2002) 925, at 927. On Walzer, see also Lang, Jr in this volume.

[21] Hathaway et al., 'War Manifestos', 85 *University of Chicago Law Review* (2018) 1139.

[22] Hathaway and Shapiro, above note 17, at xv.

[23] Ibid., at 33.

[24] Despite Anuschka Tischer's pioneer research (see also Tischer's chapter), Hathaway et al. support this argument in an interesting, partly quantitative research project on war manifestos between 1492 and 1945. According to Hathaway et al., self-defence was the central narrative in 69 per cent of these manifestos and the violations of international law in 35 per cent of these manifestos, see Hathaway et al., above note 21.

[25] Hathaway and Shapiro, above note 17, at 33.

ideal-type dichotomies between 'classical' and 'modern' international law (Habermas) respectively 'old' and 'new world orders' (Hathaway and Shapiro), they transfigure the nineteenth-century international order, allegedly shaped by the 'free right to go to war', into a dark, anarchic 'photo negative'[26] of the twentieth-century 'universal' legal order that rose from the ashes of two world wars. For the liberal teleology and its *narrative of progress*, the (anti-)thesis of nineteenth-century *liberum ius ad bellum* thus serves to emphasize the legal progress that the international order supposedly has achieved in comparison to the pre-1945 order.

As the following chapter will argue, both *realist* and *liberal* teleologies are oversimplifying the historical discourses on the justification of war and international order in the nineteenth century: the proposition of *liberum ius ad bellum* is not supported by a thorough analysis of nineteenth-century legal theory and political practice. Instead, natural law maintained a continuing importance in nineteenth-century war discourse.[27] Against the thesis of a modern *Entzauberung* (disenchantment) in the sense of a rational unification of the vocabulary in modern war discourse—away from moral theology ('just war') and towards *realpolitik (Realism)* or positive law *(Liberalism)*—, the chapter reveals the multi-normativity of the nineteenth-century discourse on war and international order. For this purpose, it offers a short genealogy of 'long nineteenth-century' war justifications, both in legal theory and political practice, from the French Revolutionary Wars to the Great War.

The chapter ties in with the contributions in Part II of this book: in her pioneer study[28] as well as in her contribution to this volume, Anuschka Tischer convincingly shows that 'just war' remained the most important normative language in official European war declarations between 1492 and 1792. By constantly justifying war, princes constituted an international 'communication community' (Tischer in this book; see also our introduction), in which multiple norms counter-balanced the rising idea of state sovereignty. Building on the early modern war discourse, the century-old communicative practice of justifying the use of force continued in the nineteenth century, the 'golden era of the state'.[29] However, in the course of the nineteenth century, important diplomatic, theoretical, and normative innovations emerged. As a consequence, the nineteenth century is more similar to the subsequent century than

[26] Ibid., at xvii, 304; on liberal peace teleologies see also Brock and Simon, 'Liberal European Peace Theories and their Critics', in K.E. Jørgensen (ed.), *The Liberal International Theory Tradition in Europe* (2020).

[27] Carty, 'The Evolution of International Legal Scholarship in Germany during the Kaiserreich and the Weimarer Republik (1871–1933)', 50 *German Yearbook of International Law* (2007) 29, at 45–55; M.-E. O'Connell, *The Power and Purpose of International Law: Insights from the Theory and Practice of Enforcement* (2008), at 38–40; Simon, 'Über das "freie Recht zum Krieg" in Politik und Völkerrecht des 19. Jahrhunderts', *Research Paper/Max Planck Institute for European Legal History* (2012); E. Jouannet, *The Liberal-Welfarist Law of Nations* (2012), at 130; Verdebout, 'The Contemporary Discourse on the Use of Force in the Nineteenth Century: A Diachronic and Critical Analysis', 1 *Journal on the Use of Force and International Law* (2014) 223; Simon, 'The Myth of *Liberum Ius ad Bellum*. Justifying War in 19th-Century Legal Theory and Political Practice', 29 *European Journal of International Law* (2018) 113; Lesaffer, 'Aggression before Versailles', 29 *European Journal of International Law* (2018) 773; see already Nussbaum, 'Just War: A Legal Concept?', 43 *Michigan Law Review* (1943) 453, at 474.

[28] A. Tischer, *Offizielle Kriegsbegründungen in der Frühen Neuzeit. Herrscherkommunikation in Europa zwischen Souveränität und korporativem Selbstverständnis* (2012),

[29] *'Das goldene Zeitalter des Staates'*, see J. Osterhammel, *Die Verwandlung der Welt* (6th ed., 2011), at 820; see also Buzan and Lawson, above note 4.

to previous ones: against the widespread proposition of *liberum ius ad bellum*, the fundamental argument of this chapter—and of Part III of this book—is that the 'long nineteenth century' is not to be understood as the anarchic converse of the modern discourse on war and international order—but as its era of birth.

2. On the Threshold of Modernity. Transformations of War and Order in Revolutionary Times? (1789–1815)

'In the beginning was Napoleon.'[30] These first words of Thomas Nipperdey's famous *Geschichte der Deutschen* are also suitable as an introduction to a genealogy of modern war justifications: the year 1789 is considered to mark the threshold of modernity.[31] This characterization does not only apply to the socio-political upheavals in France, Europe, and the world,[32] but also to international relations[33] and international law[34]—and, with this, to the justification of war and international order: the main issues of the modern war discourse—aggression, self-defence, collective peacekeeping, public sovereignty and self-determination, intervention/non-intervention, preventive/pre-emptive strikes, annexations, republican and peoples' wars, the relationship between positive and natural international law, and perhaps even total war[35]—have all been broadly discussed in the war discourse of the Revolutionary and Napoleonic Age, some for the first time in (European) history. Thus, it seems appropriate to locate the starting point of the modern war discourse in the French Revolution and its immediate aftermath.[36]

Despite its immanent historical importance, however, '1789' received comparatively little attention in the History of International Law. Randall Lesaffer explains this 'relative absence' of the French Revolution as follows: it neither fits into the common narrative of the rising, nearly absolute sovereign state (seventeenth–nineteenth centuries), nor into the narrative of a legal revolution in the first half of the twentieth century.[37] With a view to the justification of war, the Revolutionary Age questions both the *realist* and the *liberal* historiographies of the justification of war and international order, which I have mentioned above: the intense public debates in the French National Assembly on the laws of war and peace are particularly indicative of the great

[30] '*Am Anfang war Napoleon*', T. Nipperdey, *Deutsche Geschichte 1800–1866. Bürgerwelt und starker Staat* (1983), at 11.

[31] See also D. Andress, *1789. The Threshold of the Modern Age* (2008).

[32] See also B. Stone, *Reinterpreting the French Revolution: A Global-Historical Perspective* (2002); S. Desan, L. Hunt, and W.M. Nelson (eds), *The French Revolution in Global Perspective* (2013).

[33] J.-P. Bois, *De la paix des rois à l'ordre des empereurs, 1714–1815* (2003); L. Frey and M. Frey, *The Culture of French Revolutionary Diplomacy: In the Face of Europe* (2018).

[34] M. Belissa, *Fraternité universelle et intérêt national (1713–1795): les cosmopolitiques du droit des gens* (1998); B. Jacobs, R. Kubben, and R. Lesaffer (eds), *In the Embrace of France: The Law of Nations and Constitutional Law in the French Satellite States of the Revolutionary and Napoleonic Age, (1789–1815)* (2006); E.J. Kolla, *Sovereignty, International Law, and the French Revolution* (2017).

[35] D. Bell, *The First Total War: Napoleon's Europe and the Birth of Warfare as We Know It* (2007).

[36] Of course, to some extent the roots of these debates lie in the early modern war discourse, see also Tischer, above note 28, and Lesaffer, above note 27; on continuities and changes in the discourse on the justification of war and international order, see also the concluding chapter in this volume.

[37] Lesaffer, 'In the Embrace of France. An Introduction', in Jacobs et al., above note 34, 1.

public pressure to justify war both within Revolutionary France and in the European international order. Interestingly, as many of the delegates were lawyers or philosophers, the two discourse levels of political practice and legal/political theory were closely interwoven in these debates.

What is decisive for the research interest at hand is that Revolutionary/Napoleonic France constantly debated and justified the use of force on two fronts: outwards against a total of seven enemy coalitions (1792–1815), inwards against actual or supposed allies of these coalitions. The revolutionary war discourse was significantly transformed over the years—particularly in the phase between 1789 and 1793—and the prevailing voices changed distinctly. Still, the French statesmen—not even Napoleon—by no means claimed a 'free right to go to war'. Nor did they act according to such a 'right'. On the contrary, the French political elites, without exception, tried to legitimize their respective use of force with reference to moral norms, political convictions, as well as international law. So there was no transformation of the vocabulary of the European war discourse in favour of *liberum ius ad bellum* at that time.

Yet, an *actual* transformation of the European war discourse *did* occur in these years, and it was indeed founded on a new political claim: as it increasingly turned out, the appearance of 'popular sovereignty' in France challenged the understandings of international law not only as seen by Revolutionary France but also as understood by the European monarchies. This was particularly true for the justification of war and the visions of international order enshrined in it: initially, the revolutionary discourse was pacifist, as became clear in early proposals by Volney, Jallet, or the Abbé Grégoire in the National Assembly on the law of peace and war. In May 1790, the Assembly finally renounced the legitimacy of wars of conquest: '… *l'Assemblée nationale déclarant à cet effet que la nation française renonce à toute espèce de conquête, et qu'elle n'emploiera jamais ses forces contre la liberté d'aucun peuple.*' This renouncement of wars of conquest was incorporated into the French Constitution of 3 September 1791.[38]

Much in the tradition of *bellum iustum*, only defensive wars were regarded as legitimate under natural law. Grégoire put this into the formula that no justification could be given for the use of force other than to *defend* the sovereignty, freedom, and property of the people.[39] Not only should wars of aggression be outlawed, but they should also be made punishable.[40] This pacifist commitment already implied a first incompatibility with the monarchical international order based on treaties of alliance. Now alliances with monarchies, specifically the Bourbon *Pacte de famille* (1761), were fundamentally questioned because the treaties could potentially involve France in wars of aggression. Mediating between 'old' positive law and 'new' natural law, some French delegates were willed to accept alliances, but only if they were defensive in nature.[41] In rough contours—'aggression' was not narrowed to 'first-use of arms' yet[42]—the French National Assembly provided early contributions to the modern discourse of

[38] See also Steiger, 'Das natürliche Recht der Souveränität der Völker—Die Debatten der Französischen Revolution 1789–1793 (2011)', in H. Steiger (ed.), *Universalität und Partikularität des Völkerrechts in geschichtlicher Perspektive. Aufsätze zur Völkerrechtsgeschichte 2008–2015* (2015) 135, at 140.

[39] See also Grewe, above note 1, at 422.

[40] Ibid., at 423.

[41] Steiger, above note 38.

[42] See also Grewe, above note 1, at 423.

the prohibition of war. More than 130 years before the Kellogg-Briand Pact and the 'radical plan to outlaw war',[43] war was considered illegal by French revolutionary concepts of constitutional and international law. As Oliver Eberl puts it in his contribution to this volume, Immanuel Kant must have felt encouraged to pursue his *Project for a Perpetual Peace*[44] by the French Revolution which 'had enshrined the imperative of peace in its constitution'.[45]

But the peaceful course of the National Assembly did not last long: with the renunciation of wars of conquest, the revolutionaries simultaneously appealed to 'non-intervention' as a central principle of international law. Paradoxically, the principle of non-intervention, derived from popular sovereignty, became the central narrative of justification for revolutionary 'counter violence'—and finally for French conquests and annexations '*au nom de la Nation*'. Faced with the alleged threat of a military intervention by Austria and Prussia, of which the majority of the French were indeed convinced in April 1792,[46] the first French campaign was justified as a 'just defense of a free people against the aggression of a king'.[47] In fact, the revolutionaries tried to justify the war as preventive self-defence. Therefore, they resorted to basic justification principles of the doctrine of 'just war'.[48] The revolution and the constitution were to be protected from external, that is monarchical interference—even if that meant to defend the revolution offensively, argued Condorcet.[49]

The 'just-war' criteria were extended in the further course of events by war-prone revolutionaries. Anacharsis Cloots even spoke of a 'holy war'[50] and Danton praised war as the 'exterminating angel of liberty'.[51] In the sense of the 'just-war' criterion of 'right intention' (*intentio recta*), Cloots recognized the purpose of war as peacemaking, and General Charles-François Domouriez, in anticipation of H.G. Wells, spoke of the coming war as 'the last war'.[52]

After the execution of Louis XVI on 3 December 1793, the radicalization of the revolutionary vocabulary gained further ground. As Edward H. Kolla has demonstrated with reference to the French annexationism,[53] this was not a long-planned programme, but rather a discursive process of gradual argumentative escalation of violence with several turning points. On a radical ideological basis, bellicist interpretations of Jean-Jacques Rousseau's writings played an important role:[54] any actions that served the survival and preservation of the 'indivisible and universal' republic, according to Robespierre, Carnot, and Merlin, were seen as legitimate. This included looting, occupations, and annexations. The central concept of justification for the

[43] Hathaway and Shapiro, above note 17, book title.

[44] I. Kant, *Project for a Perpetual Peace. A Philosophical Essay* (1795/1891).

[45] Eberl in this volume, at 153.

[46] Belissa, above note 34.

[47] 'War Declaration of the French National Assembly, 20 April 1792', in J. Hardman (ed.), *The French Revolution: The Fall of the Ancien Régime to the Thermidorian Reaction 1785–1795. Documents of Modern History* (1982) 140.

[48] Steiger, above note 38.

[49] Ibid., at 490.

[50] Ibid., at 489.

[51] S.C. Neff, *Justice among Nations* (2014), at 208.

[52] Belissa, above note 34.

[53] Kolla, above note 34.

[54] D.W. Bates, *States of War: Enlightenment Origins of the Political* (2012); Kolla, above note 34, at 186.

expansion of the revolutionary nation became the *pouvoir révolutionnaire*.[55] With recourse to Wilhelm Grewe,[56] one could argue that war was now, in the sense of *liberum ius ad bellum*, seen as justified, if it was understood as necessary. However, one must not ignore the fact that the basic narrative for the justification of French violence remained the alleged self-defence of the republic, even if self-defence now in fact meant conquest. For the revolutionaries, war was not *just* because it was necessary—war was necessary, because it was *just*.

With this, the incommensurability of the revolutionary concepts of international order with those of the international order of monarchs became unmistakably clear: according to Heinhard Steiger, the 'internal and external destruction of the previous foundations of the European order required rhetorical, dramatic and appellative exaggerations in order to give the new order universal legitimacy, both internally and externally'.[57] However, the justification strategies were not as unequivocal as one might assume: Brissot's argument is particularly revealing here. In 1792, Brissot tried to legitimize revolutionary warfare with selective reference to classical international law, universal morality, cosmopolitanism and freedom, the French constitution, as well as to the honour and survival of the nation.[58] While he declared Leopold II an 'unjust enemy' of the French nation, he at the same time argued that the Declaration of Pillnitz had broken the defensive alliance between Austria and France established by the 1756 Treaty of Versailles. Similar combinations of 'old' and 'new' international law can be identified also in later war discourse, even when the revolutionary language became harsher.[59] The revolutionary war discourse was shaped by 'multi-normativity'.[60]

This multi-normativity continued in Napoleon's attempted self-legitimization through the combination of Royal dignity and revolutionary ideals. Obviously, this self-legitimization was a self-contradictory farce, since it neither represented the will of the people nor a purely revolutionary conception of international order. As German lawyer Georg Friedrich von Martens rightly assumed, Napoleon strove for the hegemony of France and for a universal monarchy, but was basically confirming the international order that had existed since the Peace of Utrecht (1713).[61] However, even Napoleon continued to base his war justifications on the defence of the French nation and its 'freedom'. Yet, he mixed this basic justification narrative with multiple other political and normative claims—including those of non-Christian peoples: in order to justify his invasion of Ottoman Egypt in 1798 (at that time on behalf of the Directory), Napoleon published printed proclamations distributed to the Egyptian population in

[55] Kolla, above note 34.

[56] Grewe, above note 1, at 416.

[57] Steiger, above note 38, my translation.

[58] Brissot, 'Für den Krieg (16. Dezember 1791)', in P. Fischer (ed.), *Reden der Französischen Revolution* (1974) 144.

[59] Kolla, above note 34.

[60] Vec, 'Multinormativität in der Rechtsgeschichte', in *Berlin-Brandenburgische Akademie der Wissenschaften (vormals Preußische Akademie der Wissenschaften) Jahrbuch 2008* (2009) 155, at 162–65. See also our introduction and the contribution by Miloš Vec to this volume.

[61] van Blom, 'A Very Uncertain Perspective ... The Revolutionary and Napoleonic Age and International Relations in Europe, in the Views of Georg Friedrich von Martens (1756–1821)', in Jacobs et al., above note 34, 133; on Utrecht from a perspective of International Historical Sociology, see the contribution by Benno Teschke to this volume.

which he praised not only the French nation, but also 'Allah, Mohamed and the glorious Qur'an'.[62] Furthermore, Napoleon claimed to worship the latter more than the Mamluks did, who were *de facto* ruling Egypt at that time and from whom Napoleon wanted to 'liberate' Egypt. Cairo scholar Abd al-Rahman al-Jabarti angrily rejected Napoleon's reference to Islam. However, he also reported that parts of the Egyptian rural population actually believed Bonaparte's justifications.[63]

Ironically, the 'will of the people' and the 'liberation of the nation' were also used as justification narratives to bring Napoleon down:[64] after Napoleon's invasion in 1808, his Spanish enemies published a manifesto accusing Napoleon of starting an 'unjust war' (*provocar iniquamente la guerra*) and claiming that the Spanish nation had to be liberated from '*el tirano de la Europa*'.[65] After Napoleon's disastrous Russian campaign of 1812, even the Prussian King Friedrich Wilhelm III gave in to his generals and declared war against Napoleon. He now referred publicly to the 'liberation of the nation' as a justification of war in addition to narratives of honour and religion.[66] Even if the narrative of 'national liberation' was clearly instrumentalized by Friedrich Wilhelm III and its mobilizing effect was largely limited to social and political elites in the German lands,[67] the narratives of the revolution supported the downfall of the 'Emperor of the French'. Napoleon had devoured the revolution. Its stepchildren now devoured him.

How was the revolutionary and Napoleonic discourse on war and peace judged in contemporary legal theory and in politics? *Firstly*, for conservative and positivist lawyers like Martens, the pacifist ideals of the French Revolution, as laid out in Grégoire's (unadopted) draft of a *Déclaration du droit des gens,* were too vague and utopian. In place of universal natural law thinking, Martens propagated a further development of existing positive law, which he saw threatened by the revolution. Although Martens shared some of the criticism of the 'old' diplomacy and its rules, his thinking was firmly embedded in an international order based on the balance of power. *Secondly*, with regard to the justification of war, Martens principally adhered to *bellum iustum* reasoning as mandatory.[68] Although the revolutionaries also located their war justifications in a nationally modified version of 'just-war' thinking, Martens was right about the dangers of the revolutionary war discourse for the international legal order.

This critique was also shared by Kant, who was fundamentally sympathetic to the revolution: Oliver Eberl rightly emphasizes in his contribution to this volume that the transformation of the French war discourse must have strengthened Kant's aversion against the 'just-war' theory as a theory of arbitrary violence.[69] In fact, Kant turned against both *bellum iustum* and *liberum ius ad bellum*. He refused the notion of a 'right

[62] B. Tibi, *Vom Gottesreich zum Nationalstaat. Islam und panarabischer Nationalismus* (1987), at 65f.
[63] J.R. Cole, *Napoléon's Egypt. Invading the Middle East* (2008), at 30–34.
[64] See also Kolla, above note 34.
[65] D.J.A.C., *El tirano de la Europa, Napoleon primero: manifiesto que á todos los pueblos del mundo y principalmente á los españoles* (1808).
[66] Friedrich Wilhelm III., 'An mein Volk: Se. Majeſtät der König haben mit Sr. Majestät dem Kaiser aller Reußen ein Off—und Defensiv-Bündniß abgeschlossen', *Schlesische privilegirte Zeitung* (20 March 1813).
[67] U. Planert, *Der Mythos vom Befreiungskrieg. Frankreichs Kriege und der deutsche Süden. Alltag—Wahrnehmung—Deutung. 1792–1841* (2004).
[68] G.F. Martens, *Einleitung in das positive Europäische Völkerrecht auf Verträge und Herkommen gegründet* (1796), at 297; see also van Blom, above note 61.
[69] See also H. Williams, *Kant and the End of War: A Critique of Just War Theory* (2012).

to go to war' because of 'the one-sided maxims that operate by means of force'. Instead, Kant outlined a legal prohibition of war (*ius contra bellum*) derived from practical reasoning—and with this, an international order governed by positive law.

However, there was a second theorist, on whom the violent transformation of the revolutionary discourse and the Napoleonic campaigns with its *levée en masse*[70] had an almost traumatic effect. This was one of the 'fathers of modern warfare', who was to pave the way precisely for the opposite position than the one held by Kant: Carl von Clausewitz. According to Clausewitz, war was not to be justified in the sense of international normativity, that is *bellum iustum*—it was a factual instrument of state policy, 'nothing but a duel on a larger scale', 'an act of [physical, HS] force'.[71] For the late, *conservative* Clausewitz (the *liberal-national* hopes of the young Clausewitz had been disappointed by the European monarchs after the defeat of Napoleon), war was to serve the Prussian sovereign and his political interests.

In fact, Kant and Clausewitz should become the most important references in the nineteenth-century legal discourse on *ius ad bellum*. While the non-revolutionary theory of international law of that time, such as that of Martens, still stuck to modified forms of *bellum iustum*, it was Kant and Clausewitz who developed innovative ways of thinking about war and peace: on the one hand, the doctrine of war as a means of *state policy*, 'simply a continuation of politics by other means',[72] and, on the other, the doctrine of the exclusive justification of war through *law*. Given its high impact on these two thinkers, the French Revolution had, in a dialectical way, initiated the emergence of the modern discourse on war and peace.

3. Peace through Coercion: Judging War in a Concert of Great Powers (1815–1878)

The emergence of the modern discourse of war and international order initiated by the French Revolution was to be largely challenged by the reorganization of international relations at the Congress of Vienna (1814–15). Nevertheless, by reordering Europe, the Great Powers, for their part, transformed the normative framework of the justification of war. As Paul W. Schroeder has convincingly argued, 'Vienna 1814/15' was a 'decisive turning point' in a longer transformation process of the rules, norms, and the very language of diplomacy within Europe.[73] The objective of the Congress was a procedure—formulated in 'perfect concert'[74]—to restore and secure a general European peace against any new revolutionary threats. In fact, the Congress of Vienna

[70] Lytle, 'Robespierre, Danton and the Levée en masse', 30 *The Journal of Modern History* (1958) 325.

[71] C. von Clausewitz, *On War* (1832/1976), at 75; see also Tashjean, 'The Clausewitzian Definition: From Just War to the Duel of States', 17 *Revue européenne des sciences sociales* (1979) 79, at 81; Münkler, 'Instrumentelle und existentielle Auffassung vom Krieg bei Carl von Clausewitz', 16 *Leviathan* (1988) 235.

[72] Clausewitz, above note 71, at 605.

[73] P.W. Schroeder, *The Transformation of European Politics. 1763–1848* (1994), at vii; M. Vec, *Recht und Normierung in der Industriellen Revolution. Neue Strukturen der Normsetzung in Völkerrecht, staatlicher Gesetzgebung und gesellschaftlicher Selbstnormierung* (2006), at 104; H.B. Oppenheim, *System des Völkerrechts* (1845), at 1–7.

[74] 'Treaty of Chaumont, 1 March 1814, Article I', in F.H. Hartmann (ed.), *Basic Documents of International Relations* (1951), at 3f.

inaugurated a century of cooperation and multilateral diplomacy in the form of further congresses and conferences.[75]

These congresses and conferences were political catalysts for the institutionalization and legalization of the 'Order of Vienna' and its procedures to cope with conflict in a multilateral fashion.[76] How far the normative containment of violence in Europe went, however, is controversial. For instance, the nineteenth-century legal institute of 'neutrality'[77] supports both the assumption that war had become a political affair beyond general community interests *and* the counter-assumption that the Concert was fostering the idea of international peace through law. Furthermore, the Final Act of the Congress signed on 9 June 1815 did not include a general prohibition of war—there was no written 'undeviating rule applicable to all . . . cases.' (Lord Russel)[78]

However, more recent research suggests that the significance of the normative architecture of the international order constructed in '1814/15' has so far been underestimated by those who adhere to the thesis of *liberum ius ad bellum* and who adopt the common characterization of the nineteenth-century international order as anarchic.[79] In his seminal book on *Normen und Praxis* (2009), Matthias Schulz has called *the Concert of the European Great Powers* a 'nineteenth-century Security Council'.[80] Indeed, there are some good reasons for upgrading the normative innovation going along with the Concert: the Concert passed judgement on the use of force by states and took, although selectively, actions in cases when war was not seen as justified. This included moderate measures like the withdrawal of the Concert's diplomatic support, when in 1859 Austria declared war on Piedmont-Sardinia.[81] However, it could also result in the authorization of armed measures like military interventions to maintain the international order as agreed upon in the Congresses of Troppau (1820), Laibach (1821), and Vienna/Verona (1822) and practised all over the century (see below).

As Schulz argues, the Concert developed some form of a diplomatic 'culture of peace' (*Friedenskultur*).[82] The latter was based on international moderation and behavioural norms.[83] It resulted in a regulative international order, which was shaped by power *and* norms. Accordingly, 'the preservation of the peace of Europe'[84] became a central claim in nineteenth-century war discourse—and, as I argue, even more than that, it became an emerging norm both in nineteenth-century legal theory *and* in political practice.[85] In this process, the preservation of the international order and self-defence (for the latter see also Hull's chapter) became the two central narratives of

[75] Vec, 'From the Congress of Vienna to the Paris Peace Treaties of 1919', in B. Fassbender and A. Peters (eds), *The Oxford Handbook of the History of International Law* (2nd ed., 2014) 654, at 660.

[76] Ibid. above.

[77] Abbenhuis, above note 17, at 45.

[78] M. Schulz, *Normen und Praxis. Das Europäische Konzert der Großmächte als Sicherheitsrat. 1815–1860* (2009), at 617.

[79] Wehberg, above note 11.

[80] Schulz, above note 78, at 521.

[81] Ibid., at 481.

[82] Ibid., at 4.

[83] Ibid., at 615; Vec, above note 75, at 658.

[84] Schulz, above note 78, at 615.

[85] Simon, *Der Mythos*, above note *.

justification for the modern use of force—and remained so in modified form until today (see Marauhn's chapter as well as the concluding chapter).

However, the nineteenth-century international order was by no means non-violent. To some extent, it rather was a precarious armistice or a legalized hegemony[86] of the Great Powers, which discriminated against weaker states, liberal movements, and non-European entities. Due to the lack of a full legalization of peace and war, political decisions depended on discursive acceptance in the communication community of the Great Powers. This brings us to the *Janus face* of the nineteenth-century international order governing the use of force: military interventions enabled the Great Powers to abolish, if necessary, the separation of foreign and domestic policy on which the Vienna Peace Treaty was based[87]—peace went hand in hand with violent suppression and control.[88] Furthermore, as became particularly clear in the high impact of imperialism and colonialism on the discourse of the use of force both in political practice *and* legal theory,[89] the European laws of war applied almost exclusively to the full members of the international order—the so-called 'civilized states'.[90]

Still, the normative impact of this unequal international order's coercive power can clearly be identified in nineteenth-century state practice. Contrary to the thesis of *liberum ius ad bellum*, the use of force in international relations continued to require justification. An exception to the rule was the Austro-Prussian War (1866), where no official declarations of war were made. As the war's outcome resulted largely from one decisive battle (near Königgrätz on 3 July 1866),[91] this war comes closest to the *Clausewitzian* definition of a duel of states. For Helmuth Graf von Moltke, Prussian Chief of the General Staff and a disciple of Clausewitz, it was the ideal example of a cabinet war—defined by limited, short, and decisive fighting without ideological public debates on its justification.[92]

However, already the next war, the Franco-Prussian War (1870/71), makes it highly problematic to apply the *Clausewitzian* concept of war as a duel of states. It was a hybrid war between eighteenth-century 'cabinet wars' and twentieth-century 'total wars'. As in the war of 1866, the war parties had limited goals: preventing or fulfilling German unification. But soon the war was transformed into an industrialized 'people's

[86] G.J. Simpson, *Great Powers and Outlaw States. Unequal Sovereigns in the International Legal Order* (2004).

[87] Schulz, above note 78, at 74–76; Vec 'Intervention/Nichtintervention. Verrechtlichung der Politik und Politisierung des Völkerrechts im 19. Jahrhundert', in U. Lappenküper and R. Marcowicz (eds), *Macht und Recht. Völkerrecht in den internationalen Beziehungen* (2010) 135; J. Osterhammel, *Geschichtswissenschaft jenseits des Nationalstaats. Studien zu Beziehungsgeschichte und Zivilisationsvergleich* (2001); on the colonial origins of 'humanitarian interventions', see F. Klose, *In the Cause of Humanity: Eine Geschichte der humanitären Intervention im langen 19. Jahrhundert* (2019) 256; and Lauren Benton in this volume.

[88] Koskenniemi, above note 2, at 3; A. Zamoyski, *Rites of Peace, The Fall of Napoleon and the Congress of Vienna* (2007), at 554.

[89] J. Fisch, *Die europäische Expansion und das Völkerrecht. Die Auseinandersetzungen um den Status der überseeischen Gebiete vom 15. Jahrhundert bis zur Gegenwart* (1984); A. Anghie, *Imperialism, Sovereignty and the Making of International Law* (2005); L. Benton and L. Ford, *Rage for Order. The British Empire and the Origins of International Law. 1800–1850* (2016); von Bernstorff, 'The Use of Force in International Law before World War I. On Imperial Ordering and the Ontology of the Nation-State', 29 *European Journal of International Law* 1 (2018) 233.

[90] See the chapters by Benton, Genell/Aksakal, and Hippler in this volume.

[91] J. Leonhard, *Die Büchse der Pandora. Geschichte des Ersten Weltkrieges* (2014), at 30.

[92] Ibid.

war'[93] shaped by nationalism, violence against civilians, and partisan attacks. The Franco-Prussian War heralded the end of *Clausewitzian* war and the rise of total war. 'The age of cabinet's war is behind us,—now we have only people's war', Moltke would say in his last speech in the Reichstag on 14 May 1890.[94]

Furthermore, and clearly against the proposition of a nineteenth-century *liberum ius ad bellum*, the Franco-Prussian War depended on legitimacy both in domestic and international affairs.[95] Again, as in the case of the French Revolutionary Wars, self-defence against an anticipated attack was the ultimate narrative of justification: in his war declaration, Napoleon III used the narrative of a 'defense of the weak against the strong, the reparation of great iniquities, the chastisement of unjustifiable acts'.[96] Its 'just cause'[97] *(sic!)* was a 'defence' against 'Prussian egotism', upon which Europe should pronounce. In addition, Napoleon III referred to the international order pointing at international peace, stability, and the preservation of national rights, when he scandalized Prussian hegemonic politics. For Napoleon III, the war was to be declared in the name of the international order, as 'a war of equilibrium'.[98]

Bismarck rejected the justifications given by Napoleon III for the French aggression as 'nothing but pretense and invention' and added that the German states would defend themselves against the 'unprovoked attack' ('*nicht provocierten Ueberfall*').[99] With this justification strategy, Bismarck successfully generated national and international sympathy for his war aims. Even some liberal lawyers like Bluntschli or Jhering, who were generally opposed to war as an instrument of state policy (see below), were themselves politicized by Bismarck and praised the violent German unification as just(ified).[100] However, public sympathy for Bismarck's war against France faded when Prussia decided to annex parts of Alsace and Lorraine in September 1870.[101] As an expression of *consuetudo*, the narratives of justification and critique in the context of the Franco-Prussian War also strengthen Isabel Hull's argument in her contribution on the Great War and International Law to this volume: Hull argues that wars of aggression were illegal according to nineteenth-century customary international law.

What is more, the French official account of the origin of the 1870/71 war held, 'that the rights of each nation, like the rights of each individual, are limited by the rights of others, and it is not permissible *[sic!]* that one nation, under the pretext of exercising its own sovereignty, should menace the existence or the security of a neighbouring

[93] Förster, 'Facing "People's War": Moltke the Elder and Germany's Military Options after 1871', 10 *Journal of Strategic Studies* (1987) 209, at 209.

[94] S. Förster and J. Nagler (eds), *On the Road to Total War. The American War and the German Wars of Unification, 1861–1871* (1997), at 1.

[95] C. Abel, *Letters on International Relations before and during the War of 1870* (1871), vol. 2, at 137.

[96] Ibid., at 225.

[97] Ibid., at 190.

[98] Ibid., at 225.

[99] Ibid., at 165.

[100] J.C. Bluntschli, *Das moderne Völkerrecht in dem Kriege 1870: Rede zum Geburtsfeste des höchstseligen Grossherzogs Karl Friedrich von Baden und zur akademischen Preisvertheilung am 22. November 1870* (1870); for Jhering's and Bluntschli's justifications, see Simon, 'Theorising Order in the Shadow of War. The Politics of International Legal Knowledge and the Justification of Force in Modernity', 22 *Journal of the History of International Law* (2020) 218.

[101] W. Baumgart, *Europäisches Konzert und nationale Bewegung. Internationale Beziehungen 1830–1878* (2nd ed., 2007), at 399.

nation'.[102] Contrary to the thesis of a nineteenth-century *liberum ius ad bellum*, wars waged for the extension of a states' position in the international system were seen as unjustified. State practice coincided almost ideally with the doctrine of *bellum iustum*.

Much in this vein, in mainstream legal doctrine war was *not* accepted as an instrument of state policy, but only as one of legal enforcement. *Liberum ius ad bellum* was *not* claimed in the first two thirds of the century. Instead, mainstream mid- nineteenth-century legal doctrine, as represented by German positivist August Wilhelm Heffter, held that war was only justified if it was waged in favour of the realization of *legal* purposes, '*zur Realisirung rechtlicher Zwecke*'. Heffter referred to 'just war', *gerechter Krieg*, from a positivistic perspective. Although he understood an 'unjust war' (*den ungerechten Krieg*) to be in its effects equal to a 'just war', he distinguished between 'defensive wars' (to ward off an unjust attack) and 'wars of aggression'.[103] War was, 'in other words', the ultimate self-help[104]—but its ultimate justification was the restoration of law and peace. Also for Robert von Mohl, professor for political science in Tübingen, war was an *ultima ratio regum* in order to shield the international legal order against violent anarchy.[105] Early- and mid- nineteenth-century legal discourse relied on a positivist modification of *bellum iustum*. This reflected the challenge to find a normative framework for governing the use of force in the context of an international order, in which war was not yet generally prohibited by treaty, but linked to the purpose of restoring peace. Given this vague normative constitution of the international order, moral and legal positivist claims were sometimes mixed by legal scholars.

Even stronger reference to classical, i.e. moral theological concepts of *bellum iustum* can be found in contemporary legal discourse on other forms of military force, i.e. 'measures short of war' like 'reprisals'.[106] Nowhere has this become as clear as in the discourse on the justification of (humanitarian) interventions that emerged in the nineteenth century (see also Lauren Benton and Beate Jahn in this volume). Some lawyers even understood intervention as an anticipation of a more appropriate international order (an argument which much later would be developed in a systematic way by Jürgen Habermas with regard to NATO's Kosovo Intervention in 1999).[107] The Treaty of Paris (1856) had not only made the Ottoman Empire a formal member of the European Concert (see also Genell and Aksakal in this volume), by this it had also created an international sub-system in Europe,[108] which denied a Russian Responsibility to Protect Christian citizens in the Ottoman Empire under threat of armed sanction (Articles 7 and 9). Nevertheless, in 1877 Russian foreign minister Alexander Gorchakov stated, in full agreement with the renowned Russian crown jurist Fedor Fedorovich Martens, that the independence of Turkey 'must be subordinated to the guarantees demanded by humanity, the sentiments of Christian Europe, and the

[102] Abel, above note 95, at 175.
[103] C. Calvo, *Dictionnaire de Droit International Public et Privé* (1885), at 366.
[104] A.W. Heffter, *Das Europäische Völkerrecht der Gegenwart* (1844), at 195.
[105] R. von Mohl, *Encyklopädie der Staatswissenschaften* (1859), at 453f.
[106] Neff, above note 3, at 225–27.
[107] Rougier, 'La théorie de l'intervention d'humanité', 17 *Revue Génerale de Droit International Public* (1910) 468; Habermas, 'Bestialität und Humanität. Ein Krieg an der Grenze zwischen Recht und Moral', *DIE ZEIT* (29 April 1999).
[108] W.E. Mosse, *Crimean System. 1855–71: The Story of a Peace Settlement* (1963).

general peace in Europe'.[109] This is to say that Russia, in the name of the international moral order, successfully broke a norm of the international legal order (1856 Treaty of Paris) by referring to natural law!

As illustrated here, natural law was still alive—the war discourse in the 'positivist century' was shaped by a plurality of normative arguments. A 'free right to go to war' was neither claimed in state practice nor in legal doctrine in the two first thirds of the nineteenth century. Following up on these observations, the question arises: where does the thesis of *liberum ius ad bellum* actually come from, if it was not claimed in state practice and mainstream legal doctrine?

4. The Formation of a Myth: A Forgotten War Discourse (1870–1918)

In 1913, German law professor Woldemar von Rohland wrote in a *Festschrift* on the occasion of the twenty-fifth anniversary of the reign of Emperor Wilhelm II that the twenty-five years between 1888 and 1913 represented the most significant stage in the historical development of international law.[110] That was an adequate observation. Indeed, the second half of the century was marked by a dramatic consolidation of international regulations, that is by a legalization (*eine Verrechtung* [Heinrich Rettich]) of international relations. The Hague Peace Conferences, which made the justification of war mandatory (Hague Convention III of 1907),[111] and the development of peaceful dispute settlement were particularly important in this process. The latter has been described as a transformation of international law from a law of war to a law of peace.[112]

However, this undeniable legal progress was far from sufficient to formalize the highly political field of *ius ad bellum* in positive legal terms.[113] In fact, it was possible to formulate legal rules on what was forbidden *in* war (*ius in bello*), without speaking out *against* war (*ius contra bellum*). As pointed out above, one fundamental reason for this shortcoming of legalization was the political character of the international order dominated by the European Great Powers. Against the background of the increasingly harsh tone at the end of the nineteenth century, Europe was shaped by a paradoxical simultaneity of international legalization and national *realpolitik* in combination with blunt nationalism.

It was in this context of an increasing nationalization of European politics that the thesis of *liberum ius ad bellum* emerged: in 1888, today largely forgotten German lawyer Heinrich Rettich published a 300-page treatise entitled *Zur Theorie und*

[109] E. Hertslet, *The Map of Europe by Treaty* (1891), vol. 4, at 2524; see also F. von Martens, *Die russische Politik in der Orientalischen Frage. Eine historische Studie* (1877), at 45; on Martens' justification of the war, see Simon, above note 100, at 38.

[110] von Rohland, Das 'Völkerrecht', in H.Th. Soergel (ed.), *Festschrift der Rundschau für den deutschen Juristenstand* (1913).

[111] See also Vec, above note 75.

[112] Brock, 'Frieden durch Recht: Anmerkungen zum Thema im historischen Kontext', in P. Becker, R. Braun, and D. Deiseroth (eds), *Frieden durch Recht?* (2010) 15.

[113] J. Dülffer, *Regeln gegen den Krieg?. Die Haager Friedenskonferenzen 1899 und 1907 in der internationalen Politik* (1981); Payk, above note 17, ch. 1.

Geschichte des Rechts zum Kriege, in which he attempted to reconstruct 'a positive right to go to war'. From his historical research on early modern—and nineteenth-century legal discourse, Rettich argued that there was a 'right to go to war', which, as he claimed, was rooted in interests and necessities of sovereign states.[114]

Although my arguments presented here contradict Rettich's argument fundamentally, one must acknowledge that, to some extent, Rettich's book was a pioneering work. His polemic critique, according to which prominent international lawyers like Moser, Martens, Klüber, Heffter, and Bluntschli had not come up with a shared positivist understanding of war and its functions in history,[115] was plausible to a certain extent: as mentioned above, nineteenth-century mainstream legal discourse was still using the (modified) vocabulary of *bellum iustum*. According to Rettich, these references to 'just war' were 'analogies from theological morality', not positive law.[116] Rettich saw himself in the role of an outsider who opposed a hegemonic discourse.[117]

However, it is worth noting again that in the meantime the discourse has been reversed: while in 1888 Rettich attempted to deconstruct a scholarly mainstream which he understood to be shaped by *bellum iustum* and thus natural law, today we are still faced with a mainstream which clings to the thesis of *liberum ius ad bellum*. So, what has happened to the discourse on the justification of war and international order since 1888?

As I argue, the turning point for legal theory came from late nineteenth-century state practice, more precisely from the Franco-Prussian War in 1870/71 (whose political justifications have been outlined above). In view of its new strength, gained 'overnight, so to speak',[118] the German *Kaiserreich* had to determine new preferences between *realpolitik* and normative self-commitment to the international order.[119] Particularly, the Wilhelmine period (1890–1918) created a culture of extraordinary militarism which fostered a *Clausewitzian* reorientation both in legal theory and political practice. This became evident i.a. in the hostility of the *Auswärtiges Amt* to international law in the negotiations at the Hague Peace Conferences as well as in a particular broad interpretation of *Kriegsnotwendigkeit* ('military necessity') in the contexts of the Herero and Namaqua Genocide and the First World War.[120] Accordingly, my argument is that the *Kaiserreich* is to be understood as the historical background of the *realist* turn in International Law to the philosophically founded notion of an international legal order in which there was no superior normative authority and states, at best, were bound by their 'political ethics'. This *realist* turn was driven by a respectable group of lawyers who understood war as a legally irrelevant, but politically highly effective instrument—'absolutely permitted in inter-State relations'.[121]

114 H. Rettich, *Zur Theorie und Geschichte des Rechts zum Kriege* (1888), at xiii.

115 Ibid., at 39, 67.

116 Ibid., at 41.

117 See also Carty, above note 27, at 45.

118 I. Geiss, *Zukunft als Geschichte: historisch-politische Analysen und Prognosen zum Untergang des Sowjetkommunismus, 1980–1991* (1998), at 202.

119 See also Koskenniemi, *Gentle Civilizer*, above note 2, at 204f.

120 On the Hague Peace Conference, see Payk, above note 17, at 44, 57f.; on the Herero and Namaqua Genocide, see I.V. Hull, *Absolute Destruction: Military Culture and the Practices of War in Imperial Germany* (2005), part I; on *Kriegsnotwenigkeit* and the First World War, see Hull and Vec in this volume.

121 P. Heilborn, *Handbuch des Völkerrechts*, vol. 1 (1912), at 23.

These scholars supported the idea of *liberum ius ad bellum* as the acme of state sovereignty.[122] According to German lawyer Emanuel Ullmann, every war was legal: a sovereign state accepted only its own politics as legally significant.[123]

This thesis was not only, but particularly advocated by *realist* lawyers in Imperial Germany. But how was *liberum ius ad bellum* justified in terms of legal or political theory? While conservative interpretations of Hegel were an important philosophical resource for the justification of the German *Machtstaat*, the central authority in the *realist* legal discourse of *liberum ius ad bellum* became Carl von Clausewitz: echoing his ideal-typical notion of war as a duel of states (see above), lawyers like the German Karl Lueder defined 'war' as a natural force, which often occurred as *Fortsetzung der Politik in anderer Form*, a continuation of politics by other means.[124]

Following Clausewitz's definition (and going beyond Clausewitz in a bellicist euphoria not shared by Clausewitz), for these lawyers war was not law enforcement in the sense of *bellum iustum* or *ius contra bellum*. It was a duel of states over opposing political interests. In this duel, every state simply strived 'to break down the resistance of the other to the terms which he requires for peace'.[125] Thus, war was to be freed from moral considerations. According to Lueder, war was an empirical fact of physical force.[126] In contrast to the 'natural law of philosophers', *Clausewitzians* only accepted enforceable law as 'true law'. Furthermore, law had to be positive, that is 'empirical'/'historical' law, 'a record of the habits of the species, good or bad'.[127] Natural law, by contrast, seemed inappropriate to be 'true law'. Thus, to *Clausewitzian* positivists, *bellum iustum* appeared as an outdated moral doctrine. But these scholars did not only deny the normative authority of *natural* law to regulate the use of interstate force: *Clausewitzians* also argued that the realization of Kant's project of perpetual peace based on positive international law was unthinkable and impossible, *ein unerreichbares Ideal*—and not even the right ideal at all, since they understood war as a catalyst of progress, *ein wahrer nothwendiger Culturträger*.[128]

Yet, contrary to Grewe and today's mainstream research, these *realist* lawyers, despite their respectable numbers, did not remain unchallenged at all. The Franco-Prussian war was in fact also a turning point for liberal lawyers, many of them members of the *Institut de Droit International*. With the war's brutality in mind, international lawyers like the Swiss Johann Caspar Bluntschli, the Argentinian Carlo ('Charles') Calvo, and the German August von Bulmerincq concluded that war had to

[122] C. Gareis, *Institutionen des Völkerrechts* (1888), at 192; especially Lueder, 'Krieg und Kriegsrecht im Allgemeinen', in F. von Holtzendorff (ed.), Handbuch des Völkerrechts (1889) 169, at 180; see also Ullmann, 'Völkerrecht', in E. Ullmann (ed.), *Handbuch des Oeffentlichen Rechts* (2nd ed., 1898), at 313; R.R. Foulke, *A Treatise on International Law* (1920), at 130ff.; A. Quaritsch, *Völkerrecht und Auswärtige Politik* (9th ed., 1913), at 150f.; J. Westlake, *International Law. Part II: War* (1907), at 1; K. Strupp, *Grundzüge des positiven Völkerrechts* (4th ed., 1928), at 212; L. Oppenheim, *International Law. A Treatise* (3rd ed., 1921), at 79f.; A. Lasson, *Princip und Zukunft des Völkerrechts* (1871), at 68.

[123] Ullmann, above note 122, at 313.

[124] Lueder, above note 122, at 180; Ullmann, above note 122, at 313; Foulke, above note 122, at 130ff. Lueder also quoted W. Rüstow, *Die Feldherrnkunst des neunzehnten Jahrhunderts. Zum Selbststudium und für den Unterricht an höheren Militärschulen* (1857), at 1.

[125] Westlake, above note 122, at 57; Lueder, above note 122, at 176.

[126] Lueder, above note 122, at 180–84.

[127] Westlake, above note 122, at 56.

[128] Lueder, above note 122, at 196–99, 203.

be outlawed. While they found themselves to be in total contradiction to Clausewitz's concept of war as a state duel, they accepted war only to be just (*gerecht*), if it was legal.[129] For them, war was not a political instrument, but one of law enforcement. Legal reasons for a war, argued Bluntschli, were (a serious threat of) infringements of rights or trespasses against a state, or a grave breach of world order.[130] As pointed out above, late nineteenth-century legal mainstream discourse was still marked by references to 'just war'. However, for liberal lawyers, war was not only to be prohibited in terms of morality, but of positive law. Even Franz von Liszt, who published the most important textbook in the late *Kaiserreich* (competing with the one by Ullmann), continued to describe war, in the sense of Bluntschli et al., as an 'ultima ratio for the settlement of disputes under international law'—even though von Liszt himself doubted the rationality of this 'ultima ratio'.[131]

At this point, liberal lawyers faced the same basic theoretical and methodical dilemma as *Clausewitzians*: since at that time no treaty generally prohibited war, *Liberals* had to deduce their legal opinion from other sources (as Rettich would criticize). While *Clausewitzian* positivists referred to the 'facticity' of war and thus criticized lawyers like Bluntschli for speaking *de lege ferenda* instead of *de lege lata*,[132] liberal lawyers derived their concept of law not from enforceability, but from 'historical jurisprudence, linked with liberal-humanitarian ideals and theories of natural evolution of European societies'.[133] They did not focus on codified law, but on the 'legal conscience of the civilized world',[134] that is on common European values, traditions, and norms. According to liberal lawyers, peace was a moral *and* a legal imperative for the advancement of the international normative order. A particularly important philosophical inspiration for their view of law was Immanuel Kant.[135]

These opposing opinions of *Clausewitzians* and *Kantians* did not simply coexist. They referred to each other in scientific discourse. However, this lively theoretical and methodological dispute over the legitimacy of war and international order between nineteenth-century *Realists* and *Liberals* is largely forgotten today: it is buried under the hegemonic claim of *liberum ius ad bellum*. So, why has this claim prevailed in the discourse, although, after two world wars, the idea of war in the *Clausewitzian* sense was finally replaced by a general legal prohibition of war in the *Kantian* tradition?

To answer this question, it makes sense to take a short look at the genealogy of the History of International Law as a modern academic discipline: as my analysis suggests, a *Clausewitzian* interpretation of war as a political instrument was particularly widespread in the legal discourse of the German *Kaiserreich*. This observation confirms Isabel Hull's earlier writings and her chapter in the present volume on the German war discourse in the context of the Great War: 'It is as if Imperial Germany could not

[129] J.C. Bluntschli, *Das moderne Völkerrecht der civilisierten Staaten als Rechtsbuch dargestellt* (1868), at 290; Calvo, above note 103, at 366f.; A. von Bulmerincq, *Das Völkerrecht oder das internationale Recht* (2nd ed., 1889), at 357; see also Carty, above note 27, at 49.

[130] Bluntschli, above note 129.

[131] F. von Liszt, *Das Völkerrecht* (1898), at 206.

[132] Gareis, above note 122, at 37.

[133] Koskenniemi, *Gentle Civilizer*, above note 2, at 51.

[134] Ibid.; see also Bluntschli, above note 129, at vi.

[135] Bluntschli, above note 129, at 287f.; B. Stråth, *Europe's Utopias of Peace: 1815, 1919, 1951* (2016), at 194f.

speak the same legal language as the rest of Europe.'[136] However, the *Clausewitzian* tradition in legal thinking prevailed even after the First World War. Its main exponent was Carl Schmitt, for whom a sovereign state always was authorized to use military force against its political enemy, if deemed necessary.[137] Although there was resistance against Schmitt's reanimation of the 'free right to go to war' proposition by authors like Arthur Nussbaum and Hans Kelsen,[138] the *realist, Clausewitzian* tradition prevailed in German legal doctrine. It was continued by a disciple of Schmitt, who would become an important diplomat and legal scholar in the new German Federal Republic: in 1984, Wilhelm G. Grewe published his *Epochs of International Law*, which he had written to a large extent during the 1940s.[139] Grewe's *Epochs* became *the* standard book on the subject in Germany, and was translated into English in 2000. In this way, it has remained one of the most important books on the subject across the disciplinary boundaries of the History and Theory of International Law, particularly, but not only in the German-speaking academic world. Hardly surprisingly, and as summarized above, Grewe offered only the *realist*, Schmittian reading of the nineteenth-century war discourse—and thus helped to spread the a-historical thesis of *liberum ius ad bellum* until today.

5. Conclusion: A Century of Multi-Normativity. The Justification of War and the Birth of Modern International Order

In view of the justification of war, the nineteenth-century international order was an order of 'multi-normativity'.[140] In the last third of the century a dualism evolved in international legal doctrine between viewpoints which understood war as legal enforcement (*ius contra bellum* in the sense of Kant) or as state policy (*liberum ius ad bellum* in the sense of Clausewitz). Widely ignored in the History and Theory of International Law, both narratives reflected the problem of a missing positivist codification of war *(ius ad bellum/ius contra bellum)*, and responded to this gap in contradictory ways. Although *Clausewitzians* (especially in Germany) claimed a positivist 'free right to go to war', a more precise historical analysis of the war discourses of the nineteenth century proves the critics of *liberum ius ad bellum* right. Contrary to Grewe's thesis, neither the majority of legal scholars nor state practice did claim *liberum ius ad bellum*.

Instead, within a highly political and discriminatory framework of coercion,[141] the European Concert developed a normative discourse that made the justification of war mandatory. This was confirmed by the Hague Convention III of 1907 (see also

[136] I.V. Hull, *A Scrap of Paper: Breaking and Making International Law during the Great War* (2014), at 211–39.

[137] C. Schmitt, *Die Wendung zum diskriminierenden Kriegsbegriff* (2nd ed., 1988).

[138] H. Kelsen, *Staatslehre* (1925), at 125; H. Kelsen, *Peace through Law* (1944); see also V. Neumann, *Carl Schmitt als Jurist* (2015), at 425; Nussbaum, above note 27, at 474.

[139] Fassbender, 'Stories of War and Peace: On Writing the History of International Law in the "Third Reich" and After', 13 *European Journal of International Law* (2002) 479, at 511f.

[140] Vec, above note 60.

[141] Simpson, above note 86.

Isabel Hull's chapter). Of course, this discourse only applied to Europe. From a post-colonial perspective, one has to emphasize that the norms, rules, and procedures of the European communication community were essentially limited to the latter. Non-European entities were in many cases partially or completely denied their status of sovereign political actors with reference to European 'standards of civilization'[142] in order to justify colonialism, imperial domination, unequal treaties, and violent civilizing missions.[143] Lauren Benton and Aimee Genell/Mustafa Aksakal underline this fact in their chapters in this part of the book (see also the contributions by Mallavarapu, Chimni, and Hippler in other parts). Furthermore, as Miloš Vec argues in his chapter, multi-normativity meant that legal coercion could be circumvented by justifying the use of force with recourse to other spheres of normativity. There was no normative 'centre, only many peripheries: welcome to modernity!'[144]

However, within Europe, the legitimacy of war was clearly limited. There was real normative progress, not only with regard to *ius in bello*. This means that in the nineteenth century, right was *not* might.[145] To the contrary, in the time between the Revolutionary Wars and the Great War the justification of war and international order became a complex discourse *between* right *and* might. If states wanted to go to war, they needed a legitimate cause according to the European Concert's political, moral, and legal norms. These international norms were applied and contested[146] on a case-by-case basis. They were only marginally codified, and could even contradict each other. Nevertheless, they developed a normative power, which even the most powerful actors were aware of. This normative power can clearly be reconstructed from nineteenth-century justifications of war. Thus, it seems adequate to conclude that the nineteenth century was *not* a pre-modern era of international anarchy,[147] a 'laissez-faire era for war-makers'.[148] Rather, it was the era from which the post-Second World War prohibition of the use of inter-state force slowly emerged. In other words: the modern international order governing the use of force has its roots in the political and normative debates of the nineteenth century.

142 G.W. Gong, *The Standard of 'Civilization' in International Society* (1984); see also Thomas Hippler's chapter.

143 See also Fisch, above note 89; Anghie, above note 89; von Bernstorff, above note 89; Benton and Ford, above note 89; see also Benton's contribution on *interpolity relations* to this volume.

144 Vec, above note 60, at 165.

145 For the contrary opinion that 'might was right', see Hathaway and Shapiro, above note 17, at xv.

146 For an IR-theory of contestation see A. Wiener, *Contestation and Constitution of Norms in Global International Relations* (2018); and see also our introduction.

147 For the contrary opinion cf. Wehberg, above note 11, at 162.

148 Neff, above note 3.

9

Protection Emergencies: Justifying Measures Short of War in the British Empire

Lauren Benton

1. Introduction

In March, 1848, two British traders lost their investment and narrowly escaped losing their lives on the banks of the Rio Nuñez, the river between the Gambia and Sierra Leone in West Africa. The two partners, George Martin and Joseph Brathwaite, had travelled upriver to engage in 'legitimate trade', defined at the time as any commerce other than slave trading. The men stocked a small warehouse on the riverbank with guns, ammunition, and an assortment of other goods. Africans in the area were accustomed to trade with Europeans, and at first the British merchants encountered little opposition. But then a force of locals incited by French and Belgian traders attacked and burned the Britons' compound. The British government could not claim that the attackers had violated British sovereignty—neither the British nor any other European power was claiming control over the upriver region. Instead, Lord Palmerston, British foreign secretary, asserted that the attackers had violated the merchants' right to trade anywhere in the world, and he demanded compensation from the French and British governments.[1]

In a better-known incident a decade before, another British complaint about the protection of subjects in 1839 prompted war. When British merchants were detained and refused to hand over their cargos of opium to Chinese officials in the waters off Canton, Charles Elliot, the chief superintendent of British trade in China, instructed the merchants to surrender the opium to him for delivery to Chinese officials. By seizing and turning over the opium, his critics would later argue, Elliot had committed the British government to indemnify the traders and had set the course for war. Elliot defended his actions by stating that he had merely been opposing Chinese 'aggressive measures against British life, liberty, and property'.[2] Debates in London surrounding the outbreak of the First Opium War kept the focus on the urgent need to protect British subjects and their property in China. Radical Member of Parliament

[1] Martin and Brathwaite never received a settlement; the case lingered for four years, then died, as European governments ignored the demand for compensation. R. Braithwaite, *Palmerston and Africa: The Rio Nunez Affair, Competition, Diplomacy and Justice* (1996).

[2] Quoted in Li Chen, *Chinese Law in Imperial Eyes: Sovereignty, Justice, and Transcultural Politics* (2016), at 221.

Lauren Benton, *Protection Emergencies: Justifying Measures Short of War in the British Empire* In: *The Justification of War and International Order.* Edited by: Lothar Brock and Hendrik Simon, Oxford University Press (2021). © the many contributors.
DOI: 10.1093/oso/9780198865308.003.0009

Joseph Hume declared that the rationale for war was created from 'the moment British subjects . . . were placed in prison'.[3]

These incidents exemplify what I call 'protection emergencies', a proliferating kind of crisis in the nineteenth century with many possible outcomes. The prelude to the Opium War shows that the call to protect subjects could serve as a powerful argument for war. Here and elsewhere, the imperative to protect subjects blended seamlessly with broader assertions about the need and right to protect national or imperial interests. The Rio Nuñez case reminds us that protection emergencies could also provide a mode of engagement in the politics of claim-making in places where no European power held control and no local power was recognized as fully sovereign. Diplomats might point to the mere presence of European subjects or, like French and Belgian agents in the Rio Nuñez case, might challenge the presence of other Europeans to reinforce inchoate claims to possession. Even intermittent and small-scale violence—along with the assertion of imperial agents' authority to pursue it—could project power into territories beyond imperial control, in the process pushing determinations about sovereignty into the indefinite future.

Protection emergencies multiplied with the global militarization of European empires in the nineteenth century. Wide-ranging armed patrols in the service of empire were hardly new in this period. Naval patrols had long been used to secure harbours and trade routes, and to intercept enemy shipping. In the eighteenth century Caribbean, Spanish *guarda costas* plied the waters around Cuba to intercept unauthorized foreign shipping from reaching the island and to capture enemy shipping close to shore.[4] In the early and middle decades of the nineteenth century, the British navy patrolled along the West African coast and extended patrols to other regions to capture slave ships from nations that had signed bilateral treaties to end the trade.[5] In the Pacific, European navies sent ships on vast circuits to gather information about local geographies, explore prospects for trade, and examine the nature of sovereignty in places Europeans considered politically opaque.[6]

Patrolling constituted more than a mere vehicle for extending European soft power. Increasingly, the practice also influenced definitions of legitimate use of force. Global militarization placed considerable discretion about violence in the hands of commanders on the seas. As empires delegated decisions about the recourse to violence to military officials and other imperial agents, they supplied only vague and elastic guidelines. Commanders of forces on patrol were routinely instructed to

[3] B. Semmel, *The Rise of Free Trade Imperialism: Classical Political Economy, the Empire of Free Trade and Imperialism, 1750–1850* (1970), at 153. The events leading up to the second Opium War, the so-called Arrow War, also spiralled from an emergency of protection. After a small Chinese force was reported to have pulled down the British flag aboard the *Arrow* and arrested its Chinese crew, the plenipotentiary in China, John Bowring, ordered a retaliatory attack on Canton.

[4] R. Pares, *War and Trade in the West Indies, 1739–1763* (1936); E. Schneider, *The Occupation of Havana: War, Trade, and Slavery in the Atlantic World* (2018).

[5] There is a vast literature on, and some disagreement about, the international significance of patrolling to contain the slave trade. Contrast J. Martinez, *The Slave Trade and the Origins of International Human Rights Law* (2012) and Benton, 'Abolition and Imperial Law, 1780-1820', 39 *Journal of Commonwealth and Imperial History* (2011) 355. See also J.C. Richards, *Liberated Africans and Law in the South Atlantic, c. 1839–1871* (2019) (Ph.D. dissertation on file at Cambridge University).

[6] For examples, see L. Benton and L. Ford, *Rage for Order: The British Empire and the Origins of International Law, 1800–1850* (2016), at 130–45 and 152–64.

protect subjects and, more vaguely still, to support imperial interests. Military commanders were authorized to use violence but expected not to drag empires into major wars, or create new dangers for sojourning subjects. These requirements placed the burden on colonial officials and military officers to generate justifications for particular acts of violence. This mode of operation gave shape to an implicit, unilateral right to pursue limited war. As this chapter will show, practices and vernacular legal arguments worked together to package legitimate violence in small bits, as series of brief skirmishes and lightning raids. Staccato imperial military actions were tucked inside broader, inchoate commitments to regional and global order.

In outlining this legal framework for small-scale imperial violence, this chapter builds on several methodological trends in the study of imperial legal politics and the history of international law. Recent studies call attention to the inseparability of practice from vernacular discourses and the influence of thick political contexts.[7] Such perspectives suggest that we cannot make sense of imperial agents' actions in narrating small-scale imperial violence by characterizing their normative commitments or by taking a genealogical approach that begins with later doctrine, such as humanitarian intervention, and moves backward in search of origins.[8] Recent studies of jurisdictional conflicts in empires and their formative effects—for example, their influence on ideas and practices of layered imperial sovereignty—teach us to begin instead with clusters of conflicts as sites for legal innovation.[9] Legalities of violence emerged from complex processes surrounding conflicts and the cumulative effects of ad hoc strategies of discourse and practice.

This chapter follows such an approach in examining how protection emergencies gave rise to repeating patterns of legitimate violence short of war. The first section examines pronouncements about small-scale imperial violence in European empires against a longer history of two 'jurispractices'—protection and possession—that became crucial elements of the legal and political repertoires of imperial agents as they justified violence.[10] I then discuss the way these categories were deployed in the operations of the British navy, turning to the example of legal politics on naval patrols in one region of the Pacific world. The cases reveal a mode of justifying measures short of war: the combination of high-level, broad authorization for limited war and locally produced justifications for violence. Finally, the chapter considers the broader implications of these findings. Without tracing nineteenth-century patterns forward to today's small-scale violence by hegemonic states, from drone strikes to targeted killings, we can glimpse certain structural similarities across periods in the combination

[7] For a fuller discussion of these methodological questions, see Benton, 'Beyond Anachronism: Histories of International Law and Global Legal Politics', 21 *Journal of the History of International Law* (2019) 7.

[8] My approach differs in this respect from that outlined by Hendrik Simon and Lothar Brock in their introduction to this volume, 'The Justification of War and International Order. From Past to Present', although I agree with them that the study of justifications for war can help illuminate shifts in the global legal order.

[9] L. Benton, *A Search for Sovereignty: Law and Geography in European Empires, 1400–1900* (2010); L. Ford, *Settler Sovereignty: Jurisdiction and Indigenous People in America and Australia, 1788–1836* (2011); L. Benton and R. Ross (eds), *Legal Pluralism and Empires, 1500–1850* (2013); Saksena, 'Jousting Over Jurisdiction: Sovereignty and International Law in Late Nineteenth-Century South Asia', 38 *Law and History Review* 409.

[10] The term 'jurispractice' was introduced by Hermes, 'The Law of Native Americans, to 1815', in M. Grossberg and C. Tomlins (eds), *The Cambridge History of Law in American, I* (2008) 32.

of broad authorization for war and the insistence on case-specific justifications for limited war. The decentralization of justifications for small-scale violence, moreover, may still operate as it did in the nineteenth century to cultivate indeterminacies of sovereignty. Nineteenth century imperial actions did not set such associations into stone, or even into doctrine. But they helped create durable modes of representing the use of limited force across borders as routine, and as intimately tied to visions of regional and global order.

2. Protection and Possession

Protection and possession have a long history. Both categories of discourse and practice featured prominently in interpolity law in the early modern world. Both were associated with understandings of imperial sovereignty as layered, incomplete, and uneven. And both influenced the persistence of regional formations in which competing, partial, or inchoate claims of multiple polities coexisted over long periods.[11]

Ubiquitous in the early modern world, protection arrangements typically involved the proffer of security by one political community to another, often in exchange for the payment of tribute. Participants frequently adopted positions of studied ignorance with regard to the unequal power relations that stood behind such arrangements. The ambiguities of protection talk, in fact, accounted for much of its appeal. Subordinate, tribute-paying political communities manoeuvred to get out from under obligations; colonizing powers sought to represent their right to rule as based on something other than force, such as voluntary cession of sovereignty or declarations of vassalage by new imperial subjects. In some mafia-style protection arrangements, imperial powers directly threatened weaker political communities with violence if they did not pay for protection. Such arrangements featured a purposeful lack of clarity about claims to sovereignty; protection arrangements were more profitable and politically useful when they left hierarchies of rule undefined.[12]

Possession represented another longstanding, highly flexible resource for structuring interpolity relations. Like protection, possession had a long history of particular importance for empires. In European empires, the concept of possession was defined in relation to the adaptation of the Roman law for the acquisition of property.[13] Historians have written a great deal about the Roman legal idea of *occupatio*, or occupation, which in its early modern European adaptation provided a way to secure title over territories where, it was assumed, land belonged to no one.[14] In support of claims of occupation, Europeans erected geographic markers, founded settlements, mapped new regions, and engaged in ceremonies and judicial acts to mark their presence in

[11] On interpolity law, see Benton and Clulow, 'Empires and Protection: Making Interpolity Law in the Early Modern World', 12 *Journal of Global History* (2017) 74.

[12] Benton and Clulow, 'Webs of Protection and Interpolity Zones in the Early Modern World', in L. Benton, A. Clulow, and B. Attwood (eds), *Protection and Empire: A Global History* (2017) 49 .

[13] Benton and Straumann, 'Acquiring Empire by Law: From Roman Doctrine to Early Modern European Practice', 28 *Law and History Review* (2010) 1.

[14] Ibid.; A. Fitzmaurice, *Sovereignty, Property, and Empire, 1500–2000* (2017); and K. MacMillan, *Sovereignty and Possession in the English New World: The Legal Foundations of Empire, 1576–1640* (2006).

contested regions beyond Europe. Possession was in many ways a more useful Roman legal concept than occupation for European empires. As with occupation, evidence could include settlements, fortifications, maps, ceremonies, and a range of other symbolic markers. Proofs did not need to reach an objective threshold but merely to dominate evidence of possession presented by others. The comparative framework for possession made it especially appealing in contexts where there were two rivals—as it did, for example, where there was jockeying between Spanish and Portuguese agents or settlers in Asia and the New World.[15]

The long history of protection and possession helps us to understand how these discourses were deployed by imperial agents in the nineteenth century. In the British Empire, protection gained potency as it came to be associated both with the consolidation of imperial power in colonies of varied status and with rationales for imperial expansion. The 'protection of British law' was a phrase put to use by imperial reformers advocating the amelioration of slavery and the political incorporation of indigenous peoples or subordinate subjects.[16] In the variegated rights regime of the empire, protection could be extended without implying the recognition of equal rights for all. At the same time, modes of imperial expansion proceeded under the umbrella of protection. In Ceylon (now Sri Lanka), for example, early nineteenth-century British governors cited the duty to protect subjects of the interior Kingdom of Kandy from the supposed tyranny of their king as a rationale for waging war against the kingdom and consolidating British power over the entire island.[17] In South Asia, a series of treaties between South Asian states and the British East India Company paved the way for British rule by disguising, in familiar fashion, unequal protection agreements as 'subsidiary alliances'. Discourses of possession retained their currency in nineteenth-century empires, too. Even in the context of rising British global hegemony, direct European rule claimed only a tiny portion of the globe's expanse. Intensifying inter-imperial rivalries urged stepped-up efforts to accumulate symbolic arsenals in support of implicit and future claims to territory.

In the definition of legitimate use of force, protection and possession operated in tandem. Claims of possession could create the conditions for protection emergencies, and acts taken to protect subjects could also signify proofs of possession. Punishment demonstrated jurisdiction—in many ways a more compelling proof of possession than ceremonies or other symbolic acts.[18] Significantly, discourses and practices of protection and possession had inescapably local dimensions. Protection emergencies began with detailed stories, sometimes recounted by victims, about injury to specific individuals or groups and requests for restitution or protection from future harm. Metropolitan officials, even if they ordered intervention, placed ships, soldiers,

[15] Benton and Straumann, above note 13; Benton, 'Possessing Empire: Iberian Claims and Interpolity Law', in S. Bellmessous (ed.), *Native Claims: Indigenous Law against Empire, 1500–1920* (2011) 19; T. Herzog, *Frontiers of Possession: Spain and Portugal in Europe and the Americas* (2015).

[16] On this point, and for a full discussion of protection in Ceylon and elsewhere in the British Empire, see Benton and Ford, above note 6, at ch. 4; on protection in colonial Australia and its relation to possession and sovereignty, see Bain Attwood, *Empire and the Making of Native Title: Sovereignty, Property and Indigenous People* (2020).

[17] Benton and Ford, above note 6, at 90–102.

[18] Instructions to European captains on early expeditions to the Americas were often explicit on this point. For examples, see Benton, above note 9, at ch. 1. See also Arnulf Becker Lorca's chapter in this volume.

or settlers on the scene. Once arrived, imperial agents had to interpret and respond to local events and conditions, and to gather proofs of possession, reporting on their progress to superiors and sponsors.

The paired emphases on protection and possession gave local imperial agents' communications a particular valence with regard to sovereignty. It led them to refrain from making very clear determinations about the status of sovereigns and to recognize various kinds of quasi-sovereignty. Protection, after all, denoted a relationship of dominance without the assumption of political control. Proofs of possession, similarly, offered the possibility of later moves towards territorial acquisition without leading inexorably to that step. In utilizing both discourses, often in combination, imperial agents constructed a flexible framework for imperial rule and violence.

3. Navy Captains and the Law

These modes came together in the increasing interest in, and resources devoted to, global patrolling in the nineteenth century. In the British Empire, naval operations were especially important to extending British influence and power. After the end of the Napoleonic Wars, the British navy was patrolling a sprawling, constitutionally diverse empire with porous, fluid borders.[19] Peace had delivered new colonies to the empire, and some of these—in particular Ceylon, the Cape Colony, and Malta—were strategically positioned to serve as provisioning stations for naval ships or as regional command centres for the navy. The East India Company had its own navy, which it utilized to patrol and police sea lanes and harbours across the Indian Ocean. In the Atlantic, an uneasy rapprochement with United States and French navies in the North Atlantic contrasted with growing influence of the British navy in the South Atlantic, a trend facilitated by British slave trade patrols and indirect involvement in conflicts surrounding South American independence. Around the globe, the number of men employed by the Royal Navy was at an all-time high, as was the number of mutinies and other protests against naval discipline and shipboard conditions. The global influence of the Royal Navy was being magnified, too, by the powerful new technology of steamships. Much faster than sailing ships, the new steamships could also manoeuvre more easily in riverine passageways and becalmed harbours, extending the reach of naval engagements.

Navy personnel and ships were often first responders to calls for the protection of British subjects. Those calls became increasingly frequent in the mid-nineteenth century as British merchants extended their reach and regularly sought government support for their ventures. Throughout the colonies and beyond, mid-level officials and consuls raised alarms about threats to individual British subjects and urged the use of British force. Both merchants and officials decried the seizure of trading vessels and called on the navy for protection of British shipping and retaliation against sea raiders, generating protection emergencies that swept the empire in the 1840s. Some of these

[19] Benton and Ford, above note 6 analyses this project of global legal reform.

crises acted as catalysts for war, but most gave rise to sharp and brief engagements rather than prolonged conflicts.

Between formal declarations of war and diplomatic manoeuvring lay a range of responses to protection emergencies: blockade, bombardment, patrols, and raids on land. In West Africa, British navy captains patrolling to contain the slave trade attacked coastal trading stations.[20] In Southeast Asia, naval captains bombarded Dayak settlements on the Borneo coast in 1843 and 1844, and again in 1847, causing several thousand civilian casualties.[21] Captains in both regions pointed to the need to enforce legislation against the slave trade and piracy, but they also often responded to calls to take action to protect British interests or subjects. Historians have tended to group such engagements under the heading of 'gunboat diplomacy' in the British Empire, but this label disguises important variations and hides the formative influence of imperial legal politics.[22] Situating the logic of intervention in the metropole has prevented us from investigating in depth the way justifications for the use of force were produced outside Europe and from considering with sufficient nuance the relation of limited force to configurations of sovereignty.

In the British navy, the production of justifications for violent intervention depended on the conventional pathways of the information order. Captains wrote reports that they hoped would eventually find their way, with endorsements appended, to London policy-makers. Various kinds of information jostled for space in the pages of reports, which followed a rough template in which intelligence about sea approaches, anchorages, and other geophysical properties prefaced broad-ranging commentary on political matters, from evaluations of the sovereign status of local rulers to speculation on the interests and actions of other European empires. Moral judgements filtered into passages assessing levels of 'civilization', and evaluations of resources available for naval ships blended with opinions about the potential of various sites to serve British commercial interests. Captains also recounted the details of specific engagements and expeditions on land, interspersing dry recitations of events with explanations of their decisions. The vividness of the descriptions of dangers to British subjects reported from the peripheries mattered, and so did the sophistication with which commanders appealed to hot-button political issues at home, such as abolition and free trade.

Many reports highlighted not just immediate dangers but also less well-defined goals of projecting British interests. Commanders knew that the government could not be relied upon for unanimous support for bellicose acts short of war, even when the object was to protect British subjects. Anti-imperial sentiments crossed the political spectrum in Europe. In Britain, reformers, buoyed by free trade advocates, were loudly bemoaning the costs of maintaining the empire.[23] Colonial Office lawyer James Stephen warned that there were limits to the logic of protection: 'If some 50 or 60 of the Queen's subjects ... settle themselves as a distant Community in a settlement some 100s of miles ... from any other British Colony, they have no fair right to expect the

20 P. Scanlan, *Freedom's Debtors: British Antislavery in Sierra Leone in the Age of Empire* (2017).
21 Benton and Ford, above note 6, at 139–45.
22 Benton and Ford, above note 6, at 5, 150.
23 Semmel, above note 3.

advantages of the Institutions … of other parts of the British dominions.'[24] The idea that the empire was a sink hole for resources and prestige reinforced calls for measured responses to colonial crises. Short and cheap protection engagements were politically more palatable than prolonged and costly wars.

This background gave shape to a system in which imperial agents were authorized to make judgements about violence with the understanding that their decisions might be reviewed by officials in Europe. Almost uniformly, navy captains sailed with instructions that were intentionally vague and that often included orders to protect British subjects and interests. Assessing threats and how best to respond to them required captains to answer complex questions. Were local rulers worthy treaty partners? Could they act as co-enforcers of commercial relations and property protection? Was British violence justified?

In considering such questions, captains brought their own interests into the mix as men of political intrigue who manipulated incentives for professional advancement and financial gain. Navy commanders tapped their own political views; many in the period considered themselves reformers and abolitionists.[25] They had to survey political arrangements from the seas, with the benefit of generally brief landings—a peculiar vantage point from which to discern the nature of political and legal authority on land. Familiar with symbolic proofs of possession, captains offered their actions as support for inchoate British claims against those of imperial rivals. They relied on what they knew, or could quickly learn, about risks or harm to vulnerable British subjects and the actions of local polities and other empires. Captains' talk of protection and possession allowed them to show that they were working to block the advance of rival empires, seeking to establish British influence without imposing costly rule, and identifying sovereigns with whom the British government might sign treaties. The results occasionally led them to war; much more often, they provided rationales for small-scale military actions.

We are now in sight of a widespread but little-studied mode of producing justifications for small acts of imperial aggression. When we expand the definition of aggression to include short engagements and demonstrations of force, navy captains' correspondence emerges as an important source of justifications for British imperial violence. Captains sailed with broad authorization to use force while keeping violence contained below the level of full-scale war. We find, then, both a distinctive genre—reports on interpolity relations by imperial agents and officers—and a distinctive mode of operation, the delegation of decisions about violence to 'men on the spot'.[26] The genre and practices together produced a legal framework for violence short of war, well before the concept became the subject of commentary by international lawyers.

[24] D.K. Fieldhouse, *Select Documents on the Constitutional History of the British Empire and Commonwealth* (1985), at 423. Stephen was commenting on the Rio Nuñez affair described in Braithwaite, above note 1.

[25] Opportunities for profit featured head payments on liberated slaves and on pirates captured or killed. On naval captains as reformers, see e.g. J. Samson, *Imperial Benevolence* (1998).

[26] For a good discussion of the functions of 'men on the spot' in a different context, see van Hulle, 'British Protection, Extraterritoriality and Protectorates in West Africa, 1807–80', in Benton et al., above note 11.

4. A Coastal Quagmire

British actions along the Northwest Pacific coast of North America provide a rich example of justifications advanced by imperial agents for the use of force in distant interpolity zones. In the early nineteenth century, this Pacific region was populated by multiple indigenous communities and a handful of European settlements, and it was the focus of increasingly sharp European imperial rivalry. The tentacles of Russian fur trading from the north extended into a broad area still claimed by the Spanish but infrequently visited by Spanish ships. Captain Cook's 1778 voyage to Nootka Sound, to what would later be called Vancouver Island, had revealed to the English the potential profits to be made from a trans-Pacific trade in sea otter pelts. Coastal Indian groups, including the Nuu-chah-nulth, Makah, Salish, Kwakwaka'wakw, and Haida peoples, were responding to the growing trade with Europeans and Americans by organizing and extending their fur trade connections into the interior.

Possession and protection figured prominently in early British naval actions on the coast. After British ships were intercepted and blocked by the Spanish from founding establishments in Nootka Sound, the British and Spanish governments signed the first Nootka Convention in 1790. Negotiations revolved around claims of possession: the Spanish based their claims on a series of markers erected by the Heceta and Bodega expedition of 1775, and the British pointed to the construction of several buildings by a trader named John Meares, who claimed initially to have purchased the land from an Indian polity. The British government sent Captain George Vancouver to oversee implementation of the convention on his voyage to the Pacific. His negotiations with Spanish commissioner Juan Francisco de la Bodega y Quadra produced little light. Vancouver realized that the Meares establishment—a 'small hut' that Meares had 'abandoned when he left the place'—was a slim thread from which to weave proof of British possession.[27] Vancouver also felt acutely his limited power to negotiate; his instructions did not outline the measures he should take to confirm British claims. He decided to interpret broadly his charge of 'facilitating the commercial advantages of Great Britain in this part of the world' and to leave a British subject in charge of a permanent station on the coast. The conversations of the two naval commanders, Vancouver and Bodega y Quadra, formed the basis for the second Nootka Convention, signed in 1794. It awarded the Spaniards control over the coastline south of Neah Bay and designated all points north, including Nootka Sound, as areas of free navigation and trade.[28]

British navy ships were the principal representatives of British power in the region. Captains were given vague and capacious charges to advance British interests and protect British subjects in the region, and they were expected to make reasoned and careful decisions on the spot about the need for violence. When he was sent to the Northwest coast of North America in 1813, for example, Captain Hillyar of the Royal Navy received these instructions: 'The principal object of the Service on which you are

[27] G. Vancouver, *A Voyage of Discovery to the North Pacific Ocean, and Round the World, in which the coast of north-west America has been carefully examined and accurately surveyed*, vol. I (1798), at 388.
[28] above Ibid., at 394.

employed is to protect and render every assistance in your power to the British traders from Canada.' Even in wartime, this order preceded the command 'to destroy ... any settlements which the Americans may have formed'.[29] The instructions were short on specifics. It was left to Captain Hillyar to determine what actions to take in the north Pacific to achieve these aims, and he would have to guess how to report on his exploits in a way that might reflect well on him.

Naval commanders were acutely aware of the thinness of British representation in the region, and they worried about the contrasting surge in the American presence. As American ships came to dominate the coastal trade, the centre of imperial tensions shifted south, to the mouth of the Columbia River.[30] Here competition came not just from jockeying for settlements that would serve as proofs of possession but also from the presence of overland traders, including agents of the North West Company (which merged in 1821 with the Hudson Bay Company). Over the next several decades, British naval commanders would warn that the British presence was insufficient to sustain British claims, particularly in the face of intensifying American settlement.

Tensions came to a head in the Oregon crisis of 1844–1846, a standoff between American and British forces over territorial claims. The British response was to double down on navy patrols. The Admiralty dispatched five ships to the region in the summer of 1846 with pointed instructions to protect British subjects. For example, Rear-Admiral Seymour instructed Captain Duntz of the *Fisgard* 'to afford Protection to Her Majesty's Subjects ... [and] gain intelligence of the State of Affairs in Oregon'. If Duntz encountered Americans 'employed in the erection of Forts or Strongholds to enable them to hold an adverse Military possession of the Country', he was authorized to 'remonstrate against such proceedings, and if necessary' to introduce war ships into the Columbia.[31] The treaty signed in 1846 to divide the Oregon territory along the 49th parallel did not end demands on the British navy to patrol for the protection of subjects. Tensions remained high after the British government committed to founding a colony at Vancouver Island, one that would operate under joint control of the Hudson Bay Company and the Colonial Office.[32] The frigate *Inconstant* was sent specifically for the protection of Fort Victoria, and its captain, John Shepherd, argued without success to the Admiralty that the government should station troops permanently at the fort.

In a series of conflicts with Indians, captains remained the main narrators of events and the agents mainly responsible for choosing between diplomacy and violence. The navy was effectively in charge of mapping Indian political authority in the region.

[29] Quoted in B. Gough, *The Royal Navy and the Northwest Coast of North America, 1810–1914: A Study of British Maritime Ascendancy* (1971), at 15, from Admiralty Instructions to Hillyar, 12 March 1813, most secret, ADM 2/1380, at 370–75.

[30] Indian actions had a good deal to do with this shift. In 1803, the captain of the American ship *Boston* delivered a slight to the chief of the Nuu-chah-nulth people, Chief Maquinna, whom he called a liar for claiming that the rifle given to him did not work. Maquinna returned the next day, and after a friendly interlude his followers massacred twenty-one of the ship's crew, sparing only two whom he took captive. The event intensified traders' fears and prompted a search for other trading points along the coast.

[31] Longstaff and Lamb, 'The Royal Navy on the Northwest Coast, 1813–1850', 2 *British Columbia Historical Quarterly* vol. IX 133, at 114.

[32] Vancouver Island was ceded to the Hudson's Bay Company by royal grant on 13 January 1849, under this arrangement of shared authority, which the first colonial governor found, not surprisingly, unworkable.

Although it was clear that the Oregon agreement had established British sovereignty in territories north of the 49th parallel, officials at Fort Vancouver were beginning a period of uncertain and shifting recognition of Indians' property in land and of their capacity as treaty partners.[33] The Royal Navy focused not on sorting out these questions but on enforcing British jurisdiction in cases where Indians attacked British subjects.

These events represented protection emergencies of a particular type. In 1850, Newitty Indians killed three British sailors who were deserters from a merchant ship. Captain George Wellesley travelled in the *Daedalus* to the place of the killing to arrest the Indians responsible. After a skirmish, the crew contented itself with burning the empty houses. The rear admiral of the Pacific station, Fairfax Moresby, followed the failed mission of the *Daedalus* by travelling himself to Vancouver in the imposing frigate *Portland*, accompanied by the smaller *Daphne*. Moresby wanted to punish the whole Newitty band unless the Indians surrendered those who were responsible for the murder. The Indians refused to hand over the men, then fled into the interior as naval forces destroyed their camp and killed several Indians. Moresby's brutality initially appeared no more effective than that of the previous naval commander. But a few days later, Indians returned to deliver the corpses of those identified as the culprits of the murders. Moresby concluded that his aggressive response to the protection emergency had paid off.

A similar policy was on display in British efforts to reinforce its sovereignty over the Queen Charlotte Islands against suspected American intentions to extend protection to American settlers and assert claims to British territories. The discovery of gold in the islands prompted a brief influx of prospectors. Responding to the entreaties of the chief factor of the Hudson's Bay Company and unable to wait for instructions from London, Moresby sent Captain Augustus Leopold Kuper and the frigate *Thetis* to reassert Britain's possession of the islands. Captain Kuper found that the threat of American incursions was not acute and concluded that the potential for profitable gold mining was 'overrated'.[34] Instead of containing American settlement, he became embroiled in another protection crisis when two Indians were suspected of the murder of a Company shepherd at a sheep station near Victoria. The British had demanded the surrender of the two suspects, but nothing had happened in response. The Company and the navy agreed it was important to show resolve, although Moresby was reluctant simply to take revenge. The *Thetis* provided '130 Officers, seamen and Marines' for an expedition to negotiate with (and threaten) the Indians and force them to comply.[35] The British began by notifying the Indians of the plan 'to treat them as enemies unless they submitted to the demands of justice'.[36] Eventually, one of the suspects was

[33] S. Banner, *Possessing the Pacific: Land, Settlers, and Indigenous People from Australia to Alaska* (2007), at ch. 6.

[34] Captain Kuper to Rear Admiralty Moresby, 20 July 1852, in 'Four Letters Relation to the Cruise of the Thetis', VI:3 *British Columbia Historical Quarterly*, at 192.

[35] Captain Kuper to Rear-Admiral Moresby, 4 February, 1853, in 'Four Letters Relation to the Cruise of the Thetis', ibid., at 200. The threat to treat them as 'enemies' is interesting because it does not recognize them as British subjects (and therefore rebels), despite expansive British claims to sovereignty in the region.

[36] James Douglas to Archibald Barclay, 20 January 1853, in 'Four Letters Relation to the Cruise of the Thetis', ibid., at 205.

surrendered and the other captured, and a jury 'composed of the officers present' found them guilty and sentenced them to be hanged, a punishment carried out for effect before an audience of 'the whole tribe'. The chief factor, James Douglas, regarded the incident as a very significant moment in the establishment of British authority, and the navy was happy to take credit for resolving the crisis without firing a shot.[37]

London officials did occasionally wonder whether the navy might not be getting out ahead of both the Colonial Office and the Company. Moresby's actions drew the attention of officials in London, who in 1850 requested an opinion from the Law Officers of the Crown on 'to what extent the Commanders of her Majesty's ships of War may exact from the Natives of the Pacific Islands' retribution or compensation 'when the wrongs done to British subjects extend to loss of life by unprovoked and deplorable murder'. The Law Officers' response, written in July, 1850, offered some guidance but mainly affirmed the wide discretion given to local commanders to make judgements about lawful violence. The Law Officers advised that if naval officers had been able to 'actually witness the unprovoked commission of murder or any atrocious crime against British subjects' they would be

> justified in presumptively demanding redress and the punishment of Offenders from the Chief or Chiefs to whose jurisdiction the Criminals belong; and if no reparation can be obtained, in securing and detaining for a reasonable time the persons of such Chief or Chiefs then resorting to hostilities against the Natives immediately dependent on the Same authority as the Criminals.[38]

The opinion went on to state that in cases where crimes had 'not been actually witnessed by British Officers' and they had to rely on the testimony of 'survivors' and on second-hand information about whether there had been 'previous provocation at least by some of the Europeans to the Natives', the officers should 'exercise of utmost caution and forbearance before resorting to any exercise of force whatever'.[39]

Even while urging restraint, the memorandum was careful to recognize wide latitude for judgement by British commanders. It conferred on navy commanders the capacity to decide whether acts would be considered criminal and whether they were 'unprovoked' or 'atrocious'. The Law Officers also affirmed officers' authority to determine when the exercise of force was necessary, even when they had not directly witnessed 'crimes'. Finally, in defining violence as a last resort, the formula naturalized the idea of gradual escalation in which British captains might offer peaceful resolution, only to face Natives' intransigence. The narrative blamed indigenous people's failure to comply as triggers for legitimate imperial violence.[40] It is perhaps not surprising that the guideline did not have much effect; navy captains continued to report naval violence as retaliation for Indian aggression and 'crimes' through the 1880s.

[37] James Douglas to Archibald Barclay, 20 January 1853, in 'Four Letters Relation to the Cruise of the Thetis', ibid., at 205.
[38] Letter from Law Officers, 20 July, 1850, The National Archives of Britain (hereafter TNA) ADM 172/3.
[39] Letter from Law Officers, 20 July, 1850, TNA ADM 172/3.
[40] On the long history of representing imperial aggression as a response to truce breaking, see Benton, 'The Legal Logic of Wars of Conquest: Truces and Betrayal in the Early Modern World', 28 *Duke Journal of Comparative & International History* (2018) 425.

The overall impact of the approach was to preserve the region as an interpolity zone of unsettled sovereignty. Certainly Indian sovereignty remained in question. British assertions of a right to punish Indians individually and collectively diminished their status as political interlocutors. At the same time, the British on occasion upheld Indian sovereignty against American incursions. For example, officials wrote from Fort Vancouver in January 1848 to describe 'events in Oregon that may seriously affect British subjects and British interests in that Country' and to call for a navy warship to be sent to the Columbia River to protect British traders. Agents reported that American citizens had been engaged 'for some months past in active hostilities with the Cayouse and other Indian Tribes inhabiting the upper Columbia'. After the Cayouse massacred American settlers and took sixty-four women and children captive, the British sent troops to rescue the captives, taking the opportunity to tout their better relations with, and superior understanding of, Indian polities. Oddly, having disregarded Indian sovereign rights on multiple occasions, the British extolled them to offset American power. At the same time, they pointed out that the exercise of American sovereignty lay in the indefinite future. Continued patrols were necessary, British agents asserted, precisely because the United States had yet to exert its authority in regions awarded to them by treaty. The British would need to act 'for the protection of British property in Oregon' until the United States government was 'prepared to restrain its citizens'.[41]

Historians often point to the resolution of the Oregon question as an example of inter-imperial diplomacy that successfully averted war. Decisions on the ground about the legalities of small-scale violence tell a different, more complex story. Uncertain sovereignties accompanied an ongoing right of British intervention—a right that sharpened over time as protection emergencies continued to surface. The importance of this posture is striking, given that British subjects and interests were thin on the ground. The region retained the character of an interpolity zone shaped by imperial agents' assertions of the right to use limited force.

5. Conclusion

The processes framing British imperial intervention in the early and middle decades of the nineteenth century were in some respects peculiar to the age. Global militarization of empires, commercial expansion, and sharpening imperial rivalries multiplied opportunities for conflict, both among empires and between European and indigenous polities. This context positioned naval commanders as principal actors in vast imperial projects to collect information about the sovereign capacities of distant polities, design regional orders through diplomacy, and, as this chapter has shown, render consequential decisions about when and how to use violence short of war. Captains turned to a familiar repertoire of legal languages—protection and possession—as they interpreted instructions to shelter imperial subjects and advance imperial claims.

[41] Letter of Ogden and Douglas from Fort Vancouver, 8 January 1848, TNA ADM 1/5589.

Correspondence offered justifications for series of small-scale engagements that collectively equated to a perpetual right to use force in small bursts.

The practices and statements of commanders forged an association between the right to use force and the incomplete sovereignty of local polities. Turning this statement around, we find a link between asserting a permanent right to imperial intervention and the capacity to pronounce on the nature of local sovereignties. In producing justifications for intervention—for example, by punishing Newitty Indians for violence against British subjects or in calling for a British warship to hover off American territory to protect British subjects—naval officers were also marking the incapacity or incompleteness of sovereignty by other powers.[42]

As Lisa Ford and I have shown elsewhere, this same calculus was on display in the most celebrated protection emergency of the mid-nineteenth century, the so-called Don Pacifico affair.[43] That episode responded to the appeal by a Gibraltar-born British subject, David Pacifico, to the British government after he was unsuccessful in getting the Greek government to compensate him for a raid by an anti-Semitic mob on his Athens house. In a famous, four-hour speech in the House of Commons defending the British decision to blockade Athens to force restitution, Lord Palmerston equated the British government's right and duty to protect subjects everywhere in the world to the Roman principle of '*civis Romanus sum*' ('I am [a] Roman citizen'). The speech has routinely been cited as the expression of a new doctrine of British global power backed by military force and based on a limitless prerogative to intervene on behalf of British subjects anywhere in the world. But Palmerston and the government's response closely reflected existing imperial practice. In the longer, rarely cited version of the speech, he explained that the action to protect British subjects followed a pattern of injustice by the Greek government where British subjects were concerned. It was the insufficiency of the exercise of Greek sovereignty that triggered British aggression. In this way, the blockade on Athens and Palmerston's pronouncements followed recently established patterns of the imperial use of force in protection emergencies.

As this and other cases show, justifications for imperial violence in the nineteenth century were clearly rooted in historical specificities. But the examples also suggest some interesting parallels to debates and practices much later. In the post-9/11 world, myriad small acts of aggression, from drone attacks to bombing raids, have been justified under the cover of a tightly stretched right to self-defence, including the defence of nationals anywhere in the world. The debates are too intricate and numerous to summarize here, and there is no clear consensus.[44] But we find echoes in the logic and form of justifying aggression on this basis, and in the arraying of small-scale actions under the rubric of a global war on terror. The family resemblance is as interesting as any claim (which I am not making) of historical continuity because it prompts us to ask new questions about connections suggested in the historical case: between

[42] In other places, British agents encouraged the formation of sovereign states to stabilize regions. See Benton and Ford, above note 6, at ch. 6.

[43] The discussion of the case here is drawn from Benton and Ford, above note 6, at 112–15.

[44] See C. Gray, *International Law and the Use of Force* (4th ed., 2018), at chs 4 and 5; Randall Lesaffer observes that 'we should be careful not to turn a blind eye to the elements of continuity in the history of use of force law', Lesaffer, 'Too Much History: From War as Sanction to the Sanctioning of War', in M. Weller (ed.), *The Oxford Handbook of the Use of Force in International Law* (2015), at 36; see also Thilo Marauhn's chapter.

legalities of violence and sovereignty, and between campaigns of limited violence and visions of pluri-political regional order.

Only one possibility of this juxtaposition, in other words, is to assert that imperial law haunts international law like a 'ghost in the machine'.[45] The analogy of imperial and international law allows us to ask whether the pairing of measures short of war and incomplete sovereignty is generalizable. Certainly various versions of quasi-sovereignty still exist, under different names. Pacific archipelagos harbour numerous places of quirky in-between status—a Pacific world dotted with 'dependent territories', 'external territories', 'special administrative regions', and 'autonomous collectivities', among other categories of partial sovereignty recognized by the United Nations. In today's Pacific Northwest, long after Native states appeared fully subsumed by powerful nation-states, First Nations assert rights based on nineteenth-century treaties that recognized indigenous sovereignty by degrees. The relation of such processes to legalities of violence are not the subject of this chapter, though it bears asking whether the pairing functions in similar ways beyond the nineteenth century.

Here it is enough to affirm the findings that routine responses to protection emergencies in the nineteenth century generated a framework for small-scale imperial violence. Justifications relied on longstanding imperial modes of operation of amassing proofs of possession and proffering protection. Men charged with policing and patrolling a global empire rendered decisions about small acts of jurisdiction and reprisal. In the process, they shaped regions of uncertain sovereignty and forged deep associations between the right to imperial violence and the persistence of interpolity regions under uneven imperial sway.

[45] For the description of the continued role of imperial law in international law as a 'ghost in the machine', see Benton and Ford, above note 6, at 188–92; and more generally A. Anghie, *Imperialism, Sovereignty, and the Making of International Law* (2005). See also Mallavarapu's and Chimni's chapters in this volume.

10
The Great War and International Law: German Justifications of Prevention and Pre-emptive Self-Defence

Isabel V. Hull

1. Introduction

Despite a common view that before 1914 states had a right to go to war whenever their self-proclaimed interest required,[1] the history of the nineteenth century shows that war in Europe among European states was considered by statesmen, the decision-makers on war, to be legitimate only under certain, limited circumstances. Self-defence was the most obvious of these. But what did self-defence mean? This chapter uses the German declarations of war in 1914 to examine three issues: (1) the role of customary international law (CIL) in statesmen's decision to go to war (using Germany as an example); (2) the assumptions that state actors held on *jus ad bellum*; and, more particularly, (3) how they distinguished self-defence, prevention, pre-emption, and aggression.

This chapter begins with a brief history of the nineteenth-century state system and how it regulated the right of states to go to war, before moving on to the July Crisis and Imperial Germany's declarations of war. The focus is not on the claims of jurists, but on the arguments and actions of state leaders, civilian and military.[2] Much of our (mis)understanding of nineteenth-century *jus ad bellum* is due to an overreliance on

[1] See the list of proponents of this view in Lesaffer, 'Aggression before Versailles', 29 *European Journal of International Law (EJIL)* (2018) 773, at 775. See also the contribution by Hendrik Simon to this volume.

[2] Most scholarship on *jus ad bellum* before 1914 has focused almost exclusively on the writings of lawyers, rather than on state practice. Some scholars read pre-war positivist writers as having successfully established in international-legal discourse an untrammelled 'right' to wage war, contravening the very old 'just war' tradition that limited legitimate warfare to self-defense or to avenge legal injuries. According to this reading, Germany's war in 1914 did not violate international law (see von Bernstorff, 'The Use of Force in International Law before World War I: On Imperial Ordering and the Ontology of the Nation-State', 29 *EJIL* (2018) 233, at 257). But other scholars counter that the just war tradition continued, with some modification, down to 1914, that the positivist permissivists were of very late nineteenth-century vintage, were most common in Imperial Germany, and even by 1914 did not dominate positivism: Simon, 'The Myth of *Liberum Ius Ad Bellum*—Justifying War in 19th-Century International Legal Theory and Political Practice', 29 *EJIL* (2018) 113; and Lesaffer, above note 1. Furthermore, as Lesaffer writes, 'if legal scholarship retained much of the old doctrine, nineteenth-century [state] practice adhered even closer to tradition', especially regarding justification for war, which 'remained as crucial to states as before'., see Lesaffer, above note 1, at 797, 795. In any case, the discursive-doctrinal methodology owes us an explanation for how jurists' writings alone could change customary international law. On the gap between legal theory and political practice in writing the histories of international law see also the introductory chapter by Simon and Brock and Anuschka Tischer's chapter.

Isabel V. Hull, *The Great War and International Law: German Justifications of Prevention and Pre-emptive Self-Defence* In: *The Justification of War and International Order*. Edited by: Lothar Brock and Hendrik Simon, Oxford University Press (2021).
 DOI: 10.1093/oso/9780198865308.003.0010

the writings of legal theorists, who, of course, were not responsible for decisions to go to war or usually knowledgeable about why these decisions had been made. CIL was not a pure, or closed system. It developed and changed according to diplomatic and military actions of states; decisions for or against war reflected the changing assumptions of leaders about state obligations and rights. Politics, diplomacy, and law were thus different aspects of an interactive whole, and the more we study them that way, the more accurate will be our understanding of CIL (in this case, *jus ad bellum*), and how it operated.

2. *Jus ad bellum* in the Nineteenth Century

In the wake of the tremendous disruption and destruction of the Napoleonic Wars, the statesmen of the Great Powers set up a system that they hoped would guarantee peace—in Wilhelm von Humboldt's words, 'to fix the grand principles that will produce the general happiness of Europe'.[3] The system was territorial in two senses: it applied only to Europe (not to colonies or to the Ottoman Empire), and it defined the boundaries of the units of Europe—its great, middle-sized, and small states. Although the borders were not guaranteed by obligatory force-of-arms, they were nonetheless anchored in the strongest public law treaty to date: seventeen separate treaties were contained in the Final Act (1815), which was signed by every state in Europe except the papacy. As Webster writes, 'its stipulations were thus in a special way agreed to, and in a sense guaranteed by all the sovereign States of Europe'.[4]

Statesmen understood that they were laying down new legal rules governing state behaviour in Europe. In September 1814, Humboldt (representing Prussia), wrote that the upcoming conference would remake Europe in 'the general interest and strengthen it by sanction(s), or at least by common recognition'.[5] Austria knew that to stop the cycle of wars, 'an independent European center had to derive from and rest upon a broad political consensus in Europe, underpinned by law'.[6] Charles-Maurice de Talleyrand, speaking for defeated France, agreed that peace could be assured only 'by restoring law as the basis of international life'; he insisted that the Congress declare publicly that it would conform 'to the principles of public [international] law'.[7]

The most important rule of law laid down in 1815 concerned wars of aggression and conquest. At the London pre-conference, the Allies agreed that mere occupation of territory did not mean annexation. Final disposition of territory depended on mutual consent.[8] When Prince Karl August von Hardenberg suggested to Robert Stewart Viscount Castlereagh that Prussia should simply annex Saxony, Castlereagh protested strongly against the principle 'that it should be competent for one Power to invade

[3] Cited in Sir C.K. Webster, *The Congress of Vienna, 1814–1815* (1963), at 184.

[4] Friedrich von Gentz had drawn up a document that would have obliged states to uphold the territorial status quo by force, but it failed to find approval after Russia insisted on including various measures relating to the Ottoman Empire. Webster, above note 3, at 103 f., citation: 101.

[5] Humboldt *projet* in Webster, above note 3, appendix 4.

[6] P.W. Schroeder, *The Transformation of European Politics, 1763–1848* (1994), at 527.

[7] First citation: Schroeder, above note 6, at 530; Talleyrand quote: Webster, above note 3, at 87.

[8] Webster, above note 3, at 68, 128.

another, and by force complete a recognition which was founded upon no treaty, and where no attempt had been made to disturb the possession of the invading Power in the territory to which he had laid claim'.[9] The Troppau Treaty (1820) forbade annexation resulting from Great Power intervention in states to stop revolution.[10] The whole point of the Congress, after all, was to end Napoleonic-type expansion by war. It did so by laying down all European borders in public treaty law, providing for recurrent Congresses (and after 1820, conferences of ambassadors) to mediate differences and/or create consensus on change, and by forbidding recognition of the fruits of aggression.

Subsequent state practice strongly suggests that states accepted these principles as obligatory—that is, as law. There were no wars of aggression on European soil from 1815 to 1914; there were few wars of any kind. Warfare was confined to three situations: Great Power interventions to retard or sometimes to aid revolution; unsettled questions on the periphery of Europe, particularly the fringes of the Ottoman Empire in the Balkans, where the Congress rules were not in effect; and toleration of wars of unification based on the principle of self-determination (the wars of Italian and German unification).[11] Political scientists and historians commonly conclude that 'armed force [was] effectively eliminated as a legitimate tool of statecraft among [the Concert's] members'.[12]

The European Concert (which succeeded the Congress system) operated as a society, and therefore developed its own 'rules, norms, [and] procedures'.[13] Scholars have not usually examined when rules and norms hardened into law, but it seems clear that, as F.H. Hinsley has written, the Concert was 'a coalition of leading states, founded on a public law for the defence of that law'.[14]

To understand how 1914 fitted into this system, it is convenient at this point to list the 'rules' of the Concert, as Paul W. Schroeder names them:

1. ... only the five Great Powers, the Directory of Europe, decided great European questions. Lesser powers had rights that should be protected, and they could be heard on questions that concerned them, but they had no deciding vote.
2. No power could wage war in Europe for territorial gain or promote revolution or unrest within another Great Power's territory or sphere of vital interest.

[9] Castlereagh to Lord Liverpool, 1 January 1815, cited in Webster, above note 3 at 132–33.

[10] M. Jarrett, *The Congress of Vienna and Its Legacy: War and Great Power Diplomacy After Napoleon* (2013), at 252 f., 265.

[11] See the Correlates of War Project, which, however, usually omits the smaller interventions, available at <http://www.correlatesofwar.org/> (last visited 10 January 2020). Intervention was the most controversial issue in European diplomacy before 1848. For nineteenth-century intervention and imperial ordering, see also the contribution by Lauren Benton to this volume. For the discourse on the Wars of German Unification, see Hendrik Simon's chapter.

[12] Charles Kupchan, cited in Jarrett, above note 10, at 362. See also Robert Jervis: 'war is no longer perceived as a legitimate instrument of statecraft', ibid., at 361; and P.W. Schroeder, *Austria, Great Britain, and the Crimean War: The Destruction of the European Concert* (1972), at 405.

[13] Schroeder, above note 6, at xii; 'principles, norms, rules, and decision-making procedures', Robert Keohane cited in Jarrett, above note 10, at 362; and Jarrett: 'new norms and procedures that would become enshrined as the recognized methods and traditions of diplomacy', ibid., at 354.

[14] Cited in Jarrett, above note 10, at 369.

3. No international question of vital interest to a Great Power could be raised without its consent.
4. But if a problem of genuine international import arose, no power could refuse an international conference upon it or exclude any other Great Power from it.
5. Above all, direct challenges and confrontations between individual Great Powers had to be avoided at almost any cost, mainly be referring the quarrel to the decision of the Concert. The first and great commandment of the Concert was, 'Thou shalt not threaten or humiliate another Great Power.'[15]

All of these rules were strongly reflected in state practice. No. 1, the rights of lesser powers, and No. 2, the injunction against aggressive war were settled CIL. No. 4, the mandatory holding of conferences to solve difficult issues and thus avoid war, strengthened towards the end of the century with the establishment of the Permanent Court of Arbitration (1899) and the proliferation of bilateral treaty laws requiring arbitration.

One can observe the stiffening of the norm against aggression in the rise in claims to self-defence as the primary acceptable justification for war. Prior to 1815 (and intensifying thereafter), states' explanations for why they went to war were overwhelmingly legal in nature, that is, they operated within the just war framework and therefore claimed righting a wrong: self-defence, treaty violations, violations of rights of citizens/subjects, upholding international law, rights of royal inheritance, or collecting debts.[16] Self-defence was always the most widely cited reason, because being the victim of aggression was the most serious wrong that a state could suffer. In the post-1815 state system, claiming self-defence rose steadily, and then spiked sharply upward after 1875.[17] Since 1945, of course, the United Nations Charter has made self-defence the only valid reason to go to war.[18] This pattern of development clearly shows the disappearance of aggressive war as a publicly accepted way of doing state business.

3. Hague Convention III (1907)

Prior to 1907, positive (written) law did not require states to declare war before taking up arms. In that year the states participating in the Second Hague Peace Conference agreed that signatories were obliged to precede armed conflict by a declaration or its equivalent. This step capped two, long trends in European state life. One was the virtually universal expectation that decision-makers or sovereigns owed the European community an explanation for going to war.

[15] Schroeder, above note 12, at 405, also at 410. See also the similar lists from Louise Richardson and Richard B. Elrod, quoted in Jarrett, above note 10, at 358–61.

[16] This discussion is based on the very helpful work of Hathaway et al., 'War Manifestos', 85 *University of Chicago Law Review* (2018) 1, at 1. These reasons were 58 per cent of the total reasons cited in war manifestos from 1500 to 1945, and among primary reasons given, they made up 79 per cent (my reckoning). For a summary of the first systematic evaluation of early modern war manifestos see Tischer in this volume.

[17] Hathaway et al., above note 16, Figs 5 and 7, at 66, 68.

[18] Apart, of course, from the use of force by the Security Council itself: Art. 42, United Nations Charter, available at <https://www.un.org/en/sections/un-charter/chapter-vii/> (last visited 10 January 2020); see also Thilo Marauhn's chapter.

Few wars between European states at any time in the continent's history had begun with no preceding or nearly simultaneous explanation. These explanations took many forms and performed many functions, depending on the audience. To fellow sovereigns or government leaders, they laid out the grievances or demands that might now be pursued by force of arms. To parliaments, elites, or the general public, they justified coming expenses, effort, and risk. To allies, or potential allies, they set out reasons to join in the war. To neutrals, they announced the shift from peacetime to wartime international rules, rights, and restrictions. The narrowest form of explanation tended to be the declaration of war. At its tersest, it simply registered that a war existed between X and Y as of a given date and time. Even so, declarations of war tell a great deal about how statesmen and publics interpreted the international law governing the right of states to engage in war (*jus ad bellum*).

The second trend lay in state practice since the Congress of Vienna had set up the post-Napoleonic state system based on the principle that peace was (and should be) the default condition among states. As a result of the post-1815 rules, states usually issued declarations before proceeding to war. The international lawyers gathered at the Institute of International Law meeting in 1906 almost all agreed that state practice since 1815 (some said 1848, others, 1856) had moved decidedly to requiring a declaration of war.[19] Charles Dupuis analysed fourteen wars since 1848 and discovered that only four were not preceded by a formal or informal declaration, warning, or ultimatum; and of those four, two involved the disputed Ottoman periphery, one was outside Europe altogether, and the fourth occurred during the revolutionary upheaval of 1848.[20] Albéric Rolin, the reporter at the Institute, concluded that a customary international legal obligation had clearly developed, and it was now necessary to reflect that rule in positive law.[21]

The Institute's draft articles were then debated and passed unanimously as Hague Convention III in 1907. The state representatives could not agree on the exact delay between the declaration and beginning hostilities. But otherwise, the convention was utterly uncontroversial. Hague III received the largest number of signatures (forty-four states) of any of the conventions passed at the Peace Conference, and it was the only general convention approved with no reservations.[22] It said:

> Preamble: 'Considering that it is important, in order to ensure the maintenance of pacific relations, that hostilities should not commence without previous warning;
>
> That it is equally important that the existence of a state of war should be notified without delay to neutral Powers . . . '

[19] *Annuaire de l'Institute de Droit International* 20 (1906), at 30–31, 39–40, 271–72.

[20] Ibid., at 33. Maurice's survey concluded that many wars were not preceded by declaration, but his post-1815 examples include many mere threats of force, or peacetime reprisals, and omit important diplomatic documents: J.F. Maurice, *Hostilities without Declaration of War. An Historical Abstract of the Cases in Which Hostilities Have Occurred Between Civilized Powers Prior to Declaration or Warning. From 1700 to 1870* (1883), 46.

[21] *Annuaire de l'Institute de Droit International*, above note 19, at 33.

[22] Robinson, 'Surprise Attack: Crime at Pearl Harbor and Now', 46 *American Bar Association Journal* (1960) 973, at 978.

> Article 1: 'The contracting Powers recognize that hostilities between themselves must not commence without previous and explicit warning, in the form either of a declaration of war, giving reasons, or of an ultimatum with conditional declaration of war.'[23]

The main reasons that delegates gave for supporting Hague III repeated the 1815 framework: to maintain peace (as the preamble stated) by thwarting aggression. As French General M.F.G.T. Amourel said, 'An aggressive attack in time of ordinary peace and without any plausible motive is no longer compatible with the public sentiment in the nations of the civilized world which we are representing here.'[24] The hope was to encourage arbitration or mediation and avoid war altogether. In the course of the nineteenth century, the mechanisms of arbitration/mediation grew substantially in the form of bilateral treaties pledging signatories to solve conflicts in that way, and culminating in the establishment of the Permanent Court of Arbitration at the Hague Peace Conference in 1899. From 1856 to 1914, mediation had successfully diffused thirty-three potential armed conflicts.[25]

The delegates were not utopians. They were well aware of the subterfuges states might use. They simply aimed to make it easier for 'Every one', as the Commission Report put it,

> to form a judgment on the conduct of the two adversaries. Of course this does not mean that we are to cherish the illusion that the real reasons for a war will always be given; but the difficulty of definitely stating reasons, and the necessity of advancing reasons not well substantiated or out of proportion to the gravity of war itself, will naturally arrest the attention of Neutral Powers and enlighten public opinion.[26]

As we shall see, Imperial Germany had great difficulty stating its reasons; its efforts, however, did indeed make it easier for statesmen and publics to judge why the war was breaking out.

4. Self-Defence, Pre-emption, Prevention, Aggression

Self-defence was the strongest justification in CIL for legitimate war. But how did the nineteenth century define it? The noticeable spike in claims of self-defence after 1875 was directly related to the question whether preventive war was an accepted form of self-defence. Most nineteenth-century European leaders, civil and military,

23 International Committee of the Red Cross website, available at <https://ihl-databases.icrc.org/applic/ihl/ihl.nsf/ART/190-190002?OpenDocument> (last visited 10 January 2020).

24 Amourel (1848–1908). Citation: J.B. Scott (ed.), *The Proceedings of the Hague Peace Conferences; Translation of the Official Texts* (1920), The Conference of 1907, vol. 1, Plenary Meetings of the Conference, at 162.

25 M. Hewitson, *Germany and the Causes of World War I* (2005), at 288, citing J. Dülffer, M. Kröger, and R.-H. Wippich, *Vermiedene Kriege. Deeskalation von Konflikten der Großmächte zwischen Krimkrieg und Erstem Weltkrieg* (1997).

26 Scott, above note 24, at 132.

understood preventive war specifically to mean provoking a war by attacking a power whose strength was thought to be increasing and which would therefore pose a greater threat in a future war.[27] This is the way in which Chancellor Otto von Bismarck used the term when he threatened preventive war against France on at least five occasions between 1872 and 1874.[28] Threatening again in 1875, Bismarck precipitated the 'War-in-Sight' Crisis, in which Europe debated preventive war at length in newspapers and foreign offices. Statesmen resoundingly repudiated it. The rejection of Bismarck's (and Chief of Staff Helmuth von Moltke's) tactics was so strong that the French foreign minister, arguing that 'the tacit acceptance of this principle [the right to wage preventive war] by the other European states would constitute a threat to the security of all', easily convinced Great Britain and Russia to intervene and force Bismarck to stand down, permanently.[29] In no subsequent crisis did the Iron Chancellor ever champion preventive war, which he now deprecated as 'a colossal stupidity', 'suicide for fear of death'.[30]

More tellingly, in his memoirs, Bismarck equated preventive with aggressive war. He wrote (misleadingly) that he had 'always been a principled opponent of preventive wars—that is, offensive wars (*Angriffskriegen*) undertaken because we imagine that we will later have to fight against a better armed enemy'.[31] Kaiser Wilhelm I also defined preventive war as aggression, and indicated that other European states felt the same way. He rejected preventive war in 1875, he said, because 'Germany as an aggressor would possess neither allies, nor be able to count on benevolent neutrals'—a clear statement of the strong international sanction against prevention.[32] In Bismarck's view, even successful wars 'could only be justified, if they were forced on one', that is if one were actually attacked; and in any case, no one 'could read the tea leaves so perfectly as to divine historical development according to one's own wishes'.[33] However defensive the underlying subjective aim (warding off a prospective stronger opponent), preventive war was never self-defence, which required an actual preceding attack.[34] Preventive war was a species of aggressive war.[35]

Once prevention was removed as an acceptable expansion of self-defence, there remained one other way to claim self-defence short of actually having been attacked. That way was pre-emption. A famous incident in 1838 led the United States and Great Britain to set down what they regarded as the CIL on pre-emption. In that year, incursions into Canada by American rabble-rousers led Britain one night to halt their activity permanently by attacking their vessel, the *Caroline*, without warning, setting it aflame, and sending it over Niagara Falls, killing one American. Diplomacy settled the matter short of war and produced an agreement on the customary rules governing

[27] As Strachan notes, Carl von Clausewitz outlined this possibility, Carl v. Clausewitz, *On War*, translated by J. J. Graham, introduction by A. Rapoport and Col. F.N. Maude (1968), bk. 8, ch. 9, at 398.

[28] This account follows K.-E. Jeismann, *Das Problem des Präventivkrieges im europäischen Staatensystem mit besonderem Blick auf die Bismarckzeit* (1957), at 88.

[29] J. Stone, *The War Scare of 1875: Bismarck and Europe in the Mid-1870s* (2010), at 252.

[30] Jeismann, above note 28, at 95 (February 1876).

[31] Cited in ibid., at 155.

[32] J. Janorschke, *Bismarck, Europa und die 'Krieg-in-Sicht'-Krise von 1875* (2010), at 406.

[33] Cited in Jeismann, above note 28, at 70.

[34] H. Shue and D. Rodin (eds), *Preemption: Military Action and Moral Justification* (2007), at 13.

[35] Lesaffer writes that in the nineteenth century 'aggression' became the umbrella term for all forms of war not permitted by CIL: Lesaffer, above note 1, at 796.

pre-emptive self-defence. These were that the pre-emptor must show (1) a 'necessity of self-defence, instant, overwhelming, leaving no choice of means, and no moment for deliberation'; (2) that they 'did nothing unreasonable or excessive; since the act, justified by the necessity of self-defence, must be limited by that necessity, and kept clearly within it'; (3) that 'admonition or remonstrance ... was impracticable, or ... unavailing'; and (4) that lesser measures would have been ineffective, and therefore produced the 'necessity, present and inevitable, for attacking ... '[36] The two most striking features of the rule were, first, that the threat had to be actual (real) and immediate, and second, the permission to act was extremely narrow. Pre-emption was not expected to produce protracted war; rather, the rule envisaged a single, limited operation to relieve a closely defined threat. The *Caroline* involved a threat coming from private persons, not a state, but we shall see that its requirements of real, immediate threat and proportionate response governed the European reaction to German claims in 1914.[37]

Using offensive means for (subjectively) defensive ends, that is for genuine self-defence, was thus severely limited in nineteenth-century European state practice and CIL. Despite that fact, military leaders continued to advocate preventive war. Admiral Sir John Fisher was one.[38] But the most assiduous proponents were in the general staffs of the Central Powers. Austrian Chief of Staff Franz Conrad von Hötzendorf pressed for war to prevent Austria-Hungary's decline at every possible opportunity: 1907, 1909, 1911, 1912, 1913, 1914.[39] The German general staff had a much more institutionalized tic in favour of preventive war which had begun with the wars of unification, received a check in 1875, but smouldered on, kept alive by the operative requirements of quick mobilization and deployment in the eventuality of a two-front war. Even before the Schlieffen Plan, military considerations predisposed the general staff to think in terms of prevention. In 1887–1888, the temptation re-emerged. Bismarck again thwarted considerable pressure from Chief of the General Staff v. Moltke and his quartermaster general, Alfred v. Waldersee. But their reasoning reveals the hair-trigger separating peace from military attack in their worldview: they both favoured launching a war against Russia—which was an ally!—and Waldersee pushed to attack France merely because it had adopted a new, more effective rifle![40]

After 1905 and with increasing urgency after 1911, the general staff again pressed for preventive war, still motivated by the nightmare of two-front war. But these military considerations now operated inside the larger framework of 'world policy'

[36] Daniel Webster to Lord Ashburton, 27 July 1842, cited in Jennings, 'The Caroline and McLeod Cases', 32 *American Journal of International Law* (1938) 82, at 89.

[37] Since 1946, the *Caroline* rules have been amalgamated into the rule for self-defence. Yoram Dinstein helpfully distinguishes 'interceptive' from 'anticipatory' pre-emption. The first fulfils the requirement of necessity because the victim is responding to an attack already launched; the second does not: Y. Dinstein, *War, Aggression, and Self-Defence* (3rd ed., 2001), at 172, 219–20. On the problem of private versus state threat see Ago, 'Addendum to the Eighth Report on State Responsibility', 1 *Yearbook of the International Law Commission* II (1980), at 61 f., 65.

[38] Strachan, 'Preemption and Prevention in Historical Perspective', in H. Shue and D. Rudin (eds), *Preemption: Military Action and Moral Justification* (2007), at 28.

[39] K. Canis, *Der Weg in den Abgrund: Deutsche Außenpolitik 1902–1914* (2011), at 498–99; H. Afflerbach, *Falkenhayn: Politisches Denken und Handeln im Kaiserreich* (1994), at 604. Strachan writes that some authors claim that Conrad pressed for war twenty times before August 1914: Strachan, above note 38, at 31.

[40] Jeismann, above note 28, at 139–44.

(*Weltpolitik*) that Kaiser Wilhelm II had adopted as Germany's foreign policy in 1897. The combination has tended to obscure the geo-political meaning of the general staff's proposed policy. On the one hand, preventive war was designed to foreclose the possibility that by 1917 France and especially Russia would be stronger militarily than Imperial Germany. On the other hand, 'stronger' meant not that either would attack Germany or would actually defeat it, but rather that their strength or swiftness of mobilization would make the Schlieffen Plan, Germany's 'prescription for victory', impossible.[41] That is, overwhelming German victory promised by a lightning offensive annihilating both enemies and thus destroying them as Great Powers would no longer work after 1917. Until then, the general staff estimated that Germany had a good chance at victory, which would perforce have established untrammelled German hegemony on the continent.[42] Therefore, the fear or pessimism motivating Germany's military leaders in 1914 was fear of expansion thwarted, not actual self-defence.[43]

Germany's civilian leaders were more sceptical of preventive war as a suitable method of expansion. Bernhard von Bülow rejected it in 1905, despite his aim of enhancing Germany's 'place in the sun'.[44] His successor Theobald von Bethmann Hollweg preferred 'development' to describe his policy—a purposely vague term that covered the variety of expansionary desires widespread in bourgeois and Reichstag opinion.[45] Regarding war as a means, Bethmann traversed a quick odyssey. In March 1912, he rejected it out-of-hand; by November 1912 he had 'accustomed himself' to war under the right circumstances.[46] By January 1914, the chancellor seems to have embraced Moltke's technical-military preventive war logic; he now thought war had become inevitable because of Russia's railroad expansion.[47] On 4 June, Bavaria's seasoned representative in Berlin, Hugo Graf von und zu Lerchenfeld, reported on

[41] On the Schlieffen Plan and its assumptions: I.V. Hull, *Absolute Destruction: Military Culture and the Practices of War in Imperial Germany* (2005), at 160–81.

[42] Hewitson has seen this contradiction between seeming fear and short-term confidence quite well: M. Hewitson, *Germany and the Causes of World War I* (2005), at chs 6 and 7.

[43] Moltke claimed on 5 August 1914 that he 'knew positively that Russia, England, and France had jointly agreed and prepared for an offensive war [Angriffskrieg] against Germany': (Lerchenfeld to Hertling, Berlin, 5 August 1914 in Ernst Deuerlein (ed.), *Briefwechsel Hertling-Lerchenfeld 1912–1917; dienstliche Privatkorrespondenz zwischen dem bayerischen Ministerpräsidenten Georg Graf von Hertling und dem bayerischen Gesandten in Berlin Hugo Graf von und zu Lerchenfeld* (1973), at Doc. 119. No such plan existed.

[44] In 1913, Bethmann Hollweg used the same phrase to describe his own foreign policy: G. Roth and J.C.G. Röhl (eds), *Aus dem Grossen Hauptquartier: Kurt Riezlers Briefe an Käthe Liebermann 1914–1915* (2016), at 61.

[45] On four occasions from December 1912 to December 1913, the chancellor told the Reichstag that Germany's 'development' must be the goal of foreign policy. In a major address to the Reichstag of 2 December 1914, only the second of the war, he five times used that phrase to describe Germany's aim. See Reichstag Germany, *Verhandlungen des deutschen Reichstages; Stenographische Berichte* (1915) vol. 286, at 2472; vol. 289, at 4512, 4609; vol. 291 at 6275; vol. 306 at 19. On Bethmann's preference for vague and elastic terms, see B. Sösemann (ed.), *Theodor Wolff. Tagebücher 1914–1919; Der Erste Weltkrieg und die Entstehung der Weimarer Republik in Tagebüchern, Leitartikeln und Briefen des Chefredakteurs am 'Berliner Tageblatt' und Mitbegründer der 'Deutschen Demokratischen Partei'* (1984), at 520, entry of 19 July 1917.

[46] Roth and Röhl, above note 44, at 52–53; Fritz Fischer, *War of Illusions; German Policies from 1911–1914* (1975), at 157, 164; Kaiser Wilhelm II note of circa 11 November 1912, J. Lepsius and A. Mendelssohn Bartholdy et al. (eds), *Die grosse Politik der Europäischen Kabinette, 1871–1914. Sammlung der Diplomatischen Akten des Auswärtigen Amtes (GP)* (1922–27), at 33, Nr. 12349.

[47] Wolff, above note 45, at 521–22, entry of 19 July 1917. Bethmann explained his dilemma: 'If war had come after Russia was better prepared—where would that have left us?'

Bethmann's changed position.[48] In these same days, Moltke had again assured Foreign Minister Gottlieb von Jagow that Germany could win a European war if it broke out now, but not later. He strongly recommended that 'policy should be geared to bringing about an early war'. Jagow rejected Moltke's suggestion, but not on principle. Jagow believed that prevention 'could be [considered] a defensive war' if 'a war appears inevitable'—the conclusion that the chancellor had apparently already reached.[49]

This was the situation when, on 28 June 1914, Bosnian-Serb students in league with some high-ranking, conspiratorial, nationalist officers in Serbia assassinated the Austro-Hungarian heir to the throne, Archduke Franz Ferdinand in Sarajevo. Their act set off the last of a series of European diplomatic crises that this time ended in war.

5. The July Crisis (1914)

The July Crisis is probably the most famous, and certainly the best researched, diplomatic puzzle in historiography. The long diplomatic back-and-forth was precisely what the lawyers and Hague delegates had imagined might be the prelude to declarations of war.[50] Despite the extraordinary complexity of the Crisis, the rest of this essay will try to extract just one element—Germany's declarations of war—to see what they reveal about the *jus ad bellum* in 1914.

Unbeknownst even to its ally, the Austro-Hungarian leadership had decided already in October 1913 to use the next opportunity to crush the growing Serb state whose untrammelled nationalism threatened both the territorial integrity of the Dual Monarchy and its faltering, multi-national political status quo.[51] Foreign Minister Count Leopold v. Berchtold explained to Emperor Franz Joseph (7 July 1914) that it was explicitly a preventive war; Austria-Hungary had to act 'because our situation will only become worse if we wait longer'.[52] On 5 July 1914, Germany promised its full support in an Austro-Serb war, regardless of the consequences.

There is no dispute about what military leaders, especially Chief of Staff Helmuth von Moltke (the younger) and War Minister Erich von Falkenhayn wanted in July 1914: preventive war. For them, that meant a continental war against Russia and France (with or without Great Britain—they did not care). From autumn 1913 to the assassination, Germany's military leaders had pressed for preventive war thirteen times.[53] The military was deeply involved in decision-making, and not just at the end

[48] Lerchenfeld to Georg Graf von Hertling, Berlin, 4 June 1914, Doc. 94 in Deuerlein, above note 43, at 295–96.

[49] Jagow's undated (post-war) memorandum printed in Zechlin, 'Motive und Taktik der Reichsleitung 1914', 8 *Der Monat* (1966), at 92, cited in D.C. Copeland, *The Origins of Major War* (2000), at 72; and A. Mombauer, *Helmuth von Moltke and the Origins of the First World War* (2001), at 172. English translation in J.C.G. Röhl, *From Bismarck to Hitler: The Problem of Continuity in German History* (1970), at 70.

[50] *Annuaire de l'Institute de Droit International*, above note 19, at 280; Scott, above note 24, at 132.

[51] Council of Common Ministers, meeting of 3 October 1913, in A. Schmied-Kowarzik (ed.), *Die Protokolle des gemeinsamen Ministerrates der österreichisch-ungarischen Monarchie 1908–1914* (2011), Doc. 44.

[52] Berchtold to Franz Joseph, 7 July 1914, in I. Geiss (ed.), *Julikrise und Kriegsausbruch 1914; eine Dokumentensammlung. Mit einem Vorwort von Fritz Fischer* (1963–64), vol. 1, Doc. 41, at 114.

[53] D. Hoffmann, *Der Sprung ins Dunkle oder wie der 1. Weltkrieg entfesselt wurde* (2010), at 192.

of July when Moltke's priorities usurped policy-setting. Bethmann and Jagow both knew the outline of the Schlieffen Plan; they both had become convinced of key military arguments (the closing window of opportunity); and neither ever questioned the military's expertise. The experts had convinced the chancellor and leading figures in the Foreign Office (AA) that Germany had an excellent chance of winning,[54] because, as they correctly saw, none of the three Entente powers wanted war or was prepared for it in 1914.[55] That fact, however, also refuted the claim of pre-emption. Moltke's own doubts about Germany's chances of success he kept to himself.[56]

Some historians believe that Bethmann and the AA merely risked continental war. The stronger evidence—from contemporary reports, Bethmann's and others' later admissions, and Germany's actions—supports the view that Imperial Germany and its ally aimed positively to unleash a preventive continental war.[57] Preventive war arguments were 'common currency' among civilian leaders; they appeared logical and convincing to the representatives of other German states.[58] On 5 July 5, the Kaiser and his military and civilian advisors all guaranteed their full support for whatever measures Austria might choose. Unbeknownst to his military colleagues (who worried that Austria would be satisfied with a mere diplomatic victory), Bethmann strongly pressed the Austrians to go to war with Serbia, 'even if [Austria's] action would trigger a great war', as Foreign Minister Berchtold told Conrad.[59] It was a near certainty that an actual Austro-Serbian war, as opposed to mere Austrian mobilization or threats of war as in 1912–13, would indeed cause Russia to intervene. German decision-makers had reached that conclusion already in November 1912.[60] German Ambassador Friedrich v. Pourtalès repeated that warning in February 1913, and '[f]rom the very first moment [of the July Crisis] the Russian government took up a firm attitude and made it perfectly clear that they would not allow Austria to crush Servia', as Britain's ambassador reported.[61]

The July Crisis ought to have been settled diplomatically. Mediation/arbitration was the normal method to solve European conflicts; diplomacy had last succeeded in the Balkans the previous year. Instead, German and Austrian policy aimed to thwart diplomacy, first by lulling Europe into a false sense of complacency and then by suddenly

[54] The best account is Mombauer, above note 49, at ch. 4.

[55] G.P. Gooch and H. Temperley (eds), *British Documents on the Origins of the War, 1898–1914: Arbitration, Neutrality and Security* (1926–38), vol. 11, at 194f., 361–63; Hoffmann, above note 53, at 189–90, 219, 224; Geiss, above note 52, 1, at Doc 135; Hewitson, above note 42, at 131–34, 179–80.

[56] Mombauer, above note 49, at 203, 208, 219.

[57] It is impossible to rehearse here the 'Fischer controversy' and the recently reanimated discussion of it in the historiographical literature. Readers will find a full list of works in Roth and Röhl, above note 44, at 17 n. 2, 20 n. 11, 21 n. 13, and 22 n. 17. See also G. Krumeich, *Juli 1914: Eine Bilanz: Mit einem Anhang, 50 Schlüsseldokumente zum Kriegsausbruch* (2014), at 7–14; and Mombauer, 'The First World War: Inevitable, Avoidable, Improbable or Desirable? Recent Interpretations on War Guilt and the War's Origins', 25 *German History* (2007) 78, at 78–95.

[58] 'Common currency': Mombauer, above note 49, at 190. Convincing: Lerchenfeld to Hertling, Berlin, 3 August 1914; Schoen to Hertling, Berlin, 18 July 1914, in Deuerlein, above note 43, at 326, 306; Saxon Military Representative Leuckart to Carlowitz, No. 73/3472, Berlin, 3 July 1914, cited in Geiss, above note 52, 1, at Doc. 15, at 75. Other examples: Roth and Röhl, above note 44, at 69–71.

[59] Berchtold to Conrad, 7 July 1914, cited in Hoffmann, above note 53, at 196.

[60] Roth and Röhl, above note 44, at 58–64.

[61] Pourtalès: Hewitson, above note 42, at 204; George W. Buchanan to Arthur Nicholson, private, St. Petersburg, 3 August 1914, Gooch and Temperley, above note 55, 11, Doc. 665, at 344–46.

launching the Austro-Serb war.[62] The first succeeded. Austrian slowness scotched the second, and confronted both powers with a deluge of mediation proposals from all the Great Powers, including their ally, Italy.

Mediation was an embarrassment because rejection hinted too strongly that preventive war was the real object. Germany's first line of defence was 'localization', the claim that the Austro-Serb conflict was a purely Austrian, not a European matter; mediation was tantamount to calling Austria before 'a European court of justice' (Bethmann, 27 July).[63] However, 'localization' contradicted Austria's own claim of acting as an 'agent of Europe' in punishing Serbia, and it belied the fact that Balkan crises had always been international because they concerned the stability of the Ottoman Empire with all of its strategic implications, and because the interests of two Great Powers, Austria and Russia, overlapped there.[64]

Until 30 July, Germany and Austria held fast. But the constant rejection of negotiation was making the preventive war too obvious. On 28 July, Austria declared war on Serbia and ushered in the final phase of the crisis. As war loomed, Bethmann finally transmitted, but did not recommend, a mediation proposal from Kaiser Wilhelm, which echoed a similar idea by Britain's Foreign Minister, Sir Edward Grey. It called for Austria to 'Halt-in-Belgrade', the Serbian capital, while Europe began international discussions. Some historians credit Bethmann with finally wanting to prevent war after all. But his instructions to German ambassador Heinrich von Tschirschky (28 July) admonished him 'to avoid very carefully giving rise to the impression that we wish to hold Austria back'. Instead, he explained, Germany sought a way to satisfy Austrian demands 'without at the same time bringing on a world war, and, if the latter cannot be avoided in the end, of improving the conditions under which we shall have to wage it'.[65] Improving those conditions became Bethmann's primary goal in the last days of peace. As Austria refused to budge, he wired Tschirschky (30 July), 'If Vienna declines [Grey's latest proposal] ... it will hardly be possible any longer to place the guilt of the outbreak of a European conflagration on Russia's shoulders.'[66]

Launching a successful preventive war confronted Bethmann with several problems. Domestically, Germany could only fight a continental war with the Social Democrats' support—support forthcoming only if it appeared that Germany were defending itself against attack by reactionary Russia. German leaders had agreed since November 1912 that putting Russia in the wrong was a necessary precondition of a successful war and required that Austria appear willing to negotiate while Russia

[62] The best discussion of the manipulative aspect of the plan is in Copeland, above note 49, at chs 3–4.

[63] I. Geiss (ed.), *July 1914: The Outbreak of the First World War; Selected Documents* (1968), at Doc. 96.

[64] 'Eine Art Mandat Europas': Saxon representative Frhr. v. Biedermann to Graf Vitzthum von Eckstädt, Berlin, 17 July 1914, Doc. 4 in A. Bach (ed.), *Deutsche Gesandtschaftsberichte zum Kriegsausbruch 1914; Berichte und Telegramme der Badischen, Sächsischen und Württembergischen Gesandtschaften in Berlin aus dem Juli und August 1914, Im Auftrag des Auswärtigen Amtes* (1937), at 64; and Berchtold to Austrian ambassadors to the five Great Powers and the Ottoman Empire, Vienna, 22 July 1914, The Austro-Hungarian Red-Book (official translation from 'Collected Diplomatic Documents', Cd 7860), in *The Times* (London), *The Times Documentary History of the War* (1914–18), 3, Doc. 8, at 190.

[65] Geiss, above note 63, at Doc. 115.

[66] Ibid., at Doc. 143. To German insiders, Bethmann was quite open in acknowledging that the goal of feigned negotiations was 'to put Russia in the wrong': Lerchenfeld to Hertling, 30 and 31 July 1914, in Deuerlein, above note 43, Docs. 111 and 112, at 319f. The other state representatives repeated that phrase: Bach, above note 64, Docs. 46, 49, at 52.

declined.[67] The Social Democratic sensitivity to international law also reinforced Bethmann's inclination to follow Hague III in declaring war; he intended publicly to use the forms of international law, while covertly starting a war contrary to it.[68] Internationally, Bethmann's dilemma was three-fold.

First, unlike every other European power, Germany could not mobilize its army and still engage in diplomatic negotiation. Russia at first mobilized against Austria to send the diplomatic message that the Serbian crisis was a European issue. When on 31 July impelled by technical reasons, Russia mobilized its entire forces (which included in the districts bordering Germany), that fact still did not mean war. Russia could hold its army at the border indefinitely without striking, as the Tsar, Foreign Minister Sergei Sazonov, and Russian War Minister Vladimir Alexandrovich Sukhomlinov repeatedly assured German representatives. The same was true for France and Austria. Germany's Schlieffen Plan, however, called for immediate incursion into Luxemburg (to destroy rail lines) and then invasion of France through Belgium.

And that was the second problem: in a war allegedly breaking out against Russia, Germany's war plan nonetheless required its armies to head west first. The general staff had scrapped the eastern variant of its offensive plan in April 1913. The Schlieffen Plan meant that, regardless of how successful the chancellor was in convincing German public opinion that it was a defensive war against Russia,[69] it would be almost impossible to line up the declarations of war sensibly, especially if France refrained both from declaring war and from attacking.

Finally, Germany's only military plan meant that it must first attack neither of the two Great Powers at which it aimed, but instead two tiny states whose neutrality was guaranteed by major international treaties to which Germany was itself a signatory and guarantor. The contradiction between military planning and the requirements of civilian diplomacy saddled Chancellor Bethmann Hollweg with the impossible task of solving all four problems at once. Not surprisingly, he failed. The task of cloaking a preventive war inside the form of international law caused Germany's explanations (its declarations of war) to oscillate incoherently between military pre-emption and self-defence against actual attack. Significantly, Germany never publicly admitted to prevention. On the contrary, the AA's aim, as Lerchenfeld reported to Munich, was 'to make Germany not appear the aggressor (*Angreifer*)'.[70]

6. Germany's Declarations of War

Just as the decision (5 July) to back Austria even at the cost of continental war was taken by Germany's civilian and military leaders, together with the Kaiser, the declarations

[67] Kaiser Wilhelm II note of circa 11 November 1912, *GP* 33, at No. 12349.

[68] Hoffmann, above note 53, at 245.

[69] Admiral Georg von Müller (1 August 1914) was pleased that 'the government has managed brilliantly to make us appear the attacked', cited by Röhl, 'Admiral von Müller and the Approach of War, 1911–1914', 12 *The Historical Journal* (1969) 651, at 670. Bethmann's secretary, Kurt Riezler, thought 'the staging [*Inszenierung*] was very good': Riezler to Käthe Liebermann, 22 August 1914, in Roth and Röhl, above note 44, Doc. 6, at 124.

[70] Lerchenfeld to Hertling, Berlin, 3 August 1914, in Deuerlein, above note 43, Doc. 116, at 326.

were discussed and drafted jointly. In fact, Chief of the General Staff von Moltke began the process himself on 26 July by drafting the note to be handed to Belgium. It was not a declaration of war—the general staff fervently hoped that Belgium would not impede German troops on their way to France. Therefore, Moltke's draft tried to lure Belgium to 'come on the side of Germany' (changed by Director of the Political Division of the Foreign Office, Wilhelm von Stumm, to 'observe benevolent neutrality') by promising to quit Belgian territory and guarantee its independence after the war, pay for damages and requisitions, and even compensate Belgium, 'at the expense of France' (added by Stumm).[71] Stumm's addition admitted that Germany's was going to be a war of conquest that would reorganize the map of Europe.[72] Only if Belgium refused, would the two countries be at war. This extraordinary offer was justified by 'reliable information about the intended deployment of French troops on the Meuse at Givet-Namur. That leaves no doubt about France's intention to proceed through Belgium against Germany.' Since Moltke drafted these words six days before France even mobilized, they were obviously a mere cover for the Schlieffen Plan. More important for our purposes, it is also clear that mere purported intentions of foreign powers were sufficient in Moltke's view to justify going to war (pre-emption). While the AA emended parts of Moltke's draft (e.g. removing his claim that British troops would join France in violating Belgium, and Stumm's offer of territorial compensation), it retained the justification by divined intent and shortened the deadline to twelve hours for 'an unequivocal answer'.[73]

Thus, at the beginning of the cascade of declarations, Moltke like Bethmann favoured using international forms for military reasons. At the first (31 July) of four meetings that the military and civilian leadership held to discuss and draft the declarations, Moltke also favoured the declaration of war against Russia, which he considered necessary to launch the Schlieffen Plan.[74] But as he grew more nervous that Russian military preparations would diminish or even cancel Germany's advantage in deployment speed, Moltke argued that no declarations were necessary. At the three subsequent joint meetings (1 August, 2:30 AM; 2 August, 10:00 AM; and the morning of 3 August), he adduced various fake border violations by Russia and France, based on rumours whose accuracy he did not investigate and to which he clung, even after they had been disproved.[75] He argued that these violations meant that war already existed, making declarations superfluous. He seems variously to have thought that (together with Navy Minister Admiral Alfred von Tirpitz and War Minister Erich von Falkenhayn) no declaration was necessary 'before we actually marched into France', or that, if Germany waited, surely France would violate Belgian neutrality before Germany did and cancel out Germany's own impending breach.[76] At base, Moltke

[71] K. Kautsky (ed.), *Die Deutschen Dokumente zum Kriegsausbruch* (*DD*) (1919), vol. 2, at Doc. 376.

[72] Germany's diplomatic preparations in July included offering territory to potential allies, including Romania, Bulgaria, Greece, and possibly Montenegro: Deuerlein, above note 43, at Doc. 104; Kautsky, above note 71, vol. 3, at Docs. 506, 548, 567, 673. Moltke proposed offering territorial inducements to Sweden and Japan: ibid., at Doc. 662.

[73] Kautsky, above note 71, vol. 3, at Doc. 648.

[74] L. Albertini, *The Origins of the War of 1914* (1952–57), vol. 3, at 192, and citing Bethmann Hollweg's memoirs, vol. 1, at 156.

[75] Ibid., vol. 3, at 204–19.

[76] Ibid., vol. 3, at 195, 197.

feared that diplomatic activity risked holding up military operations.[77] He complained to the Kaiser that the AA 'refused to believe that a tremendous avalanche like the one now starting was impossible to stop; they still went on thinking that notes were still of use'.[78] None of the military leaders comprehended Bethmann's domestic and diplomatic reasons to abide at least with the forms of international law. When Bethmann tried to explain that 'without a declaration of war on France he could not present the demand on Belgium', Tirpitz confessed, 'I cannot understand this reasoning.'[79] Moltke snapped at the AA's chief legal counsel, Johannes Kriege, when Kriege tried to explain to him the requirements of international law.[80]

In the end, the military leaders' claims of necessity won the day regarding the timing of the declarations. Just as the ultimata to France and Russia had been sent out simultaneously (about which, more below), Bethmann had intended to declare war on both Entente allies, and then hand Belgium its note, on 1 August. It turned out to be too late to stop Ambassador Pourtalès from handing in the declaration against Russia, though Berlin did not receive confirmation for a day, but Tirpitz managed to delay the Belgian note (1 August); the following day, Moltke stopped the declaration of war against France. Bethmann gave in to military pressure both times, joining Moltke's hope that 'the most possible French violations would pile up'.[81] The result was that the Belgian note preceded the declaration against France (3 August), making Germany's legal claims even less coherent and believable than they were already.[82]

The chancellor and foreign office had envisioned following the format suggested in Hague III: first an ultimatum, then a declaration of war. The ultimata to France and Russia went out at the same time, a fact not lost on Entente statesmen.[83] The ultimatum to Russia (31 July, 7 PM Russian time), which Bethmann drafted himself, demanded that Russia cease all military measures against both Germany and Austria within twelve hours, or else Germany would mobilize, which for the *Kaiserreich* meant launching war.[84] In both the ultimata and the subsequent declarations, Germany was careful to avoid the phrase 'declare war', and instead to say 'considers itself in a state of war', which sounded more innocent.[85] Germany's claim was security, but the inclusion of Russian measures directed against Austria meant that Germany was still demanding that Russia accept Austria's war against Serbia—that is 'localization' and thus an enormous Russian diplomatic loss in the Balkans and beyond.[86] The basic claim,

[77] On 30 July Moltke told his adjutant Major Hans von Haeften that 'Germany can only maintain peace now through a tremendous national humiliation, because any negotiation under the pressure of Russian mobilization equals national humiliation.' J.C.G. Röhl, *Wilhelm II. Der Weg in Den Abgrund 1900–1941* (2008), at 1151.

[78] Tirpitz on Moltke, cited in Albertini, above note 74, vol. 3, at 197.

[79] Ibid., vol. 3, at 197.

[80] Ibid., at 195.

[81] Bethmann told this to the state representatives: Bach, above note 64, at Doc. 87; and Kautsky, above note 71, vol. 3, at Doc. 629.

[82] The AA's claims did not even convince German state representatives. Württemberg's envoy Axel von Varnbüler found them festooned with 'slender, it seemed to me overly slender nuances': Varnbüler to Weizsäcker, Berlin, 2 August 1914 in Bach, above note 64, at Doc. 83.

[83] Albertini, above note 74, vol. 3, at 181 f.

[84] Kautsky, above note 71, vol. 3, at Doc. 490.

[85] A. von Wegerer, *A Refutation of the Versailles War Guilt Thesis* (1930), at 272.

[86] On 2 August, Kaiser Wilhelm minuted his goal of 'freeing the Balkans from Russia forever': Kautsky, above note 71, vol. 3, at Doc. 702.

however, was that Russian mobilization by itself was a sufficient threat to trigger self-defence. No European state accepted that claim. During the Balkan Wars, Austria and Russia had remained mobilized for months without declaring war.

The ultimatum to France (drafted by Jagow) informed Paris of the ultimatum to Russia and demanded that France declare within eighteen hours whether it intended to remain neutral or to join its ally.[87] In a secret addendum, Bethmann told Ambassador Wilhelm Frhr. von Schoen that if, 'unexpectedly', France agreed to neutrality, Germany would occupy the forts of Toul and Verdun as guarantees, promising to return them at war's end. Verdun was the last fortress before Paris; its cession would have stripped French defences naked and offered up the Republic to German noblesse oblige.[88]

Both ultimata required of the recipients that they de facto give up their long-standing military alliance (since 1892), their diplomatic prestige, and in France's case, its security as a Great Power—all of this without war.[89] That is why it was 'unexpected' that either would comply. Moltke, however, continued to fear that they might, and thus ruin Germany's chances of winning the military victory that would guarantee Germany's European hegemony. He reacted to the (incorrect) news that Britain might remain neutral, by muttering 'now it only remains for Russia to bale out as well'.[90] As it happened, Moltke's fears were groundless. Both Tsar Nicholas II and Foreign Minister Sazonov informed Ambassador Pourtalès that Russia could not stop mobilization, and French Foreign Minister René Viviani told Schoen that France would be guided by its 'interests'.[91]

As the deadlines for both ultimata ran out, the next step was to declare war—Bethmann later explained that the ultimata were mere 'formalities' to ensure the war.[92] In the Russian case, the declaration was consistent with the ultimatum. The declaration repeated that Russian measures were 'threatening' to the 'security and existence' of Germany, and, since no reply to the ultimatum had reached Berlin, Germany 'accepted the challenge and regards itself as in a state of war with Russia'.[93] Efforts by Moltke and Falkenhayn to stop the declaration and rely instead on putative border violations came too late. Nevertheless, when Foreign Minister Jagow informed the navy of the state of war, he cited the border violations.[94]

As we have seen, military leaders intruded into the smooth functioning of the step from ultimatum to declaration, holding up the French declaration, mixing up the timing of the French and Belgian notes, and inserting false allegations of border

[87] Ibid., vol. 3, at Doc. 491.

[88] On 1 August, Moltke claimed that, if France surrendered Toul and Verdun—he added Epinal and Belfort—'then its power is broken'., Röhl, above note 77, vol. 3, at 1159.

[89] The Kaiser told the Austrian ambassador that he was 'determined to finish off France [mit Frankreich abzurechnen], which will hopefully succeed'., ibid., vol. 3, at 1163.

[90] Admiral von Müller's diary entry of 1 August 1914, cited in Geiss, above note 63, at 336.

[91] Kautsky, above note 71, vol. 3, at Docs 535, 536, 571.

[92] Wolff diary of 19 July 1917: Wolff, above note 45, vol. 1, at Doc. 578.

[93] Kautsky, above note 71, vol. 3, at Doc. 542. In order to guarantee the outbreak of war, Pourtalès's instructions contained two variants: one for an 'unsatisfactory' reply, and the other for no reply at all. The *White Book* suppressed the full language, presumably because it disclosed too nakedly the aggressive intention, , at 50 n. 2.

[94] Ibid., vol. 3, at Doc. 623.

violations into both. The French case is particularly interesting, because it reveals the real reasons behind the declarations, and the official efforts to hide them from Social Democratic critics and international public opinion.

The first draft of the declaration against France was written prophylactically, before the French reply was known, by A. von Rosenberg, a counsellor in the AA. After hearing from Paris, Bethmann, Jagow, and interestingly, the press chief of the AA, Otto Hammann, heavily edited the draft. It explained that Russian military measures 'threatened German security'. France's 'vague reply' meant that the Republic might join its ally and 'attack Germany from behind'. In the absence of Russian assurances that it would stop its measures, 'Germany cannot leave to France the choice of the moment when the threat to Germany's western border becomes a reality, but must, threatened from two sides, immediately launch its defense.'[95] This is the clearest claim of pre-emptive war (based on military reasoning) in the documents. But it was never sent.

At the morning meeting on 3 August, Bethmann, Moltke (for part of the time), Jagow, and Undersecretary of the AA Arthur Zimmermann decided to suppress the claim of pre-emption. Instead, Jagow's substitute draft followed Moltke's pretexts and accused French troops of various border violations in Alsace, violation of Belgian air-space, and bombing Karlsruhe and Nuremberg. This was all 'false news', as Viviani explained to London.[96] Bethmann removed Jagow's assertion about violating Belgian neutrality. Zimmermann added the conclusion: 'France has thus [on account of the border crossings and bombings] put us in a state of war.'[97] In short, the claim was now self-defence, not pre-emption. The reason for this change is clear. As Bethmann had explained already on the preceding day to Ambassador Prince Karl von Lichnowsky in London, 'It is desirable that England should regard this step [Germany's invasion of Belgium] as an act of self-defence against French threat.'[98] Hence, also, the indirect language: 'put us in a state of war'. The edited second draft arrived in Paris in a garbled version, but nonetheless, Ambassador Schoen correctly pieced together the argument and claimed self-defence.[99] The German ultimatum to Belgium (2 August, 7 PM) cited 'reliable information' of 'the intention of France to march through Belgian territory against Germany'.[100] Doubting Belgium's ability to offer 'an adequate guarantee against danger to Germany', the note said that 'It is essential for the self-defence of Germany that she should anticipate any such hostile attack.' This claim amounted to pre-emption based on putative intent. The next day, however, German Ambassador Claus von Below Saleske shifted the argument. His new assertion reflected Bethmann's acquiescence to Moltke's and Falkenhayn's insistence on holding back the declaration of war on France 'in hope that France [will] attack us [first]'.[101] Below now repeated to

95 Ibid., vol. 3, at Doc. 608.

96 France. Ministère des affaires étrangères, *Diplomatic Correspondence Respecting the War Published by the French Government*, [Gt. Brit. Foreign Office] Miscellaneous. vol. 1914 Gt. Brit. Parliament. Papers by command. Cd. 7717 (1914), at Doc. 146.

97 Kautsky, above note 71, vol. 3, at Doc. 734.

98 Albertini, above note 74, 3, at 202.

99 Kautsky, above note 71, vol. 3, at Doc. 734b. Wilhelm Eduard von Schoen, *Erlebtes; Beiträge Zur Politischen Geschichte der Neuesten Zeit* (1921), at 182–84.

100 Belgium. Ministère des affaires étrangères, *Diplomatic Correspondence Respecting the War*, published by the Belgian government, Great Britain. Foreign Office. Miscellaneous, vol. 1914 (1914), at Doc. 20.

101 Kautsky, above note 71, vol. 3, at Doc. 629.

the secretary-general of the ministry of foreign affairs Moltke's claims about French attacks on Germany, adding that they constituted 'violation[s] of international law, seeing that war had not been declared'.[102] Below said that the supposed violations led 'to the supposition that other acts, contrary to international law, would be committed by France', presumably, a French incursion into Belgium. The AA directed an even more remarkable argument to the British public. It offered as proof of Germany's *bona fides* regarding restoring Belgian territory, the fact that Germany had promised to observe the neutrality of the Netherlands; 'It is obvious that we could not profitably annex Belgian territory without making at the time territorial acquisitions at the expense of Holland.'[103] This assurance rested on practical, military grounds, not on the fundamental legal-diplomatic basis that no European state was permitted simply to annex parts of another, for whatever reason.[104] Sir Edward Grey found this reasoning damning enough that he quoted this document to the House of Commons, as it deliberated on going to war.[105]

Such arguments were utterly unconvincing. Belgian leaders were fully aware that France was honouring its pledge not to invade Belgium first. More important, they knew the international law on neutrality, which forbade a neutral state from permitting through-passage to belligerent troops, as Germany requested. And they also knew that the international guarantee of Belgian neutrality (1831, 1839) was the basis of international recognition of Belgium as a state altogether. Rejecting Germany's 'amazing proposal', as Foreign Minister Davignon called it, was a painful, but easy decision.[106] The Belgian reply labelled Germany's attack 'a flagrant violation of international law'.[107] Belgium explicitly rejected the military reasoning behind it: 'No strategic interest justifies such a violation of law.'

After the German invasion of Belgium (4 August), the *Kaiserreich* tried once more to convince the small country to stop fighting. However, this second offer (9 August) guaranteed only Belgium's territory, not its independence, and the earlier promise to pay for damages and requisitions had disappeared.[108] This note, too, was easy to reject.

Alone among the four countries which Germany invaded in August 1914, Luxemburg received no preceding diplomatic note. Neither the chancellor nor his diplomats had any inkling that Moltke's emended Schlieffen Plan called for seizing the Luxemburg railway on the first day of mobilization, even before entering Belgium. Kaiser Wilhelm signed the mobilization order at 5 PM on 1 August.[109] Just three hours later, a small detachment of German troops entered Luxemburg and tore up the tracks at Trois-Vièrges (Ulflingen). Early on the morning of the 2 August, the main force of troops occupied the entire country. Foreign Minister Davignon entered an 'energetic

102 Belgium, above note 100, at Doc. 21.

103 Ibid., at Doc. 36.

104 The statement referred to the Dutch province of Limburg, which protrudes to the south and thus hindered a broad, German sweep westward into France. Earlier versions of the Schlieffen Plan had proposed violating Dutch neutrality for this reason.

105 Grey's speech to Parliament consisted largely of reading diplomatic documents; see I.V. Hull, *A Scrap of Paper: Breaking and Making International Law During the Great War* (2014), at 38f.

106 Belgium, above note 100, at Doc. 38.

107 Ibid., at Doc. 22.

108 Ibid., at Doc. 60.

109 Mombauer, above note 49, at 219.

protest' against 'the flagrant violation of [Luxemburg's] neutrality', which was guaranteed by Germany (Prussia) in 1867.[110] Moltke drafted a possible reply for the AA in which he cited 'necessity' (*Zwangslage*) and 'French reports that French troops are marching on Luxemburg'.[111] At 11:30 AM, hours after the main attack, the chancellor instead sent a similar message of his own denying hostile intent; the measures were meant 'for security against French attack', and Germany promised payment for damages. Less than three hours later, Jagow telegraphed Prime Minister Paul Eyschen, reverting to Moltke's language about 'reliable information that French troops are marching on Luxemburg. We had to take measures to protect our army and secure the railroads.'[112] All of these documents claimed pre-emption. Eyschen replied that there were no French troops in Luxemburg, no threat of such, and indeed the French had destroyed the rail tracks on their own side of the border, proving they had no intention whatever of disturbing the Grand Duchy's neutrality.[113] However, having virtually no troops at its disposal, Luxemburg bowed to the inevitable.

In February 1918, Bethmann Hollweg admitted privately that, 'Yes, by God in a certain sense it was a preventive war.' He went on to describe the preventive schema of the general staff: 'But if war was hanging over us anyway, if it had come in two years even more dangerously and unavoidably, and if the military says, now it's possible without losing, but in two years no longer! Yes, the military!'[114] This was a backhanded admission that Bethmann himself had followed the military's logic and launched war, not just risked it, not to defend against an actual threat or attack, but in order to have the greatest chance of military victory. But as we have seen, none of Germany's diplomatic documents cited preventive war. Germany's civilian leaders knew that since 1875 European states regarded preventive war as aggression and thus unlawful. AA press director Hammann furiously denied rumors that 'Germany wanted a preventive war and had used the present circumstances to begin one'. 'One must repeat, Russia alone is forcing Europe into war.'[115]

Instead, Germany argued two versions of self-defence. The logic of the ultimata and most of the behind-the-scenes diplomatic exchanges was pre-emption. Germany argued that Russian mobilization, even against Austria, was an existential threat. France could join its ally at any time. Reliable news foretold impending French attacks on Belgium and Luxemburg. But, it seems clear, even Imperial Germany suspected that the Great Powers would not accept such argumentation. Self-defence against actual attack was obviously a stronger claim, so Moltke instrumentalized rumours of border violations to make it appear that Russia and even France had attacked the *Kaiserreich*. Moltke's ulterior motive was to cut through the Gordian knot of diplomatic red tape and launch operations while Germany still enjoyed an advantage. But both Bethmann and Jagow accepted his claims and the stronger legal protection they seemed to afford.

[110] Kautsky, above note 71, vol. 3, at Doc. 644.
[111] Ibid., vol. 3, at Doc. 639.
[112] Jagow wrote his message himself: ibid., vol. 3, at Doc. 649.
[113] Ibid., vol. 3, at Doc. 730.
[114] Cited in F. Fischer, *Krieg der Illusionen; Die deutsche Politik von 1911 bis 1914* (1969), at 671.
[115] Hammann to Ballin, 1 August 1914 in Kautsky, above note 71, vol. 3, at Doc. 572.

To bolster its case, Germany published selected diplomatic documents for public consumption. The *German White Book* was presented on 4 August 1914 to the Reichstag and thereafter made widely available.[116] The subtitle proclaimed its mostly domestic purpose, to show *How Russia Went Behind Germany's Back and Unleashed the War*. It held Russia indirectly responsible for Serbia's threat to Austria-Hungary, and finally blamed Russian mobilization for cutting short diplomatic talks. Though it labelled Russian military measures as 'menacing the security of our country', it did not admit to Germany's own declaration of war. Instead, 'Russian troops crossed our frontier ... Thus, Russia began the war against us.' 'On the morning of the next day, France opened hostilities.'[117] In short, it claimed self-defence.

Although he characterized it as a gloss on the *White Book*, Bethmann's own speech to the Reichstag on 4 August was more revealing of the confused argumentation justifying Germany's four wars.[118] He began by claiming a military version of pre-emption. Regarding Russian mobilization, he asked the Reichstag rhetorically, 'should we wait until the powers, between whom we are wedged in, choose the moment to attack?' The chancellor then shifted to self-defence, apparently feeling that Germany's attack in the west was more difficult to justify. He used the only quotation of his speech to cite extensively from Moltke's report of alleged border violations. He concluded: 'Gentlemen, we are now in a condition of necessary self-defence *(Notwehr)*; and necessity knows no law.' Necessity was therefore the basis of his next admission: 'Our troops have occupied Luxemburg and have perhaps already entered Belgian territory.' As critical jurists later pointed out, *Notwehr* was out of the question, since neither Belgium nor Luxemburg had attacked or even threatened Germany. In German civil law the concept of *Notstand* existed to excuse harm done to innocent third parties in the course of a legitimate effort to defend oneself, but that concept did not exist in international law.[119] Press secretary Hammann inserted the next passage in Bethmann's speech, in order to mollify the Social Democrats: 'Gentlemen, that [the incursions into Luxemburg and Belgium] contradicts international law.'[120] But the chancellor reverted to pre-emptive logic: 'we however knew that France was ready to attack. France could wait, but we could not.' Germany was 'forced' to override the neutrality of both countries. 'This injustice [*Unrecht*]—I speak openly—this injustice that we are doing, we will seek to make good as soon as our military goal is reached.' It could hardly be clearer that the 'necessity' at the bottom of this train of arguments was military necessity, not self-defence or even pre-emption. Worse,

[116] Germany. Auswärtiges Amt, *Das Deutsche Weissbuch; Wie Russland Deutschland hinterging und den Europäischen Krieg entfesselte* (1914); English translation available in *The Times* (London), *Times Documentary History* (1914–1918), vol. 2: 123–64.

[117] *German White Book*, in *The Times* (London), *Times Documentary History*, above note 116, vol. 2, at 123f., 134f.

[118] Germany, above note 45, Legislaturperiode XIII, 2. Session, vol. CCCVI (1914), at 5–7.

[119] Hull, above note 105, at 43–47.

[120] Hammann: Auswärtiges Amt, Politisches Archiv, Nachlaß Jagow, vol. 8, pt. 1, 'Politische Aufsätze', 'Der Durchmarsch durch Belgien', at fols 52f.

military necessity, a concept within the laws of war (not *jus ad bellum*), was being adduced *before* war had broken out.

7. Conclusion: Judging the Law via International Reaction

There were many reasons why other European states doubted Germany's (and Austria's) *bona fides*—claiming localization for an obviously European issue; adamant rejection of mediation; Austria's harsh ultimatum, and Germany's complete support of its ally while claiming it knew nothing of Austrian plans; and both nations' rush to war, among others. Nevertheless, the unusual language of seasoned diplomats makes clear how far removed from normal diplomatic and international-legal expectations Germany's actions and justifications were. Localization seemed to them a 'sophism'; Austria's ultimatum left them 'dumbfounded'; Germany's mobilization 'amazed' them; its ultimatum to Belgium, that 'amazing proposal', left the Dutch foreign minister unable 'to hide my astonishment'; the political director in the French foreign ministry simply could 'not comprehend' Germany's ultimatum to Russia.[121] In a 'personal and private' conversation on 26 July, French political director Philippe Berthelot-de Margerie told Ambassador Schoen that 'to any simple mind Germany's attitude was inexplicable if it did not aim at war'.[122] Italy, Germany's and Austria's ally, concluded the same thing. It judged the Austro-Serbian war to have 'an aggressive quality, even according to the words of the German ambassador', which contravened the defensive nature of the Triple Alliance.[123] Italy remained neutral, it announced, on account of the 'voluntary aggression initiated by' Germany and Austria.[124]

In London, the Foreign Office had concluded 'that the German and Austrian plan of a general war was definitely made some time ago and is now deliberately carried out, this being considered the best moment for German success'.[125] That is, it was a preventive war, and therefore aggression. The argument of pre-emption remained unpersuasive because none of the other European Great Powers believed, as Germany's

[121] Sophism: Philippe Berthelot-de Margerie to ambassadors in London, Berlin, Rome, Vienna, Constantinople, Belgrade, 29 July 1914 in France. Ministère des affaires étrangères, *Diplomatic correspondence*, above note 96, Doc. 85, at 100. Dumbfounded: Yov. M. Yovanovitch to Prime Minister Nicholas Pašić, Nish, 16 August 1914 in Serbia and Ministry of Foreign Affairs, *Collected Diplomatic Documents (1914)*, in *The Times*, above note 64, vol. 2, at Doc. 52, at 117. Amazed: Ambassador Alexander Izvolsky to Foreign Minister Sazonov, Paris, 1 August 1914, in Russia, *Documents Respecting the Negotations Preceding the War Published by the Russian Government*, in ibid., vol. 1, Doc. 74, at 491. Amazing proposal: Belgian Minister of Foreign Affairs J. Davignon to ambasssadors in Paris, London, and St. Petersburg, Brussels, 4 August 1914, in Ministry of Foreign Affairs Belgium, 'Diplomatic Correspondence Respecting the War (1914)', in ibid., vol. 2, Doc. 38, at 42. Hide my astonishment: Baron Fallon to Minister of Foreign Affairs Davignon, 10 August 1914, The Hague, citing Jonkheer Loudon, Belgium, 'Belgian Grey Book', in ibid., vol. 2, Doc. 62, at 57. Do not comprehend: Ambassador Francis Bertie to Edward Grey, No. 111, 1 August 1914, Paris, citing de Margerie, in Gooch and Temperley, above note 55, vol. 11, at 254.

[122] France, *Diplomatic Correspondence*, above note 96, Doc. 57, at 68.

[123] Enclosure to the communication of the French embassy, in Gooch and Temperley, above note 55, vol. 11, Doc. 406, at 244.

[124] Rennell Rodd to Grey, No. 299, 4 August 1914, Rome, in ibid., vol. 11, Doc. 668, at 348.

[125] Ambassador Walter Hines Page to Bryan, London, 3 August 1914 in Department of State United States, *Papers Relating to the Foreign Relations of the United States; 1914–1918; Supplement: The World War* (1928–33), Supplement 1914, at 38.

Schlieffen Plan forced it to do, that 'mobilization means war'. Thus, mobilization without incursion was not justification for attack. And the argument from self-defence was risible. The Italian foreign minister rejected as 'absurd' Germany's contention that French border violations justified war.[126] 'There would always be irregularities on the frontier. It was perfectly clear on which side the provocation had been .... ' Moltke's own draft declaration to the Netherlands had admitted that some violations of the Dutch border were possible, but called them 'accidental'.[127] The preeminent student of the July Crisis, Luigi Albertini, emphasizes how remarkable it was that Germany would resort to 'trivial border incidents as pretexts' for a major world war.[128] Germany ultimately gave up trying to find 'proof' of its allegations.[129]

The July Crisis confirms the consensus of opinion among European states that, on the continent, preventive war was illegal, pre-emptive war was severely restrained, and genuine self-defence—meaning both fending off armed attack against one's territory, independence, or sovereignty, and defending the treaty-structure that guaranteed the state order—was the only justification for war acceptable to the community of states. Imperial Germany's struggles with its declarations of war show that its civilian leaders were well aware of these rules and went to desperate (and transparent) efforts to cover up its aggressive actions. Despite agreement among its civilian and military decision-makers that the *Kaiserreich* 'needed' to go to war for its own future development (euphemized by many as 'security'), it nonetheless never actually made that argument; it never publicly admitted to 'preventive war in a certain sense'.[130] It is unclear whether any German leader was aware of the consequences that German policy would surely have for the European state order and legal system, or even if they considered the problem. With a less divided government, the pre-war *Kaiserreich* might have made a better case for itself. But it is hard to see how it could have made one that would not have violated the basic political *and* legal foundations of post-1815 Europe. Those foundations had evolved and strengthened by 1914, particularly regarding the evolution of international law and institutions for arbitration and mediation; but Imperial Germany's assumptions, foreign policy methods, and sense of self had diverged from the European consensus. As Fritz Fellner writes, 'a deep alienation had grown up between Germany and the other Great Powers'; 'German statesmen stood in contradiction to their times.'[131] He understands that contradiction as moral, but not yet legal. This view would be true if one considers only codified international law; not until 1928 did European states sign the Kellogg-Briand Pact outlawing war in international relations.[132] But diplomatic

[126] Rodd to Grey, No. 160, 3 August 1914, Rome, in Gooch and Temperley, above note 55, vol. 11, Doc. 579, at 308.

[127] Kautsky, above note 71, vol. 3, at Doc. 674. The German Foreign Office removed the section on potential border violations.

[128] Albertini, above note 74, vol. 3, at 212.

[129] Bundesarchiv Berlin-Lichterfelde, 'Die Verletzung der Neutralität Belgiens aus Anlaß des Krieges, 1914–1920', R 901 86587–86593.

[130] See Fischer's excellent account of the disappearance of the phrase 'preventive war' from German public and official discourse after September 1914: Fischer, above note 46, at 462–68.

[131] Fellner, 'Die "Mission Hoyos"', in H. Maschl and B. Mazohl-Wallnig (eds), *Vom Dreibund zum Völkerbund; Studien zur Geschichte der internationalen Beziehungen 1882–1919* (1994) 112, at 133.

[132] The latest account: O.A. Hathaway and S.J. Shapiro, *The Internationalists: How a Radical Plan to Outlaw War Remade the World* (2017).

institutions, nineteenth-century state practice, and the expectations of European statesmen (including Germany's civilian leadership) all show that a state's 'right' to resort to arms in Europe had been significantly limited in CIL to genuine self-defence against actual or truly imminent, not imagined or theoretical attack. Preventive war, in whatever sense, was aggression.

11
Salvation through War? The Ottoman Search for Sovereignty in 1914

Aimee Genell and Mustafa Aksakal

1. Introduction

> Only the nation armed with national feelings can participate in the struggle and gain as a result of it the right to remain alive. But let us not spend much time on the word 'right,' for there is no one left who does not believe that 'right' is nothing other than 'might.' The most obvious truth is that those who do not crush will be crushed. The inevitable place of those who do not heed this proverb is the cemetery and history. Thus in order to live we must not strive for 'right' but for power ... Right can only be derived from power, civilization only from power, happiness only from power. Power is everything.[1]

Why did the Ottomans not stay out of the First World War? It is not far-fetched to imagine that, instead of entering the war on 29 October 1914, the empire's leaders might have seized upon this intra-European conflict as a moment of respite. By the first decade of the twentieth century, after all, European settlement or control worldwide extended to some 85 per cent of the globe's surface, rendering the Ottoman Empire one of the last holdouts. China, the Ottomans' fellow agrarian empire, had been partitioned into spheres of influence under the Open Door agreements, Iran had gone a similar way shortly thereafter, and the Ottomans themselves had been subjected to economic and political restrictions imposed by the European Concert. Swimming in a sea of military crises, the Ottomans may have come up for air and taken a reprieve from a century of defeats and territorial losses whose most recent chapters included the Austro-Hungarian annexation of Bosnia-Herzegovina (1908), the Bulgarian declaration of independence (1908), the Italian grab for Tripoli in Libya and the Dodecanese islands (1911), and the attack by the Balkan League and ensuing wars (1912–13). With such a record of defeat, it seems, the leadership should have opted for less war, not more.

The generation at the helm of the Ottoman state, however, welcomed the July Crisis not as a pause from international politics but as an opportunity to end the empire's diplomatic and military isolation. Whereas the Ottoman leaders of the nineteenth

[1] 'İntikam Duygusu [Feelings of Revenge]', *Büyük Duygu* [The Great Longing], Sayı 2, 16 Mart 1329 [29 March 1913], at 17.

Aimee Genell and Mustafa Aksakal, *Salvation through War? The Ottoman Search for Sovereignty in 1914* In: *The Justification of War and International Order*. Edited by: Lothar Brock and Hendrik Simon, Oxford University Press (2021).
 DOI: 10.1093/oso/9780198865308.003.0011

century had launched a broad programme of reforms in their search for a place in the European Concert, by the twentieth century a new crop of radical leaders had taken charge of the state, first in June 1908 in the so-called Young Turk Revolution, and then more firmly in the January 1913 *coup d'etat*, the March on the Sublime Porte. Many of these new leaders came from borderlands, from territories now lost, and they believed that the survival of their state in the international system driven by the European Great Powers rested on the creation of a 'nation in arms' capable of going to war at the necessary moment. During the July Crisis, with the stock of all neutrals rising rapidly, they were able to trade *promises* of military intervention for immediate military aid and a long-term alliance with the Great Power of their choice, Germany. The Ottoman leaders were not expansionists eager to follow the Germans into the trenches, but they were fully capable of responding to an opportunity, however risky, when they saw one. They viewed the German alliance not as a ticket to the battlefield in a war they expected to be over within months, but as their key to international security, the first step towards regaining sovereignty—in practice and not just theory—and thus full membership in the European system. However anomalous in the context of the ambitions of the parties to Europe's alliance system—and contrary to what it has come to stand for in the historiography—the secret 2 August agreement with Germany was conceived by Ottoman policy-makers as an instrument of peacetime, *post-war* diplomacy.

Today the process that led to the Ottoman entry into the First World War lies buried under various heaps of rubble: decades of inaccessibility of the Ottoman archival material; the (apparent) paucity of first-hand sources documenting the decision-making process; an entrenched historiography that celebrates the birth of the Turkish nation-state while taking for granted a catastrophic mismanagement by their Ottoman predecessors; and, finally, a high-stakes politicization of the late Ottoman period, especially scholarship on policies towards Christian minorities, of which the Armenian case is emblematic—and one whose legacy continues to affect Turkey's domestic politics and foreign relations. Here we examine the international environment in which the Ottomans found themselves, or, more accurately, in which they *thought* they found themselves, in 1914. The main Ottoman justification for war, as we shall see, was to reassert sovereignty domestically and to attain a position of equality within the European state system.

2. The Ottoman Empire in the International Order before 1914

In the aftermath of Italy's unprovoked invasion of Libya in 1911,[2] Ottoman politicians and writers increasingly regarded international law as a scheme rigged in Europe's favour, without moral force and impractical as an instrument to preserve the empire. Yet this was a drastically new position. For much of the Hamidian period (1876–1909) and even the early part of the Second Constitutional Era (1908–1918), the Foreign Ministry had relied on international law, especially the 1856 Treaty of Paris, to assert

[2] See also the chapter by Thomas Hippler in this volume.

territorial rights. Despite the fact that the Great Powers treated the empire as an unequal partner within the international arena, officials in Istanbul employed international law as a weapon of the weak to shore up sovereignty in negotiations with European states from the 1850s down to the Balkan Wars. How did international law come to play such an important role in Ottoman–European relations?

The Ottoman Empire was very much part of the European state system in the early modern era—the empire made war and peace, conducted diplomacy and trade, including law governing warfare and the exchange of prisoners.[3] Starting in the middle of the nineteenth century, military weakness, accompanied by civilizational arguments about non-European state capacities, relegated the Ottomans to the margins of the European system. At the same time, the Great Powers gradually imposed a variety of legal, administrative, and financial restrictions, which limited the full exercise of sovereignty within the Ottoman Empire—the most pernicious of which were the Capitulations, but also included European-imposed autonomous provinces such as in Egypt and Mount Lebanon.

Dating back to the sixteenth century, the Capitulations were privileges the sultan had extended unilaterally to European merchants trading within the empire. By the nineteenth century, however, these privileges had been incorporated into bilateral treaties governing the relations between the empire and nineteen other states; for Istanbul these former privileges had turned into a derogation of sovereignty.[4] The Capitulations not only exempted foreign subjects from taxation, search of domicile, and conscription, but also local law.[5] European consular courts and offices sprang up in cities across the empire. What's more, Europeans extended their capitulatory privileges to Ottoman subjects, known as *protégés*, who were employed by embassies and consulates. This system was rife with corruption—Ottoman subjects who wished to evade taxation and local law could purchase *protégé* status from European consuls.[6] This not only deprived the state of significant tax revenue but also undermined the authority and sovereignty of Istanbul. Europeans, meanwhile, justified the Capitulations by arguing that the Ottomans lacked the capacity to carry out justice according to European standards.

The 1856 Treaty of Paris, which concluded the Crimean War, explicitly admitted the Ottoman Empire in Article 7 'to participate in the advantages of the Public Law and System *(Concert)* of Europe'. The same article recognized the empire's independence and committed the Great Powers to guarantee the Ottomans' 'territorial integrity'—a phrase that would be repeated in Ottoman–European treaties and negotiations down to 1923.[7] But even in the moment where it seemed that the first non-Christian state had been invited to enter the European Concert, equality and 'non-intervention' in

[3] W. Smiley, *From Slaves to Prisoners of War: The Ottoman Empire, Russia, and International Law* (2018).

[4] W. Hanley, *Identifying with Nationality Europeans, Ottomans, and Egyptians in Alexandria* (2017), at 11–13; M. van den Boogert, *The Capitulations and the Ottoman Legal System: Qadis, Consuls and Beraths in the 18th Century* (2005), at 1–15; Ahmad, 'Ottoman Perceptions of the Capitulations, 1800–1914', 11 *Journal of Islamic Studies* (2000) 1, at 2–4.

[5] Ahmad, above note 4; Özsu, 'Ottoman Empire', in B. Fassbender and A. Peters (eds), *The Oxford Handbook of the History of International Law* (2012) 429.

[6] Hanley, above note 4, ch. 9; van den Boogert, above note 4, chs 1–2.

[7] E. Hertslet, *Map of Europe by Treaty* (1875–91), at 1250–65. In particular, see Article XII, at 1254 f.

Ottoman affairs were qualified and made dependent on the sultan's guarantee to ameliorate the condition of the Christian populations of the Ottoman Empire.[8] Starting in the 1870s, European international lawyers debated whether the empire had been invited to join the 'European family of nations' too soon. James Lorimer, the Scottish international lawyer, was particularly brutal in his assessment: 'In the case of the Turks we have had bitter experience of the consequences of extending the rights of civilisation to barbarians who have proved incapable of performing its duties.'[9] By that time, however, Ottoman legal advisors and Foreign Ministry officials had already used the language of territorial guarantees and non-intervention to defend the imperial domains against the Great Powers. While some European lawyers and diplomats argued that the Ottoman Empire was not fully equal, Ottoman legal experts appealed to international law and pointed to the long history of treaties signed between European states and themselves as evidence of sovereign equality.[10] For instance, the Ottoman Foreign Ministry responded to the Russian declaration of war in 1877 by arguing that it was Russia who was destabilizing the European order by ignoring the Treaty of Paris:

> History will take note of this astounding fact, that in this age of enlightenment, of civilization, and of justice, a great Power has carried into a neighbouring State fire and sword, because that State demanded that international engagements, that international law, the independence of internal administration, and, above all, the honour and dignity of its Sovereign, should be respected in her case, as it respected them in that of other States.[11]

In the aftermath of a series of political and financial catastrophes in the 1870s and early 1880s, the Ottoman Empire formally incorporated international law into its diplomatic apparatus as a strategy to prevent further European intervention. The Russo-Ottoman War[12] of 1877–1878 resulted in territorial losses in the Balkans. More important perhaps, the war created a massive refugee crisis. Muslim refugees from the Balkans and the Caucasus arrived in the empire and dramatically altered the religious and ethnic composition of the state. In 1881, the European-controlled Ottoman Public Debt Administration (*düyûn-u umûmiye-i osmaniye varidât-ı muhassasa idaresi*)

[8] See Reform Edict of February 1856 (*ıslahat fırmanı*) and Treaty of Paris, Article IX, ibid., at 1255. A number of Ottoman historians of international law have considered how the Treaty of Paris shaped Ottoman–European relations. See Özsu and Skouteris, 'International Legal Histories of the Ottoman Empire: An Introduction to a Symposium' and especially Aral, 'An Inquiry into the Turkish 'School' of International Law', Hanley, 'International Lawyers without Public International Law: The Case of Late Ottoman Egypt,' all in 18 *Journal of the History of International Law/Revue d'histoire du droit international* (2016); see also Palabıyık, 'International Law for Survival: Teaching International Law in the Late Ottoman Empire (1859–1922)', 78 *Bulletin of the School of Oriental and African Languages* (2015) 271; Esmeir, 'On Becoming Less of the World', 8 *History of the Present* (2018) 1.

[9] J. Lorimer, *The Institutes of the Law of Nations: A Treatise of the Jural Relations of Separate Political Communities* (1883-84), vol. 1, at 120.

[10] Genell, 'Ottoman Autonomous Provinces and the Problem of "Semi-Sovereignty" in International Law', 18 *Journal of Balkan and Near Eastern Studies. Special issue: 'Autonomy and Federation in the Ottoman Empire'* (2016) 6; Smiley, above note 3.

[11] 'Turkish Manifesto in answer to Russian declaration of War, Constantinople, 26 April, 1877', in Hertslet, above note 7, vol. 4, Doc. No. 495, at 2601.

[12] For Russian justifications of this war as a 'humanitarian intervention', see Simon and Robinson/Antonov in this book.

began operations to collect debt payments on behalf of European creditors—it later financed major infrastructure projects in the empire that were controlled by European investors.[13] These affronts to domestic sovereignty were further compounded by the French protectorate established in Ottoman Tunisia in 1881 and the British military occupation of Egypt in 1882.

In response to these disasters, the Foreign Ministry established the Office of Legal Counsel (*istişare odası*) to deal with international legal questions.[14] The Office of Legal Counsel, which was headed by two chief legal advisors, wrote opinions that became the foundation of negotiations with Europe. Although it cuts against the grain of older histories of the Eastern Question—European intervention into Ottoman affairs—this was a strategy that worked surprisingly well. Even as they established a string of permanent military occupations across the Eastern Mediterranean, European powers had to find new ways of justifying intervention in Ottoman affairs. For example, during the British invasion and occupation of Ottoman Egypt in 1882, the British War Office and Cabinet had a lengthy debate about whether or not laws of war applied to Egypt. The British Cabinet determined that there was no way around considering Egyptians as anything other than Ottoman subjects. The Cabinet then instructed the War Office 'to treat rebels [in Egypt] according to recognized rules of civilized warfare, including exchange of prisoners'.[15] Egyptians, in other words, had to be treated as Ottoman subjects, and as such, at least in theory, were entitled to the protections of international law.

In addition to using international law to protect the boundaries of the state, the empire embarked upon a series of domestic legal reforms.[16] These reforms aimed at ending European restrictions on Ottoman sovereignty at home. During the great imperial reordering of the mid-century known as the *tanzimat*, reformers sought to present Ottoman legal and political institutions in language and form more easily accessible to European audiences. They took up an ambitious project of codifying Islamic law into a new Civil Code (*mecelle*), completed in 1876. The Imperial Law School opened in 1874 and offered courses in Islamic legal theory as well as international law.[17] Such institutions paved the way for the creation of the Office of Legal Counsel as more and more trained lawyers staffed it. Reconciling legal norms with European conventions gave ammunition to the Ottoman Foreign Ministry to argue that the Capitulations and other privileges for Europeans—and Ottoman subjects enjoying legal status as Europeans—should be repealed, allowing the state to reclaim its authority throughout the empire. The First World War provided the first real opportunity to abrogate the

[13] M. Birdal, *The Political Economy of Ottoman Public Debt: Insolvency and European Financial Control in the Late Nineteenth Century* (2010).

[14] The Ottoman Empire was the first state in Europe to incorporate international lawyers into their Foreign Ministry—Germany, France, the United Kingdom, and the United States created similar offices in the late 1880s and 1890s. See Genell, 'The Well-defended Domains: Eurocentric International Law and the Making of the Ottoman Office of Legal Counsel', 3 *Journal of Ottoman and Turkish Studies* (2016) 2.

[15] British National Archives hereafter *NA/CAB/37* General Mediterranean Telegrams. 21 August 1882. To Adm. Seymour, Port Said.

[16] Rubin, 'Modernity as Code: The Ottoman Empire and the Global Movement of Codification', 59 *Journal of Economic and Social History of the Orient* (2016) 828.

[17] E. İhsanoğlu, *Darülfünun: Osmanlı'da Kültürel Modernleşmenin Odağıı* (2010); Bedir, 'Fikh to Law: Secularization through Curriculum', 11 *Islamic Law and Society* (2004) 378.

Capitulations, which the ruling regime, the Committee of Union and Progress, did on 9 September 1914—a move not recognized by the Allies until 1923.

During the Second Constitutional Era (1908–1918), critics of Sultan Abdülhamid II cast his reign as an era of despotism and unmitigated failure in the international arena. It is true that those years witnessed catastrophic defeat at the hands of Russia in 1878 and subsequent British occupation of Cyprus and Egypt, as well as massive ethnic violence against Ottoman Armenians along the Russian border in the 1890s. And yet, over the course of Abdülhamid II's thirty-three-year rule, the empire achieved relative stability within the international system. It saw the expansion of railway and telegraph networks, innovations in agriculture alongside the beginnings of industrialized manufacturing, an increasing number of schools opening their doors to students, and ever-widening international trade and commercial relations. While Europeans continued to intervene in Ottoman domestic affairs, their actions had to be justified in new ways that adhered to the text of existing treaties. European governments were generally hesitant to recognize Ottoman territorial losses. The Second Constitutional Era, in contrast, opened with immediate vulnerability for the empire. The Great Powers refused to prevent and later condemn Austria-Hungary for annexing Bosnia and Herzegovina, nor Bulgaria for declaring independence. The years following the Young Turk Revolution between 1908 and 1914 were marked by political instability at home and Great Power unwillingness to honour international legal commitments. Ottoman lawyers and Foreign Ministry officials initially fought tooth and nail to obtain European territorial guarantees to shore up sovereignty. But in the years leading up to the First World War, after decades of undermining of Ottoman sovereignty, European declarations to recognize the 'territorial integrity' of the empire were now regarded as little more than empty promises without force.

3. Libya and the Laws of War

Italy's wanton aggression in Ottoman Libya transformed the role of international law in the empire's relations with Europe. When Italy bombarded the port of Tripoli in 1911 and began its unprovoked year-long war, Istanbul's diplomatic response was largely crafted by the members of the Office of Legal Counsel. On the eve of Italy's invasion, Tevfik Pasha, the Ottoman ambassador at London, repeatedly requested that the British government intervene to prevent war. Once Rome launched its attack, he issued a plea to the Great Powers to end the war 'in the name of civilization' and for the security of Europe.[18] Britain declined to apply the brakes—even though British diplomats admitted privately that the Italian action would place Britain in 'a very embarrassing position' precisely because Britain had guaranteed the territorial integrity of the empire.[19] Despite fierce Ottoman resistance and Italian brutality, European powers refused to censure or stop Italy.

On 7 November 1911, Rome announced its formal annexation of Libya. In response, Ottoman legal advisors argued that Italy violated the territorial guarantees of

[18] *NA/FO/407/177*, No. 135 Tewfik Pasha to Sir Edward Grey, 1 October 1911.

[19] *NA/FO/407/177*, No. 88 Sir Edward Grey to Sir R. Rodd, 29 September 1911 and enclosure No. 87.

the Treaty of Paris (1856) and the Treaty of Berlin (1878).[20] In London, Tevfik Pasha formally protested the annexation and announced that the declaration was 'In law, as in fact, effectively null, because the Ottoman Empire and Italy are still in the midst of war and the Government intends to retain and defend by arms its sovereign and inalienable rights over these two provinces.'[21] In Istanbul, Grand Vezir Said Pasha threatened European ambassadors that if European powers recognized the Italian annexation of Libya, Istanbul would consider itself released from any obligations emanating from either treaty.[22]

British officials in the Foreign Office anticipated Ottoman legal arguments and asked its own legal advisors to outline the consequences of accepting the Italian annexation of Libya. The Law Office of the Crown conceded only that Italian annexation violated the territorial guarantees enshrined in the Treaty of Paris.[23] The Foreign Office, in turn, requested that the Italians supply a legal rationale for its seizure of Libya. The Italian ambassador at London argued that the territorial provisions of the Treaty of Paris had been superseded by the Treaty of Berlin when Europe recognized the independence of Montenegro, Romania, and Serbia. Moreover, the fact that none of the powers went after Austria-Hungary for annexing Bosnia and Herzegovina proved that the 1856 treaty was a dead letter. Britain responded that although the Treaty of Paris had been modified, no single power could decide that the treaty had been abrogated without the permission of the other powers.[24] Italy was pointing to the Great Powers' previous violations of Ottoman sovereignty to justify its own in Libya in 1911.

In addition to arguing that Italy's invasion of Ottoman territory was illegal and unprovoked, the lawyers in the Office of Legal Counsel also focused on *jus in bello* (laws in war) questions. In a series of legal opinions, Foreign Ministry advisors claimed, correctly, that Italy had violated the Hague Conventions of 1899 and 1907 through the use of prohibited weapons of war. In a formal protest they decried the Italian use of poisonous and asphyxiating gases as well as the use of exploding projectiles, all methods of warfare prohibited by the Hague Conventions.[25] They addressed the formal protest to the Foreign Ministry of the Netherlands with a view to bringing Italy before the Permanent Court of Arbitration at the Hague. Just as the European powers refused to acknowledge Ottoman protests, the Dutch Foreign Ministry maintained they could not intervene on behalf of Ottoman claims and that the government should instead jointly address the signatories of the Hague Conventions.

While Istanbul's protests were roundly ignored, the international press was much more sympathetic to Ottoman arguments. German papers ran a series of articles warning that allowing Italy to attack Libya would erode peace in Europe and open up the entire Eastern Question and all the potential dangers to peace that came with it. Edwin Mead, the American reformer and pacifist, speaking in Paris, worried that ignoring the Hague Conventions would undermine international peace: 'If the parties

[20] Başbakanlık Osmanlı Arşivi/ hereafter *BOA/HR. HMŞ.IŞO.* 203/15 1329 L. 9. The guarantees can be found in Articles VII and IX of the Treaty of Paris (1856) and Article LXIII of the Treaty of Berlin (1878).

[21] *NA/FO/407/177*, No. 483, Tewfik Paşa to Sir Edward Grey, 8 November 1911.

[22] *BOA/HR. HMŞ.IŞO.* 203/15 1329 L. 9.

[23] *NA/FO 412/100/* Law Officers of the Crown and Reports/ 1911, No. 10026.

[24] *NA/FO/407/177*, No. 462, Sir Edward Grey to Sir R. Rodd, 6 November 1911.

[25] *BOA/HR. HMŞ.IŞO.* 203/25. 1329 Za. 12; *BOA/HR. HMŞ.IŞO.* 203/15 1329 L. 9.

to The Hague Convention are silent while thus insulted and ignored, then those conventions are a farce. If the peace party in Italy itself and the world is dumb at such a juncture, then it might as well not exist.'[26]

The Japanese had shown in the nineteenth century that 'negotiating with imperialism' was possible as a tactic for fending off the Great Powers in the short term. But Japan's case had also demonstrated that only a military powered by the modern forces of industrialization, mass education, and social mobilization could make diplomacy effective.[27] Following Japan's defeat of Russia in 1904–1905, Ottoman leaders, in their more optimistic moments, referred to their empire as the 'Japan of the Near East', holding up Japan as a model to be followed. They had learned the bitter lessons of the primacy of military power, and they had done so mostly at the hands of the Entente powers: the French in North Africa; the British in Cyprus, Egypt, and the Gulf; and the Russians in the Caucasus and the Balkans.[28]

The Ottoman press depicted European diplomacy—especially diplomacy espousing 'humanitarian intervention'—as scarcely disguised tools of imperialist expansionism. These external pressures, moreover—whether coming directly from the Great Powers themselves or indirectly through their support of this or that Balkan state or independence movement inside the empire—could be countered only with military power. To the press, national recovery was not attainable 'by those old books of international law, but only by war'.[29] This outlook did not mean that the Ottoman leaders were war-hungry in 1914 and that they jumped into war at the first opportunity, however. Indeed, for three long months after signing the German alliance they did everything they could to stay out of the fighting.

4. 1914

Ottoman entry into the First World War extended the conflict deep into the Eastern Mediterranean and South Asia, opening up several new battlefronts. For the Ottomans, the points of contiguity to belligerent powers had been the flashpoints over territorial disputes in the recent past. The Ottomans' geopolitical situation, moreover, delivered them their own version of encirclement, with Russia looming across the Black Sea and in the Caucasus, the British ensconced in Cyprus, Egypt, and the Persian Gulf (and

[26] 'Judging the Seizure of Tripoli', *The Literary Digest*, XLIII, 16 (14 October 1911), at 616.

[27] M.R. Auslin, *Negotiating with Imperialism: The Unequal Treaties and Culture of Japanese Imperialism* (2004); C. Aydin, *The Politics of Anti-Westernism in Asia: Visions of World Order in Pan-Islamic and Pan-Asian Thought* (2007), ch. 4; Worringer, *Ottomans Imagining Japan East, Middle East, and Non-Western Modernity at the Turn of the Twentieth Century* (2014), chs 5–7. On Japan's use of international law in diplomacy, especially the law of war, see D. Howland, *International Law and Japanese Sovereignty: The Emerging Global Order in the 19th Century* (2016), chs 5–6. For the various ways in which the early Meiji approached European international law see J. Stern, *The Japanese Interpretation of the 'Law of Nations', 1854–1874* (1979).

[28] H.N. Akmeşe, *The Birth of Modern Turkey: The Ottoman Military and the March to World War I* (2005), at 28–31, 72–79; P.J. Brummett, *Image and Imperialism in the Ottoman Revolutionary Press, 1908–1911* (2000).

[29] Z. Arıkan, 'Balkan Savaşı ve Kamuoyu [The Balkan Wars and Public Opinion]', in Bildiriler: Dördüncü Askeri Tarih Semineri [Proceedings of the Fourth Military History Seminar]. *Ankara: Genelkurmay Basımevi*, at 176, quoting the daily newspaper *Ahenk* [Harmony], 13 October 1912.

both in occupation of neighbouring Iran since 1911), and the French extending their Mediterranean encroachments across North Africa to Morocco. It is not surprising, therefore, that Germany became the Ottoman ally of choice. Calculating that the guns of August would fall silent as quickly as they had erupted, the Ottomans hoped to make the German alliance the foundation on which to construct a new international security and national recovery. The alliance, as we now know, did not bring about the empire's security but its dissolution, and the subsequent Anglo-French mandate regimes under the new League of Nations re-instituted the old imperialist encirclement as a full-fledged colonial occupation after the First World War.

That the Ottomans were interested in a long-term German alliance can be seen in Grand Vizier Said Halim Pasha's insistence that the alliance treaty be concluded for seven rather than the four years proposed by Berlin, with immediate renewal for an additional seven years. He also proposed the creation of a German naval base in the Sea of Marmara just off the capital, pointing again to the Ottomans' international vulnerability.[30] And yet, once they had secured the alliance, the Ottoman leaders, including Minister of War Enver Pasha, who has played the role of the pro-German war hawk in the historiography, endeavoured to stay out of the war. Their most spectacular effort, perhaps, were the Ottoman-Bulgarian negotiations during the first half of August 1914, in which both states secretly acknowledged their preference for neutrality.[31] Berlin had hoped that the two could agree on a joint operation. Instead Istanbul and Sofia, in a piece of previously agreed-upon theatre, each feigned (to the Germans) a desire to intervene thwarted only by their opposite number's hesitation.[32]

For Berlin, the alliance had a very different significance. The German chancellor, Theobald von Bethmann Hollweg, instructed his ambassador to sign the 2 August alliance only with the guarantee that 'in the event of war with Russia, Turkey will take direct and significant action'.[33] Since Germany had already declared war[34] on Russia (and Russia launched its first engagement on 17 August), for the Germans 'in the event of war' meant 'immediately'. Ottoman intervention had the potential to divert and tie up Entente troops, providing the Central Powers with relief in European theatres. There was also some cause for hope that Ottoman entry would bring in other neutrals, especially Bulgaria and Romania, on the side of the Central Powers, and perhaps even Greece and (Germany's wayward alliance partner) Italy. The fact that Berlin looked to Istanbul, the seat of the sultan-caliph, as the match that could set ablaze a global Muslim revolution scorching Entente rule in Africa and Asia handed the Ottomans an additional important lever, and they proved skilful in employing it. Finally, alliance with the Ottomans served the Central Powers in another crucial function: the

[30] *BA-MA, Bundesarchiv-Militärarchiv Freiburg i.Br.* [Federal Republic of Germany, Military Archives], Freiburg i.Br., RM 40-456, folios 310–12, *Memorandum*, Hans Humann, 1 October 1914.

[31] Yasamee, 'Ottoman Empire', in K. Wilson (ed.), *Decisions for War, 1914* (1995), at 242.

[32] M. Aksakal, *The Ottoman Road to War in 1914: The Ottoman Empire and the First World War* (2008), at 123.

[33] *PAAA, Politisches Archiv des Auswärtigen Amts* [Political Archives of the Foreign Office], Berlin, Bethmann to Wangenheim, tel. 296, 1 August 1914, R 22402.

[34] For the German justifications of war in 1914, see the chapter by Isabel V. Hull; for German justifications of the use of poison gas, see the contribution by Miloš Vec to this volume.

complete closure of the Straits in late September meant the end of Anglo-French supply and communication lines through the Mediterranean to Russia.

Thus throughout August–October, with the treaty signed, the Ottomans hoped to sit *on* the alliance and to sit *out* the war. But with German armies bogged down on the Western front and the news of heavy Habsburg losses arriving from Galicia, Berlin's demands for intervention intensified. With a military and naval mission, two battleships, and over 1,000 German officers and personnel in the Ottoman capital already, the top German military leaders on 10 September called for the immediate stop to any further aid to the Ottoman Empire.[35]

We may ask whether at this point, as the Germans began threatening to withdraw from the alliance, the Ottomans could have walked away from their German allies and joined the Entente, or gone back to a policy of true neutrality. Turning their back on Germany meant not only the rupture of the alliance but also, in the event of the war's rapid conclusion, its settlement at the Ottoman Empire's expense. Defecting to the other alliance bloc seemed equally out of the question, since it was the Entente powers, with their large imperial footprint in the Middle East, who posed a mortal threat to the empire's territorial integrity. That the choice of ally fell on Germany, therefore, should not come as a surprise, and there is stunning evidence for the extent to which an alliance with Germany had been discussed publicly even before 28 June. When the German battlecruiser *SMS Goeben* visited Istanbul in May 1914, already at that point it was rumoured that the powerful ship might be sold to the Ottoman navy, or that in the event of war between the Triple Alliance and the Triple Entente the *SMS Goeben* would assume duties on behalf of the empire. Hence, about six weeks prior to Franz Ferdinand's assassination in Sarajevo, participation in a general war alongside Germany was presented in a major Istanbul daily as a reasonable and acceptable course of action.[36]

The conviction that international law could hardly provide the empire's security set the tone for all pre-war political discussions. Deputies in the Ottoman Chamber of Deputies, military and political leaders, and intellectuals argued time and again that international relations should be seen as entirely determined by military success. As a result they cast the functions of society and military as one and the same. In the Chamber a deputy argued, for example, that

> it does not matter however many books we write on international law or however many human rights laws we implement. In order to get states to respect these we must still possess additional means, means of force. Every state has adopted this position and for that reason builds up its [military] strength. [A state] will use all of its defensive or offensive strength in order to defend and protect its rights. We are a state, too, and we therefore cannot escape this truth.[37]

[35] *PAAA*, Jagow to Wangenheim, tel. 121, 10 September 1914, R 22402.
[36] *Tanin*, 7 May 1914 and 17 May 1914.
[37] Ferhad Bey in *Meclis-i Mebusan Zabıt Ceridesi* [Proceedings of the Ottoman Chamber of Deputies], 30 Haziran 1330 [13 July 1914].

The fact that the First Balkan War of 1912 had ended in terrible defeat, that the adversary was not a Great Power but a coalition of the small Balkan states, that the loss of territory came within 100 miles of the capital and included such prominent places as Salonica, and that so many in the leadership itself hailed from the Balkans, all accentuated the gravity of the situation.[38]

Ottoman fears concerning Entente intentions were not unfounded. When the war began, the French foreign minister had suggested offering the Istanbul government a guarantee of the empire's territorial integrity in order to 'calm' Ottoman nerves—a guarantee, he continued ominously, would not prevent the Entente 'from solving the Straits question in line with our thinking at war's end'. Others in the French Foreign Ministry took yet a more aggressive line, arguing bluntly that it would 'be more advantageous for us to include Turkey on the side of our enemies and in that way to finish her off' for good.[39] Similarly, throughout the autumn of 1914, Britain demanded demobilization and strict neutrality from the Ottoman Empire in exchange for maintaining the political status of Egypt.[40] The Foreign Office, too, tied any post-war decisions in the Near East to this condition, noting that Britain would continue to guarantee the territorial integrity of the Ottoman Empire in return for non-intervention and demobilization. By this point, Foreign Office promises fell on deaf ears.

Intercepted correspondence between the Russian ambassador and Petersburg confirmed Ottoman suspicions that the Entente's offers of friendship were a trap. On 6 August the Russian ambassador in Istanbul, Giers, proposed to his government keeping the Ottomans neutral until 'that point in time when circumstances permit our own firm entrance into the Straits'.[41] Evidence such as this, coming at the end of what the Ottomans perceived as a long history of violation suffered at the hands of Britain, France, and Russia and the international system they dominated, told the Ottomans that they could expect little from any agreements with the Entente powers. Ottoman intervention and the Ottomans' justification for war, is best understood in terms of the deep sense of defeat and resentment at the relentless violation of Ottoman international rights and sovereignty. It was for this reason, therefore, that the Ottomans

[38] Zürcher, 'The Young Turks—Children of the Borderlands?', in K.H. Karpat and R.W. Zens (eds), *Ottoman Borderlands: Issues, Personalities and Political Changes* (2003), at 5f.

[39] 'Soviet Union [IBZI]', in O. Hoetzsch (ed.), *Die Internationalen Beziehungen im Zeitalter des Imperialismus [International Relations in the Age of Imperialism]* (1931–43), Izvolskii to Sazonov, tel. 65, 11 August 1914, Series II, vol. 6/1, at 44, and n. 2. Izvolskii reported these French attitudes a day after the two German battleships, the *Goeben* and the *Breslau*, made their famous escape into the Ottoman Straits. The French talks took place before the news had reached Paris, however, and hence they had not been prompted by the crisis that ensued over the German ships' obtaining Ottoman sanctuary.

[40] Periodically the Foreign Office accused Enver Paşa and the 'military party' of preparing a campaign against Britain in Egypt and warned Said Halim and Ottoman ambassadors that armed neutralization posed a threat to Egypt's political status. For example, in one of the many episodes, the British Ambassador to the Ottoman Empire, Louis Mallet, linked Ottoman mobilization in Syria to a plan to attack Egypt, based in part on articles published in *Tanin*. BOA/HR.SYS./4/5 / 10 October 1914. 'Ingiltere Sefareti'nin Tanin'de neşredilen Mısır'la ilgili paragrafa dair açıklama isteğine verilen cevap', Said Halim Paşa to L. Mallet, 8 October 1914. Mallet wrote to Grey that 'I think that it would be right to remind the Grand Vizier that we have always made it perfectly clear that undertaking not to change the status of Egypt was conditioned on Turkey maintaining strict neutrality.' [cd. 7628] No. 13 (1914) 'Correspondence Respecting Events Leading to the Rupture of Relations with Turkey', No. 117, Mallet to Grey, 12 October 1914.

[41] *IBZI*, Giers to St. Petersburg, tel. 631, 6 August 1914 and *BA-MA*, RM 40-457, folio 254. Also in Giers to Sazonov, 5 August 1914, Series I, vol. 2, No. 9, at 6 f.

hoped to seize the opportunity for an alliance with Germany in 1914, even at the risk of entering a world war.

After three months of Ottoman foot-dragging, with the German armies bogged down on the Marne and Habsburg troops held back in the Serbian campaign, Berlin's patience with its Ottoman alliance partner ran out. As a result, on 29 October 1914, an Ottoman navy went out into the Black Sea and opened fire on Russian ships and port cities. When the fleet returned to Istanbul/Constantinople, however, the government claimed that Russian forces had shot first. As this chapter has argued, the Ottoman encounter and experience with international law and the Great Powers, especially Britain, France, and Russia, throughout the nineteenth century, not only made the fiction of a Russian attack in 1914 appear plausible to an Ottoman public but it also rendered an alliance with Germany the most promising long-term, post-war course of action for the empire's leaders. Nonetheless, Ottoman leaders justified the empire's entry into war in 1914 on the grounds of self-defence, much as other European states had done throughout the nineteenth century (see Simon's and Hull's contributions).

5. Conclusion

Ottoman observers became increasingly disappointed in the promises of international law and the Concert's ability to provide international security. The cover of a textbook on diplomatic history used to train Ottoman officials featured these lines: 'The great questions of the time will not be resolved by parliamentary speeches and majority decisions, but by iron and blood.'[42] The words were Bismarck's, of course, and dated back to 1862, but they were put there by the book's author, Ahmed Salahaddin, a professor of international law, because they captured, in his view, the way the world appeared from Istanbul in the aftermath of the Balkan Wars. By 1914, Ottoman newspapers and books cast the Great Powers, and especially those of the Entente, as rapacious states eager to undercut if not terminate Ottoman sovereignty. The French and now the Italian invasions of North Africa, British rule in Cyprus and Egypt, and Russian occupation of eastern Anatolia and support of the Balkan states made that charge appear highly plausible. The Ottomans, in 1914, faced their own version, they believed, of the German *Einkreisung*, or encirclement.

For the ruling Committee of Union and Progress the justification for war became enshrined in the Ottomans' decades-long pursuit for sovereignty and self-defence. These arguments also served as effective mobilizers of Ottoman society, and army conscripts, at least in the opening stages of the war. The fact that it had been an *Ottoman* fleet, led by the *SMS Goeben*, that opened fire on *Russian* forces, and not vice versa, thus had to be hidden from the public. In order to claim justification of war on defensive grounds, the leaders of the Committee of Union and Progress propagated the fiction of a Russian attack. There was no declaration of war, therefore, ahead of the 29 October 1914 Ottoman entry into war.

[42] A. Salahaddin, *Makedonya Meselesi ve Balkan Harb-i Ahiri* (1331/1915).

The Ottoman search for sovereignty and self-defence did not die with the signing of the Armistice in October 1918. Rather, diverse factions throughout the imperial domains fought to determine their post-imperial lives on their own terms.[43] By autumn 1919, power within the Ottoman state was increasingly divided between two factions—one under Allied military occupation in the imperial capital, headed by the sultan, and the other gathering around Mustafa Kemal (Atatürk) in Ankara. While the Istanbul government opted for an older diplomatic approach—one that sought accommodation with the Great Powers and appealed to international law and treaty rights—the Ankara government took a different tack. In the midst of this power struggle, with the capital and the Arab provinces under Allied occupation and the Greek invasion of İzmir/Smyrna, the Ankara government promulgated a new constitution in 1921. The very first article proclaimed that 'sovereignty resides unconditionally in the people'.[44] Article 1 became the cornerstone and focal point of Turkish politics under Atatürk, and remains so to this day. In the Arab lands of the empire, contrary to wartime agreements, the League of Nations denied the right to sovereignty explicitly. Article 22 of the League of Nations Covenant maintained that

> Certain communities formerly belonging to the Turkish [sic] Empire have reached a stage of development where their existence as independent nations can be provisionally recognized subject to the rendering of administrative advice and assistance by a Mandatory until such time as they are able to stand alone.

Arab subjects who in the Ottoman Empire had voted and sent elected deputies to Istanbul, now had lost those rights under British and French mandate rule.

When Ottoman leaders finally chose to abandon the empire in favour of the Turkish nation, the new Republic maintained suspicion towards what international law and European diplomacy could offer the state. In his famous, thirty-six-hour long speech in October 1927—the foundational text of official historiography on the establishment of the Turkish Republic—Mustafa Kemal, echoing decades of Ottoman international lawyers and political observers, laid bare the idea that force supplanted law:

> Sovereignty is never transferred for some theoretical reason, or through debate or talk. Sovereignty is seized with strength and power and violence. The Sons of Osman [i.e., the founders of the Ottoman state c.1299] seized the Turkish people's sovereignty by force and held this power they usurped for six centuries. And now the Turkish nation, through rebellion (*isyan*)—after warning these invaders [the Sons of Osman] to know their limits—has taken its sovereignty into its own hands.[45]

[43] Aksakal, 'The Ottoman Empire', in R. Gerwarth and E. Manela (eds), *Empires at War: 1911–1923* (2014) 32 f.

[44] *Ceride-i Resmiye*, No. 1, 7 Şubat 1337 [7 February 1921].

[45] Taken from a speech by Mustafa Kemal on 1 November 1922, following the Allied invitation to Lausanne, which was sent both to the government in Istanbul and the government in Ankara. Mustfa Kemal gave the speech before the Grand National Assembly voted to abolish the Sultanate. M. Kemal (Atatürk), *Nutuk: Gazi Mustafa Kemal tarafından* (1927), at 494.

After the war, in Atatürk's Turkey, justification of war had become an article of faith. A military high school textbook made clear that every member of society was a soldier:

'Wars have now become wars fought by the entire nation. This is a new age. Wars are no longer fought by the army alone. All women and men, each member of the nation, must take on a duty and a role according to his or her age and skills that will serve the war.'[46] The experience of Ottoman partition and the false promises of international law now demanded absolute military preparedness in the future.

[46] C. Tahir, *Askerliğe Hazırlık Dersleri* (1926), at 1.

12

Juridification, Politicization, and Circumvention of Law: (De-)Legitimizing Chemical Warfare before and after Ypres, 1899–1925

Miloš Vec[*]

1. Introduction: Chemical Weapons as a Subject of Multi-Normativity

The chapter that follows deals with the interaction of technology, politics, and law of war. It discusses the changing interdependencies between them, focusing on the creation and implementation of legal norms, political discourse, and legal scholarship before and after Ypres, 1899–1925. The German chlorine gas attack in the Second Battle of Flanders of 22 April 1915, took place immediately after intense efforts in international law to make war more civilized and to restrict poisonous weapons, particularly at the Hague Conferences of 1899 and 1907 as a milestone.[1] The Hague Conferences were characterized by, among other things, political activism, which was supported by the pacifist movements together with some parts of the community of international law scholars.[2] The public response was eminent, and it reflected the high expectations of international law at the time. This attitude was an expression of the optimism towards international law at the turn of the twentieth century. As can be seen here, 'communication communities' (see Tischer in this volume; and Simon and Brock: Introduction) had developed at that time. They assessed the use of force and of specific weapons on the basis of normative standards and criteria.

With regards to international legal sources, the nineteenth century was marked by a landslide shift to treaties and customs. The nineteenth century produced more treaties than there were ever before. This *Verrechtlichung* (juridification) of international relations culminated in a 'treaty-making revolution of the 19th century'.[3] Certain types

[*] This text is based on Vec, 'Challenging the Laws of War by Technology, Blazing Nationalism and Militarism: Debating Chemical Warfare before and after Ypres, 1899–1925', in B. Friedrich, D. Hoffmann, J. Renn, F. Schmaltz, and M. Wolf (eds), *One Hundred Years of Chemical Warfare: Research, Deployment, Consequences* (2017) 105.

[1] J. Dülffer, *Regeln gegen den Krieg? Die Haager Friedenskonferenzen von 1899 und 1907 in der internationalen Politik* (1978).

[2] S.C. Neff, *Justice among Nations. A History of International Law* (2014), at 323–28.

[3] Keene, 'The Treaty-Making Revolution of the Nineteenth Century', 34 *The International History Review* (*Int. Hist. Rev.*) (2012) 475.

Miloš Vec, *Juridification, Politicization, and Circumvention of Law: (De-)Legitimizing Chemical Warfare before and after Ypres, 1899–1925* In: *The Justification of War and International Order*. Edited by: Lothar Brock and Hendrik Simon, Oxford University Press (2021). © the many contributors. DOI: 10.1093/oso/9780198865308.003.0012

of treaties functionally compensated for the lack of codification of international law, and they were therefore called 'law-making treaties'.[4] International law never had it as good as it did in the final years before the First World War.

What is more, as Martti Koskenniemi has emphasized it, 'the laws of war have perhaps never before nor since the period between 1870 and 1914 been studied with as much enthusiasm'.[5] Some contributions to this anthology plead with good arguments for a stronger focus of historical research on the political practice of war justification (Hull, Tischer; for the combination of theory and practice see Simon and Brock: Introduction). While the present text also deals with political and military discourse on the use of chemical weapons, it is firmly anchored in the tradition of the history of international legal scholarship. After all, the nineteenth-century 'treaty revolution' came late in the field of weapons regulation; in the preceding years and in light of the lack of treaty law, essentially legal scholars were the ones formulating norms on the legitimacy of weapons. Thus, they were establishing international legal standards.

Within an overall friendly-sounding panorama of legal internationalism the law of war was a particularly touchy sphere. The use of violence was, on the one hand, restricted, but, on the other hand, legitimized, in that international law itself supplied a normative order to its employment.[6] Unlike economic and administrative regulation, the law of war concerned questions that were understood to be more delicate and more political. All of them addressed competencies of state sovereignty, but restrictions on the law of war were subject to clearer bounds than those of the other fields of regulation. The evolution typically took the form of transforming customary law into international treaty law (e.g. the prohibition of poison in the law of war).

However, as the political and legal discourse around Ypres 1915 will illustrate, neither international law nor international lawyers were isolated and completely independent from the influence of politics or other spheres of normativity. Different communication communities interacted. This is due to the inner logic of justification discourses. Justifications of a specific type of weapon and warfare are, like most social interactions, regulated by plural norms at the same time. In other words: they were subject to 'multi-normativity'[7]; regarding war discourses see the editors' introduction). In the writings of international lawyers, complex interaction between legal approaches, moral convictions, and social conventions crystallized. Studying these scientific discourses can therefore help us to understand how modern international law as a legal discipline emerged in the 'era of its birth'[8]—and how it interacted in a certain situation with extra-legal spheres such as politics and morality.

[4] M. Vec, *Recht und Normierung in der Industriellen Revolution. Neue Strukturen der Normsetzung in Völkerrecht, staatlicher Gesetzgebung und gesellschaftlicher Selbstnormierung*, (2006), at 111–26.

[5] M. Koskenniemi, *The Gentle Civilizer of Nations* (2001), at 87.

[6] Simon, 'Das Recht des Krieges', 24 *Rechtsgeschichte—Legal History* (*Rg.*) (2016) 508, at 508; see Bernstorff, 'The Use of Force in International Law before World War I: On Imperial Ordering and Ontology of the Nation-State', 29 *The European Journal of International Law* (*EJIL*) (2018) 233; Lesaffer, 'Aggression before Versailles', 29 *The European Journal of International Law* (*EJIL*) (2018) 773.

[7] Vec, 'Multinormativität in der Rechtsgeschichte', in Berlin-Brandenburgische Akademie der Wissenschaften (vormals Preußische Akademie der Wissenschaften) (ed.), *Jahrbuch 2008* (2009) 155.

[8] Vec, 'Principles in 19th Century International Law Doctrine', in L. Nuzzo and M. Vec (eds), *Constructing International Law—The Birth of a Discipline* (2012) 209.

The crucial turning points in the interaction were treaties on international law and declarations, technological innovations, the outbreak of the war in the summer of 1914, the aforementioned use of poison gas in April 1915, and finally the end of the war in 1918. The various phases of the international law approach to poison gas can perhaps be characterized by three terms: juridification, politicization, and circumvention of law. In my view, the discussion of international law around the First World War followed this periodization, and for extended periods, at least, it held few surprises with unpredictable points. The years immediately following the use of poison gas in the First World War were perhaps the exception—they bore surprises.

2. Codifying War Technologies in International Law around 1900

Four sets of international regulations are particularly relevant to the discussion of the poison gas attacks near Ypres in 1915. Their historical genesis is characterized by comprehensive first steps to the transformation of the civilizing ambitions of the time into the legal regulation of war.[9] The legal bases for restrictions on chemical weapons, the validity of which was debated in the context of the use of poison gas at Ypres and beyond, can be identified precisely. In this section, they will first simply be presented, in their original wording. The heated debates over their validity and interpretation will be addressed later.

2.1 From the Principle of Humanity to the First Poison Prohibition in International Treaty Law (1868–1907)

The first of these legal bases was the Saint Petersburg Declaration of 29 November (11 December) 1868. 'Considering that the progress of civilisation should have the effect of alleviating as much as possible the calamities of war', the declaration regulated and formulated not only the special, explicit prohibition of certain weapons but also general principles of the law of war. According to Thilo Marauhn, it prohibited specific types of weapons (projectiles), on the one hand, and, on the other, established the 'fundamental obligation to avoid unnecessary suffering and referred to the laws of humanity'.[10]

These same principles were applied again in the 'Project of an International Declaration concerning the Laws and Customs of War, Brussels, 27 August 1874', which, in Article 13, included the following provision: 'According to this principle are especially "forbidden": (a) Employment of poison or poisoned weapons.' The Brussels Declaration on Land Warfare of 1874 never came into effect. It did, however, lead to analogous resolutions by the Institut de Droit International in 1880 in

[9] Dülffer, above note 1.
[10] T. Marauhn, *Der deutsche Chemiewaffen-Verzicht. Rechtsentwicklungen seit 1945* (1994), at 46.

the form of the 'Manuel des lois de la guerre sur terre',[11] unofficially known as the 'Oxford Manual', which in turn was the model for the later positive law regulations of 1899 and 1907.

On 29 July 1899, the concluding act of the First Peace Conference in The Hague followed. It included the Hague Declaration (IV, 2) concerning asphyxiating gases. The declaration proper is as brief as possible: 'The contracting powers agreed to abstain from the use of projectiles the sole object of which is the diffusion of asphyxiating or deleterious gases.'[12] It goes back to Article 13 of the draft declaration of the Brussels Conference of 1874. Although a prohibition of poison and poisoned weapons already existed under customary international law, the Hague Declaration of 1899 was the first time international treaty law expressly referred to 'poison' weapons.[13]

In 1899, the 'Convention with Respect to the Laws and Customs of War on Land', was passed.[14] Section II, Chapter 1, Article 22 reads as follows: 'The right of belligerents to adopt means of injuring the enemy is not unlimited.'[15] Definitions and specifications followed in Article 23, in which the parties were expressly denied unlimited rights in the choice of means to harm the enemy. It reads in part:

Article 23.

Besides the prohibitions provided by special conventions, it is especially prohibited:—

(a) To employ poison or poisoned arms;
(b) To kill or wound treacherously individuals belonging to the hostile nation or army;
...
(e) To employ arms, projectiles, or material of a nature to cause superfluous injury;

Articles 22 and 23 established minimum standards for humanitarian warfare under existing international law. Furthermore, a general principle prohibiting the use of poison in a certain sector was codified by treaty.

The final act of the Second Hague Conference of 1907 contained in Article 22 of the Fourth Convention (with identical wording) and in 23 a, b, e of the regulation the almost identical definitions of the Hague Convention on Land Warfare of 1899 (with minor changes compared to the 1899 version).

[11] J.L. Kunz, *Gaskrieg und Völkerrecht*, Erweiterter Sonderabdruck aus 'Zeitschrift für öffentliches Recht' Band VI, Heft 1 (1927), at 13.
[12] *Declaration (IV,2) concerning Asphyxiating Gases* (1899), available at <https://ihl-databases.icrc.org/applic/ihl/ihl.nsf/Article.xsp?action=openDocument&documentId=2531E92D282B5436C12563CD00516149> (last visited 10 January 2020).
[13] Marauhn, above note 10, at 46.
[14] Kunz, above note 11, at 13.
[15] *Convention with Respect to the Laws and Customs of War on Land*, available at <http://avalon.law.yale.edu/19th_century/hague02.asp> (last visited 10 January 2020).

2.2 No Notion of What Lay Ahead: The Intense Legal Discourse on the Hague Convention prior to 1915

The Hague Conferences marked a new era in the history of international law on warfare. Following the gas attacks at Ypres in 1915, the definitions established therein would become the central points of reference and interpretation in the discussion of the legality or illegality of the German actions. At the time of their enactment, no one could have foreseen in which form poison gas would be employed on the battlefields of the war to come.[16] For that reason, Bothe argues that the intentions of the norm creators are interesting but not truly helpful when it comes to interpreting Article 23(a).[17] However, for the years between the passing of the norms quoted above and the German gas attack near Ypres, a lively, widespread juridical discourse on the law of war can be documented among jurists as well as military. The first and formal feature is that a considerable number of treatises appeared that addressed various subjects from the large field of regulating new warfare technologies.

If one considers only this thematic subset of the discussion of international law in this period, it is noticeable that the prohibition of poison is mentioned repeatedly. Nowhere, however, is this discussion focused on poisonous gas in the sense of the later events. For example, the German international law specialist Hans Wehberg, who had a reputation as a pacifist, only reproduces the original wording of the norm; there are no explanations of it.[18] The Swiss-German jurist and partisan of peace Otfried Nippold mentions poison but does not interpret or expose the problems of the term.[19] The same is true of Ernest Nys,[20] professor and historian of international law in Brussels and member of the Permanent Court of Arbitration. Both Karl Strupp and Henry Bonfils associated the prohibition of poison with the pollution of rivers, wells, and water pipes by infectious materials.[21]

To my knowledge, Albrecht Tettenborn deviates most from this widespread brevity in such accounts; he interprets the term '*Gift*' (poison).[22] In this case, his intense preoccupation with the wording of the norm that takes up the term '*Gift*' leads to a discussion of possible regulatory gaps. They are filled by the traditional historical practice of broad interpretation. The *ratio legis* of weapon prohibitions is extended to other fields, resulting in an expanded protection by the law of war.

In the final years before the First World War, this sort of treatment that exposes the problems of the subject was the exception, despite the codification of the prohibition

[16] M. Bothe, *Das völkerrechtliche Verbot des Einsatzes chemischer und bakteriologischer Waffen. Kritische Würdigung und Dokumentation der Rechtsgrundlagen* (1973), at 8.

[17] Ibid., at 88.

[18] H. Wehberg, *Die Abkommen der Haager Friedenskonferenzen, der Londoner Seekriegskonferenz nebst Genfer Konvention. Mit einem Vorwort von Prof. Dr. Zorn* (1910).

[19] O. Nippold, *Die zweite Haager Friedenskonferenz. 2. Teil. Das Kriegsrecht, unter Mitberücksichtigung der Londoner Seerechtskonferenz* (1911), at 10.

[20] E. Nys, *Le droit international. Les principes, les theories, les faits*, vol. 3 (2nd ed., 1912), at 144.

[21] H. Bonfils, *Manuel de droit international public (Droit des gens). Destiné aux étudiants des Facultés de Droit et aux aspirants aux fonctions diplomatiques et consulaires* (5th ed., 1908), at 660; K. Strupp, *Das internationale Landkriegsrecht* (1914), at 58.

[22] A. Tettenborn, *Prinzipien und Richtungen der Kriegsmittelverbote des Landkrieges* (1909), at 22–24.

of poison in treaties. The hesitant discussions that would follow in 1915 are even less evident. In the years since the prohibition of poison had been passed, modern technology had not yet posed a challenge to international law and its protective regulations. If one compares the tone and style of the comments of scholars of international law, something other than their brevity is striking as well: scholarship on international law was not very politicized in a nationalistic sense. There were rarely nationalistic undertones, no accusations or stereotypical discussions when it came to the establishment and application of norms of the law of war. On the contrary, the quest for juridification of international relations and foreign policy was a transnational phenomenon, shared by international lawyers from many countries.[23]

3. Militarization and Circumvention of Law: Debates on International Law during the Continental War with Gas, 1915–1918

On the one hand, all of that is to be expected and not very surprising. On the other hand, academics working in the area of international law after 1915 will take another look at the positions previously taken by their colleagues and interpret them in light of recent events. This impression of the brevity, even inadequacy, of international law scholars' interpretations of the prohibition of poison, combined with the not very politicizing tone of jurists' interpretations, would change after 22 April 1915, with a slight delay but a lasting effect. It was not just that norms were read anew in the face of the ongoing or recently occurred gas warfare; previous positions were cited as well. The old interpretations were reinterpreted. They took on a new weight and became in turn not only arguments, but potentially sources of international law as well.[24]

3.1 International Law: Alive, but Not Kicking

The international law of these years was often intensely preoccupied with the events of the First World War. Many academics in international law adopted passionate stances on the legality or illegality of specific actions; others behaved more guardedly.

It should be recognized first that the panorama of publications on international law during the First World War is quite rich. However, they did not comment on or even mention the ongoing gas warfare.[25] The German *Zeitschrift für Völkerrecht*, for example reflected new thematic trends and changes in international law.[26] It is interesting, however, that I was unable to find a single article in *Die Zeitschrift für*

[23] M. Payk, *Frieden durch Recht? Der Aufstieg des modernen Völkerrechts und der Friedensschluss nach dem Ersten Weltkrieg* (2018), at 27–78.

[24] Vec, 'Sources in the 19th Century European Tradition. The Myth of Positivism', in S. Besson and J. d'Aspremont (eds), *Oxford Handbook on the Sources of International Law* (2017) 121.

[25] H. Pohl, *Deutsches Landkriegsrecht. Quellensammlung mit Sachregister* (1915).

[26] Hueck, 'Die Gründung völkerrechtlicher Zeitschriften in Deutschland im internationalen Vergleich', in M. Stolleis (ed.), *Juristische Zeitschriften. Die neuen Medien des 18.–20. Jahrhunderts* (1999) 379.

Völkerrecht concerning the subject of the gas attack. German scholars on international war seem to have been completely silent on the subject during the First World War.[27]

The connection between the German decision in favour of the use of poison gas and international law also remained murky. According to Isabel Hull,[28] the precise motive for the German employment of gas near Ypres is unclear. Only subsequent justifications of the action with references to international law can be found. In these justifications the Germans argue that their use of poison gas was a reprisal for the French use. Modern scholars emphatically reject this justification as self-protection.[29] Instead, the Germans independently wished to employ poison gas. Nevertheless, the argument that the use of poison gas was supposedly a reprisal under international law shows that the Germans assessed its use based on the upstanding validity of the norm, since they asserted thereby that its use was a sanction for an injustice committed by the other side. That makes it clear that they assumed, at least ex post, that the use of poison gas violated international law. Other historical sources also suggest that the use on the part of the Germans had to overcome resistance that assumed it was illegal.[30] Fritz Haber,[31] a chemist and the scientist responsible for the planning of German gas warfare, claimed in the 1920s that the military had conducted a legal review. Colonel General, head of the *Generalstab* (General Staff), and Minister of War Erich von Falkenhayn had, according to Haber in a report to the investigating committee of the German Reichstag on 1 October 1923, 'apparently personally reviewed the permissibility of gas weapons under international law'. Haber stated: 'He was convinced beyond any doubt that his orders in the area of gas warfare did not violate international law.'[32]

The striking absence of public evaluations based on international law or any other normative considerations raises questions. What does this say about the German attitude to international law? To what can it be traced back? Hull is very emphatic here:

> the pattern of decision making seems clear. Civilian leadership, which was chiefly in charge of applying legal considerations, faced especially strong undertow from military institutions: from the junior, and then quickly from the senior naval officer corps regarding submarines; from the war ministry, which had already bought flamethrowers without advance discussion; and from OHL [Oberste Heeresleitung], which had already bombed civilian targets from the air.[33]

[27] W. Zecha, *'Unter die Masken!' Giftgas auf den Kriegsschauplätzen Österreich-Ungarns im Ersten Weltkrieg* (2000), at 27.

[28] I.V. Hull, *A Scrap of Paper. Breaking and Making International Law during the Great War* (2014), at 232.

[29] Ibid., at 233; Zecha, above note 27, at 22.

[30] Otto Hahn reports a conversation with Haber, see D. Martinetz, *Der Gaskrieg 1914–1918. Entwicklung, Herstellung und Einsatz chemischer Kampfstoffe. Das Zusammenwirken von militärischer Führung, Wissenschaft und Industrie* (1996), at 104.

[31] Haber, 'Zur Geschichte des Gaskrieges. Vortrag, gehalten vor dem parlamentarischen Untersuchungsausschuß des Deutschen Reichstages am 1. Oktober 1923', in F. Haber (ed.), *Fünf Vorträge aus den Jahren 1920–1923* (1924) 76, at 76.

[32] J. Bell and W. Schücking, *Das Werk des Untersuchungsausschusses der Deutschen Verfassungsgebenden Nationalversammlung und des Deutschen Reichstages 1919–1928, Völkerrecht im Weltkrieg*, vol. IV (1927), at 13; Haber, above note 31, at 76–77.

[33] Hull, above note 28, at 238.

In this view, international law was disdained by the very parties who made the military decisions during the First World War and bore responsibility for them. Hull's verdict is even more harsh from a comparative international perspective: Germany is said to have expressed particular disdain for international law.[34] With regard to the legal history, it is not easy to determine from the available sources whether this was in fact the case.

A first approach would be to ask how the other European powers reacted to the German poison gas attack and to what extent international law played a role in that discourse. But my thesis is that considerations of international law continue to be largely absent. Legal assessments of the German use of poison gas are found only here and there: the English international law professor Coleman Phillippson noted in 1915, in an addendum to his account of the law of war during the First World War, that '[the Germans] diffused asphyxiating gases among their enemy; such conduct being not only unlawful under the international declaration made in 1899, but contrary to humanity and civilization'.[35] A similar verdict is also found in Hall and Higgins.[36] French law professor at the Sorbonne Antoine Pillet wrote in a book published towards the end of the First World War that the prohibition of gas in Article 23(a) applied only to fluid or solid poison, since the prohibition of gas had been regulated elsewhere.[37]

Hence the surprise that international law played hardly any role in the German decision continues to some degree with Germany's enemies as well. As will be demonstrated in what follows, there was hardly any normative discourse on gas warfare, whether among politicians or in the general public.

3.2 The Lack of Protest: Political Voices and Official Reports

One first point concerns the question of the extent to which political voices and official reports articulated protest over the German poison gas attack. The findings are quantitatively sparse. Zecha observed in 2000 'that neither the warring nor the neutral countries, for example, the United States until 1917, protested the use of poison gas or chemical weapons'.[38] In the period immediately following the First World War, not only Haber[39] but also the Germans Johannes Bell and Walther Schücking (with apologetic intent) asserted with satisfaction that no warring or neutral power had protested at all.[40] Jaschinski condensed this into the memorable formulation 'the silence of the Allied forces'.[41]

By contrast, Garner mentions a charge from the British War Office dated 21 April 1915, that 'Germans had violated the laws of civilized warfare during their recapture

[34] See also F.C. Partridge, *The Future of International Law* (1917), at 6.
[35] C. Phillippson, *International Law and the Great War. Introduction by John MacDonald* (1915), at 217.
[36] W.E. Hall, *A Treatise on International Law* (7th ed., 1917), at 569 n. 2.
[37] A. Pillet, *Les conventions de La Haye du 29 juillet 1899 et du 18 octobre 1907. Étude juridique et critique* (1918), at 218.
[38] Zecha, above note 27, at 26.
[39] Martinetz, above note 30, at 114.
[40] Bell and Schücking, above note 32, at 9.
[41] H. Jaschinski, *Neuartige chemische Kampfstoffe im Blickfeld des Völkerrechts* (1975), at 115.

of hill 60 east of Ypres, by employing shells which emitted asphyxiating gases'.[42] Moreover, 'the Belgian commission of inquiry investigated the use of asphyxiating gases at Ypres'.[43] Both Garner and Kunz report that the British Field Marshal Sir John French (later Earl of Ypres) denounced the gas attack in a battle report: 'the enemy ... by the use of an entirely new war method, one contrary to engagements entered into by them at the Hague Convention'.[44] In the House of Lords, Lord Kitchener would have protested this kind of warfare on 18 May 1915.[45] *The Times* of London reported a number of times.[46] Hull in turn quotes four statements by politicians that express outrage at the use of poison gas weapons.[47] These four statements were not, however, official protests but only personal remarks, which Hull cites as evidence of the authenticity of the outrage of those who made them.

The picture in France is similar. In contrast to Britain, where official protest with underpinnings in international law was supposed to have been articulated, no one would ascribe that to France. Olivier Lepick, the scholar with the best knowledge of this material (1998),[48] responded to a request by the present author by saying that no French protest could be found.[49] Hull notes that there were no discussions worth mentioning in France of the legality of poison gas weapons.[50] Interest focused on France's own capacities: to catch up with Germany and to deploy gas weapons.

In the end, official protest did not follow until near the end of the First World War—albeit not by the countries involved but by the International Committee of the Red Cross (ICRC). On 6 February 1918, it appealed to the warring powers to renounce poison gas weapons.[51] The use of chemical warfare agents was said to violate international law. The ICRC evoked the risk of the escalation of gas warfare and proposed a treaty on the renunciation as a return to the Hague Convention.[52] In their reply to this note of protest, the Allied countries first referred to their own use of poison gas as a 'reprisal'. By doing so they implicitly admitted (as the Germans had previously) that the use of this weapon was illegal but justified it with reference to earlier violations of international law by the enemy.

The warring powers thus demonstrated their awareness of the abstract legal standard; several of them (including the United States) nevertheless employed this weapon to obtain military advantage.[53] It is even more interesting, of course, that the

[42] J.W. Garner, *International Law and the World War*, vol. I (1920), at 284–85.

[43] Ibid., at 272.

[44] Abridged quotation in Kunz, above note 11, at 3, 14; A. Palazzo, *Seeking Victory on the Western Front. The British Army and Chemical Warfare in World War I* (2000), at 43; Zecha, above note 27, at 27; with different wording in Garner, above note 42, at 276.

[45] Palazzo, above note 44, at 43.

[46] Garner, above note 42, at 275–76.

[47] Hull, above note 28, at 235.

[48] O. Lepick, *La Grande Guerrre Chimique* 1914–1918 (2nd ed., 1998).

[49] Oral communication to the author on 21 April 2015 at Harnack-Haus in Berlin.

[50] Hull, above note 28, at 237.

[51] Protest by the International Committee of the Red Cross, published in *Papers relating to the Foreign Relations of the United States. 1933. 1918, Supplement II: The World War*, at 779–81, available at <https://history.state.gov/historicaldocuments/frus1918Supp02/pg_779> (last visited 10 January 2020); Jaschinski, above note 41, at 60; A. Overweg, *Die chemische Waffe und das Völkerrecht. Eine rechtshistorische und rechtskritische Studie* (1937), at 64.

[52] Jaschinski, above note 41, at 60.

[53] Ibid., at 116.

first use did not produce an official outcry that mobilized international law as a normative basis for a complaint against the enemy in question. International law was almost inaudible in the discourse on poison gas with regard to the years of the First World War. Other violations, such as that of Belgian neutrality,[54] were publicly denounced much more strongly.

3.3 The Daily Press: Restraint, Disinformation, and Loud Silence

Another surprise is the almost complete lack of discussion in the daily press. Here too astonishingly few traces of and references to the poison gas attack can be found. Even when they did occur, they occurred with strange distortions. The daily newspapers of various countries reported little about their own use; publication was subject to the conditions of wartime censorship. Sometimes readers would first learn about it when the domestic press reacted to accusations from abroad or took positions justifying its own government. It remains surprising nevertheless that even the enemy's use of poison gas was not treated very prominently.

Several examples can be cited briefly for this assessment and they are based on intense archival studies conducted by students in the summer semester of 2014–15. According to these studies, the German use of gas weapons near Ypres was mentioned several times in the reporting on Ypres.[55] Often, however, the reader only learned about the use of gas from biased official reports from the Major Headquarters of the German Reich. It is even more curious that numerous reports refer first to French or English papers in order to reject their presentation of the facts or their legal views.[56] Moreover, such reports referred to official German announcements, according to which the enemies had been using such means for several months. In addition, there is a conspicuous silence where one would have expected reports. The first gas release by Austria-Hungary went unmentioned in the *Arbeiter-Zeitung*, the *Reichspost*, the *Neues Wiener Journal*, and the *Pester Lloyd*.[57] By contrast, there were lengthy reports on the attack on 24 October 1917, as part of the Twelfth Battle of the Isonzo; the victory was celebrated by the newspapers, but the gas that made it possible in the first place was not mentioned anywhere!

But the poison gas attacks of the enemies also had a relatively weak journalistic response. The first English release of gas was not mentioned in any of the four Austrian-Hungarian newspapers studied. The same was true of the reporting on the first French release of gas on 25 February 1916, near Reims: the attack was not mentioned in any edition (21–28 February 1916); nor was the attack using phosgene gas shells on 21

[54] Hull, above note 28, at 16–50; Norris, 'Uninvited and Unwelcome: The SS Appam and the US Law of Neutrality', 29 *The European Journal of International Law* (*EJIL*) (2018), 477; Payk, above note 23, at 82–90, 109, 496.

[55] S. Spitra, *Rechtfertigungsnarrative & Gaskrieg. Eine theoretische Betrachtung des Begriffs 'Rechtfertigungsnarrativ' & eine praktische Untersuchung seiner Anwendung beim Gaskrieg bei Ypern 1915* (2015) (seminar paper, University of Vienna), at 19.

[56] D. Bischof, *Reaktionen und öffentliche Debatten zum Gaskrieg in vergleichender Perspektive (Deutsches Reich—Österreich-Ungarn)* (2015) (seminar paper, University of Vienna), at 13–20.

[57] A. Herzog, *Die Darstellung des Giftgaskrieges anhand zeitgenössischer österreichisch-ungarischer Tageszeitungen* (2015) (seminar paper, University of Vienna), at 11.

February 1915, near Verdun mentioned, even though this involved a new weapon.[58] In addition to the complaint of violations of international law, which by this time had become problematic, concerns about making one's own soldiers and nationals uncertain probably played the primary role in omitting such reports.

3.4 Possible Interpretations: *Raison de Guerre* as Its Own Form of Normativity?

If we sum up the observations thus far, we can conclude that international law played an astonishingly peripheral role not only in the Germans' decision whether to employ poison gas but also in the public discourse. Justifications often followed only in response to opponents' accusations of illegality. In what follows I will attempt to explain this attitude, in particular on the German side. Three possible factors seem to me worth mentioning.

First, the disdain for international law during the First World War in general can be traced back to older traditions. The keyword here is the so-called 'deniers of international law'.[59] This collective name brings together heterogeneous doubts about international law as a genuine normative order. The flipside of such denial was often a complementary over-emphasis on the sovereignty of the nation-state. In this view, the normative order between states could be subordinated to national law and national interests. This mixed explosively with the interests of certain military and political figures, and precisely in the situation of military conflict it was able to become acute that certain regulations of behaviour between states would be regarded as non-binding.[60] In this view, which is represented in the chapters by Hull and Simon, traditions of dismissing and denying international law were particularly strong in Germany.

The second factor that can be identified is a particular dismissal of international law on war that took the form of the German military circumventing the law. For example, Hull suspects there was a specifically German mentality that easily thrived in the ideological soil described above.[61] It found its classical expression in the notorious text *Kriegsbrauch im Landkriege* (translated as *The Usages of War on Land*), which was published by the German General Staff in 1902. It states that *raison de guerre* permits any warring state to employ all means that make it possible to achieve the aim of the war. This was restricted only by '*customs, traditions, or manner of war*', but not by international law.[62] It is hardly surprising that such publicly expressed positions caused outrage from Germany's enemies in war and were exploited for propaganda. The text

[58] Ibid., at 11.

[59] See the dissertation project by Paul Hahnenkamp, *The Deniers of International Law*, at the University of Vienna.

[60] See also Hull, above note 28, at 239; Simon, 'The Myth of *Liberum Ius ad Bellum*: Justifying War in 19th-Century Legal Theory and Political Practice', 29 *European Journal of International Law (EJIL)* (2018) 113.

[61] I.V. Hull, *Absolute Destruction. Military Culture and the Practices of War in Imperial Germany* (2005).

[62] Großer Generalstab (ed.), *Kriegsbrauch im Landkriege*, Kriegsgeschichtliche Einzelschriften 31 (1902), at 2–3.

was quoted critically by international law scholars[63] and was translated into several languages.[64]

Andreas Toppe came to a similar conclusion, identifying a lack of implementation in the German military and a 'radicalization of military doctrine'.[65] The systematic location of this thinking is the idea that war has its own mechanisms. They are in a position to annul law and especially international law. It is condensed in the phrase 'Necessity knows no law.' Evidence of this normative stance can also be found here and there in the late nineteenth century.[66] The results of research of recent years and decades has shown that military figures—not only in Germany—often rejected the limitations on their means of warfare that were sought or agreed upon at international conferences on international law.[67] The argument was that while humanity may be a feasible principle for modern international law on war, it is alien to the true nature of war.[68] Other military figures argued that one could better pursue humanity by creating new, more effective weapons than by prohibiting them.[69]

Hence the limits should themselves be limited.[70] Not international law but military utility should have the last word. Prohibitions in international law were therefore eyed with distrust because the military wanted to keep open the option of better weapons in the future.[71] Interestingly, this attitude of criticizing or annulling international law is expressly shared by some experts in international law.[72] Even Hans Wehberg, who is regarded as a pacifist, writes: 'In the extreme case ... every principle of the law of war can be breached.'[73]

Other, non-German scholars on international rights made comparable arguments.[74] Thus around 1900 there was an unholy alliance in which the military and international law scholars placed the validity of the law for certain extreme cases under the proviso of necessity. It is hardly surprising that as the war continued, the aspect of the political utility of international law was emphasized more strongly.[75]

[63] J. Westlake, *International Law. Part II: War* (1907), at 91; Garner, above note 42, at 280.

[64] German General Staff, 'Kriegsgebrauch—The Customs of War', in German General Staff, *Britain as Germany's Vassal. Together with Kriegsbrauch—The Customs of War, 1902 and Extracts from Regulations adopted by the Hague Conference 1907 and subscribed to by Germany* (1914); Grande Stato Maggiore Germanico, *Le Leggi della guerra (Kriegsbrauch im Landkriege)* (1915); Grand État-Major Allemand, *Les lois de la guerre continentale (Kriegsbrauch im Landkriege)* (1916).

[65] A. Toppe, *Militär und Kriegsvölkerrecht. Rechtsnorm, Fachdiskurs und Kriegspraxis in Deutschland 1899–1940* (2008), at 28, 30, 105.

[66] Vec, 'All's Fair in Love and War or the Limits of the Limitations. Juridification of Warfare and its Revocation by Military Necessity', in M. Killingsworth (ed.), *Who Do the Laws of War Protect? Civility, Barbarity and IHL* (2020 forthcoming).

[67] Messerschmidt, 'Völkerrecht und "Kriegsnotwendigkeit" in der deutschen militärischen Tradition seit den Einigungskriegen', 6 *German Studies Review* (*GSR*) (1983) 237, at 240.

[68] Dülffer, above note 1, at 150.

[69] Ibid., at 150.

[70] Huber, 'Die kriegsrechtlichen Verträge und die Kriegsraison', 7 *Zeitschrift für Völkerrecht* (1913) 351, at 359.

[71] Dülffer, above note 1, at 76.

[72] Cybichowski, 'Notwendigkeit und Völkerrecht', in S. Cybichowski (ed.), *Studien zum internationalen Recht* (1912) 21, at 68–69; F. Lentner, *Das Recht im Kriege. Kompendium des Völkerrechtes im Kriegsfalle* (1880), at V.

[73] Wehberg, above note 18, at 14-15.

[74] A. Rivier, *Principes du Droit des Gens*, vol. II (1896), at 241–42.

[75] Koellreutter, 'Kriegsziel und Völkerrecht. Betrachtungen aus der Front. Abhandlungen', 10 *Zeitschrift für Völkerrecht* (1917/1918) 493, at 500.

A third and last approach to explaining the absence of discourse on the use of gas as a weapon and its permissibility under international law is the discomfort in ethos of broad swaths of the military to poison gas. This argument focuses on the perception of this weapon and contrasts it with the disposition of military actors. Gas and poison were perceived as cruel, unmanly, and unchivalrous.[76] All poison was considered a 'womanish weapon', which is in keeping with the attribution of this way of killing to women in criminal law and criminology.[77] Gas and poison were not used in man-to-man close combat; they did not cause bleeding wounds in physical battle. Rather, it was a weapon that only worked at a distance. This had already been an argument for considering medieval crossbows to be illegitimate or even illegal. All these elements underscore the asymmetry of the debate, in which they ran counter to a chivalrous ethos that preferred beating, stabbing, and shooting weapons.[78] The military had little sympathy for this new weapon, its use was nearly considered a 'moral taboo',[79] and that can be seen as another reason for its silence on the subject.

4. The Continuing Politicization of International Law: The Legal Assessment of War Crimes, 1918–25

After the war ended in 1918, there followed a third phase in the discourse on international law concerning the use of poison gas weapons. It was an intense, retrospective debate on the legality of their use under international law embedded in the general public assessments of the events of the war years. The not very surprising form taken by this third and final phase is characterized by stark nationalization and an irreconcilable polarization of the political and legal standpoints. All the events were evaluated in terms of domestic and international law. The parties sought the tribunal of the public and tried to win over public opinion. Between the wars, especially, the subject of gas warfare reached a climax in the public debate.

Perspectives of history and of international law thus went hand in hand politically, and they were also associated with narratives of proper conduct.[80] For that reason it is right to speak of 'war innocence research'.[81] The public fight over the question of war guilt and the politically tinged legal discourse on war crimes followed the battle with weapons. Gas warfare was also assessed in a number of publications after 1918. The major international scholarly works on international law published between the wars

[76] J. Encke, *Augenblicke der Gefahr. Der Krieg und die Sinne* (2006), at 197–218; Encke, 'Die gemeine Waffe. Interview mit Herfried Münkler', *Frankfurter Allgemeine Sonntagszeitung* (12 April 2015) 41.

[77] I. Weiler, *'Giftmordwissen und Giftmörderinnen.' Eine diskursgeschichtliche Studie* (1998); R.M. Price, *The Chemical Weapons Taboo* (2007), at 23.

[78] Encke, 'Waffe', above note 77.

[79] Price, above note 77, at 19; see also the title from R.M. Price, *The Chemical Weapons Taboo* (2007).

[80] See also S. Zala, *Geschichte unter der Schere politischer Zensur. Amtliche Aktensammlungen im internationalen Vergleich* (2001).

[81] Große Kracht, *Kriegsschuldfrage und zeithistorische Forschung in Deutschland. Historiographische Nachwirkungen des Ersten Weltkriegs* (2004), available at <http://www.zeitgeschichte-online.de/thema/kriegsschuldfrage-und-zeithistorische-forschung-deutschland>, at 8 (last visited 10 January 2020).

mention gas warfare frequently. Monographs, brochures, and essays devoted exclusively to poison gas were published.[82]

A number of such works were published in Germany, investigating the conduct of the war by Germany and other countries, some expressly discussing gas warfare. *Die deutsche Kriegführung und das Völkerrecht* was published already in 1919;[83] in the 1920s followed a five-volume work.[84] The former work claims the legality of the use of gas on 22 April 1915. The poison gas attack at Ypres was placed in historical and international contexts that were intended to relativize it. The specific claim of permissibility under international law resulted from additional arguments that the German authors presented to their readers with great care. Their conclusion defines the guilt and innocence of the parties to the war in a clear black-and-white schema:

> The factual and legal procedures in gas warfare are presented by applying the critical probe; they certainly justify the conduct of Germany, while to France falls the burden of having violated a global treaty. The reproaches made against us are thereby revealed to be part of the battle of lies by which the enemy league unceasingly strives to disparage us in the public opinion of the world.[85]

Compared with that of the war years, this discourse on international law was intense and decidedly detailed. It is characterized by a nationalist polarization of standpoints. The attributions of guilt to the respective wartime enemy were expressed in the language of law and morality. The preferred politics of international law could often be derived simply from the nationality of the authors.

The Germans loudly and energetically defended themselves against what seemed to them a form of victor's justice and corresponding assessments of international law.[86] They appealed to a number of counter-arguments purporting to justify the use of poison gas. By contrast, the assertion that the use of poison gas was illegal could not be found in a single publication from Germany. The authors presented a whole arsenal of arguments for its legality under international law: The Hague Declaration of 28 July 1899 is said not to have applied.[87] The Hague Convention of 1907 was also said not to be applicable in the First World War.[88] Alternatively, it was claimed that Article 23(a) was not relevant because gas is not a 'poison'.[89] It was claimed that the Hague

[82] Ewing, 'The Legality of Chemical Warfare', 61 *American Law Review* (1927) 58; Van Eysinga, *La guerre chimique et le movement pour sa repression (Academie de Droit International. Rec. des Cours 1927)*, vol. 16 (1928), at 329; J.L. Kunz, *Kriegsrecht und Neutralitätsrecht* (1935), at 85–88.

[83] Deutsches Kriegsministerium and Oberste Heeresleitung (eds), *Die deutsche Kriegführung und das Völkerrecht. Beiträge zur Schuldfrage* (1919), at 20–26.

[84] J. Bell and W. Schücking, *Das Werk des Untersuchungsausschusses der Deutschen Verfassungsgebenden Nationalversammlung und des Deutschen Reichstages 1919–1928, Völkerrecht im Weltkrieg*, vols I, II, III,1, III,2, IV (1927).

[85] Ibid., at IV, 42.

[86] Payk, above note 23, at 393–429; J. Leonhard, *Der überforderte Frieden. Versailles und die Welt 1918–1923* (2018), at 813, 899–902.

[87] Kunz, above note 11, at 20, 28; J. Meyer, *Der Gaskampf und die chemischen Kampfstoffe* (2nd ed., 1926), at 296.

[88] Meyer, above note 87, at 296-297.

[89] Deutsches Kriegsministerium and Oberste Heeresleitung (eds), above note 83, at 24; R. Hanslian, *Der chemische Krieg* (2nd ed., 1927), at 5; Kunz, above note 11, at 33; Meyer, above note 87, at 298; Overweg, above note 51, at 48–51.

Declaration of 1899 did not apply to the 'blue and yellow cross shells of the world war' or to artillery gas shells since 'the spreading of poisonous gases is not their only purpose; rather, their main purpose was to render the enemy harmless'.[90] Article 23(e) was also said not to be relevant because no 'unnecessary suffering' was caused.[91]

Furthermore, it was claimed in the alternative that the Hague Declaration of 28 July 1899 had been violated first by France; Germany merely followed suit and retaliated.[92] It had been a case of a 'state of emergency as recognized by international law' (*Deutsches Kriegsministerium und Oberste Heeresleitung 1919*).[93] In general, it was claimed that the 'Hague Accords had barely touched on the essence of gas warfare', since they had failed to recognize its humane essence.[94] The prohibitions of asphyxiating gas were said to have been annulled in the world war since both sides had made use of gas, claimed the international law scholar Josef Kohler in 1918; thus gas warfare had been legal.[95] The German international law professors Julius Hatschek and Arthur von Kirchenheim claimed between the wars that the use of poison gas conformed to international law.[96]

It is probably not oversimplifying too much to say that the standpoints of German jurists were primarily legitimizing, affirmative, defensive, and militaristic. One of the few exceptions was Mendelssohn-Bartholdy. He published a critical essay in 1927 with a revealing title.[97] In this text he expressed criticism of Haber and rejected the argument that the German use of poison gas had been a reprisal. He did, however, concede that the ambiguities of the rules had left many backdoors open.

The positions of academics in international law from Allied countries, by contrast, cannot be characterized so simply. There seems to have been a greater diversity of opinion and not so much thinking in camps. For example, there are French assessments in which poison gas as such is said to violate international law (but could be justified as a reprisal).[98] The Allied use had been justified if and to the extent it had been reprisal.[99] The Archbishop of Canterbury and the Bishop of London (on 16 May 1915), by contrast, appealed to their government not to employ poison gas and descend to the level of the enemy.[100] Articles 23(a) and 23(e) had been violated by the Germans to the extent that chlorine gas had been employed; by contrast, it was argued, other types of gas should be judged less harshly under international law.[101]

[90] Kunz, above note 11, at 26; K. Strupp, *Grundzüge des positiven Völkerrechts* (2nd ed., 1922), at 201.

[91] Deutsches Kriegsministerium and Oberste Heeresleitung (eds), above note 83, at 24; Hanslian, above note 89, at 5; Kunz, above note 11, at 32; Meyer, above note 87, at 298-299.

[92] Deutsches Kriegsministerium and Oberste Heeresleitung (eds), above note 83, at 23; Haber, above note 31, at 83; Kunz, above note 11, at 3-4; Meyer, above note 87, at 301.

[93] Deutsches Kriegsministerium and Oberste Heeresleitung (eds), above note 83.

[94] Meyer, above note 87, at 302.

[95] J. Kohler, *Grundlagen des Völkerrechts. Vergangenheit, Gegenwart, Zukunft* (1918), at 212–13.

[96] J. Hatschek, *Völkerrecht als System rechtlich bedeutsamer Staatsakte* (1923), at 316; Von Kirchenheim, 'Geschosse', in K. Strupp (ed.), *Wörterbuch des Völkerrechts und der Diplomatie, Erster Band* (1924) 403, at 405–06.

[97] Mendelssohn-Bartholdy, 'Kriegsnotwendigkeiten und Repressalien: Zwei Feinde des Völkerrechts', 6 *Europäische Gespräche: Hamburger Monatshefte für auswärtige Politik* (1927) 319.

[98] A. Rolin, *Le Droit moderne de la Guerre. Les Principes—Les Conventions—Les Usages et Les Abus* (1920), at 326–27.

[99] W.E. Hall, *A Treatise on International Law* (8th ed., 1924), at 637, n. 2; Garner, above note 42, at 262, 271–92.

[100] Garner, above note 42, at 273, n. 1.

[101] T.J. Lawrence, *The Principles of International Law* (7th ed., 1923), at 531.

Finally, there were positions not formulated by any of the countries that had been involved in the world war. Authors of such texts include Franz Carl Endres and Gertrud Woker. The latter published her book on behalf of the Women's International League for Peace and Freedom. Both authors emphasized not only that the German use of poison gas violated international law, but also the ongoing threat and the alarming role of a transnationally active armaments industry.[102] Woker frontally attacked Haber's argumentation, polemicizing against the 'magnificent militaristic logic' she saw in the arguments that gas was effective and in conformity with international law.[103] Given the highly politicized climate and the intent to legitimize past actions, it was not to be expected the scholarly discourse would get closer to the subject matter, and in fact it did not. The camps remained hostile to each other, and there was no discernible mediation by third positions or even any softening of tone during the period between the wars.[104] This thinking as part of a camp was particularly stark in German scholarship on international law.

Thus the scholarly community in international law was not unified and did not come together on a view of past events. It was, however, better disposed to shaping such a view in the future. After the experiences of the First World War, whether or not one judged certain past acts as in conformity with or as illegal under international law, all sides appeared agreed that poison gas should not be used in the future. Legal reforms and further standardization served to affirm the prohibition of poison under international law from the Paris Peace Treaties of 1919 (prohibiting the production and possession of weapons in Germany) and the League of Nations to the Geneva Protocol on Poison Gas of 1925. The general prohibition was in that sense a kind of self-reassurance and affirmation of the content of treaties and of customary law that had been valid prior to the First World War but had often been breached in praxis between states during the war.

5. Summary: Expectations Regarding International Law

The Geneva Protocol on Poison Gas seems a potential terminus for the narratives about the events of Ypres in 1915. It is still today valid and has been judged 'probably the most significant special standard thus far prohibiting the use of chemical weapons'.[105]

The prohibition of perfidious means and unnecessary suffering was violated many times in April 1915 and up until the end of the war in 1918, in the face of express agreements and despite the ongoing validity of customary international law. Even if one were to regard the detailed special prohibitions under international law as somehow inapplicable, there would remain the spirit of the norms as a further point of reference. After all, alongside the specific prohibitions there were also older general legal principles proscribing perfidious means and unnecessary suffering under international

[102] F.C. Endres, *Giftgaskrieg die grosse Gefahr* (1928), at 38–39.
[103] G. Woker, *Der kommende Giftgaskrieg* (1925), at 16–19.
[104] Payk, above note 23, at 127, 314, 403–17, 543; Leonhard, above note 86, at 1046.
[105] Marauhn, above note 10, at 47.

law deriving from the principle of humanity. That general principles of international law were disregarded to such an extent in the First World War relates to the particular moral values expressed in these principles. By this expression, legal principles can endow values with more vigour and prompt surprising decisions, but are also particularly fragile. It is no coincidence that certain legal disciplines in which the norm-setting is not yet complete tend towards legal principles. The international law of the nineteenth century was one such area. Alongside the juridification of international relations, it relied heavily on general legal principles as well as on natural law.[106]

The conflict surrounding poison gas in the law of war shows how fragile certain rules can become in practice. Under the conditions of the world war, under pressure from blazing nationalism and militarism, international law is drawn into the undertow of politicization, an undertow that weakened it so severely that one is inclined to speak of its almost total absence from certain areas of regulation in these years. Certain actors, including both military figures and international lawyers, practised the circumvention or even denial of the law of war. Whether one is disappointed by this depends on the expectations of a contemporary reader contemplating the historical events of earlier epochs. While in recent discourses on the normative justification of the use of force one can sometimes observe a veritable 'lawfare' in the area of International Humanitarian Law, in the years 1914–18 it seemed more reasonable to the actors to circumvent the law.[107] Whether this refers to an increased authority of international law in the trias of war, norms and technology must remain open at this point. However, what has to be highlighted from a legal-historical perspective here is the following: the law of war in the First World War was, especially when compared to other matters, a highly political area of regulation. In any event, the consideration of this time should always take into account the political circumstances and the larger context when it accentuates the *Realpolitik* and the relativism of this international law.[108] In this sense, political practice and its context always matter in the History of International Law.

[106] Vec, above note 8; Vec, above note 24.

[107] For later, but similar developments in the twentieth and twenty-first centuries, see T.M. Fazal, *Wars of Law. Unintended Consequences in the Regulation of Armed Conflict* (2018); see also the concluding chapter in this volume.

[108] Orakhelashvili, 'The 19th-century Life of International Law', in A. Orakhelashvili (ed.), *Research Handbook on the Theory and History of International Law* (2011) 441, at 454-455.

PART IV
FROM THE LEAGUE TO THE UN

The Universe of Western International Legal Order Revealing its Self-Contradictions

13
Peace through Law: Lessons from 1914

B.S. Chimni

1. Introduction

The theme 'peace through law' has engaged the attention of states and international legal scholars for more than a century.[1] As the chapters in the previous Part III of this book have shown, this also applies to the international legal and political discourses of the nineteenth century. But despite the longstanding debates about the legitimacy and legality of war and the enormous changes in the normative and institutional structures concerning this issue since the beginning of the twentieth century, wars are still with us. It is in this context that this chapter revisits, a hundred years after the end of the First World War (WWI), some reflections of those times on the root causes of war. It briefly discusses three texts either published or adopted after the beginning of WWI, two of them written by Sigmund Freud and Rosa Luxemburg. The third comprises resolutions adopted by the International Congress of Women that met at The Hague in 1915. These texts offered different perspectives on the causes of the war. The Freud text suggested that humankind had yet to come to terms with the destructive instincts that inhabit the human mind. Rosa Luxemburg used the lens of political economy to trace the war to the nature of capitalism and its internal relationship with imperialism. She argued that the war was primarily a conflict over colonial territories manifesting inter-imperialist contradictions. The resolutions of the Hague Congress of Women stressed the significance of enfranchising women, increasing their participation in decision-making, and subjecting foreign politics to democratic control. The chapter then proceeds to ask whether WWI could have been prevented by more, and more effective, international law and organizations, a question that has been asked with respect to innumerable conflicts and wars since. While much has changed since 1914 this has not diminished the power of many of the insights of Freud, Luxemburg, and the Hague Congress. Yet the tendency of much international legal scholarship over the past century, dominated as it is by positivism and pragmatism, has been to place faith in the 'peace though law' project to the general neglect of extra-legal realities.[2]

[1] This chapter is a modified and extended version of my 'Peace through law: Lessons from 1914', 3 *London Review of International Law* (2015) 245.

[2] A. Anghie, *Imperialism, Sovereignty and the Making of International Law* (2004), chs 2 and 3. On 'positivism', see further Simma and Paulus, 'The Responsibility of Individuals for Human Rights Abuses in International Conflicts: A Positivist View', in S.R. Ratner and A.M. Slaughter (eds), *The Methods of International Law* (2004) 23, at 25–32. See also the chapters by Siddharth Mallavarapu and Benno Teschke in this book. On different narratives of political and normative war discourses, see Hendrik Simon's and Lothar Brock's Introduction.

B.S. Chimni, *Peace through Law: Lessons from 1914* In: *The Justification of War and International Order.* Edited by: Lothar Brock and Hendrik Simon, Oxford University Press (2021). DOI: 10.1093/oso/9780198865308.003.0013

The response is epitomized by the work of jurists like Hans Kelsen, who despite his knowledge of WWI having served 'as legal adviser to the war minister' of Austria,[3] and despite also the subsequent occurrence of the Second World War (WWII), still contended that 'peace through law' was possible. The spirit of Kelsen continues to inform much mainstream international legal scholarship (MILS) which overlooks in the process the factors of human nature, gender, and political economy.[4] It has therefore been unable to appreciate for instance that international laws contribute to the creation of imperial structures or to understand the full import of the exclusion of the voices of women for achieving 'peace through law'.[5] The disregard of deeper sources of international conflicts has also meant that MILS has not seriously considered ways of addressing them.

It is, however, not the aim of the chapter to argue that law and legal institutions do not have an important role to play in the peace project. They undoubtedly have. The objective is merely to stress the elementary point that the deeper sources of conflict among nations cannot be ignored, especially at a time when the outcome of another conflict—the conflict with Nature (or war against Nature)—gains global attention, complicating the idea of 'peace' which now indisputably includes what may be called 'Peace with Nature'.

On the face of it the canvas of the present essay is vast. But the intent is an extremely modest one—to make certain *bare* connections that are ignored by MILS such as those between human nature, gender, capitalism, imperialism, and war (both between nations as well as against Nature). The essay does not contest the view that in explaining the origin of wars 'the particular matters'.[6] It only suggests that it would be imprudent to ignore the deeper causes of war, of which, admittedly, there need not necessarily be one account. For its part the essay identifies, and no more than that, one set of deep causes that can possibly explain the innumerable wars between nations, as also the war against Nature, over the century since WWI. The hope is that international lawyers will reflect on deep structures to see how these bear upon the 'peace through law' project and in turn upon legal methodology.

It is time to turn to the three texts mentioned beginning with that of Freud.

[3] Ladavac, 'Hans Kelsen (1881–1973): Biographical Note and Bibliography', 9 *European Journal of International Law* (1998) 391, at 391. See also 'Hans Kelsen', *New World Encyclopedia*, available at <http://www.newworldencyclopedia.org/entry/Hans_Kelsen> (last visited 29 November 2019).

[4] For a blueprint approach, see G. Clark and L.B. Sohn, *World Peace through Law* (3rd ed., 1966). But the subject of critique here is mainstream international law scholarship (MILS) in general. While MILS may not subscribe to the blueprint mode it shares the essential premise that war can be prevented through appropriate legal and institutional structures.

[5] See generally Chimni, 'Third World Approaches to International Law: A Manifesto', in A. Anghie, B.S. Chimni, O. Okafor, and K. Mickelson (eds), *The Third World and International Order: Law, Politics and Globalization* (2003) 47; H. Charlesworth and C. Chinkin, *The Boundaries of International Law: A Feminist Analysis* (2000).

[6] Lebow, 'What Can International Relations Theory Learn from the Origins of World War 1?', 28 *International Relations* (2014) 387, at 388.

2. Sigmund Freud on Human Nature and War (1914)

In an essay entitled 'Thoughts for the Times on War and Death' written in 1914, Freud made a number of observations that, despite the passage of time, remain germane to understanding war as a human condition. He began, however, by expressing his puzzlement that the war had occurred:

> We had expected the great ruling powers *among the white nations upon whom the leadership of the human species has fallen*, who were known to have *cultivated world-wide interests*, to whose creative powers were due our *technical advances in the direction of dominating nature*, as well as the artistic and scientific acquisitions of the mind—*peoples such as these we had expected to succeed in discovering another way of settling misunderstandings and conflicts of interest.*[7]

He lamented that instead of providing leadership, the white civilized nations set 'at naught all those restrictions known as International Law, which in peace-time the states had bound themselves to observe'.[8] Freud was also perplexed, or so he suggested, by the paradox of civility in internal relations and brutality abroad. He wrote of:

> the destitution shown in moral relations externally by the states which in their interior relations pose as the guardians of accepted moral usage, and the brutality in behavior shown by individuals, whom, as partakers in the highest form of human civilization, one would have not credited with such a thing.[9]

He then attempted to provide an answer to the question whether 'the white nations' could do something to set right the state of affairs that led to war and brutality. From this perspective Freud noted 'the two cooperating factors', one internal and the other external, which could possibly transform 'bad instincts' responsible for war.[10]

> The internal factor consists in an influence on the bad—say, the egoistic—instincts exercised by erotism, that is, by the human need for love, taken in its widest sense. By the admixture of *erotic* components the egoistic instincts are transmuted into *social* ones. We learn to value being loved as an advantage for which we are willing to sacrifice other advantages. The external factor is the force exercised by up-bringing, which advocates the claims of our cultural environment, and this is furthered later by the direct pressure of that civilization by which we are surrounded.[11]

[7] Freud, 'Thoughts for the Times on War and Death', in *Collected Papers*, vol. IV, edited by M.M.R. Khan (1925) 288, at 289 (emphasis added).
[8] Ibid., at 292.
[9] Ibid., at 295.
[10] Ibid., at 297.
[11] Ibid.

But even as he noted the possibility of transforming 'bad instincts' he warned that 'the primitive mind is, in the fullest meaning of the word, imperishable.'[12] He went on to make the unsettling statement that the idea of civilized society was based on an 'illusion', a view he elaborated upon later in *Civilization and its Discontents* (1929). 'In reality', he wrote, 'our fellow-citizens have not sunk so low as we feared, because they had never risen so high as we believed.'[13] Or, as he put it afterwards, 'the fateful question of the human species seems to me to be whether and to what extent the cultural process developed in it will succeed in mastering the derangements of communal life caused by the human instinct of aggression and self-destruction.'[14]

In singling out the dark and evil forces that reside in the human mind, Freud did not seriously explore other possible ways of conceptualizing human nature or other reasons for conflict among nations. He did not pay attention for instance to the kind of understanding contained in Karl Marx's sixth thesis on Feuerbach that 'human essence is no abstraction in each individual. In its reality it is the ensemble of the social relations.'[15] Freud therefore did not attach significance to economic causes of war and thus did not see rivalries over colonial possessions as a possible cause of WWI.[16] Indeed, his reference to the leadership of 'white nations' and 'cultivated world-wide interest' suggests that he accepted the distinction between civilized and uncivilized nations and saw nothing terribly wrong in the colonial project.

If he had paid heed to Marx and capitalism, as we shall see Rosa Luxemburg did, he would not only have better grasped the complex causes of war, but would also have understood the social basis of competing nationalisms that subjected him to the same passions as others: nationalism. It is interesting that while Freud condemned the war and the violation of international law he wrote that

> [w]e live in the hope that the impartial decision of history will furnish the proof that precisely this nation [i.e., Austria-Hungary], this in whose tongue we now write, this for whose victory our dear and near ones are fighting [his immediate family was involved], was the one which least transgressed the laws of civilization—but at such a time who shall dare present himself as the judge of his own cause?[17]

It may not be out of place to mention the views of his other illustrious contemporary Max Weber. Weber had viewed 'the military arms race among the superpowers [preceding the war] as an unfortunate but inevitable side-effect of a welcome state of

[12] Ibid., at 297, 301. The idea of 'primitive mind' is for the purposes of this essay treated as distinct from the idea of 'primitive peoples' or 'primitive cultures' that underlies the Western civilizing mission. It is seen as a reference to inherent and unchanging human nature.

[13] Ibid., at 300.

[14] Freud, *Civilization and its Discontents* (2000) [1930].

[15] K. Marx and F. Engels, *Selected Works*, vol. 1 (1969), at 14.

[16] Freud criticized Marxist thinking in *Civilization and its Discontents* observing that 'the psychological premises on which the [communist] system is based are an untenable illusion ... Aggressiveness was not created by property.' Freud, above note 14, at 25. But he did add later: 'I too think it unquestionable that an actual change in men's attitude to property would be of more help in this direction than any ethical commands; but among the Socialists this proposal is obscured by new idealistic expectations disregarding human nature, which detract from its value in actual practice.' Ibid., at 39.

[17] Freud, above note 7, at 293.

affairs'.[18] Indeed, he favoured the German naval build-up.[19] For, as he wrote, the division of world power ultimately means the control of the nature of future culture. Future generations will hold us responsible in these matters, and rightly so, for we are a nation of seventy and not seven millions'.[20] We look in vain in Weber, one of the great masters of social science, for a serious analysis of the war beyond nationalist rhetoric. Wolfgang Mommsen is therefore not perhaps altogether wrong in concluding that 'Weber was an imperialist out of conviction'.[21]

Be that as it may, it would be myopic to be dismissive of the view that 'bad instincts' is at least one root cause of war, a view that classical *Realists* like Hans Morgenthau came to adopt.[22] Therefore, along with the legal positivist programme for peace which, as we shall see, Freud also endorsed, ways have to be found to address 'bad instincts' in so far as is possible.

3. Women and WWI: The Hague Congress of Women (1915)

About a year after Freud wrote his essay, the International Congress of Women met in The Hague and called for an immediate end to the war.[23] The Congress also agreed with the need for a legal positivist programme to maintain peace, endorsing particular proposals that had been advanced in the period.[24] But it also emphasized dimensions that were unique for the time. The Congress adopted a resolution stating that 'Foreign Politics shall be subject to democratic control' immediately followed by the demand

[18] W.J. Mommsen, *The Political and Social Theory of Max Weber: Collected Essays* (1989), at 28; I. Dronberger, *The Political Thought of Max Weber* (1971), at 29f.

[19] Dronberger, above note 18, at 152.

[20] Cited in H.H. Gerth and C. Wright Mills (eds), *From Max Weber: Essays in Sociology* (1970) 40.

[21] Mommsen, above note 18, at 28. In response to the suggestion, for which Mommsen admits there is some basis, that Weber's support was tactical, as a way to advance the 'liberalization of German society', Mommsen writes: 'In the light of the available data one cannot but conclude that for Max Weber an active German imperialistic policy was more than just a clever strategy to give life to German liberalism.' Ibid., at 29. See also Dronberger, above note 18, at 155f.

[22] H. Morgenthau, *Politics among Nations: The Struggle for Power and Peace* (4th ed., 1967), at 4.

[23] For text of resolutions, see 'Resolutions Adopted', International Congress of Women, The Hague, 28 April–1 May 1915, available at <http://www.ub.gu.se/kvinndata/portaler/fred/samarbete/pdf/resolutions_1915.pdf> (last visited 29 November 2019) ('Resolutions Adopted'). See generally Baetens, 'International Congress of Women (1915)', *Max Planck Encyclopedia of Public International Law* (2010), available at <http://opil.ouplaw.com/home/epil> (last visited 29 November 2019).

[24] There were several resolutions that indicated elements of a positivist programme. Some of these may be worth mentioning. Resolution 6 stated: 'This International Congress of Women, believing that war is the negation of progress and civilization, urges the governments of all nations to come to an agreement to refer future international disputes to arbitration and conciliation.' 'Resolutions Adopted' 7. In Resolution 9, the Congress urged 'that a third Hague Conference be convened immediately after the war'. Ibid., at 8. In Resolution 11(a), the Congress called for the 'development of the Hague Court of Arbitration, a permanent International Court of Justice to settle questions or differences of a justiciable character, such as arise on the interpretation of treaty rights or of the law of nations'. Ibid. Finally, Resolution 14 stated 'This International Congress of Women demands that all secret treaties shall be void and that for the ratification of future treaties, the participation of at least the legislature of every government shall be necessary'. Ibid., at 9. For a discussion of the contribution of the Hague Congress to the positivist programme of international law, see Baetens, above note 23.

that women should be granted 'equal political rights with men'.[25] The stated assumption was that 'the combined influence of the women of all countries is one of the strongest forces for the prevention of war'.[26] Further, in Resolution 11(b) the Congress called for 'a permanent International Conference holding regular meetings in which women should take part' to deal 'with practical proposals for further International Cooperation among the States'.[27] The International Congress of Women also passed a resolution opposing 'the assumption that women can be protected under the conditions of modern warfare'.[28] It protested 'vehemently against the odious wrongs of which women are the victims in time of war, and especially against the horrible violation of women which attends all war'.[29] The resolution anticipated contemporary debates on the gaps in international humanitarian law when it comes to protection of women in times of conflicts. Finally, the Congress adopted a resolution expressing solidarity with the peoples of all nations involved in the war. It stated: '[t]his International Congress of Women of different nations, classes, creeds and parties is united in expressing sympathy with the suffering of all, whatever their nationality, who are fighting for their country or laboring under the burden of war'.[30] This resolution offers testimony to the different way in which women can think about war and suffering even in times of conflict.[31] While there is a need to avoid the essentialist trap, it can be said that women, as a consequence of their position in the sexual and social division of labour, or of social relations in general, are able to bring to bear on conflicts experiences and perspectives that may be more favourable to their resolution.[32] It has been argued for instance that women can help promote cooperation among nations as they tend to, among other things, 'see issues in contextual rather than in abstract terms'.[33] To be sure, the relationship between feminism and war is 'contested and complex'[34]

[25] '[T]he Congress declared that women ought to have equal political rights, not only the right to vote but also to be elected and equally represented on all levels of government, as well as in the judiciary and academia. Seen in its historic context, this is a spectacular demand in light of the fact that in 1915 only five countries in the world gave fully equal voting rights to all women.' Baetens, above note 23, para. 11. The call for enfranchisement of women reminds that the war had little to do with the defence of 'western civilization', 'liberal values', or 'democracy' as it is often made out to be by the victorious powers. As Michael Faulkner points out, at this point of time 'only about 40% of the male electorate in Britain had voting rights' and 'women's suffrage campaigners were still fighting for their rights and going to jail for their principles'. Faulkner, 'World War One 1914–18: One Hundred Years of a Failure to Learn from "The War to End Wars"', *No Glory*, 12 February 2014, available at <http://noglory.org/index.php/articles/132-world-war-one-1914-18-one-hundred-years-of-failure-to-learn-from-the-war-to-end-wars> (last visited 29 June 2015).

[26] Resolution 9, in 'Resolutions Adopted', above note 23, at 8.

[27] Ibid.

[28] Resolution 2, in ibid., at 5.

[29] Ibid.

[30] Resolution 3, entitled 'The Peace Settlement', in ibid., at 5f.

[31] It may however be noted that women from most Western countries involved in the war were present at the Congress: Austria, Belgium, Canada, Denmark, Germany, Great Britain (from which 180 women were prevented from attending), Hungary, Italy, the Netherlands (from which 1,000 women participated), Norway, Sweden, and the United States (from which 47 women participated). Ibid., at 4.

[32] See generally Holmstrom, 'A Marxist Theory of Women's Nature', in N. Holmstrom (ed.), *The Socialist Feminist Project: A Contemporary Reader in Theory and Politics* (2011), at 360.

[33] Tickner, 'Hans Morgenthau's Principles of Political Realism: A Feminist Reformulation', 17 *Millennium: Journal of International Studies* (1988) 429, at 433.

[34] Riley, Mohanty, and Pratt, 'Introduction: Feminism and US Wars—Mapping the Ground', in Riley et al. (eds), *Feminism and War: Confronting U.S. Imperialism* (2008), at 1f. They go on to add that 'few feminist theorists continue to accept the claim that women are natural pacifists, yet there is much debate over how to

but it does not take away from the need for different standpoints on questions of peace and war.[35]

The troubling aspect of the Congress was that no women from the colonies were invited.[36] In fact, delegates 'spoke in rather patronizing terms about the people in the colonies; "protecting the locals" who are not capable of governing their lands themselves',[37] albeit the Congress also spoke of justice for subject communities. One of the resolutions adopted by the Congress stated that:

> This Conference should be so constituted that it could formulate and enforce those principles of justice, equity and good will in accordance with which the struggles of subject communities could be more fully recognized and the interests and rights not only of the great Powers and small nations but also those of weaker countries and primitive peoples gradually adjusted under an enlightened international public opinion.

The reference to 'primitive peoples' merely showed that women are also a product of their place and times for otherwise the Congress could have linked the oppression of women with the oppression of colonies, and in turn with capitalism.[38]

4. Rosa Luxemburg on Capitalism, Imperialism, and War (1915)

Rosa Luxemburg made the connections that had eluded the International Congress of Women. At the same time as the meeting of the International Congress of Women was taking place, Rosa Luxemburg (1871–1919) wrote a most remarkable essay known as the 'Junius Pamphlet' that repays reading even today.[39] Like Freud she began by noting how 'treaties and alliances, the holiest words and the highest authorities have been torn into scraps'.[40] However, her analysis of WWI differed radically from both that of

understand what is in women's interests and how to advance those interests'. Ibid.; see also the contribution by Anthony Lang, Jr who refers to Cynthia Enloe's take on the justification of war.

[35] See Zalewski and Runyan, 'Taking Feminist Violence Seriously in Feminist International Relations', 15 *International Feminist Journal of Politics* (2013) 293.

[36] Baetens, above note 23, para. 4.

[37] Ibid., para. 16.

[38] Johanna de Groot, for instance, has demonstrated how 'nineteenth century representations and discourses of sexual identity and difference drew upon and contributed to comparable discourses and representations of ethnic, "racial", and cultural identity and difference'. de Groot, '"Sex" and "Race": the Construction of Language and Image in the Nineteenth Century', in C. Hall (ed.), *Cultures of Empire: A Reader* (2000) 37, at 57.

[39] R. Luxemburg, *The Crisis in German Social-Democracy: (The 'Junius' Pamphlet)* (1919). An edition is also at *Marxists Internet Archive*, available at <https://www.marxists.org/archive/luxemburg/1915/junius/> (last visited 29 November 2019). The pamphlet is said to have been written between February and April 1915 from her prison cell. Luxemburg was incarcerated for opposing the war. The essay contains exceedingly rich analysis of the times. However, this is not the occasion to offer detailed comments on other matters she addressed including the strategy and tactics of social democrats.

[40] Ibid., at 8.

Freud and the Congress of Women. Luxemburg traced the war to the nature and development of capitalism. In a powerful and insightful passage she wrote:

> The events that bore the present war did not begin in July 1914 but reach back for decades. Thread by thread they have been woven together on the loom of an inexorable natural development until the firm net of imperialist world politics has encircled five continents. It is a huge historical complex of events, whose roots reach deep down into the Plutonic deeps of economic creation, whose outermost branches spread out and point away into a dimly dawning new world, events before whose all-embracing immensity, the conception of guilt and retribution, of defense and offence, sink into pale nothingness.[41]

Luxemburg noted how capitalist nations 'were competing in their expansion toward the non-capitalist countries and zones of the world'.[42] She referred to the 'strong tendency toward colonial expansion' and detailed the efforts of Britain, France, Russia, Germany, and the United States in this direction.[43] It may be recalled here that the last quarter of the nineteenth century witnessed a tremendous drive towards carving out the still independent sectors of the world: that is, nearly all of Africa, a large part of Asia, and a number of Pacific islands. In fact, as Harry Magdoff has pointed out:

> the rate of new territorial acquisitions of the new imperialism was almost three times that of the earlier period. Thus, the increase in new territories claimed in the first seventy-five years of the nineteenth century averaged about 83,000 square miles (210,000 square kilometres) a year. As against this, the colonial powers added an average of about 240,000 square miles (620,000 square kilometres) a year between the late 1870s and World War I (1914–1918).[44]

Luxemburg pointed out that these colonial expansions,[45] and the ensuing struggle for colonies, had generated 'competition of armaments in all nations' and concluded that

[41] Ibid., at 97.

[42] Ibid., at 34. She observed that colonial expansion had:

> created new, extra-European antagonisms on all sides: between Italy and France in Northern Africa, between France and England in Egypt, between England and Russia in Central Asia, between Russia and Japan in Eastern Asia, between Japan and England in China, between the United States and Japan in the Pacific Ocean—a very restless ocean, full of sharp conflicts and temporary alliances, of tension and relaxation, threatening every few years to break out into a war between European powers. It was clear to everybody, therefore, (1) that the secret underhand war of each capitalist nation against every other, on the backs of Asiatic and African peoples must sooner or later lead to a general reckoning, that the wind that was sown in Africa and Asia would return to Europe as a terrific storm, the more certainly since increased armament of the European states was the constant associate of these Asiatic and African occurrences; (2) that the European world war would have to come to an outbreak as soon as the partial and changing conflicts between the imperialist states found a centralized axis, a conflict of sufficient magnitude to group them, for the time being, into large, opposing factions. This situation was created by the appearance of German imperialism.

Ibid., at 35.

[43] Ibid.

[44] H. Magdoff, *Imperialism: From the Colonial Age to the Present Times* (1978), at 34f.

[45] Luxemburg, above note 39, at 35.

'the fact that each country will strive to occupy the colonial possessions of its opponent, or at least to create disturbances therein, automatically turns every war into an imperialistic world conflagration'.[46]

It is important to note that Luxemburg is making two distinct points here. First, she is speaking of the inherent tendency for capitalism to expand and occupy non-capitalist spaces. Indeed, she was amongst the first to argue, in her book *The Accumulation of Capital*, that imperialism is linked to the very survival of capitalism.[47] As the economist Joan Robinson noted, she drew attention to the historical fact that 'the extension of capitalism into new territories' was the mainspring of the 'vast secular boom' between the seventeenth and the nineteenth centuries.[48] Luxembourg grasped the essence of imperialism in stating that 'imperialism is the political expression of the accumulation of capital in its competitive struggle for what remains still open of the non-capitalist environment'.[49] This insight remains valid even today, albeit that the nature of imperialism has undergone transformation in the post-colonial era.

Second, Luxemburg was pointing to the phenomenon of inter-imperialist rivalry, that is, the tendency of imperialist nations to compete with each other for non-capitalist spaces. It is this competition that led to the armaments race. In the words of Luxemburg, 'militarism is a weapon in the competitive struggle between capitalist countries for areas of non-capitalist civilization'.[50]

She summed up the two trends by observing that in the process of acquiring non-capitalist areas, 'imperialism grows in lawlessness and violence both in aggression against the non-capitalist world and in ever-more serious conflicts among the competing capitalist countries'.[51] The two world wars and innumerable conflicts since have revealed the importance of her insights. International laws have been cast aside as so many irrelevancies, underlining the need for the 'peace through law' project to come to terms with the political economy of imperialism.

But in so far as inter-imperialist rivalry went, for decades after the end of WWII the views of Luxemburg's senior colleague in the left movement, Karl Kautsky, who spoke of ultra-imperialism, that is, the peaceful exploitation of the world by collaborating

[46] Ibid., at 100. It is worth noting here that the Hague Conference of Women also passed a resolution on 'general disarmament' that traced the problem to 'private profits' and to that extent coincided with the Luxemburg view:

> The International Congress of Women, advocating universal disarmament and realizing that it can only be secured by international agreement, urges, as a step to this end, that all countries should, by such an international agreement, take over the manufacture of arms and munitions of war and should control all international traffic in the same. It sees in the private profits accruing from the great armament factories a powerful hindrance to the abolition of war.

Resolution 12, in 'Resolutions Adopted', above note 23, at 9. Luxemburg's analysis of the political economy of war was later captured in the language of international law by Evgeny Pashukanis who wrote that the essence of international law in his times was 'the struggle of capitalist states among themselves'. Pashukanis, *Pashukanis: Selected Writings on Marxism and Law*, edited by P. Beirne and R. Sharlet (1980), at 169.

[47] R. Luxemburg, *The Accumulation of Capital* (1951 [1913]).

[48] J. Robinson, *Introduction to Luxemburg* (1951), at 28.

[49] Luxemburg, above note 47, at 446.

[50] Ibid., at 454. She went on to aptly observe that 'in addition, militarism has yet another important function. From the purely economic point of view, it is a pre-eminent means for the realization of surplus value; it is in itself a province of accumulation.' Ibid.

[51] Ibid., at 446.

imperialist nations, seemed to have greater validity.[52] However, that understanding is challenged today by the events in Ukraine. At least one possible way of understanding the annexation of Crimea is as a manifestation of inter-imperialist competition between NATO countries and Russia. This theme is reverted to presently.

5. Could More International Law Have Prevented WWI?

There is at first the need to address the significant question whether WWI could have been prevented with more international law and better organizations. Could the acceptance of the international legal positivist war prevention programme advanced by the United States at that time have succeeded? One contemporary American scholar, Francis Boyle, has observed that 'a good historical argument could be made that the First World War occurred in substantial part because there was too little … international law and organizations'.[53] The key elements of the proposed agenda between 1898 and 1917 have been conveniently summarized by Boyle:

> (1) The creation of a general system for the obligatory arbitration of disputes between states; (2) the establishment of the international court of justice; (3) the codification of important areas of customary international law into positive treaty form; (4) arms reduction, but only after, not before, the relaxation of international tensions by means of these and other legalist techniques and institutions; and (5) the institutionalization of the practice of convoking periodic conferences of all states in the recognized international community.[54]

After spelling out the proposed agenda, Boyle asks a series of questions:

> What if Germany had not objected to the principle of obligatory arbitration at the First Hague Peace Conference, or to the conclusion of a multilateral obligatory arbitration treaty at the Second? What if the Latin American states had not opposed the formation of the Court of Arbitral Justice at the Second Hague Peace Conference over the issue of its composition, which did not impede adoption of the plan for the International Prize Court? What if the House of Lords had not rejected the Declaration of London and the International Prize Court in 1911? What if the nations of the world had proceeded on schedule in 1913 to enter into preliminary preparations of the convocation of the Third Hague Peace Conference in 1915? Would there have been a First World War in 1914 if any or more of these international legal positivist developments had occurred prior thereto?[55]

Boyle concludes:

[52] Kautsky, 'Ultra-Imperialism', I/59 *New Left Review* (1970), at 41.
[53] F.A. Boyle, *World Politics and International Law* (1985), at 43.
[54] Ibid., at 28.
[55] Ibid., at 43.

> The historical record will substantiate the proposition that with just a little more support from a few obstreperous actors at key moments in time, the elements of the pre-World War I American legal positivist war prevention program for world politics could have fallen into place soon enough to create a reformed structure of international relations in which conditions propitious for the outbreak of a general systemic war in Europe could have been substantially ameliorated.[56]

The writings of the times published in *The American Journal of International Law* reflected many of the assumptions and sentiments that inform Boyle's questions and response on the positivist programme. There was little appreciation of the limits of the role of international law. Neither human nature, nor gender, nor capitalism, nor imperialism figured in them. Instead stress was laid on education in international law.[57] In April 1914, just three months before the war broke out, the American Society of International Law sponsored the Conference of American Teachers of International Law 'to increase and broaden the instruction of international law'. In the circumstances 'the outbreak of war in Europe', according to Boyle, 'was a shock to many international lawyers'.[58] Elihu Root, Secretary of State at the time of Second Hague Conference and later president of the American Society of International Law was disillusioned. He noted that the structure of international law had been 'rudely shaken' and that nations will 'observe ... rules only so far as their interest at that time prompts them'.[59] Root came to recognize that 'little reliance can be placed upon unrestrained human nature, subject to specific temptation to commit forcible aggression in the pursuit of power and wealth'.[60] He went on to observe that 'the acquired habits of civilization are weak as against the powerful, innate tendencies which survive from the countless centuries of man's struggle for existence against brutes and savage foes'.[61] Yet he felt that the only way out was to make the restraint of law more effective[62] through having 'sanctions behind them'.[63] He went on to propose 'a court of international justice, with a general obligation [on states] to submit all justiciable questions to its jurisdiction and to abide by its judgment' as 'a primary requisite to any real restraint of law'.[64] Thus, while 'innate tendencies' and 'unrestrained human nature' along with the 'pursuit of power and wealth' received mention (as yet the issue of enfranchisement and representation of women did not) faith was still placed in the 'peace through law' project.

This echoed the fact that in the Allied declarations of WW I, there had been quite active references to international law, and after their defeat even the Germans actively referred to legal norms in order to demand a peace treaty based on Wilson's Fourteen

[56] Ibid.

[57] The lead article in the very first issue of *The American Journal of International Law* was written by Elihu Root entitled 'The Need of Popular Understanding of International Law', 1 *The American Journal of International Law* (1907) 1. See also H. Shinohara, *US International Lawyers in the Interwar Years: A Forgotten Crusade* (2012) 20, at 15.

[58] Boyle, above note 53. at 21.

[59] Root, 'The Outlook for International Law', 10 *American Journal of International Law* (1916), at 1f.

[60] Ibid., at 4.

[61] Ibid., at 4.

[62] Ibid.

[63] Ibid., at 5.

[64] Ibid.

Points. The emergence of modern international law of war and peace in the nineteenth century[65] had raised expectations that could no longer be ignored. But this also meant that the defeated Germans were to be punished uncompromisingly according to a rigid formal legalism. Taking up this observation, German historian Marcus M. Payk asks himself whether *less* international law could ultimately have meant *more* peace.[66]

6. Ignoring Lessons: Hans Kelsen's *Peace through Law* (1944)

In fact the belief in 'peace through law' prevailed both among leading thinkers and international lawyers in the inter-war period.[67] Thus, for instance, in response to a letter from Albert Einstein, Freud returned fifteen years later to the question of how war can be avoided. Despite his scepticism about the possibility of reining in 'bad instincts' he saw no alternative to supporting the 'peace through law' project:

> There is but one sure way of ending war and that is the establishment, by common consent, of a central control which shall have the last word in every conflict of interests. For this, two things are needed: first, the creation of such a supreme court of judicature; secondly, its investment with adequate executive force. Unless this second requirement be fulfilled, the first is unavailing.[68]

He then went on to applaud the League of Nations experiment as one which had 'never before, perhaps, on such a scale been attempted in the course of history'.[69] Needless to add, the nature of the mandate system did not engage his attention.[70]

The 'peace through law' thinking even survived WWII. It will suffice for the present purposes to draw attention to some features of the work of that great legal mind, Kelsen, encapsulated in his book *Peace through Law*.[71] In it, he contended that peace can be realistically achieved only through 'a slow and steady perfection of the international legal order'.[72] The underlying premise was that 'the idea of law, in spite of everything [i.e., the two world wars] still seems to be stronger than any other ideology of power'.[73] Kelsen did briefly turn to the causes of war, in a section titled 'Economic or

[65] See I.V. Hull, *A Scrap of Paper: Breaking and Making International Law during the Great War* (2014); Simon, 'The Myth of *Liberum Ius ad Bellum*: Justifying War in 19th-Century Legal Theory and Political Practice', 29 *European Journal of International Law* (2018) 113. See also the contributions by Lauren Benton, Isabel Hull, and Hendrik Simon to this volume.

[66] M.M. Payk, *Frieden durch Recht? Der Aufstieg des modernen Völkerrechts und der Friedensschluss nach dem Ersten Weltkrieg* (2018); Payk, '"What We Seek Is the Reign of Law": The Legalism of the Paris Peace Settlement after the Great War', 29 *European Journal of International Law* (2018) 809.

[67] This is a trend that classical *Realists* like E.H. Carr and Hans Morgenthau subjected to scathing critique. See for instance E.H. Carr, *The Twenty Years' Crisis, 1919–1939* (2001).

[68] Freud, 'Why War?', in P. Mayer (ed.), *The Pacifist Conscience* (1966) 238, at 242.

[69] Ibid. Of course he noted that '[i]f the propensity for war be due to the destructive instinct, we have always its counter-agent, Eros, to our hand. All that produces ties of sentiment between man and man must serve us as war's antidote.' Ibid., at 246.

[70] On the mandate system of the League of Nations, see Anghie, above note 2, at 115–95.

[71] H. Kelsen, *Peace through Law* (1944).

[72] Ibid., at ix.

[73] Ibid., at 21.

juridical approach?', but privileged the juridical approach, as many American scholars had done in previous decades, rejecting what he termed the Marxist view of war that traced it to the capitalist system.[74] Kelsen admitted that wars have economic causes but observed that 'they are not the root cause'.[75] The colonial project or the struggle over colonies did not figure in the book at all. And despite having interacted with Freud and even written on aspects of his work, Kelsen did not speak about human nature. From all accounts, it appears that he did not endorse the Freud thesis on human instincts of aggression and self-destruction.[76] The issue of gender also, despite the Hague Congress, and the women's movement, did not receive a mention. In Kelsen's view the ultimate cause was the anarchic international system.[77] In the event he concluded that the elimination of war 'is a problem of international policy, and the most important means of international policy is international law'.[78]

In *Peace through Law*, Kelsen went on to advance 'a complete legal-institutional strategy to pursue a stable and universal peace among nations' to address the problem of an anarchic international system.[79] Annex I of the book contained a 'Covenant of a Permanent League for the Maintenance of Peace'[80] and Annex II 'Treaty Stipulations establishing Individual Responsibility for Violations of International Law (International Criminal Jurisdiction)'.[81] Kelsen assigned a central role to an international court of justice, to individual responsibility for breaches of international law, and the creation of an international police force.[82] In order to maintain the integrity of the legal order furthermore Kelsen endorsed the idea of just war. In this respect, a recent critic has aptly noted that 'it is undoubtedly paradoxical for an author who lays claim to pacifist and anti-imperialist ideals—and makes peace the ultimate end of law—to assume (just) war as the condition for the legal nature of the international system'.[83] Besides the fact that the 'practical applicability of the theory is problematic in the absence of a neutral higher authority invested with the power to determine whether acts of war are just or unjust' there was the serious objection that 'only a state which is stronger than its adversary state is in a position to use war as a legitimate instrument of coercion'.[84] However, in Kelsen's view, the idea of just war, along with the notion of self-help, was essential if international law were to serve as instrument of peace in an anarchic international system. Thus, an ironic outcome of the 'peace through law' project was the justification of the use of force by powerful imperialist

[74] Ibid., at 17.
[75] Ibid.
[76] Jabloner, 'Kelsen and his Circle: The Viennese Years', 9 *European Journal of International Law* (1998) 368, at 384.
[77] Kelsen, above note 71, at 17.
[78] Ibid., at 18.
[79] Zolo, 'Han Kelsen: International Peace through International Law', 9 *European Journal of International Law* (1998) 306, at 317: 'Kelsen borrows from Kant both the ideal of perpetual peace and the federalist model, as well as the idea of a ... "world citizenship" which includes as its subjects all the members of the human species'; on Kant, see Oliver Eberl's contribution to this volume.
[80] Kelsen, above note 71, at 127–41.
[81] Ibid., at 141f.
[82] For how 'Peace through Law' was appraised by his contemporaries, see von Schmadel, 'Kelsen's Peace through Law and its Reception by his Contemporaries', 39 *Hitotsubashi Journal of Law and Politics* (2011) 71.
[83] Zolo, above note 79, at 315.
[84] Ibid.

states. It is therefore not surprising that Kelsen was forgetful about the fact that Asia and Africa were still under colonial rule. In fact the nature and character of international law in general—that is, its colonial character—was almost of no concern. Kelsen would therefore have been unable to understand why someone like Mahatma Gandhi, the apostle of non-violence, was critical of the Kellogg-Briand Pact that sought to prohibit the use of war as an instrument of national policy. For Gandhi, the Pact talked of a peace that coexisted with colonialism.[85] Gandhi would have expressed the same sentiment about contemporary attempts to bring about 'peace through law' without addressing the phenomenon of neo-colonialism that manifests itself in, among other things, imperialist wars against weak nations.

7. Inter-Imperialist Competition Today: Annexation of Crimea

The possibility of realizing 'peace though law' has in recent years been brought into question by a range of armed interventions and wars waged by the United States and its NATO allies against the former Yugoslavia, Iraq, Afghanistan, and Libya. In these instances the international rule of law has not been respected.[86] While a range of legal justifications have been offered it can be said that the dominant view is that these do not withstand scrutiny. These interventions and wars have again brought to the fore the relationship between capitalism, imperialism, and war. The Russian annexation of Crimea and events in Eastern Ukraine even take us back to WWI. NATO countries have cried foul over the Russian violation of the international rule of law.[87] The question is whether it is gaps in international law or flaws in the structure of international bodies (such as the Security Council or the International Court of Justice (ICJ)) that encouraged Western interventions and wars against weak nations or the Russian boldness in Ukraine? Or do we need to go back to Freud and Luxemburg to understand these developments? Are the innumerable wars that the world has witnessed in some ways about the imperishable dark and destructive forces that reside in the human mind? Or are the intervention and wars, including the Russian annexation of Crimea and aggression against Eastern Ukraine, to be traced to the workings of global capitalism that leads to inter-imperialist rivalry that cannot be contained today? It is pertinent in this context to note that the first foreign policy document of President Putin spoke of restoring the 'Great Power' status of Russia, revealing a concern of being hemmed in by NATO powers.[88] The British Marxist scholar Alex Callinicos suggests

[85] Gandhi criticized 'the Briand-Kellogg Pact of 1929 [1928, BSC], because the Pact was still exploitative if through peaceful means, and because the Pact recommended avoiding violence to those who had already been deprived of the use of violent means. As he was also to point out, "A person who has never hurt a fly will fail to understand the meaning of an appeal made to him not to spill blood"'. Bartolf, 'Gandhi and War: The Mahatma Gandhi/Bart de Ligt Correspondence', *Satyagraha Foundation for Nonviolence Studies*, 22 August 2013, available at <http://www.satyagrahafoundation.org/gandhi-and-war-the-mahatma-gandhi-bart-de-ligt-correspondence/> (last visited 29 November June 2019).

[86] See the contribution by Anna Geis and Wolfgang Wagner to this volume.

[87] See the contribution by Paul Robinson and Mikhail Antonov to this volume.

[88] For a comparative table of the 2000, 2008, and 2013 official Russian foreign policy documents, see Gonzalez, 'The Foreign Policy Concept of the Russian Federation: A Comparative Study', Instituto Español de Estudios Estratégicos, Framework Document No. 06/2013, April 2013, available at <http://www.ieee.es/

that the annexation of Crimea is best seen as the sharpening of inter-imperialist rivalry once again. In his words:

> The classical theory of imperialism is, more than anything else, a theory of intra-capitalist *competition*. Imperialism is a system, the form taken by capitalism when the concentration and centralization of capital bring about the fusion of economic competition among capitals and geopolitical competition among states. Putin's actions express exactly this imperialist logic, combining geopolitical preoccupations (above all, blocking NATO expansion) with economic motivations (fear that Russian firms will be squeezed out of the Ukrainian market by European rivals).[89]

There is certainly merit in the Callinicos analysis. However, from the perspective of the 'peace through law' imagination it is interesting to note that in the instance of the annexation of Crimea the ICJ, the centrepiece of all 'peace through law' projects, may have actually contributed to the conflict through its Advisory Opinion in the *Unilateral Declaration of Independence by Kosovo* case. While the ICJ clarified in this case that it was not concerned with 'remedial secession' it provided implicit legitimacy to the idea by recognizing that the unilateral 'declaration of independence' by the Provisional Institutions of Self-Government of Kosovo 'did not violate international law'.[90] In his address to the Russian parliament, President Putin invoked the case to justify the annexation of Crimea.[91] Thus the *Kosovo* case proved to be a dangerous precedent.

8. Capitalism Today: War against Nature

Meanwhile, even as international laws and institutions have not been able to prevent wars between (and inside) nations, the very idea of 'peace' has come to be transformed. In contemporary times the issue of war against Nature has become as central as war between (or within) nations complicating thereby the idea of 'peace'. The concept of

en/Galerias/fichero/docs_marco/2013/DIEEEM06-2013_Rusia_ConceptoPoliticaExterior_FRuizGlez_ENGLISH.pdf> (last visited 29 November 2019).

[89] Callinicos, 'Imperial Delusions', *International Socialism*, 31 March 2014, available at <http://isj.org.uk/imperial-delusions/> (last visited 29 November 2019).

[90] *Accordance with International Law of the Unilateral Declaration of Independence in Respect of Kosovo* (Advisory Opinion), ICJ Reports (2010) 403, at 453.

[91] President Putin noted in his address that:

> the Crimean authorities referred to the well-known Kosovo precedent—a precedent our western colleagues created with their own hands in a very similar situation, when they agreed that the unilateral separation of Kosovo from Serbia, exactly what Crimea is doing now, was legitimate and did not require any permission from the country's central authorities. Pursuant to Article 2, Chapter 1 of the United Nations Charter, the UN International Court agreed with this approach and made the following comment in its ruling of July 22, 2010, and I quote: 'No general prohibition may be inferred from the practice of the Security Council with regard to declarations of independence,' and 'General international law contains no prohibition on declarations of independence'. Crystal clear, as they say.

'Address by President of the Russian Federation' Moscow, 18 March 2014, available at <http://en.kremlin.ru/events/president/news/20603> (last visited 27 May 2015).

peace today has perforce to include the idea of Peace with Nature.[92] In fact the war against Nature has raised the possibility of 'a world without us' in the same way as nuclear weapons have done.[93] Thinking of living in peace without ensuring peace with Nature has become impossible. The ecological crisis returns us to the idea of human nature on the one hand and global capitalism on the other. Dipesh Chakrabarty insightfully observes, in this regard, that the ecological crisis 'requires us to put global histories of capital in conversation with the species history of humans'.[94] It calls for 'thinking simultaneously on both registers, to mix together the immiscible chronologies of capital and species history'[95] for 'the whole crisis cannot be reduced to a story of capitalism'.[96] In other words, the crisis compels us to think of ways in which the destructive instincts of both humankind and capital, this time directed against Nature, can be addressed.

It is worth noting here that even in this instance the ICJ could not rise to the occasion. The advisory opinion of the Court in the *Nuclear Weapons* case is evidence of the limits of law in bringing about both peaceful relations among nations and peace with Nature. The Court, by a majority, advised the United Nations General Assembly that 'in view of the current state of international law, and of the elements of fact at its disposal, the Court cannot conclude definitively whether the threat or use of nuclear weapons would be lawful or unlawful in an extreme circumstance of self-defense, in which the very survival of a State would be at stake'.[97] The agnosticism of the Court, here, reflects above all upon the inability of legal discourse to come to grips with the reality of a 'world without us'.

9. Conclusion: The Paradox of Peace and Law

It is time for some concluding remarks. In plain terms, mainstream international lawyers, using particular legal methodologies, have rarely engaged with the question of root causes of war. These causes may be conveniently categorized as internal and external causes. Freud and Luxemburg represent these two alternative perspectives of looking at the causes of war. While there is a need to interrogate the essentialist ontology that underlies the theses of both Freud and Luxemburg, the role of human drives and the political economy of global capitalism help us to understand the limits of the 'peace through law' project.[98]

In so far as external causes of war are concerned it appears to be the case that without fundamental changes in the workings of global capitalism peace between nations, and peace with Nature cannot be achieved. It can also be safely said that the exclusion of the voices of women in the process of taking decisions about war is 'a great strategic

[92] See among earlier formulations Galtung, 'Cultural Violence', 27 *Journal of Peace Research* (1990) 291, at 292, 294.

[93] It is the title of Alan Weisman's path-breaking book *The World Without Us* (2007).

[94] Chakrabarty, 'The Climate of History: Four Theses', 35 *Critical Inquiry* (2009) 197, 212.

[95] Ibid., at 220.

[96] Ibid., at 221.

[97] *Legality of the Threat or Use of Nuclear Weapons* (Advisory Opinion), ICJ Reports (1996) 226, at 266.

[98] See also the critique in Benno Teschke's contribution to this book.

mistake' as they bring with them a distinct set of experiences and perspectives that is often conducive to promoting peace.[99] While the relationship between capitalism, imperialism, and patriarchy is complicated, as manifested in strands of 'imperial feminism', the significance of the greater participation of women in decisions relating to war and its conduct cannot be underestimated, as The Hague Congress pointed out a century ago.

But the question remains whether attempts to attend to the external causes of war are enough. Freud cautioned us against taking this view. It cannot of course be denied, as Marx pointed out, that human nature is shaped in fundamental ways by the 'ensemble of social relations' and therefore the transformation of global capitalism is essential to achieving global peace. On the other hand, there is arguably a surplus of 'human nature' which manifests instincts of aggression and destruction. One possible way of addressing it is through work on the self, an appealing understanding of which was advanced by Mahatma Gandhi who saw work on the self as an integral part of a process of creating a world in which men can live in harmony with fellow men and with Nature.[100] He thus linked work on the self to non-violent resistance to all forms of oppression and injustice. The ethics and spirit of such resistance should inform the work of international lawyers as well.

The talk about root causes of war is not meant in any way to underestimate the role of international law and international institutions in preventing and regulating war. They provide the shared criteria, processes, and mechanisms through which states can prevent war or actively intervene on the side of peace when war does break out. In other words, while the belief that 'peace through law' can be achieved is false, it is also not possible to bring about peace without law. But if international laws and institutions are to contribute to establishing peace there is an urgent need to reform those legal regimes that are the carriers of 'deep causes' of wars and conflicts in the domain of law (e.g. in the area of investment, environment, and human rights law). If such reforms are to be undertaken MILS must address the tyranny of methodological prejudices that prevent it from seriously engaging with deep structures. We simply cannot keep using failed legal methods and worn-out vocabularies and hope to contribute to achieving global peace.[101]

[99] Charlesworth and Chinkin, 'Editorial Comment: Sex, Gender and September 11', 96 *American Journal of International Law* (2002) 600, at 604.

[100] See Chimni, 'The Self, Modern Civilization and International Law: Learning from Mohandas Karmchand Gandhi's *Hind Swaraj*', 23 *European Journal of International Law* (2012) 1159.

[101] See also Mallavarapu's chapter as well as my chapter on 'humanitarianism' in Part VII of this book.

14

Re-Ordering the World from the Skies? The Emergence and Justification of Aerial Warfare

Thomas Hippler

1. Introduction

Although ruled out both by classical concepts of *jus in bello* and by modern law of armed conflict, deliberate targeting of civil populations has nevertheless been a common feature in war.[1] This chapter will focus on strategic bombing as one of the most salient forms that targeting of civilians took during the first half of the twentieth century. In order to explain, and to justify these practices, various interpretations have been brought forward. A classic is the distinction between wars in the 'civilized' and wars in the 'uncivilized' world.[2] Early modern *ius publicum europaeum* already distinguished between European and extra-European theatres of conflict. This dichotomy between the 'civilized' and 'uncivilized', in other words, was spatially defined.[3] Since the Enlightenment, and more prominently during the nineteenth century, the spatial distinction is increasingly transposed onto a temporal scale.[4] 'Civilization' is thought to be unfolding in history which means that some regions are more advanced than others. Especially in the twentieth century, however, this linear thinking was replaced by a more complex one inasmuch as it became conceivable that formerly civilized regions regressed towards barbarity. Yet another interpretation insists on the difference between *norms* and *practices*. that is the unavoidable difference between what actors claim to do and what they do.[5]

This chapter will follow a different path. Inspired by Nietzsche's 'genealogical' method revived by Foucault, it will try to historically situate the emergence of specific discourses of justification,[6] thus distancing itself from any dualistic structure of reasoning. The question is how narratives of justification arise within complex historical

[1] The following relies on material previously published in different papers and books, in particular on *Bombing the People: Giulio Douhet and the Foundations of Air Power Strategy, 1884–1939* (2013) and *Governing from the Skies: A Global History of Aerial Bombing* (2017).

[2] See also the contributions by Mallavarapu and Chimni to this volume.

[3] See Tischer and Becker Lorca in this book.

[4] See the contributions by Genell/Aksakal and Benton in this volume.

[5] See M. Schulz, *Normen und Praxis. Das Europäische Konzert der Großmächte als Sicherheitsrat 1815–1860* (2009).

[6] See M. Saar, *Genealogie als Kritik: Geschichte und Theorie des Subjekts nach Nietzsche und Foucault* (2007); see also Hendrik Simon's and Lothar Brock's remarks on a *genealogy of justification* in this book's introductory chapter.

Thomas Hippler, *Re-Ordering the World from the Skies? The Emergence and Justification of Aerial Warfare* In: *The Justification of War and International Order*. Edited by: Lothar Brock and Hendrik Simon, Oxford University Press (2021).
 DOI: 10.1093/oso/9780198865308.003.0014

situations in which strategic competition and different motivational settings interact. Justifications, in other words, are but one particular form of political practice.[7]

If looked upon from this angle, some aspects of an otherwise well-known military history appear in a different light. Take the most obvious example, the strategic bombing campaigns during the Second World War. Especially German historiography commonly used to argue that carpet bombing of civil populations was employed by the Luftwaffe in the first place. Historiographic debate seemed focused for a long time on the question of knowing whether the German civilian population can be viewed as victims of the Allied air war, or whether strategic bombing had been simply a response to German aggression, violent but necessary. The question of knowing 'who started it', Germans or the British, is not only puerile, it also hides the real problems. In reality, the first bomb was dropped not from a German nor from a British airplane, but by an Italian pilot in Libya.

During the interwar period aerial bombing became a core component of territorial control for all colonial powers. Aerial bombing, in other words has been part of the military arsenal of European expansion against the colonized. It is not, however, because of a lesser number of victims that these facts are less commonly known. To stick with the Libyan example, the population of the Italian colony diminished between 1928 and 1931 from 225,000 to 142,000.[8] British Historian David Omissi even estimates that three quarters of the tribal population died, not only through bombs dropped from the air and the extensive use of poison gas, but also by hunger and disease following the colonial conquest.[9] The actual reason was that the perception of aerial bombing was focused mainly on the European—and to a lesser extent far-eastern—theatre of the Second World War. As a consequence, the history of aerial bombing has been seriously misconstrued. A 'colonial matrix' of aerial warfare existed well before it reached Europe.

Therefore, it is necessary to explain the breakdown of a world-order that was capable of exporting total enmity to the periphery of the world system. This also entails an explanation of how and why total war 'returned', as it were, from the periphery to the centre. Moreover, the uneven proneness of European powers to employ indiscriminate bombing against civil populations needs equally to be explained. It is particularly unpleasant for supporters of liberal democracies to recognize that carpet bombing of civil populations has never been a predominant strategic option either for Nazi Germany nor for the Soviet Union, but rather for liberal democracies like the United Kingdom or the United States of America.

2. The Colonial Origins of Aerial Warfare

It was in the context of the colonial war in Libya in 1911 that an Italian plane dropped the first bomb from the air. The next year, air power was used during the First Balkan

[7] See Hippler and Vec, 'Peace as a Polemic Concept: Writing the History of Peace in Nineteenth Century Europe', in T. Hippler and M. Vec (eds), *Paradoxes of Peace in Nineteenth-Century Europe* (2015) 3, at 13.

[8] Entry 'Libya' in P.V. Cannistraro (ed.), *Historical Dictionary of Fascist Italy* (1982), at 306.

[9] D.E. Omissi, *Air Power and Colonial Control. The Royal Air Force, 1919–1939* (1990), at 199–201.

War, in which the Ottoman Empire was opposed by a coalition between Montenegro, Serbia, Bulgaria, and Greece. Immediately before the First World War, both France and Spain began to use aircraft in their colonial wars in North Africa.[10] The use of aircraft in colonial wars was a constant theme in pre-1914 treatises of future warfare and in science fiction literature—if the two genres can be effectively distinguished. And there is indeed some evidence that military science and science fiction are intimately interwoven. In this respect it is interesting but not astonishing to note the affinity of early aviation science fiction before the First World War with racist themes: novelists who wanted to depict the fascination and the military usefulness of aeroplanes for dropping bombs or hunting and machine-gunning were naturally led to situate the plot in an extra-European context.[11] On the one hand, this introduced an effect of asymmetry to European readers, since Europeans were never victims of these kinds of military practices. On the other hand, these horrifying practices seemed to be more legitimate if they were performed in a context of colonial warfare against 'uncivilised races'.[12] The first target of literary science fiction was the 'yellow peril', as in Matthew Phipps Shield's greatest success *The Yellow Danger*. The Chinese villain Dr Yen How is a source for the fictional character of Dr Fu Manchu. The same holds true for French writer Émile Driant, whose novel *L'aviateur du pacifique* (The Aviator of the Pacific) of 1909 'foresaw the Japanese attack on Pearl Harbor over fifty years later, and divined the critical role that flying machines would play in the conflict for control of the Pacific'.[13] Two years later, he published *Au-dessus du continent noir* (Above the Black Continent), another aviation novel in which aircraft are employed for missions of colonial control, such as punitive missions against barbarous rebels against French colonial rule. Science fiction literature thus endorsed a theme that Englishman R.P. Hearne had developed during these same years from a military point of view. In his 1910 *Airships in Peace and War* he declares that 'in savage lands the moral effect of such an instrument of war is impossible to conceive.... The appearance of the airship would strike terror into the tribes.... It will enable an expedition to be made with astounding rapidity, it will create the most terrifying effect on savage races, and the awful wastage of life occasioned to white troops by such expeditionary work would be avoided, whilst the cost would be considerably reduced.'[14]

The early discussions about the military use of air power—especially among the colonial powers—were indeed centred on the idea of the colonial possibilities of air power and the employment of the new device in 'small wars'. And the 'moral effect' of air power is an important argument in this respect. As already seen, the first military employment of aircraft did indeed take place in the colonies, during the Libyan

[10] Paris, 'The First Air Wars—North Africa and the Balkans, 1911–13', 26 *Journal of Contemporary History* (1991) 97, at 100–03.

[11] To be sure, this characteristic changes after the First World War, and in particular during the 1930s, when European consciousness begins to conceive of itself as a potential victim of air warfare. See S. Lindqvist, *A History of Bombing* (2001); and Hippler 'Krieg aus der Luft: Konzeptuelle Vorüberlegungen zur Entstehungsgeschichte des Bombenkrieges', in W. Hardtwig (ed.), *Ordnungen in der Krise. Zur politischen Kulturgeschichte 1900–1933* (2007) 403.

[12] The following paragraph relies on Lindqvist, above note 11.

[13] R. Wohl, *A Passion for Wings: Aviation and the Western Imagination, 1908–1918* (1994), at 86.

[14] R.P. Hearne, *Airships in Peace and War, Being the Second Edition of Aerial Warfare with Seven New Chapters* (1910), at 183–85.

war of 1911, and in subsequent conflicts virtually all colonial powers made use of air power. In their conquest of Morocco in 1912–14, the French used aircraft for reconnaissance and bombing.[15] According to René Martel, author of a history of the French Bombing Fleet up to 1918, bombing—and not reconnaissance—was the most important contribution of aircraft to the conquest of Morocco and the 'projectiles coming from France make a great impression on the adversary'.[16] Lieutenant Armand Des Prez de La Morlais wrote thirty years later in an internal report that these bombardments 'produced a considerable moral effect on the native rebels. . . . We ourselves were aware of the fear we were inspiring due to the rapidity with which groups of fighters were dispersing as the plane approached.' De La Morlais adds that in several villages the local markets did not take place any more for fear of aerial attack.[17] In a memorandum, French commander Lyautey spoke of the employment of aviation in support of troops acting against 'rebel tribes'.[18] It can be inferred from this that attacks on basic economic and social structures such as marketplaces were intentional and systematic.

As for the British Commonwealth, Winston Churchill, then First Lord of the Admiralty, commissioned a report in 1914 on the possible use of aircraft in Somaliland and proposed employing an airship in the colony to the Cabinet;[19] and in a memorandum entitled 'The Future of Air Power', dated 27 June 1918, Frederick Sykes developed the idea of the imperial use of air power.[20] The real breakthrough in colonial air policing, however, was made after the First World War. Formed on 1 April 1918, the Royal Air Force (RAF) was the first independent air force in the world. After the war the service was drastically cut, mainly for budgetary reasons, and even threatened in its institutional independence. In this context air power advocates in London came back to Churchill's idea in 1914 of employing aircraft against a rebellion which had been going on for some twenty years in northern Somalia, the Dervish resistance movement led by Muhammed Abdullah Hassan, called the 'Mad Mullah'.[21] The mission started in 1919 and was accomplished in January 1920.[22] According to one British author describing the operation, the Somalian fighters had been ignorant of the existence of aeroplanes and 'conjectured that the airplanes were chariots of Allah coming to take the Mullah to paradise, or a Turkish intervention come to tell of the Turkish Sultan's victory in the First World War'.[23] Almost killed by the first bomb, the Mullah fled to the desert where he was hunted by the RAF until he gave up.[24] This success was

[15] Niessel, 'Le role militaire de l'aviation au Maroc', 33 *Revue de Paris* (1926) 508.

[16] R. Martel, *L'aviation française de bombardement des origines au 11 novembre 1918* (1939), at 11.

[17] 'Les origines du bombardement par avion', Service historique de l'Armée de l'Air, fonds Des Prez de la Morlais Z 35 342', cited Pernot and Villatoux, 'L'aéronautique militaire au Maroc avant 1914', *Revue Historique des Armées* 3 (2000) 89, at 96.

[18] 'Rapport du général Lyautey au ministre de la guerre, 4 septembre 1913, Service historique de l'Armée de l'Air', C 34, cited in ibid., at 96f.

[19] Killingray, '"A Swift Agent of Government": Air Power in British Colonial Africa, 1916–1939', 25 *Journal of African History* (1984) 429.

[20] The memorandum is reproduced as Appendix V in F.H. Sykes, *From Many Angles: An Autobiography* (1942), at 544.

[21] D. Jardine, *The Mad Mullah of Somaliland* (1923).

[22] Gray, 'Bombing the "Mad Mullah", 1920', 125 *Journal of the Royal United Services Institute for Defence Studies* 41.

[23] Jardine, above note 21, at 239.

[24] Hess, 'The "Mad Mullah" and Northern Somalia', 5 *Journal of African History* (1964) 415, at 432.

of great importance for the RAF and actually ensured its survival as an independent service. Aircraft could be promoted as a swift and cheap means to perform missions of colonial policing, and this was indeed the main task of the air force in the interwar period.[25] The most common feature of colonial air policing was what was called an air blockade. The operations started with heavy bombing from the air for a couple of days. In a second step, the intensity would be lowered, but bombing would remain persistent in order to keep the rebels far from their villages, fields, grazing grounds, and water supply. In other words, the tribe was bombed into starvation or they resigned to colonial rule. In other cases, aircraft directly attacked the rebels or opened machine-gun fire on their cattle. Sometimes aircraft were used to support the infantry, 'attacking carefully defined areas and their flanks, front and rear'.[26]

The first territory in which the practice of colonial bombing from the air was systematically employed was Iraq, but similar features were also employed on the North-West Frontier of India, that is, Afghanistan, and in Sudan, Syria, and Morocco by the French and Spanish, and by the Americans in Nicaragua.[27] The archetypical concept of 'police bombing', however, was developed and employed in Iraq.[28] An effective occupation of the formerly Ottoman territory was too expensive for the tight British budget, and so both the forces and the budget became smaller with 'air control'. They included eight RAF squadrons, four armoured-car companies, a modest indigenous force, and a single British infantry brigade.[29] In cases of insurrection, the regulations required that at first, warning leaflets should be dropped, and then villages would be bombed no more than forty-eight hours later.[30] However, in reality the poor navigation systems often caused confusion and the wrong villages were bombed.[31] The advocates of air policing argued that it was 'both economical and humane since it inflicts neither great nor permanent suffering upon the people against whom it is used nor heavy casualties among those who have to wield it'.[32] Critics of air policing, on the other hand, insisted on the inability of the device to control the populations effectively. Lord Lloyd thus stated in the House of Lords on 9 April 1930 that 'whatever air control can do, it can never civilize people nor pacify people'.[33]

The method of indirect control through air policing may indeed be considered as a major historical shift. Within Europe, traditional international law had established a link between legitimate forms of warfare and the occupation of territory as its scope. Occupation establishes a pacified relationship between occupying power and civil

[25] Beaumont, 'Policing the Empire', *Aerospace Historian* (1979) 84.

[26] J. Ellis, *From the Barrel of a Gun: A History of Guerrilla, Revolutionary and Counter-Insurgency Warfare, from the Romans to the Present* (1995), at 137.

[27] Omissi, above note 9.

[28] Paris, 'Air Power and Imperial Defence 1880–1919', 24 *Journal of Contemporary History* (1989) 209.

[29] Paschall, 'The Strategic View. Air Control: Iraq, 1920–30', 3 *The Quarterly Journal of Military History* (1991), at 20–22.

[30] A. Harris, *Bomber Offensive* (1947), at 22f.

[31] J.S. Corum and W.R. Johnson, *Air Power in Small Wars: Fighting Insurgents and Terrorists* (2003), at 62–66; and P.A. Towle, *Pilots and Rebels: The Use of Aircraft in Unconventional Warfare, 1918–1988* (1989), at 9.

[32] 'Air control in undeveloped countries, Air Staff Memorandum N. 41, 1 January 1929', AIR 20/674, cited by Killingray, above note 19, at 440.

[33] Cited in Townshend, 'Civilization and "Frightfulness": Air Control in the Middle East Between the Wars', in C. Wrigley (ed.), *Warfare, Diplomacy and Politics, Essays in Honour of A.J.P. Taylor* (1986) 142, at 157.

population. To the extent to which the populace does not engage in a national resistance movement, the occupying power acts as executive sovereign on the occupied territory. There is, in other words, a relation of obedience and protection between occupiers and occupied. In the extra-European sphere, on the other hand, this model has never had any normative value. Colonization aimed at permanent control of the subjected territories, and this permanent dominion was ideologically justified by a civilizing mission on the part of the colonial powers. In order to be able to civilize, however, the social structure of the colonized country had to be substantially modified. The method of indirect air control, as employed since the 1920s, cancelled this social interventionalism. Colonial rule thus relied on existing tribal structures and it was these tribal entities that owed submission to their colonial yet distant masters. And in case of rebellion, these tribal structures were bombed in an exemplary manner.

The employment of air power was generally acceptable in the context of warfare against non-white populations. In 1919, the South African government under Jan Christiaan Smuts—author of the famous report that had led to the creation of an independent Royal Air Force[34]—requested aircraft from the imperial government, reassuring London that

> the machine will not be used against the white population of Johannesburg but will be held in reserve mainly for demonstration purposed and consequent moral effect against the natives on the Reef should they become out of hand.[35]

Despite this promise the South African government did not hesitate to employ air power in order to crush white miners on the Rand in March 1922. Unconditionally forbidden according to the rules of civilized warfare, the employment of indiscriminate bombing against civil populations was reserved to non-white 'barbarians' outside of Europe.

3. The Legitimation of Aerial Warfare

The use of aerial warfare in colonial contexts largely follows the narratives of justification of colonial warfare in general.[36] And this justification relies on a crucial difference between Europe and the rest of the world. In Europe war has been an affair between states since the seventeenth century.[37] It was the state that put an end to the 'state of nature', establishing on its territory a power capable of containing civil war. Internal armed conflicts steadily came to an end. The corollary of turning war into an inter-state affair was a limitation of warfare: once war was defined as a relation between states, it ceased to denote a relationship between individuals. It followed from

[34] C.J. Luck, *Interpreting and Misinterpreting the Promise of Air Power* (Masters Thesis, Air University, Maxwell Air Force Base 2007).

[35] Killingray, above note 19, at 432.

[36] For these narratives in general, see the chapters by Mallavarapu and Chimni; for other case studies in this volume, see the contributions by Becker Lorca, Benton, and Lange.

[37] For the debates on early modern state-building and the justification of war in Europe see the chapters by Tischer, Teschke, and Eberl in this volume.

this that the latter had the right to be protected from warlike violence. Rousseau only repeated a common opinion when he wrote that, even in war, states were bound to respect the persons and goods of citizens.[38] In the colonies, however, things were fundamentally different. Since colonial wars did not oppose two states monopolizing legitimate violence, as embodied in an army, the distinction between 'defenders of the state' and 'ordinary men' was not applicable. Thus, a regular war 'may be terminated by the surrender or capitulation of the hostile sovereign or chief, who answers for his people; but in the suppression of a rebellion the refractory subjects of the ruling power must all be chastised and subdued' as Charles Edward Callwell put it famously in his 1896 classic *Small Wars: Their Principle and Practice*.[39] 'The main points of difference between small wars and regular campaigns ... are that, in the former, the beating of the hostile armies is not necessarily the main object ... that moral effect is often far more important than material success, and that the operations are sometimes limited to committing havoc which the laws of regular warfare do not sanction.'

It is interesting to compare these lines with the ideas of American 'airpower' prophet William Mitchell, according to whom 'the advent of air power which can go straight to the vital centres and entirely neutralize and destroy them has put a completely new complexion on the old system of war. It is now realized that the hostile main army in the field is a false objective and the real objective are the vital centres. The old theory that victory meant the destruction of the hostile main army is untenable.'[40] Colonial war on the one hand, and aerial warfare on the other thus share one common feature since their main objective is not to beat a hostile army. In other words, distinct features of colonial warfare are now being applied to the European 'centre' in which traditionally the rules of civilian immunity were supposed to apply. As a consequence, the traditional distinction between 'barbarity' and 'civilization' was being reconfigured. This reconfiguration could take two different directions. On the one hand, within Europe the internal conflict led to a notable extension and evolution of the notion of the barbarian. On the other hand, the perception of extra-Europeans and their relation to war equally paved the way for a rapprochement between the formally distinct spheres of war in Europe and war in the colonies.

With regard to the first point, the employment of air power—real or only in strategic planning—is a good indicator of the extension of the notion of the barbarian to internal conflicts. In December 1917, in the immediate aftermath of the Bolshevik Revolution in Russia, British aircraft were first employed on the British home front in a non-violent but heavily threatening way: aircraft dropped leaflets urging workers in Coventry to end their strike.[41] During the 1919 rail strike and the 1926 General Strike a significant number of aircraft were employed to ensure essential transportation within Britain and to provide the provinces with daily copies of conservative newspapers

[38] 'The purpose of war being to destroy the enemy state, its defenders may rightfully be killed so long as they are carrying arms; but as soon as they lay them down and surrender, ceasing to be enemies or agents of the enemy, they become simply men again, and there is no longer any right over their lives.' J.-J. Rousseau, *The Social Contract*, vol. 1 (2008), ch. 4, at 52.

[39] C.E. Callwell, *Small Wars: Their Principle and Practice* (1996), at 41; see also S. Lindqvist, *A History of Bombing* (2003), sections 46f.

[40] W. Mitchell, *Skyways: A Book on Modern Aeronautics* (1930), at 255.

[41] Omissi, above note 9, at 40–43.

like *The Times* and, in 1926, the governmental *British Gazette*. In early 1920 Hugh Trenchard, the 'father of the Royal Air Force', had already drafted a memorandum which discussed the possible use of aircraft to suppress 'industrial disturbances or risings' in 'settled countries' such as India, Egypt, Ireland, and England. If 'murder and arson' were being committed in cities in which 'a majority of the inhabitants are definitely hostile' to the government, it should be made possible to fight the insurrection with 'a limited amount of bombing and machine gun fire'. Fearing a bad reaction in the press if it became known that the employment of the RAF against British workers was under discussion, Churchill urged the military authorities to withdraw the references to the British Isles—at least in the written communication.

Nonetheless, the RAF was instrumental during the 1926 strike in transporting volunteer strike breakers, 'who manned steam engines, drove underground trains, maintained power stations and arrested prominent communists'. 'Trenchard had good reason to be pleased with the work of the air force during the strike, and he received letters of fulsome congratulation from several railway managers'.[42] After 1926 there was no further need to defend Britain's internal order with the air force, one exception being the discussions about the bombing of warships held by mutineers at Invergordon in the autumn of 1931. In contrast to England, however, the 'Celtic periphery' and Ireland in particular, has always been a different case in point. During the Irish revolt of 1920, aircraft were employed for reconnaissance missions and for the protection of armoured cars from ambush.[43]

It thus becomes clear that the 'barbarians' who can legitimately be fought by bombs dropped from aircraft needed no longer to be extra-Europeans. In the face of what German conservative historian Ernst Nolte has termed the 'European Civil War', insurgents, rebels, or communists within Europe were subject to treatment that was formerly reserved for extra-European space. And the example of the use of air power in different forms of conflict—traditional wars between Western powers, colonial wars, or revolutionary war within Europe—gives credibility to Hannah Arendt's famous insight on the continuity between colonialism and totalitarian dominion. However, it is puzzling to see that the ideas of employing air power against workers on strike were precisely not used in a totalitarian setting but were, on the contrary, mostly developed in Britain, a country with a long and well-established democratic and liberal tradition.

It was, however, not only the extension of methods of colonial warfare to internal conflict that paved the way for the breakdown of a world-order that was capable of exporting total enmity to the periphery of the world system. The shift equally concerned the perception of the colonized 'others' by European powers. As pointed out before, the Arab lands in general and the formerly Ottoman territory of Mesopotamia in particular, were the primary theatres of colonial 'police bombing'. It is therefore no coincidence that discourses of justification were foremost developed with regard to this region. The concepts used by the British defence and intelligence establishment were to be particularly influential in this respect.[44] After the First World War, the Arab

[42] Ibid., at 41.

[43] Ibid., at 42f.

[44] P. Satia, *Spies in Arabia: The Great War and the Cultural Foundations of Britain's Covert Empire in the Middle East* (2008).

warrior was styled as the exact opposite of the 'tommy'.[45] For the Arabs, war was still romantic, an 'excitement' whose tragic outcomes they accepted as a fatality inherent to life. In short, the Arabs were different from us, so different that 'they have no objection to being killed', as Hugh Trenchard, head of the RAF general staff, explained to the sensitive souls of the British Parliament.[46] Arabs loved war precisely because it involved a confrontation with death, and as opposed to the effeminate Europeans, they did not make the flabby distinction between combatants and non-combatants. Thought about properly, not bombing them would almost amount to insulting their values.[47]

Besides, bombing was particularly appropriate for a population whose 'fatalism' was inherent to their religious and cultural heritage. Fatefully impersonal by its nature, the bomb falling from the sky was like divine anger, requiring not a response but immediate submission. The life of an Arab was thus permanent war, but a war of a quite other nature than that which had just ravaged Europe. Arabian war had not yet abandoned the notions of honour and courage. The individual fought freely, proudly, and for his independence. The noble savage was more than a warrior, he was indeed a knight of the desert. It thus comes as no surprise that aviation was seen as the weapon most suitable to the Arab art of war, quite simply because air combat presented the same romantic characteristics: the airman was a knight of the sky as the Arab was a knight of the desert. It was not surprising, therefore, that 'a kind of mutual sympathy existed between these Arab nomads and the air force', as the RAF officer Robert Brooke-Popham remarked. Thomas Edward Lawrence, known as 'Lawrence of Arabia', even saw the 'Bedouin way of warfare' as the antidote to the anonymous massacres of the Great War: 'What the Arabs did yesterday, the air forces will most likely do tomorrow.'[48]

It is perhaps not entirely by coincidence that it was foremost in Germany that this orientalizing discourse was adopted and used to make sense of her own geopolitical situation. A good example is Hans Ritter's *Der Zukunftskrieg und seine Waffen* (Future War and its Weapons) of 1924, which discusses the employment of air power in some detail. At the beginning of human history, war was 'a war of annihilation' (*Ausrottungskrieg*) of one tribe against another.[49] The subsequent development of absolute monarchies introduced a split between warriors and populace and thus a barrier between two different political classes. Ritter argues that the whole nineteenth century had been characterized by a paradoxical development: on the one hand, the legacy of the French Revolution had mobilized the whole male population for the national war effort while, on the other hand, international law upheld and systematized the distinction between combatants and non-combatants. This intrinsic paradox had to be overcome.[50] According to Ritter it was 'the superior logic and the cold vision of the purpose of English politics which made the necessary first step towards

[45] The paragraph relies on Satia, 'The Defence of Inhumanity: Air Control and the British Idea of Arabia', 111 *American Historical Review* (2006) 16.

[46] Townshend, above note 33, 148–51.

[47] Glubb, *The Story of the Arab Legion* (1948), at 149; and *Arabian Adventures: Ten Years of Joyful Service* (1978), at 148, quoted by Satia, above note 45, at 37–39.

[48] Satia, above note 45, at 29.

[49] H. Ritter, *Der Zukunftskrieg und seine Waffen* (1924), at 5.

[50] See also the chapter by Miloš Vec in this book.

the rupture of the sacrosanct law of humanity, when enacting a hunger blockade, regardless of age or sex, on the inhabitants of the besieged fortress of Germany'.[51] In this way war was returning to its former and more original condition of an existential struggle of one community against another. Following this logic, it was legitimate to bomb civil populations from the air. French authors also adhered to this idea, arguing that only the part of 'the civilian population not taking part in the manufacture of military equipment' was immune from being targeted.[52] The author of a 1923 doctoral dissertation in international law held the same point of view when concluding that international law had not provided clear criteria as to what determines the military character of a given target. Consequently, he bluntly states that 'aeroplanes should not be forbidden to attack these necessary auxiliaries of the enemy army' such as railway workers, workers in ammunition factories, and anybody whose economic role is of direct or indirect military interest.[53] In the lectures he delivered at the British school of air warfare at Andover, Robert Brook-Popham cited 'democratization', 'industrialization', and 'trade unionism' as the three decisive factors that had led to a closer integration of the people into the war effort. As a consequence, the distinction between combatants and non-combatants was de facto obsolete. His Italian colleague Giulio Douhet held the same opinion, arguing that 'the distinction between belligerents and non-belligerents no longer exists now, since all are working for the war, and the loss of a worker may well be more serious than the loss of a soldier'.[54]

We thus see how the themes of industrialization and the perceived threat of social disorganization and insurrection paved the way towards the applicability of methods of colonial warfare to Europe. These themes were to become prominent in the interwar period, especially in German military literature. Erich Ludendorff, for instance, remarks that total war, the origins of which could be found in universal conscription and thus on the reorganization of politics upon a popular basis, had features similar to colonial war. Limited wars were actually deemed to be morally suspect, since the existence of a community is not at stake. The morality of fighting, in other words, stems from the existential danger to which a community is exposed. And in this sense, colonial wars are total wars, at least for the non-Europeans who wage war for 'ethical reasons'.[55] In contrast, 'cabinet wars' are considered more as organized theft than as morally legitimate existential struggles. Ludendorff's argument is thus underpinned by the idea that the total mobilization during the First World War had the effect of overcoming the split between the state and the population; state and population had in fact become united. And in this respect the situation in Europe after the First World War increasingly resembled the situation in the colonies. After the Second World War, this idea was perhaps best expressed by Sir Arthur Harris according to whom 'as in all the R.A.F.'s many operations over savage countries, including Germany, capture by the natives was one of the worst things we had to fear'.[56]

[51] Ritter, above note 49, at 7.
[52] L. Rolland, *Les pratiques de la guerre aérienne dans le conflit de 1914 et le droit des gens* (1916), at 70.
[53] J. Bouruet-Aubertot, *Les Bombardements Aériens* (1923), at 63–65.
[54] Douhet, 'La grande offensive aerea', in G. Douhet, *Scritti inediti* (1951), at 127.
[55] E. Ludendorff, *Der totale Krieg* (1988), at 6.
[56] Harris, above note 30, at 20.

4. War, Democracy, and the International Order

It can thus clearly be seen that the distinction between the European centre and the colonial periphery is being perforated. This discursive strategy could be used for very different purposes. Harris used it to justify the use of strategic bombing against civil populations in Europe, while Ludendorff insisted in the contrary on the higher moral value of an existential struggle as exemplified in colonial warfare, thus opposing monarchical forms of war to the new wars of the twentieth century that are characterized as *völkisch*.[57] These views are certainly not incongruent with his being an early supporter of the Nazi movement and of Hitler's putsch in Munich in 1923. What is puzzling, however, is that what Ludendorff called '*völkisch*' bears some resemblance to what others called 'democratic'. On a conceptual level, this should not surprise us, given the fact that the Greek *demos*—'the people' as opposed to the elite (*aristoi*)—is one of the meanings of the German word *das Volk*. *Volk*, however, as with most of the equivalents in modern European languages—like the French *le peuple* or the Italian *il popolo*—also included what the Greek language had called *ethnos*, that is, a community of descent.[58] The modern vocabulary denoting both the people as a social category and the people as an ethnic community implies that the two meanings are indeed close. In both cases the emphasis is put on the idea that the political form, the state, had to be a consequence of the social organization of the population and that no split has to separate the population from its political form.

Another dimension, however, has to be added to this picture: the gendered imagination of 'the people'. In the European theatre, the people is a fundamentally unstable entity. The notion designs, on the one hand, the subject of politics and conveys, as such, a male connotation. The exemplification is the citizen-soldier fighting for his fatherland. On the other hand, however, the people is also a permanent threat to social stability. In this respect the people has an essentially female connotation. This latter connotation is perhaps best exemplified in Giulio Douhet's *Come finì la Grande Guerra* (How the Great War Ended) of 1919, a text situated somewhere between a novel, counter-factual history, and a work of military strategy. The Italian general had become increasingly famous as a military writer on air power and is best known as the author of *The Command of the Air*, where he recommends ruthless bombing attacks on urban populations. In his 1919 military fantasy, Douhet hints at the same nexus between aerial bombing and its revolutionary impact on the urban populations. In addition, the description of the social disorder induced by aerial bombing, according to this view, always involves the presence of women.[59] Women perform here and elsewhere the function of a metonymy of the people, and more particularly of the undisciplined rabble, in contrast to the male discipline that constitutes a coherent body politic out of an anonymous mass. Air warfare thus has a gendering impact on the imagination of warfare. The coherent nation of which the male citizen-soldier is the

[57] Ibid., at 61.
[58] See also the contribution by Lange in this volume.
[59] G. Douhet, *Come finì la grande guerra: La vittoria alata* (1919), at 62.

perfect example is shattered and becomes an anonymous and disorganized rabble with a female face.

Quite similar ideas could already be found in Herbert George Wells' 1908 science fiction novel *The War in the Air*. Wells was both a Fabian socialist and a member of the Air League of the British Empire.[60] The novel is plotted around the story of Bert Smallways, a 'vulgar little creature', who finds himself embarked on a German airship on a bombing mission against New York City.[61] Confronted with the selective bombardment of the city's infrastructure, the authorities decide to hoist the white flag. However, public opinion reacts against this decision and an insurrection movement for national defence breaks out. Given the German inability to land troops and to occupy and pacify the city, they have to bomb New York into submission 'because she was at once too strong to be occupied and too undisciplined to surrender in order to escape destruction. Given the circumstances, the thing had to be done.'[62]

In Wells' view, aerial warfare had a decisive shortcoming: its inability to occupy territory. There was a fundamental contradiction between its enormous power of destruction and its helplessness in controlling the ground. The particular strength of Wells' analysis lies in the connection he established between technological means, forms of social organization, and warfare. According to him, the specific political contradiction of his time was that between democracy and war.[63] But, contrary to today's fashionable theory of Democratic Peace, Wells did not argue that democracy was inherently in contradiction to war. According to him, the nation had abandoned the thought of war to the professional military: 'War had become a matter of apparatus of special training and skill of the most intricate kind. It had become undemocratic.'[64] Democracy is linked to some forms of patriotism and the needs of the scientific age to a certain professionalization of the armed forces. Yet the highly specialized technological arm of aviation is unable to contain the insurrectional movement of a population driven by its patriotic feelings. Wells sums up the difficulties in the following paragraph:

> The difficulty of the Germans ... came from the impossibility of landing any efficient force or, indeed, any force at all from the air-fleet.... From above they could inflict immense damage; they could reduce any organised Government to a capitulation in the briefest space, but they could not disarm, much less could they occupy, the surrendered areas below. They had to trust to the pressure upon the authorities below of a threat to renew the bombardment. It was their sole resource. No doubt, with a highly organised and undamaged Government and a homogeneous and well-disciplined people that would have sufficed to keep the peace. But this was not the American case. Not only was the New York Government a weak one and insufficiently provided with police, but the destruction of the City Hall and Post-Office and other central ganglia had hopelessly disorganised the co-operation of part with part.... The Germans had

[60] S. Hynes, *The Edwardian Turn of Mind* (1968), at 15–52; and D.C. Smith, *H. G. Wells. Desperately Mortal: A Biography* (1986), at 86.

[61] H.G. Wells, *The Works of H. G. Wells, The Atlantic Edition*, vol. 20 (1926), at 162f.

[62] Ibid., at 200f.

[63] Ibid., at 100f.

[64] Ibid., at 176.

> struck at the head, and the head was conquered and stunned—only to release the body from its rule. New York had become a headless monster, no longer capable of collective submission.[65]

With admirable clear-thinking Wells thus hints at a decisive dilemma of democratic government: the 'constitutive power' of the population and the 'constituted power' of the government.[66] Moreover, the depiction of the population of New York City as a 'headless monster' hints at resemblances between urban populations under warlike stresses and the colonial imaginary of 'savage people'.

The vision of Wells is a very sombre one, as it is indebted to a vision of a crisis of European democracy in the years around 1900. The sharp awareness of the 'democratic paradox' potentially leads him to anti-democratic and racist positions.[67] In Wells' novel, the Japanese actually join the war between Germany and America, and their swifter planes rapidly destroy the more vulnerable German dirigibles. And the British protagonist Bert Smallways, who had been embarked by the Germans, even ends up feeling sympathetic to his captors in the face of the 'yellow peril': when the airship Hohenzollern was brought down, 'the visible world [was left] to Asia, to yellow people beyond Christendom, to all that was terrible and strange!'[68] In his description of the bombing of New York, Wells thus depicts 'men, women, and children mixed together as though they had been no more than Moors, or Zulus, or Chinese'.[69]

> The special peculiarities of aerial warfare were of such a nature as to trend, once it had begun, almost inevitably towards social disorganisation.... Nothing comparable to this state of affairs had been known in the previous history of warfare, unless we take such a case as that of a nineteenth century warship attacking some large savage or barbaric settlement... Then, indeed, there had been cruelties and destruction that faintly foreshadowed the horrors of the aerial war. Moreover, before the twentieth century the world had had but one experience, and that a comparatively light one, in the Communist insurrection of Paris, 1871, of the possibilities of a modern urban population under warlike stresses.[70]

Similar ideas can be found in the pre-1914 political writings of H.G. Wells. Wells casts doubt on democracy as a viable form of political organization and to envisage more authoritarian forms of government.[71] Commenting on the implications of Blériot's crossing of the Channel in the *Daily Mail* on 27 July 1909, Wells claims that 'the days of natural democracy' were coming to an end.[72] In these political writings, which are a

[65] Ibid., at 195f.

[66] For the conceptual history of '*pouvoir constituant*' see A. Negri, *Insurgencies: Constituent Power and The Modern State* [*Il potere costitutente*] (1999).

[67] For the paradox of most recent democratic wars, see also Anna Geis and Wolfgang Wagner in this volume.

[68] Wells, above note 61, at 266.

[69] Ibid., at 201f.

[70] Ibid., at 238–89.

[71] See his texts *Democracy under Revision* (1927) and *After Democracy* (1932).

[72] Wells above note 61, at 422. See also R. Wohl, *A Passion for Wings: Aviation and the Western Imagination, 1908–1918* (1994), at 89.

useful complement to his highly influential science fiction novels, he draws the logical conclusions from these views.[73] As he points out in his autobiography, his impression of the 'complete insufficiency of the current parliamentary methods of democratic government', had convinced him, from 1900 on, of the necessity of a world state.[74] During the First World War, he developed the idea of a 'League for Peace', that is, a supranational organization capable of effectively controlling the non-proliferation of modern weaponry.[75] An 'International Tribunal' for the 'discussion and settlement of international disputes' should be set up. In order to be able to prevent war effectively, the 'International Tribunal' 'would need also to have power to intervene in the affairs of any country or region in a state of open and manifest disorder'.[76]

The idea is actually almost as ancient as the discussions about the military uses of air power. Victor Hugo had already formulated the idea that air power would bring about universal peace in 1864.[77] In two of Rudyard Kipling's science fiction stories, *With the Night Mail* (1905) and *As Easy as A.B.C.* (1912), an 'Aerial Board of Control'—the A.B.C. of the second story's title—has developed from the technical necessity of regulating aerial traffic into a form of world government. With sovereign states on the verge of extinction, technocrats have seized power and keep the world running smoothly. The second of these stories revolves around three aviators called up to deal with social unrest in 2065, which clearly shows that the A.B.C.'s world government ultimately relied on its air power.[78] In his 1933 work *The Shape of Things to Come*, H.G. Wells was to pick up the ideas of 'World Air Control' and an 'Air Force for Mutual Assistance' as executive organs of a world government.

It is at this point that the writer's politically informed science fiction converges with the perspectives that can be found in the writings of military strategists. The most surprising example is certainly the Italian general Giulio Douhet who was not just content to recommend attacks from the air against civilian populations, by use of bombs and toxic gas, but who championed at the same time a key idea of pacifism, that of an 'international tribunal' that would prevent war by having its decisions sanctioned by air forces.[79] 'The pilot as gendarme and the bomb as truncheon'—that is the precise point at which the colonial practice of police bombing links up with humanist cosmopolitism.[80] It is perhaps in this last respect that the discourses of justification of aerial warfare appear to be the most topical. While the colonial idea is today largely discredited, that of a military air force with cosmopolitical ends continues to prosper, so much so that it is found in Article 45 of the United Nations charter: 'Members shall hold immediately available national air-force contingents for combined international

[73] M.R. Hillegas, *The Future as Nightmare: H. G. Wells and the Anti-Utopians* (1967).

[74] H.G. Wells, *Experiment in Autobiography: Discoveries and Conclusions of a Very Ordinary Brain (since 1866)*, vol. 2 (1934), at 651.

[75] H.G. Wells, *War and the Future: Italy France and Britain at War* (1917), at 275.

[76] Ibid., at 278.

[77] I.F. Clarke, *Voices Prophsysing War: Future Wars 1762–3749* (1992), at 4.

[78] M. Paris, *Winged Warfare: The Literature and Theory of Aerial Warfare in Britain, 1859–1917* (1992), at 39; and C. Carrington, *Rudyard Kipling: His Life and Work* (1955), at 375.

[79] Douhet, 'Incursione in Utopia' (5 March 1915), in A. Curami and G. Rochat (eds), *Giulio Douhet. Scritti 1901–1915* (1993), at 492. See also R. Beaumont, *Right Backed by Might: The International Air Force Concept* (2001).

[80] S. Lindqvist, *A History of Bombing* (2003), section 74.

enforcement action.' Air strikes as not only a 'democratic' practice, but also humanist, cosmopolitan, and even pacifist?

5. Conclusion

The strategic use of air power is not just an issue for the ethics of warfare but its genealogy provides insights into the problematic link between war, justification, and democracy in modern history. This chapter argues that the 'pacifist' impetus which paradoxically accompanies the waging of war throughout history did not come to an end with the introduction of aerial warfare. Arial warfare rather added a new dimension to the war and peace discourse first in the colonial setting of the early twentieth century and later in the European theatre of war. Aerial warfare added to utopian thinking on overcoming war the idea of controlling the impulse to wage war by establishing a universal police system that would provide for international order. This idea came to bear in the form of police bombing, the reality of which, however, betrayed any hopes for turning war into a civilizing endeavour. As seen from Europe, the notion of strategic air power as an instrument for enhancing global civilization did not survive the aerial bombings during the Second World War. But it had died much earlier in the attempts of the Europeans to economize on the suppression of anti-colonial resistance. In recourse to Michel Foucault, one could even argue that aerial warfare, if anything, made possible a form of violent global governance in the worst sense of the word.[81] This form of attempted governance is still with us.[82] As such it signals the breakdown not just of a spatial order in which Europe stood against the rest of the world, but also the breakdown of linear teleologies of civilization.

[81] M. Foucault, *The Government of Self and Others: Lectures at the Collège de France 1982–1983* (2010).
[82] See also Geis and Wagner in this volume.

15

The Justificatory Potential of International Law: National Socialists' Dreams of African Colonies

*Felix Lange**

1. Introduction

For some observers, international law has the progressive potential to protect humanity.[1] Instead of leaving international relations to Hobbesian anarchy, the legalization of international politics promises to tame the powerful and create a more peaceful global order.[2] Ideally, international law exerts its constraining power even when it comes to the use of force. Both the Kellogg-Briand Pact of 1928 prohibiting the initiation of war as a measure of foreign policy[3] and the UN Charter are premised on the idea that aggression is not only unethical, but illegal. The UN Charter offers only two exceptions: self-defence and enforcement action under the collective security scheme of Chapter VII of the Charter.[4]

However, for long *Realists* and *critical scholars* alike asserted that international law can be instrumentally used by the dominant hegemon to advance his respective objectives[5] or to apologize all kinds of international conduct by states.[6] This is particularly true for the use of force. For instance, the intervention in Iraq in 2003 was legitimized by the Bush administration and the Blair government on the basis of international legal argument, while the majority of international lawyers regarded it as a violation of the UN Charter.[7] The potential use of international legal argument in the context of justifying war, thus, seems obvious for the contemporaneous observer.

Nonetheless, the editors convincingly refer us to a mismatch in the historical analysis of the role of international law in relation to war and annexation. While the

* Revised version of 'The Dream of a Völkisch Colonial Empire. International Law and Colonial Law during the National Socialist Era', 5 *London Review of International Law* (2017) 343; contributed upon invitation by the editors. All translations are by the author.

1 R. Teitel, *Humanity's Law* (2011)

2 See G. Clark and L.B. Sohn, *Peace Through Law* (1958).

3 For a positive interpretation of the effects of the pact, see O. Hathaway and S. Shapiro, *The Internationalists: How a Radical Plan to Outlaw War Remade the World* (2017). For a critical perspective see Chimni's 'Peace through Law' in this volume.

4 Articles 2(4), 39–42, and 51 UN Charter.

5 For a classical *realist* account see W. Grewe, *The Epochs of International Law* (2000).

6 For a classical *critical* account see M. Koskenniemi, *From Apology to Utopia* (1989).

7 On this see C. Gray, *International Law and the Use of Force* (4th ed., 2018), at 253–56; 367–76; see also Michael Stohl's and Anna Geis'/Wolfgang Wagner's contributions to this volume.

Felix Lange, *The Justificatory Potential of International Law: National Socialists' Dreams of African Colonies* In: *The Justification of War and International Order.* Edited by: Lothar Brock and Hendrik Simon, Oxford University Press (2021).
 DOI: 10.1093/oso/9780198865308.003.0015

scholarship on just war theory seems abundant, not as many studies exist on the justificatory uses of international law in concrete conflicts.[8] While we know a lot about who developed which arguments on the preconditions for just war, we do not know as much about how this relates to historical practice.

This contribution looks for a historical example in an unexpected place. It examines the use of international law in the context of the German debate about Germany's right to retake its colonies after the First World War. It focuses on the years between 1933 and 1945—the period when German National Socialist leaders tried to conquer Europe and create a world based on the ideology of German racial superiority.

To study this period as an exemplar for justificatory uses of international law might be surprising. One might assume that National Socialists' foreign policies of imperial war and terror are distinct and distant from international law. In fact, the international legal regime after the Second World War was partly developed as a departure from National Socialist imperial and genocidal policies. Moreover, one might wonder what National Socialists had to say about African colonies. It is well known that Hitler's foreign policy focused on the East rather than Africa as the potential German *Lebensraum* for the future. Why then should German international lawyers during National Socialism resort to legal argumentation in relation to 'uncivilized' African territories in the first place?

The example, however, seems to be telling for various reasons. First, under National Socialism international legal arguments continued to play some role, even after the start of the Second World War. While some international lawyers developed elaborate critiques of the Versailles Treaty, others aimed at inventing National Socialist conceptions of international law.[9] It is well known that Carl Schmitt published his notorious theory of Great Space (*Großtraumtheorie*) in the context of the German invasion of *Rest-Czechoslovakia* in March 1938.[10] Based on the American Monroe Doctrine of 1823 and the Roosevelt Corollary to the Doctrine of 1904, Schmitt implicitly demanded that the Western powers had to tolerate German expansion in Europe because it belonged to the German sphere of influence.[11] German legal scholars thus attempted to shape the general discussion on international law.

Second, even though Hitler's government never successfully gained colonies overseas, it developed far-reaching legal plans for a German-controlled Central African empire. As one historian has argued: 'It may well be questioned whether at any time in history a non-existent empire had been so well administered!'[12] In the *Kolonialpolitische Amt der NSDAP* (KPA) (Colonial Policy Office of the NSDAP) and the *Ausschuss für Kolonialrecht der Akademie für Deutsches Recht* (Committee on Colonial Law of the Academy for German Law) politicians, civil servants, and academics developed elaborate legal arguments about the right to colonial possessions,

[8] Simon and Brock, Introduction, at 8–13.

[9] See below.

[10] C. Schmitt, *Völkerrechtliche Großraumordnung und Interventionsverbot für raumfremde Mächte: Ein Beitrag zum Reichsbegriff im Völkerrecht* (1939).

[11] On this, see U. Herbert, *Best: Biographische Studien über Radikalismus, Weltanschauung und Vernunft 1903–1989* (1996), at 271–75.

[12] W. Schmokel, *Dream of Empire: German Colonialism, 1919–1945* (1964), at 160.

and produced detailed drafts of laws governing the envisioned colonial structures.[13] For some groups in the National Socialist regime, the colonial quest was a key objective.

Third, the German case also deserves attention because of its somewhat extraordinary nature. Instead of developing the concept of '*terra nullius*', as in the nineteenth century, and instead of only using administrative law to govern colonies like other Western powers, Germany resorted to legal arguments to demand the colonies it had lost after the First World War. How did German politicians and international legal scholars justify this claim? To what extent were the legal arguments advanced in line with more traditional critiques of the international legal system?

This contribution addresses these issues by contextualizing the academic discourse on international law and the colonial quest in the political debates of the time. It studies the writings of German international lawyers and relates them to the political conceptions of a National Socialist colonial empire. The contribution starts by considering the sceptical reaction of the German public and German international lawyers to the loss of the German colonies during the interwar period. It then addresses how politicians and international lawyers demanded the return of the lost German colonies in the context of National Socialist colonial policy. The contribution concludes with remarks on the justificatory potential of international law in the National Socialist era.

One further premise needs to be mentioned. For the purpose of this chapter, I understand colonialism as applying only to overseas territories outside of Europe. Some historians today argue that Hitler's *Lebensraum* policy in Eastern Europe was nothing other than colonialism on the European continent.[14] Despite some obvious similarities, there are, however, important differences.[15] Even though the National Socialists at times used the language of colonialism to refer to their Eastern policies, Hitler regarded the colonialism of the *Kaiserreich* as a failure, and intended to start something new with his turn to the East.[16] Moreover, the legal arguments for Eastern expansion were based on

[13] On this see Schubert, 'Das imaginäre Kolonialreich. Die Vorbereitung der Kolonialgesetzgebung durch den Kolonialrechtsausschuss der Akademie für Deutsches Recht, das Reichskolonialamt und die Reichsministerien (1937–1942)', 115 *Zeitschrift der Savigny-Stiftung für Rechtsgeschichte: Germ anistische Abteilung* (1998) 86, 88–90; Sippel, 'Rasse, Recht und neues deutsches Kolonialreich in Afrika: Rassendiskriminierung als wesentlicher Bestandteil eines intendierten nationalsozialistischen Kolonialrechts', in W. Wagner (ed.), *Rassendiskriminierung, Kolonialpolitik und ethnisch-nationale Identität* (1992) 119. For the political and socio-economic dimensions of colonial policy during the Third Reich, see K. Linne, *Deutschland jenseits des Äquators? Die NS-Kolonialplanungen für Afrika* (2008); K. Hildebrand, *Vom Reich zum Weltreich. Hitler, NSDAP und koloniale Fragen 1919–1945* (1969); D. van Laak, *Imperiale Infrastruktur. Deutsche Planungen für eine Erschließung Afrikas 1880 bis 1960* (2004); A. Kum'a Ndumbe III, *Was wollte Hitler in Afrika? NS-Planungen für eine faschistische Neugestaltung Afrikas* (1993); H.A. Jacobsen, *Nationalsozialistische Außenpolitik 1933–1938* (1968), at 429–32.

[14] See in particular J. Zimmerer, *Von Windhuk nach Auschwitz. Beiträge zum Verhältnis von Kolonialismus und Holocaust* (2007); Zimmerer, 'Colonialism and the Holocaust: Towards an Archeology of Genocide', 50 *Development Dialogue* (2008) 95; Conrad, 'Rethinking German Colonialism in a Global Age', 41 *Journal of Imperial and Commonwealth History* (2013) 543.

[15] See Kundrus, 'Continuities, Parallels, Receptions. Reflections on the "Colonization" of National Socialism', 4 *Journal of Namibian Studies* (2008) 25; Gerwarth and Malinowski, 'Hannah Arendt's Ghosts: Reflections on the Disputable Path from Windhoek to Auschwitz', 42 *Central European History* (2009) 279; Gerwarth and Malinowski, 'Der Holocaust als "kolonialer Genozid"? Europäische Kolonialgewalt und nationalsozialistischer Vernichtungskrieg', 33 *Geschichte und Gesellschaft* (2007) 439; I.V. Hull, *Absolute Destruction: Military Culture and Practices of War in Imperial Germany* (2005), at 5–90.

[16] Kundrus, above note 15.

the idea of annexation during wartime,[17] and thus varied significantly from the legal arguments advanced in relation to the Southern colonies before the war. Since Southern and Eastern policies were regarded as distinct in the understanding of contemporaries, and since the context of the legal argumentation was completely different, this chapter focuses only on colonial claims directed towards overseas colonies.[18]

2. Weimar, the League's Mandate System, and the 'Unjust' Treaty of Versailles

After the First World War, Germany had to give up its claims to the pre-war colonies. Article 119 of the Treaty of Versailles stated: 'Germany renounces in favour of the Principal Allied and Associated Powers all her rights and titles over her overseas possessions.'[19] Germany thus lost its rather short-lived colonial empire (1884–1919) consisting of German East Africa, German South-West Africa, Cameroon, Togo, Samoa, German New Guinea, and Kiautschou. In a Note of 16 June 1919, the Allied Powers justified the confiscation of the German colonies:

> Germany's dereliction in the sphere of colonial civilization has been revealed too completely to admit of the Allied and Associated Powers consenting to make a second experiment and of their assuming the responsibility of again abandoning thirteen or fourteen millions of natives to a fate from which the war has delivered them.[20]

In the eyes of the Allied powers, Germany had lost the right to lead a colony as a 'civilized' nation because of its cruel treatment of the local populations. This more political than legal argument was based on the British *Blue Book*, which had collected material on the mistreatment of colonial inhabitants by Germany, while ignoring Britain's own colonial shortcomings.[21]

The former colonies of Germany and the Ottoman Empire soon became part of the mandate system of the newly established League of Nations. As mandatories of the League, the victorious powers were supposed to educate the mandated territories towards independence. According to Article 22 of the Covenant of the League, they had to spur the development of those 'not yet able to stand under the strenuous conditions of the modern world', whose 'well-being and development' would 'form a sacred trust of civilization'.[22] The article also explained that 'the tutelage of such peoples should be

[17] On the discussions on Poland, see Messerschmitt, 'Revision, Neue Ordnung, Krieg. Akzente der Völkerrechtswissenschaft in Deutschland 1933–1945', 9 *Militärgeschichtliche Mitteilungen* (1971) 61, at 90f.

[18] However, a comparative research project on the history of colonial law in different empires should include the National Socialist domination policies in the East. For an endorsement of a comparative 'globalization' of Germany's colonial history, see Conrad, above note 14, at 543.

[19] Treaty of Peace with Germany (Treaty of Versailles), signed 28 June 1919, available at <https://www.loc.gov/law/help/us-treaties/bevans/m-ust000002-0043.pdf> (last visited 2 April 2020).

[20] See A. Crozier, *Appeasement and Germany's Last Bid for Colonies* (1988), at 14.

[21] See Union of South Africa, *Report on the Natives of South-West Africa and Their Treatment by Germany* (1918).

[22] Covenant of the League of Nations, signed 28 April 1919, available at <http://avalon.law.yale.edu/20th_century/leagcov.asp> (last visited 2 April 2020).

entrusted to advanced nations who by reason of their resources, their experiences or their geographical position can best undertake this responsibility'.[23] Prior to the agreement on this scheme, the discussions among the Allies had been controversial. While the British and the French aimed for annexation of the former German colonies, the American President Woodrow Wilson insisted on the mandate system. As a compromise, the territories were categorized into 'A', 'B', and 'C' territories depending on their alleged stage of development.[24] In the individual mandates, the Council of the League spelled out the duties of the mandatory powers towards the populations concerned.[25]

The prevailing reaction in Germany was not directed at the provisions of the Mandate system as the post-First World War continuation of colonialism, but rather at Germany's loss of the colonies. An influential colonial revisionist movement quickly built up.[26] One of its protagonists, the former governor of German East Africa, Heinrich Schnee, invented the slogan of the 'colonial guilt myth'. Inspired by the metaphor of the 'war guilt lie' directed towards Article 231 of the Treaty of Versailles, he maintained that Germany had treated its colonies in a benign way, and that the mandate system was based on a lie.[27] Moreover, Hans Grimm's novel *Volk ohne Raum*, which was published in 1926 and dedicated to the colonial cause in Africa, came to be a bestseller and later a byword for expansion into the East under the National Socialists.[28] Even the member of the Weimar-friendly Catholic *Zentrum* party and mayor of Cologne, Konrad Adenauer, argued, as vice-president of the German Colonial Society, that colonial possessions were a necessity for Germany because of the limited space in the *Reich*.[29] (Adenauer was dismissed from political office following the National Socialist takeover and became chancellor of the German Federal Republic after the Second World War.) With the exception of the communists, some intellectuals like Albert Einstein,[30] and sections of the Social Democratic Party, the public, and the political parties of Weimar broadly agreed on the goal of regaining the colonies.[31]

23 Ibid.

24 On the background of the mandate system see A. Anghie, *Imperialism, Sovereignty and the Making of International Law* (2005), at 119–23; S. Pedersen, *The Guardians: The League of Nations and the Crisis of Empire* (2015).

25 Article 22 (VIII) of the Covenant of the League of Nations.

26 See Linne, above note 13, at 21–25.

27 See H. Schnee, *German Colonization, Past and Future—The Truth About the German Colonies* (1926). See also W.W. Schmokel, *Dream of Empire, German Colonialism, 1919–1945* (1964), at 65.

28 H. Grimm, *Volk ohne Raum* (1926).

29 See 'Soll Deutschland Kolonialpolitik betreiben? Eine Umfrage', 12 *Europäische Gespräche—Hamburger Monatshefte für Auswärtige Politik* (1927) 611; Horstmann, '"Das deutsche Volk braucht Kolonien"—Konrad Adenauer und der Kolonialrevisionismus', in M. Bechhaus-Gerst and A.K. Horstmann (eds), *Köln und der Deutsche Kolonialismus. Eine Spurensuche* (2013) 197. In the leading biographical studies of Konrad Adenauer, his role as vice-president of the German Colonial Society does not play a major role. See H.P. Schwarz, *Adenauer. Der Aufstieg: 1876–1952* (1986); H. Köhler, *Eine politische Biographie* (1994).

30 See 'Soll Deutschland Kolonialpolitik betreiben?', above note 29, at 626.

31 See Rüger, 'Das Streben nach kolonialer Restitution in den ersten Nachkriegsjahren', in H. Stoecker (ed.), *Drang nach Afrika. Die deutsche koloniale Expansionspolitik und Herrschaft in Afrika von den Anfängen bis zum Verlust der Kolonien* (2nd ed., 1991) 262; Heyn, 'Der kolonialkritische Diskurs der Weimarer Friedensbewegung zwischen Antikolonialismus und Kulturmission', 5 *Wiener Zeitschrift für kritische Afrikastudien* (2005) 37; Hildebrand, above note 13, at 56–62.

The German government also defended Germany's colonial policies, and demanded colonial equal treatment with the other Great Powers. The short-lived *Reichskolonialministerium* rapidly produced accounts of the German, French, and British treatment of the colonized people which called into question the claim that the German colonial reign had been particularly cruel.[32] In 1924, the German government sent a memorandum to the Council of the League in which it demanded to become a mandatory power in the near future.[33] After Germany's entry into the League in 1926, the government kept the German claim alive. Even though officials in the German Foreign Office doubted the feasibility of the colonial goals and the government instead supported self-determination in the colonies as a means to strengthen German political and economic influence, there was reluctance to give up the colonial claims.[34]

The loss of the colonies was also addressed in the writings of Germany's international lawyers. During the 1920s, the German discipline of international law was preoccupied with two major developments: the foundation of the League of Nations, through which institutionalization at the global level reached a new peak; and the Treaty of Versailles, which sanctioned the territorial losses of Germany, and laid the basis for reparation claims against the country. While some academics subscribed to the project of international institutionalization and developed theories challenging the sovereign state dogma,[35] almost the entire discipline regarded the Treaty of Versailles as a 'dictate', and expended considerable energy attacking the provisions of the agreement.[36] At the same time, most of these critics favoured the German colonial enterprise and despised the mandate system of the League. In the 1925 edition of the most popular German textbook on international law, Max Fleischmann denounced the Treaty of Versailles for its 'military emasculation', 'military defamation', and the 'robbery' of Germany's territory. In losing all its colonies to the victors, Germany would be '*entrechtet*' (disenfranchised). In Fleischmann's view, the mandate system was only invented 'to calm the conscience of the American President'.[37] The Breslau Professor Axel von Freytagh-Loringhoven criticized the disregard of the Wilsonian promise of a 'free, open-minded, and absolutely impartial adjustment of all colonial claims', which the American president had included in his Fourteen Points. Von Freytagh-Loringhoven, an anti-Semitic and nationalistic founder of the right-wing bourgeois *Deutsche Nationale Volkspartei*,[38] also argued that the powers granted to the victor states in the mandates would go much further than foreseen in Article 22 of the

[32] See Reichskolonialministerium, *Deutsche und französische Eingeborenenbehandlung* (1919); Reichskolonialministerium, *Die Behandlung der einheimischen Bevölkerung in den kolonialen Besitzungen Deutschlands und Englands* (1919).

[33] See Wehberg, 'Friedensbewegung und Kolonialpolitik', 26 *Die Friedens-Warte* (1926) 35.

[34] See Pedersen above note 24, at 195–203.

[35] On the writings of Hans Kelsen, Walter Schücking, and Hans Wehberg, see, for instance, J. von Bernstorff, *The Public International Law Theory of Hans Kelsen: Believing in Universal Law* (2010).

[36] On the reaction of German international legal scholarship to the Treaty of Versailles, see M. Stolleis, *Geschichte des öffentlichen Rechts in Deutschland, Dritter Band, Staats—und Verwaltungsrechtswissenschaft in Republik und Diktatur 1914–1945* (1999), at 86–88.

[37] F. von Liszt, *Das Völkerrecht. Systematisch dargestellt*, edited by M Fleischmann (12th ed., 1925), at 56–58.

[38] See Stolleis, above note 36, at 162.

League Covenant. As long as Germany could not regain its colonies, it would have to try to prevent the annexation of the mandated territories by the mandatories.[39]

Yet national-conservative international lawyers were not the only international lawyers to reject Versailles and its colonial provisions. Even though the pacifists Walther Schücking and Hans Wehberg provided a rather positive account of the League's mandate system in their famous League commentary,[40] and even though Wehberg welcomed the mandate system as progress in comparison to the colonial suppression of the past,[41] both lawyers in general supported Germany's colonial claims. For Schücking, the Treaty of Versailles was an 'egregious injustice' that triggered a 'right to revision'.[42] In relation to the colonies, he argued that

> as one of the greatest and most capable *Kulturvölker* (civilized peoples) of the world, we cannot renounce the right to take an active part in the colonisation of overseas countries. For colonisation is a cultural mission, and our colonies have been stolen from us under false pretences.[43]

Wehberg stressed that

> the placing of the German colonies under the Mandate of the League on the basis of the exclusion of Germany as mandatory power ... while the British, French etc. colonies stay in possession of those respective powers, is ... an injustice which typifies the spirit of Versailles.[44]

The German academy was thus sceptical of the new international law governing the colonies. In line with the general critique of the Treaty of Versailles, the League's mandate system was considered discriminatory. What is interesting to note, however, is that despite the general opposition, German international lawyers did not develop arguments about a legal entitlement to the former colonies. Rather than making an elaborate legal claim, they merely expressed political concerns about unjust treatment. Because in the political context of the 1920s, the recovery of the colonies seemed to be too unrealistic, it was regarded as futile to develop sophisticated legal arguments. Instead, as Susan Pedersen has shown in her recent study, Germany tried to prevent the annexation of the mandated lands after its entry into the League in 1926 and advocated self-determination for the colonized, in order to increase its political and economic influence on the African continent.[45] As Wehberg contended, 'whatever one

[39] von Freytagh-Loringhoven, 'Die Mandatsherrschaft des Völkerbundes', in H. Schnee and H. Dräger (eds), *Zehn Jahre Versailles, II. Band. Die politischen Folgen des Versailler Vertrages* (1929), at 171–200.

[40] W. Schücking and H. Wehberg, *Die Satzung des Völkerbundes* (2nd ed., 1924), at 680–711.

[41] See Wehberg, above note 33, at 35–37.

[42] W. Schücking, *Die nationale Aufgabe unserer Politik* (1926), at 8f.

[43] Ibid., at 12.

[44] Wehberg, 'Die Revision des Versailler Vertrages', 25 *Die Friedens-Warte* (1925) 150. Later Wehberg emphasized that after the accession to the League, Germany should calmly support the interests of the natives in the Mandate Commission without pushing for its own colonies in order to gain credibility. See Wehberg, 'Deutschland in der Mandatskommission', 27 *Die Friedens-Warte* (1927) 259.

[45] See Pedersen, above note 24, at 195–203. On 'counter-imperial' legal thought in Germany, see Rietzler, 'Counter-Imperial Orientalism: Friedrich Berber and the Politics of International Law in Germany and

might think of the problems of colonial policy', it seemed prudent to regard Germany's task on the Mandates Commission as an international responsibility, rather than as an opportunity for German propaganda.[46]

3. The National Socialist Era and International Law as Justification for Colonial Claims

The National Socialist takeover of 1933 spurred the imagination of many colonial advocates. Given the rhetoric of a strong, powerful German *Volk*, which would no longer feel bound by an unjust Treaty of Versailles, the colonial movement pinned high hopes on the National Socialist government. Over and over again, the government violated the 'dictate' of Versailles, in particular by introducing compulsory military service in March 1935 and by remilitarizing the Rhineland one year later.[47] In the eyes of some observers, this indicated that the colonial system too would be rearranged in the near future.[48]

The colonial policies of the National Socialist government were, however, ambivalent. Hitler did not regard the African colonies as a political priority. Already in his propagandistic autobiography *Mein Kampf*, written in the mid-1920s, he had claimed that Germany should rather look to the East than to the South. As a historic destiny, Germans would need *Lebensraum* in the East that would provide raw materials for the German people. He emphasized:

> We must stop the endless German movement to the South and West, and turn our gaze toward the land in the East. We must finally terminate the colonial and trade policies of the pre-war period and turn to the *Bodenpolitik* (land policy) of the future.[49]

As Hitler saw it, African colonies were only a vague project for the distant future. Germany would have to eschew all colonial claims for 100 years in order to concentrate on the European continent. Only then could it start to achieve world domination.[50]

After coming to power, Hitler continued to concentrate on *Lebensraum* in the East. However, the *Führer* used the colonial claims as a strategic instrument for potential bargains with Great Britain, and for securing the support of the old colonial elites in Germany. In spite of his *Lebensraum* policy towards the East, he therefore stated in the press, and in public speeches, that Germany needed its colonies back. He also

India, 1920s–1960s', 11 *Journal of Global History* (2016) 113. For a critique of such attitudes see also the contributions by B.S. Chimni and Siddharth Mallavarapu to this volume.

[46] See Wehberg, 'Deutschland in der Mandatskommission', above note 44, at 260f.

[47] See U. Herbert, *Geschichte Deutschlands im 20. Jahrhundert* (2014), at 353f.

[48] See van Laak, above note 13, at 291f.

[49] See A Hitler, *Mein Kampf. Zwei Bände in einem Band* (9th ed., 1932), at 742, cited in I. Kershaw, *Hitler: 1889–1936: Hubris* (2001).

[50] On which, see Hildebrand, above note 13, at 79, 86–88, 767f. For the historical debate on whether Hitler in his foreign policy intended to follow a step-by-step plan to world domination or whether he was focused on the European continent, see I. Kershaw, *The Nazi Dictatorship: Problems and Perspectives of Interpretation* (2015) [1985] 157–86.

supported the founding of a NSDAP-dominated *Reichskolonialbund*, which absorbed the earlier colonial movement and became the new organ of colonial propaganda. The *Reichsbank* president and minister for economic affairs, Hjalmar Schacht, likewise embraced the colonial plans for economic reasons.[51] Furthermore, the Foreign Office favoured the colonial goals, which were more in line with the *Wilhelminean* tradition than the idea of expanding to the East. Joachim von Ribbentrop used his *Dienststelle Ribbentrop* and later the Foreign Office (he became foreign minister in February 1938) to push for German colonies.[52] The press, theatres, and cinemas soon embraced the colonial cause, and produced the image of a heroic German colonial fighter as a role model for the public.[53] Moreover, the newly founded *Kolonialpolitische Amt* (KPA) and Committee on Colonial Law of the Academy of German Law came to be the two most important institutions in charge of the legal implementation of the colonial plans.[54] The KPA was created in 1934 to study the benefits of potential colonies from a political, economic, and scientific perspective.[55] The Committee on Colonial Law was established in 1937 to discuss pressing academic colonial law questions.[56] These institutions later produced detailed drafts on the legal norms for governing a potential German Central Africa (*Deutsch-Mittelafrika*), which was supposed to supply the *Reich* with raw materials and colonial products.[57] However, whenever Great Britain and France considered returning some of the mandated colonies to Germany in an attempt at 'colonial appeasement',[58] Hitler refused to make concessions. His political vision remained focused on the East.[59]

In the meantime, international legal scholarship in Germany had changed significantly since 1933. The new leadership drove numerous international lawyers of Jewish origin and pacifist convictions into exile. Hans Kelsen, the famous author of the *Pure Theory of Law*, had been dismissed from his chair in Cologne and fled via Geneva and Prague to the United States.[60] Schücking, a judge at the Permanent Court of International Justice, decided not to return from The Hague to Germany and was

[51] See Schacht, 'Germany's Colonial Demands', 15 *Foreign Affairs* (1937) 223. On this generally, see van Laak, above note 13, at 296–300; Pedersen, above note 24, 331, 339–43.

[52] On which, see Wendt, 'Außenpolitik', in W. Benz, H. Graml, and H. Weiß (eds), *Enzyklopädie des Nationalsozialismus*, (5th ed., 2007), at 66, 75–77; Jacobsen, above note 13, at 296–98; Hildebrand, above note 13, at 357–73, 771–72. Alfred Rosenberg as head of the *Außenpolitisches Amt* of the NSDAP and Ernst Wilhelm Bohle as head of the *Auslandsorganisation* of the NSDAP were opponents of the idea of African colonies: Hildebrand, above note 13, at 288f. The *Außenpolitisches Amt* focused on Britain, the Balkans, and the East, while the *Auslandsorganisation* of the NSDAP established National Socialist associations abroad. On these institutions, see Jacobsen, above note 13, at 45–89, 90–160.

[53] See Pedersen, above note 24, at 330.

[54] See Linne, above note 13, at 49f. who also refers to the Ministry of Justice as an institution which developed legal plans on the colonies. It is important to stress that to date the role of international law in the *Kolonialreferat* of the Foreign Office has not been studied.

[55] Linne, above note 13, at 7–9, 26–33.

[56] Schubert, above note 13, at 90.

[57] Linne, above note 13, at 7–9, 26–33. For '*Mittelafrika*' as a war aim during the First World War, see Neitzel, '"Mittelafrika". Zum Stellenwert eines Schlagwortes in der deutschen Weltpolitik des Hochimperialismus', in W. Elz and S. Neitzel (eds), *Internationale Beziehungen im 19. und 20. Jahrhundert* (2003) 83.

[58] On 'colonial appeasement', see Pedersen, above note 24, at 325–47; van Laak, above note 13, 296–300.

[59] See Schmokel, above note 12, at 113–26.

[60] See R.A. Métall, *Hans Kelsen: Leben und Werk* (1969), at 60–77; von Bernstorff, above note 35, at 278–81.

stripped of his position at the University of Kiel. He died in 1935.[61] Meanwhile, the national-conservative Erich Kaufmann had to leave for the Netherlands in 1939 because of his Jewish heritage.[62]

At the same time, the critique of the Treaty of Versailles came to be advanced more forcefully. Carl Schmitt lambasted that hidden behind the 'thin veneer of juridical legalisations' was the 'grimace of a menial and cruel kind of rape and suppression'.[63] The director of the Kaiser-Wilhelm-Institute of Comparative Foreign Law and International Law, Victor Bruns vehemently attacked the Treaty of Versailles as a violation of Germany's equality.[64] In addition, *völkisch*-oriented authors dismissed the 'abstract-formalist' thinking of the League of Nations era, and tried to integrate terms like *Volk* and *Reich* into a theoretical account of international law.[65] Friedrich Berber, a leading propagandist for Foreign Minister Joachim von Ribbentrop, formulated the programme for international legal research in light of imminent war:

> [The] exceedingly successful foreign policy of the Führer [has] led German international legal scholarship to the recognition of its task as reality-based and present-enrooted responsible political science *(wirklichkeitsverbundener und gegenwartsverantwortlicher politischer Wissenschaft)*. Instead of dead formulas and abstract terms, the politics of international law come to the fore as the academic observation of the concrete political international law, as the treatment of international law under the dynamic aspect of transformation, the struggle of new ideas with old formulas. It has not only the task of finding and unmasking the political, historical and ideological backgrounds of the Western European and Anglo-Saxon international law, to supply German foreign policy in its struggle for freedom and greatness of the German people with weapons in international law and to find new forms, new vessels for new policy ideas and creations; it has above all to work out the system of a real international legal order (Völkerrechtsordnung) that no longer is a summation of more or less random formal rules.[66]

The international lawyers continued the Weimar critique of the mandate system. Alfred Verdross, famous for his *Verfassung der Völkerrechtsgemeinschaft*, published in 1926, emphasized, in his 1937 textbook *Völkerrecht*, that the mandate system was a disguised expropriation of the German colonies without compensation.[67] Similarly, G.A.

[61] See F. Bodendiek, *Walther Schückings Konzeption der internationalen Ordnung. Dogmatische Strukturen und ideengeschichtliche Bedeutung* (2001), at 77–81.

[62] See F. Degenhardt, *Zwischen Machtstaat und Völkerbund. Erich Kaufmann (1880–1972)* (2008), at 126–30.

[63] C. Schmitt, *Nationalsozialismus und Völkerrecht* (1934), at 28. On Schmitt's take on international law, see Carty, 'Carl Schmitt's Critique of Liberal International Legal Order between 1933 and 1945', 14 *Leiden Journal of International Law* (2001) 25, at 34–47.

[64] See V. Bruns, *Deutschlands Gleichberechtigung als Rechtsproblem* (1934).

[65] See Wolgast, 'Konkretes Ordnungsdenken im Völkerrecht', 4 *Völkerbund und Völkerrecht* (1937/1938) 74; N. Gürke, *Volk und Völkerrecht* (1935); Walz, 'Völkerrecht und Nationalsozialismus', 1 *Völkerbund und Völkerrecht* (1934/1935), at 473–79. See further Diner, 'Norms for Domination: Nazi Legal Concepts of World Order', in D. Diner, *Beyond the Conceivable: Studies on Germany, Nazism, and the Holocaust* (2000) 49; Vagts, 'International Law in the Third Reich', 84 *American Journal of International Law* (1990) 661.

[66] Berber, 'Deutsche Völkerrechtswissenschaft', in B. Rust (ed.), *Deutsche Wissenschaft: Arbeit und Aufgabe* (1939) 62, at 63

[67] A. Verdross, *Völkerrecht* (1937), at 62.

Walz described the mandates as the 'legal form, in which the Allied and Associated Powers dressed up the robbery of the German colonies',[68] while Ulrich Scheuner referred to the 'robbery' of the colonies and to the 'great achievements' of German colonization.[69] In the *National Socialist Handbook for Law and Legislation*, the chapter on the mandate system declared that the system was based on the 'colonial guilt lie' and a violation of Wilson's promises.[70]

The main institution addressing questions of international law and the colonies was the Academy of German Law and its Committee on Colonial Law. According to its notorious head, Hans Frank,[71] the Committee had a threefold task: to assess the policies and administration of the mandates from a legal perspective; to gather arguments for a German entitlement to the colonies; and to develop the foundations of a future German colonial law.[72] At the fourth annual meeting of the Academy in 1937, its members discussed the legal arguments for retaking the former colonies. It was the political face of the colonial movement, Franz Xaver von Epp, who provided the framework for the debate.[73] Von Epp was an early National Socialist (since 1928) with strong ties to the traditional colonial movement. Born in 1868, he had made his career as a military man and a right-wing politician who knew the former colonies from his colonial adventures. In 1900, he had been part of the East Asian Expeditionary Corps, which was deployed to crush the Boxer Rebellion in China.[74] He was also involved in the fighting against the Herero in German South-West Africa. After serving as a German military commander in Europe during the First World War, he established the *Freikorps Epp*, which fought the communist Bavarian Soviet Republic in 1919 and various communist uprisings in the Ruhr area. After 1933, von Epp became the leading National Socialist concerned with colonial policies. As head of the KPA and (from 1936) leader of the *Reichskolonialbund*, he linked the old colonial elites with National Socialist institutions.[75]

In his presentation, *Germany's Legal Entitlement to its Colonial Property*,[76] von Epp relied on four arguments. *First*, he accused the Allied Powers of a violation of international law prior to the conclusion of the Treaty of Versailles. He stressed that the Congo Act—agreed at the Berlin Conference with no African participants—stipulated in Article 11 that the colonies in the traditional Congo Basin area should be treated as neutral, even if the colonizer was involved in a war somewhere else.[77] According to

[68] G.A. Walz, *Handbuch des Völkerrechts. Vierter Band. Völkerrecht und internationales politisches Staatensystem. Zweite Abteilung. Das System der politischen Staatsverträge seit 1918* (1937), at 58.

[69] Scheuner, 'Zur Geschichte der Kolonialfrage im Völkerrecht', 22 *Zeitschrift für Völkerrecht* (1938) 442.

[70] The article was written by a former member of the Mandates Commission, Julius Ruppel. Ruppel, 'Das Mandatsystem', in H. Frank (ed.), *Nationalsozialistisches Handbuch für Recht und Gesetzgebung* (1935), at 266.

[71] On Hans Frank, see P. Sands, *East West Street: On the Origins of 'Genocide' and 'Crimes against Humanity'* (2016).

[72] See Schubert, above note 13, at 97.

[73] On von Epp, see Hildebrand, above note 13, at 113–18; Linne, above note 13, at 29; van Laak, above note 13, at 286; K. Wächter, *Die Macht der Ohnmacht. Leben des Franz Xaver Ritter von Epp (1868–1946)* (1999).

[74] However, the German troops arrived after the uprising already had been put down.

[75] See Hildebrand, above note 13, at 113–18.

[76] See F.X. von Epp, *Deutschlands Rechtsanspruch auf sein koloniales Eigentum* (1937).

[77] The article held that the territory should be 'considered as belonging to a non-belligerent State, the belligerents thenceforth abstaining from extending hostilities to the territories thus neutralized, and from

von Epp, this provision had to be applied to the German colonies in East and South-West Africa. In his view, the Allied Powers had violated Article 11 by attacking the German Empire in the colonies.[78] Von Epp thus indicated that the Allied incorporation of the German colonies had begun with a violation of neutrality. In a *second* step, von Epp stressed that the League of Nations had been built on this violation of neutrality. According to him, the 'Dictate of Versailles' sanctioned the violation of the Congo Act, while the League of Nations became the institution for monitoring injustices. The mandate system reflected the 'political-juridical deformity of Versailles' and 'manipulation of Versailles', and was nothing other than a 'recycling of earlier covert British annexation methods'.[79] Von Epp thus implicitly invoked the principle of *ex injuria ius non oritur*, according to which an unjust act cannot create law. According to him, the violation of neutrality could not justify the deprivation of German colonies, as sanctioned in Article 119 Treaty of Versailles. *Third*, von Epp indicated that the Treaty of Versailles violated the promises of Woodrow Wilson's famous Fourteen Points, in particular, the fifth point on colonies. Instead of pushing for 'a free, open-minded, and absolutely impartial adjustment of all colonial claims',[80] the mandate system, in von Epp's view, was driven by annexationist interests of the colonial powers.[81] Like many scholars in German academia at the time, von Epp thus took the view that Woodrow Wilson's Fourteen Points constituted the offer of a preliminary peace treaty, which Germany had accepted in a Note reacting to Wilson's proposal. This preliminary peace treaty had then been violated by the much stricter conditions imposed by the Treaty of Versailles.[82] *Fourth*, von Epp claimed that the mandate system of the League was based on an erroneous factual account, the 'atrocities lie' ('*Greuellüge*') regarding 'the colonial incapacity of Germany'.[83] As British and French politicians would have conceded, the narrative from the German 'colonial guilt' was a myth. Hence, the 'basis for the forced renunciation of colonial property no longer exists, and there is therefore no legal justification for the continued existence of the mandates'.[84]

Subsequent (and partly prior) to von Epp's statements, German legal academics developed these positions in various articles and monographs. The head of the Committee on Colonial Law, von Freytagh-Loringhoven (who had almost become a minister in Hitler's first cabinet)[85] argued that 'the idea that bad administration of a colony involved its loss, was totally foreign to pre-war international law'.[86] Even the

using them as a base for warlike operations'. For an overview on the Berlin Conference, see Craven, 'Between Law and History: The Berlin Conference of 1884–1885 and the Logic of Free Trade', 3 *London Review of International Law* (2015) 31.

[78] Von Epp, above note 76, at 6–8.
[79] Ibid., at 3–5, 8, 11.
[80] This was the Wilson's fifth point.
[81] Von Epp, above note 76, at 9–13.
[82] See e.g. von Freytagh-Loringhoven, 'Wer entscheidet über die Rückgabe der Kolonien?', 7 *Deutsches Recht* (1937) 485, at 490; H. Brenner, *Wem hat Deutschland seine Kolonien aufgrund des Versailler Diktats überlassen?* (1938), at 82–89; F. Winkelmann, *Die deutsche Kolonialfrage als Völkerrechtsproblem* (1936), at 17–36.
[83] Von Epp, above note 76, at 11.
[84] Ibid., at 15f.
[85] See Stolleis, above note 36, at 162.
[86] A. von Freytagh-Loringhoven, *Das Mandatsrecht in den deutschen Kolonien. Quellen und Materialien* (1938), at xii.

mandate system would not give rise to such legal consequences. Furthermore, he decried the Allies' accusations as 'an injury to Germany's honour' giving rise to a satisfaction claim in form of the return of the colonies.[87] He stressed that since the mandatories had not lived up to their obligations under Article 22 of the League's Covenant to protect the interests of the inhabitants, the mandates system would not hold as a justification for taking the colonies from Germany.[88] Similarly, Norbert Gürke, who later became professor at the University of Vienna, criticized Article 22 as a 'disguise for geopolitical interests'.[89] Articles, monographs, and dissertations elaborated how French and British colonial troops had (allegedly) violated the Congo Act during the First World War,[90] underlined the principle of colonial equal status between states,[91] emphasized that the whole Treaty of Versailles was null and void because of illegal coercion,[92] criticized the breach of Wilson's promises as a breach of a preliminary peace treaty,[93] and attacked the mandate system as concealing international law violations.[94] Hence, these arguments elaborated and intensified the criticism of the League's mandate system of the 1920s. In other words, after 1933, the voices became louder. With the political back-up of the National Socialists, newly established institutions like the Committee on Colonial Law provided the space to entrench the Weimar positions. Protagonists like von Epp or von Freytagh-Loringhoven, who had already subscribed to the colonial cause in the Weimar years, now seized the opportunity to advance their colonial programme in a politically more promising environment.

While the argumentation was to a large degree in line with the traditional critiques made in Weimar times, some scholars also became more radical in tone. One suggested that 'if Germany, in view of this clear legal position, takes its colonies by force, then the annulment of the wrong created at Versailles concerning the colonial question could be considered permissible self-help. Resistance against such force, however, would be wrongful under international law.'[95] Another one put forward that the 'juridical, state law conception of the world of the Versailles model' had led to the separation of peoples, who by blood would otherwise belong together. In contrast, the international law of the future would have to govern an order of different peoples instead of different states. Also, colonial law would have to be built on a racial hierarchy, which would recognize the superior capability of the German race.[96] Furthermore, it was argued that the many Jews in German international legal scholarship had contributed to producing a statist, anti-*völkisch* and anti-German law reflected in the injustices of the fraudulent mandate system.[97] For these scholars, it was not only an option to take back

[87] Ibid., at xvi.

[88] Ibid., at xviii.

[89] Gürke, 'Völkerrechtslehre und koloniale Ordnung', 7 *Deutsches Recht* (1937), at 513f.

[90] Hashagen, 'Der Kolonialkrieg im Weltkriege und das Völkerecht', 12 *Zeitschrift für Völkerrecht* (1938), at 71–76.

[91] See Karlowa, 'Das Recht der Deutschen auf Kolonien', 7 *Deutsches Recht* (1937) 482; Winkelmann, above note 82, at 1–16.

[92] W. Mellem, *Der deutsche Kolonialanspruch* (1938), at 14–25; Karlowa, above note 91, at 482.

[93] Brenner, above note 82, at 82–89; Winkelmann, above note 82, at 17–36.

[94] L. Kühne, *Das Kolonialverbrechen von Versailles. Die Räuber von Versailles, die Mandatare und das etatistische Völkerrecht: Die Totengräber der weißen Kolonialherrschaft* (1939), at 42–56.

[95] Mellem, above note 92, at 78.

[96] Leistritz, 'Wandel im Völkerrechtsdenken und Kolonialfrage', 3 *Der Kolonialdienst* (1938) 1.

[97] Kühne, above note 94, at 20, 80, 92f.

the colonies by force; they also openly incorporated the racism of National Socialism into their legal claims.

These radical critics, however, were a minority and did not have an influential position in the academy.[98] Most of the scholarship followed traditional patterns of international law argumentation. In line with the discursive rules of the discipline, German lawyers accused the Allied powers of treaty violations, unlawful coercion, and violations of the principle of sovereign equality. Through these arguments, they hoped to win potential supporters abroad.[99] Von Freytagh-Loringhoven edited a book on sources and materials of the mandate system, which printed all sources in French or English and included a preface in German, English, French, and Italian depicting Germany's legal position.[100] International law's justificatory power was supposed to persuade other states that Germany's colonial claims had a fair legal basis.

The German attempts did not bear fruit. Even though the idea of winning back the former colonies gained new political momentum with the German attack on Poland and the outbreak of the Second World War, legal argumentation lost its central role. Military force instead of law came to be the instrument of the colonial quest. After Germany had proven its military strength with the speedy victories against Poland and France, German expansion into the former overseas colonies seemed only a matter of time. For the German navy and the Foreign Office under anti-British Ribbentrop, a German Central Africa (*Deutsch-Mittelafrika*) seemed closer than ever. In May 1940, Ribbentrop asked his staff to concretize the plans for a German *Kolonialreich*. In light of the *Endsieg*, officials of the Foreign Office argued that because Britain would give in to German ambitions, a *Reich* consisting of the former German colonies, Belgian-Congo, French Equatorial Africa, and potentially also British-Nigeria would be possible. This territory could provide *Großdeutschland*, and even the Greater European economic area, with tropical and subtropical raw material.

Hitler's position, too, somewhat shifted. From 1940 to 1941, Hitler temporarily regarded the colonies in Central Africa as part of the war aims because he hoped that a peace with Britain would enable him to 'divide the world' and thus share the African continent with Britain. Soon the administration prepared for governing the imagined empire.[101] The KPA now drafted many legal regulations on how to administer the African territories while the Committee on Colonial Law of the Academy of German Law provided legal expertise. The objective was to develop a National Socialist colonial law dominated by *völkisch* ideas.[102] As one protagonist claimed a National Socialist colonial law 'must respect the racial character of the natives ... and refuse education as pseudo-European'.[103] It would be wrong 'to force European civilisation

[98] None of them was a professor at a German university.

[99] See Schmokel, above note 12, at 64f.

[100] von Freytagh-Loringhoven, above note 86, at xii–xviii. Similarly, in order to advance economic arguments for the retaking of the colonies, the German Society of International Law and World Policy, founded by Heinrich Schnee, produced a report in English stressing the importance of the German colonies for Germany as suppliers of raw materials. See Deutsche Gesellschaft für Völkerrecht und Weltpolitik, *Raw Materials and Colonies. Report of Committee Appointed to Investigate the Question of Raw Materials* (1940).

[101] See Hildebrand, above note 13, 652–773; Linne, above note 13, at 70f.; van Laak, above note 13, at 307–09, 315.

[102] Lange, 'The Dream of a Völkisch Colonial Empire. International Law and Colonial Law during the National Socialist Era', 5 *London Review of International Law* (2017) 343.

[103] See Karlowa, cited in Schubert, above note 13, at 99.

on the indigenous ... because this method has led to the dislocation and decay of the natives and has driven them into the hands of African bolshevism'.[104] The 'supremacy of the white race' would be better advanced if National Socialism concentrated on the' '*arteigene*' ('race-specific') upward development' of the natives.[105] The colonized population was profoundly different from the German people by birth, and lacked the capacity to become a pillar of German culture.[106] 'Every biological connection between whites ... and other races has to be forbidden.'[107] As the international lawyer Gürke proposed, the 'political incapacity' of some races legitimized their governance by the superior race: 'History proves that the natural racial and *völkisch* feeling of every strong people regards colonial acquisition and possession as lawful.'[108]

However, after the German invasion of the Soviet Union in 1941 and the failure of the concept of the *Blitzkrieg*, Hitler lost all interest in the reacquisition of colonies (which, as indicated, he had never prioritized anyway). With German troops unable to reach Moscow, resources needed to be concentrated on winning the war in the East. While some colonial enthusiasts still hoped that the war in Africa would open the door for a German-controlled African continent, they were quickly disappointed. With the loss at El Alamein and the Soviet counter-offensive in Stalingrad in November 1942, the idea of a German colonial empire completely collapsed. Hitler repeatedly stressed that the raw materials in the East would suffice to satisfy Germany's needs.[109] The colonial institutions were liquidated shortly afterwards. The KPA, which had already been asked to dismantle some of its operations by the end of 1941, was shut down in March 1943. Discussion of the laws to govern renewed colonial administration did not reach its conclusion.[110] The dream of a *völkisch* colonial empire came to an abrupt end.

4. Conclusion

Already during the Weimar Republic, the loss of the German colonies had led to fierce criticism of the Treaty of Versailles. The German government, pushed by a strong colonial revisionist movement, stated on numerous occasions that it intended to retake its former colonies. German international lawyers likewise voiced concerns about the 'unjust' mandate system of the League of Nations. Yet the redistribution of colonial territories seemed unrealistic during the 1920s, and the discipline did not devote attention to Germany's legal position on the former colonies. After the National Socialist takeover, however, the pro-colonial voices became louder. Even though Hitler's political design looked predominantly towards the East, newly established institutions—like the KPA and the Committee on Colonial Law—developed ideas for governing

[104] Ibid., at 99.
[105] Ibid., at 103.
[106] See von Hecht, cited in W. Schubert, *Akademie für Deutsches Recht. Protokolle und Ausschüsse, 12. Ausschuss für Rechtsfragen der Bevölkerungspolitik 1934–1940 und Ausschuss für Kolonialrecht zusammen mit den Entwürfen des kolonialpolitischen Amtes 1937–1941* (2001), at 492f.
[107] See Weber, cited in ibid., at 496.
[108] Gürke, above note 89, at 514f.
[109] See Linne, above note 13, at 139f., 147–51.
[110] See Schubert, above note 13, at 94; Linne, above note 13, at 154–62.

the potential German colonies. Politicians and international lawyers now embraced and elaborated the claim of a legal entitlement to the former colonies. Building on the argumentative schemes of the interwar period, they accused the Allies of violations of the Congo Act, Wilson's Fourteen Points, and the principle of sovereign equality. In their view, the 'unjust' Treaty of Versailles and the Covenant of the League of Nations masked, but could not condone, these violations of international law.

So what does this story tell us about the justificatory uses of international law in the context of colonialism? Perhaps most obviously, this study can be read yet again as exposing the 'dark sides of international law'.[111] The chapter describes how international legal arguments were advanced by National Socialist politicians and international lawyers in order to support a particular foreign policy agenda of the German *Reich*, namely, the aim of regaining the former colonies. The justificatory potential of international law was used by a dictatorial and racist regime for the political purpose of colonial expansion.[112]

At the same time, this study also points to something else. If German international lawyers wanted to be heard, and to convince foreign statesmen and officials of their ideas, they had to advance their arguments in accordance with the discursive rules of the discipline. As the editors of this volume highlight, actors who justify their policies with legal argumentation are part of a 'communication community'.[113] Therefore, German international lawyers needed to rely on established patterns of international legal argumentation, referring to violations of treaty obligations and making arguments about sovereign equality. The introduction of the *völkisch* world-view into these discussions would not work. It was only when the governance of the colonies was at issue and no other power had to be persuaded that National Socialist ideology left its imprint on the legal drafts. Envisioning the German *Reich* as the unrestricted sovereign over the colonial territories, the German legal theorists infused colonial law with their ideas of racial superiority. While international law was supposed to pave the way for regaining the colonies, colonial law was framed according to the political principles of the rulers in the motherland.

[111] On this line of thinking in the history of international law, see Becker Lorca, 'Eurocentrism in the History of International Law', in B. Fassbender and A. Peters (eds), *The Oxford Handbook of the History of International Law* (2012) 1034; see also the chapters by Becker Lorca, Chimni, Hippler, and Mallavarapu in this volume.

[112] On the dark sides of international law see D. Kennedy, *The Dark Sides of Virtue: Reassessing International Humanitarianism* (2004); A. Anghie, *Imperialism, Sovereignty and the Making of International Law* (2005).

[113] See Simon and Brock, Introduction, and Tischer in this volume.

PART V
'DEMOCRATIC WARS' AND THE POST-COLD WAR INTERNATIONAL ORDER

Rise and Decline of the 'Liberal Peace'

16

'What We Are Fighting For': Democracies' Justifications of Using Armed Force since the End of the Cold War

Anna Geis and Wolfgang Wagner

1. Introduction: Liberal Democracies and War—An Uneasy Relationship

As Simon and Brock state in their introduction to the present volume: 'The history of war is also a history of its justification.'[1] This chapter deals with changing justification practices of liberal democracies after the end of the Cold War. At that time, it seemed that both liberalism and democracy emerged as the undisputed 'winners' of the Cold War. Thirty years later liberal euphoria and hubris have largely vanished. Regional orders have developed in quite diversified directions; an increasingly multipolar world order, comprising democratic and non-democratic powers, puts current global governance arrangements under pressure. A shift in global power relations directs enhanced attention to justificatory practices of Russia and China, for example in the UN Security Council.[2] Other countries from the Global South[3] and their regional organizations are also gaining in agency. However, this chapter will focus on liberal democracies' justification practices over the last decades, drawing on research on the so-called 'Democratic Peace' and its flipside, 'democratic wars'.

Since 1990, liberal hegemony under US leadership has manifested itself in many forms. Core principles of liberal thought such as market economy, promotion of democracy, and human rights migrated into numerous international organizations and institutions beyond the liberal 'core' and thus helped to stabilize the global hegemony of 'the West'. The discourses and practices of institutions such as the UN Security Council, NATO, or the European Union—the two latter forming 'democratic clubs' that both enlarged their membership substantially—demonstrate that democracies have played a major role in governing central security matters: they have shaped the international ordering practices via their interpretations of international law, their justifications of the use of military force, their modes of warfare, their (dis-)armament and arms control policies, and their alliance politics.[4]

[1] See Simon and Brock, 'Introduction', this volume, at 3.
[2] See H. Niemann, *The Justification of Responsibility in the UN Security Council. Practices of Normative Ordering in International Relations* (2018).
[3] A. Acharya, *Constructing Global Order. Agency and Change in World Politics* (2018).
[4] See e.g. M. Evangelista, H. Müller, and N. Schörnig (eds), *Democracy and Security: Preferences, Norms and Policy-Making* (2008).

Anna Geis and Wolfgang Wagner, *'What We Are Fighting For': Democracies' Justifications of Using Armed Force since the End of the Cold War* In: *The Justification of War and International Order.* Edited by: Lothar Brock and Hendrik Simon, Oxford University Press (2021). © the many contributors. DOI: 10.1093/oso/9780198865308.003.0016

Such a focus on liberal democracies seems warranted given their material and normative international ordering resources: first, a number of democracies possess enormous military and economic capabilities which have enabled this group of states to shape international politics to a large extent since 1990. Second, liberal democratic states often consider themselves as global guardians of human rights, a rule-based international order, and a cosmopolitan international community while refraining from subjecting themselves fully to procedural obligations under international law.[5] Hence, states (or more specifically: regimes/governments) that violate human rights on a large scale are regarded as outsiders to the international community that are sanctioned by the liberal guardians by punitive action. Stigmatizing labels such as 'rogue', 'pariah', or 'evil' are part and parcel of liberal 'ways of war'.[6] While democratic elites might justify the use of force by referring to 'good' or even 'just' reasons and derive *substantial legitimacy* from such a practice, they face serious deficits in attaining *procedural legitimacy* in the form of a clear UN mandate (Kosovo 1999, Iraq 2003, the airstrikes against Syria in 2017 and 2018). In the case of Libya (2011) such a mandate was given but it was promptly transgressed to serve regime change.

Democratic justification patterns for using military force have also changed over time during the period of liberal hegemony since those justifications always reflect the interplay between changing domestic constellations of interest and power, hegemonic discourse within the state and the surrounding normative order. The international establishment of the notion of 'human security' after the end of the Cold War highlights the strong relevance of human rights for the normative international order. Whereas the 1990s saw a higher congruence between democracies' domestic liberal norms and the normative principles of world order, the US-led 'Global War on Terror' following the terrorist attacks of '9/11' ushered in a period of militarized counter-terrorism and a concomitant macro-securitization of global governance, resulting in the blending of liberal-cosmopolitan with national security justifications. For example, some of the current international military interventions in Africa rather follow an 'intervention as stabilization' frame than a distinctly humanitarian outlook.[7]

In this chapter, we will draw on a large body of literature that highlights the distinctiveness of liberal democracies' security policies and which has, in turn, been inspired by the debate about the so-called Democratic Peace.[8] This literature argues that democracies are both powerful *and* hesitant warriors: 'powerful' because many of them possess effective military capabilities and resource mobilization strength;[9] 'hesitant'

[5] Brock, 'The Use of Force in the Post-Cold War Era', in M. Bothe, M.E. O'Connell, and N. Ronzitti (eds), *Redefining Sovereignty: The Use of Force after the Cold War* (2005) 21.

[6] Wagner and Werner, 'War and Punitivity under Anarchy', 3 *European Journal of International Security (EJIS)* (2018) 310.

[7] See the contribution by Nina Wilén to this volume and Karlsrud, 'From Liberal Peacebuilding to Stabilization and Counterterrorism', 26 *International Peacekeeping* (2018) 1.

[8] Proponents of the so-called Democratic Peace argue that democracies rarely if ever wage war against each other. See for the early debate B. Russett, *Grasping the Democratic Peace: Principles for a Post-Cold War World* (1993). For a review of the democratic distinctiveness programme that emerged in the wake of the Democratic Peace debate see Geis and Wagner, 'How Far Is it from Königsberg to Kandahar? Democratic Peace and Democratic Violence in International Relations', 37 *Review of International Studies (RIS)* (2011) 1555.

[9] D. Reiter and A. Stam, *Democracies at War* (2002).

because democratic leaders face numerous domestic constraints: they are accountable to domestic constituencies, bound by constitutional law, face political opposition and media control. Costly foreign policy decisions such as deploying troops usually require at least *some* extent of public justification, although parliamentary oversight powers in security policy vary significantly across democracies.[10]

In line with the theme of the anthology, the remainder of this chapter will discuss continuity and change in the justifications of the use of armed force by liberal democracies since the end of the Cold War. We will first discuss why liberal democracies have a distinct need for justifying the use of military force (section 2). We will then outline the actual justifications that liberal democracies have brought forward in legitimizing military missions since the end of the Cold War (section 3). In doing so, we distinguish a period of liberal democratic dominance and euphoria in the 1990s (section 3.1) and a period since '9/11', characterized by the macro-securitization of transnational terrorism (section 3.2). For the first period, we primarily draw on the findings from the project 'Wars of Democratic States since 1990' that examined parliamentary debates in seven liberal democracies on the 1991 Gulf War, the 1999 Kosovo intervention, and the 2003 Iraq War.[11] For the second period, we present new data on parliamentary debates in Canada, Germany, and the United States about the continued use of force in Afghanistan between 2001 and 2016 and on the fight against *Daesh* in Iraq and Syria.[12] We find a remarkable pluralism and concomitance of justifications: whereas liberal norms such as the protection of human rights are prominent in the justificatory discourses of liberal democracies, such normative references come to bear alongside traditional concerns for national security and are often seconded by arguments about international law, multilateralism, and alliances. We can also show that '9/11' and the subsequent 'Global War in Terror' have led to a renaissance of national security arguments. However, humanitarian justifications have persisted and remained prominent in debates about military interventions in Afghanistan and the fight against *Daesh*.

2. Hesitant Warriors?: The Need for Justification

Commenting on the use of force by democracies and its justification, C. Lipson observed in 2003:

> (I)t is clear that democracies go to war often. In the twentieth century, for instance, Britain went to war more frequently than Germany. The US has obviously been willing to use force repeatedly, both to defend its material interests and to extend

[10] Peters and Wagner (2010), 'Between Military Efficiency and Democratic Legitimacy: Mapping Parliamentary War Powers in Contemporary Democracies, 1989–2004', 64 *Parliamentary Affairs* (2010) 175.

[11] See A. Geis, H. Müller, and N. Schörnig (eds), *The Militant Face of Democracy. Liberal Forces for Good* (2013). This project studied justifications in the United States, Australia Canada, Germany, France, the United Kingdom, and Sweden.

[12] The data stem from Wolfgang Wagner's work on party politics and foreign policy; see W. Wagner, *The Democratic Politics of Military Interventions. Political Parties, Contestation and Decisions to Use Force Abroad* (2020).

> its values. Moreover, democracies do not merely respond to provocations by others. They often seize the initiative and sometimes attack first. We need to understand, then, how norms and beliefs could lead democracies to make war against some states but not against others.[13]

According to 'Democratic Peace' scholarship, the relationship between democracy and war is delicate: (some) democracies possess overwhelming material capabilities to fight wars, but the use of force is restrained by numerous legal provisos and by norms and values deeply rooted in their societies. The preference for peaceful conflict resolution, respect for human rights, and aversion to casualties are among the norms inherent in political liberalism. However, the phenomenon of 'liberal interventionism' or 'liberal' and 'democratic wars'[14] after 1990 has warranted more attention for the 'militant' face of democracies, in particular to the question of to which norms and values belligerents refer in justifying their use of force. It should be emphasized that democracies engage in military actions with widely differing frequency. A small group of democracies such as the United States, the United Kingdom, France, Australia, India, and Israel have been involved comparatively often in military conflicts, while some democracies rarely take part in military action and others abstain altogether.[15]

Research on 'democratic wars' since the 1990s suggests that liberal norms which promote peacefulness in interaction with other democracies can under certain circumstances also legitimate violence in interactions with some non-democracies. To derive this theoretical argument, an analogy to the use of force *within* democratic polities can be drawn: using force is not ruled out per se in the domestic sphere of democracies, but its legality and legitimacy are governed by the respective legal order and politico-cultural norms, which vary between democracies. Transferred to the external use of force, decisions on war participation will have to meet certain domestic and international legitimacy standards. First, they should be in accordance with procedural requirements, that is fulfil standards of domestic and international lawful decision-making, and, second, they should be justified with substantial reasons acceptable to the domestic public.[16] Political elites and citizens alike can also develop preferences favouring the use of force in a given conflict. This is to say that there are some justifications that are accepted by the general public as legitimate reasons for the use of force while others are not.

[13] C. Lipson, *Reliable Partner. How Democracies Have Made a Separate Peace* (2003), at 3.

[14] D. Chandler, *From Kosovo to Kabul and Beyond. Human Rights and International Intervention* (2006); Freedman, 'The Age of Liberal Wars', in D. Armstrong, T. Farell, and B. Maiguashca (eds), *Force and Legitimacy in World Politics* (2005) 93; Vasquez, 'Ethics, Foreign Policy, and Liberal Wars', 6 *International Studies Perspective (ISP)* (2005) 307; Heinze, 'Failed Interventions and the Inherent Contradictions of Liberal Internationalism', in A. Hehir and R. Murray (eds), *Protecting Human Rights in the 21st Century* (2017) 51; Geis, Müller, and Schörnig, above note 11.

[15] See e.g. Müller, 'The Antinomy of Democratic Peace', 41 *International Politics* (2004) 494, at 494–97; and P. Mello, *Democratic Participation in Armed Conflict: Military Involvement in Kosovo, Afghanistan, and Iraq* (2014).

[16] Müller and Wolff, 'Democratic Peace. Many Data, Little Explanation?', in A. Geis, L. Brock, and H. Müller (eds), *Democratic Wars: Looking at the Dark Side of Democratic Peace* (2006) 41, at 61.

Such requirements explain why *justification* is such an important dimension of analysing the use of force employed by democracies. Even if democratic leaders might try to 'bypass' legal procedures or do not come up with 'good' reasons before engaging in military actions, at some point they will probably be confronted by other members (or by institutions) of the democratic community, for example via parliamentary or expert inquiry or via media scandalizing. If one follows 'Democratic Peace' research arguments, one can expect a *special* need for democratic leaders to justify military actions in comparison to non-democratic leaders. The proto idea of this expectation was developed by Immanuel Kant, who is often cited as the founding father of 'Democratic Peace' research,[17] in his famous essay 'On Perpetual Peace' (1795). Kant argues that if

> the agreement of the citizens is required to decide whether or not one ought to wage war, then nothing is more natural than that they would consider very carefully whether to enter into such a terrible game, since they would have to resolve to bring the hardships of war upon themselves ... By contrast, in the case of a constitution where the subject is not a citizen of the state, that is, in one which is not republican, declaring war is the easiest thing in the world, because the head of state is not a fellow citizen, but rather the owner of the state, and hence forfeits nothing of his feasts, hunts, summer residences, court festivals, and such things due to the war. The head of state can decide to wage war for insignificant reasons as a kind of game for amusement and can, for the sake of decency, indifferently leave its justification up to his diplomatic corps, which always stands ready for such tasks.[18]

Modern advocates of the 'Democratic Peace' claim that democratic institutions and (liberal) norms have a pacifying influence on the foreign policies of democratic states. Citizens are considered to be sensitive to the risks and costs of war and hence rather reluctant warriors, except for situations of self-defence; democratic leaders, interested in being re-elected, will respect their citizens' preferences. Hence, if a leader or a part of the elite intend to engage in interventions or wars, they will have to argue for this, make a case, provide reasons to their 'audience(s)' in order to gain legitimation for such a decision. These audiences can be quite different: domestic voters of a party, oppositional parties, the general public; elites or populations of other countries (partners or adversaries); the so-called 'international community'; an emerging 'global' public in the era of enhanced globalization etc.

Content analyses of justificatory discourses in democracies can help to identify empirical patterns over time and across countries. Such analyses do not necessarily provide reliable information about the *motivation* behind the speaker's policy choice, as actors can always act strategically and simply lie about their motives. However, public justifications are an important indicator of what is considered as '(un)acceptable', '(un)

[17] Instead of many see B. Russett and J. Oneal, *Triangulating Peace. Democracy, Interdependence, and International Organization* (2001). For a critical discussion of such Kant interpretations see O. Eberl, *Demokratie und Frieden: Kants Friedensschrift in den Kontroversen der Gegenwart* (2008). See also Oliver Eberl's contribution to this volume.

[18] Kant, 'Perpetual Peace', in P. Kleingeld (ed.), *Perpetual Peace and Other Writings on Politics, Peace and History* (2006) 75; see also the chapter by Oliver Eberl in this volume.

sayable' in a specific community at a specific time. Politicians 'belong to a community whose constitutive values and norms they share'[19]—or at least are familiar with. Hence, if actors have private motives or interests, they will at least try to relate their action to an accepted argument within the range defined by the constitutive values and norms of the respective community. It is very likely that democratically elected agents who have to justify their actions or intentions publicly in order to reach consent put forward 'resonant' arguments, that is arguments which they think will correspond best to the value orientations and interests of their constituents.

3. Changing Patterns of Justification since the End of the Cold War

Whenever liberal democracies used military force after the end of the Cold War, they typically intervened in manifest violent conflicts (the Iraq war in 2003 being a partial exception). Such interventions are not directed towards acquiring territory, but towards stopping violence and long-term transformation of societies and political systems, that is enforcing a new normative political order.[20] Since the 1990s, with the emergence of 'liberal interventionism' and the paradigm of 'liberal peacebuilding', the future political orders of pacified territories were to be modelled after 'Western' conceptions of statehood, human rights, democracy, and market economy.[21] During the last decade, we have been observing a gradual reframing of a number of interventions, particularly in Africa towards the 'stabilization' of existing regimes, that is rather restoring the sovereignty of ruling governments, no matter whether autocratic or else, vis-à-vis violent non-state actors, and thus deemphasizing human rights and democracy as pillars of a normative order. The failures of Western-inspired interventionists and other liberal peacebuilders[22] to establish democracy, human rights, and statehood in quite a number of military missions has partly—together with the gradual normative shift towards a 'Global War on Terror' and doctrinal changes at UN level[23]—contributed to a toning down of ambitious transformative goals of interventions.

[19] Schimmelfennig, 'The Community Trap: Liberal Norms, Rhetorical Action, and the Eastern Enlargement of the European Union', 55 *International Organization* (2001) 47, at 62.

[20] On some occasions, more limited air strikes might be designed to basically 'punish' leaders that are held accountable for atrocities and violations of international law such as in the case of the Assad regime and the use of chemical weapons in the Syrian civil war, see Wagner and Werner, above note 6.

[21] For critical treatments of 'liberal peacebuilding' see e.g. Paris, 'Peacebuilding and the Limits of Liberal Internationalism', 22 *International Security* (1997) 54; Chandler, above note 14.

[22] Heinze, above note 14; Grimm, 'External Democratization after War: Success and Failure', 15 *Democratization* (2008) 525; L. Brock, H.H. Holm, G. Soerensen, and M. Stohl, *Fragile States. Violence and the Failure of Intervention* (2012); Gromes and Dembinski, 'Practices and Outcomes of Humanitarian Military Interventions: A New Data Set', *International Interactions* (2019), 1032.

[23] Karlsrud, 'For the Greater Good?: "'Good States" Turning UN Peacekeeping towards Counterterrorism', 74 *International Journal* (2019) 65; Wiuff Moe and Geis, 'From Liberal Interventionism to Stabilization: Towards a New Consensus on Norm-Downsizing in Interventions in Africa', 9 *Global Constitutionalism* (2020), 387.

3.1 Militant Democracies as 'Forces for Good'? Justifying Military Missions in the 1990s

The normative framing of the use of force by democracies after the end of the Cold War has inspired different labels for this phenomenon: 'liberal' wars and 'democratic' wars are the two most prominent, and they overlap to a large extent. Lawrence Freedman regards 'liberal wars' as wars 'conducted in pursuit of a humanitarian agenda.... The ideal type for a liberal war is that it is altruistic in inspiration and execution.... It is liberating and empowering while involving as few casualties as possible.'[24] In a similar vein, John Vasquez considered the Kosovo War as 'the quintessential liberal war of our time.'[25] Liberal wars are, according to Vasquez' definition, 'wars that are based on moral claims to do what is right rather than claims about the national interest. These wars include wars for democracy or humanitarian wars arising out of armed humanitarian interventions.'[26] The concept of 'democratic wars' is in many respects congruent with that of 'liberal wars' but addresses democratic wars as the flip-side of the Democratic Peace.[27] The underlying claim is that democracies have special reasons to go to war against non-democracies which non-democracies do not have because governments in the former are accountable to their electorate.[28]

Liberal or democratic interventionism in practice faces many contradictions since the benign self-image of democracies as 'forces for good' is not easy to maintain:[29] political judgements on the use of force have to cope with enormous complexity, imperfect knowledge, and moral/political ambivalences. People are killed in order to save people; the rule of law is disregarded in order to install it; international security is undermined in order to strengthen it. The controversial public debates within democracies as well as disputes between democracies about the appropriateness and legitimacy of the use of force indicate that democratic publics are quite aware of these problems. Different interventions/wars are also not treated as new phenomena in war discourses but speakers often refer to prior experiences, failures, and historical analogies in order to make their case and derive 'lessons learnt.'

Comparative empirical analyses of justificatory discourses in liberal democracies since the 1990s show that democratic leaders and members of parliaments usually refer to quite a number of different norms in order to justify the use of force. In our own empirical research of pre-intervention discourses,[30] we have found evidence for the

[24] Freedman, 'The Age of Liberal Wars', in D. Armstrong, T. Farell, and B. Maiguashca (eds), *Force and Legitimacy in World Politics* (2005) 93, at 98.

[25] Vasquez, 'Ethics, Foreign Policy, and Liberal Wars: The Role of Restraint in Moral Decision Making', 6 *International Studies Perspectives (ISP)* (2005) 307, at 311.

[26] Ibid., at 307.

[27] Doyle, 'Kant, Liberal Legacies, and Foreign Affairs', 12 *Philosophy and Public Affairs* (1983) 205, at 230. E.O. Czempiel, *Friedensstrategien* (1986).

[28] For a detailed explanation of 'democratic wars' see Geis, Brock, and Müller, above note 16; and see Geis, Müller, and Schörnig, above note 11.

[29] Heinze, above note 14; Macmillan, 'Introduction: The Iraq War and Democratic Politics', in A. Danchev and J. MacMillan (eds), *The Iraq War and Democratic Politics* (2005) 1, at 7.

[30] Geis, Müller, and Schörnig, above note 11; Wagner and Werner, above note 6.

phenomenon that Hendrik Simon and Lothar Brock describe as 'multi-normativity' in the introduction to this book:

> Thus, war discourses may not only be seen as the sum of disputes over 'justifying reasons', but also as the sum of disputes between different narratives and traditions of justification. The simultaneity of different normative frameworks of reference may be summarized as 'multi-normativity'.[31]

In terms of political science theories of international relations, both speakers who argue in favour and those who argue against the use of force refer to arguments that can be attributed to a 'realist' frame (such as national interest, balance of power, regional stability), to 'institutionalist' frames such as alliance obligations and international organizations, to 'human rights' frames such as stopping a humanitarian disaster or avoiding casualties, to 'international law-based order' frames' such as enforcing a UN mandate, or to arguments that are derived from the foreign policy role/identity conception of the democracy in question. Our case studies of seven democracies in the run-up to the Gulf War 1990/91, the Kosovo War 1998/99, and the Iraq War 2002/03[32] showed that democratic actors do invoke a large spectrum of liberal norms in justifying military actions but that normative-liberal motivations always mix with more 'traditional', narrow national interests—whether these are prominently debated in public or not. The selectivity of Western military actions indicates that normative goals such as protection of human rights have to be weighed against the costs and risks for one's own polity such as accepting casualties.

Which frames will dominate a pre-intervention/war discourse depends on domestic norms, interests, and specific situations but also on the type of intervention/war in question. The Kosovo war as 'the quintessential liberal war of our time' (Vasquez, see above) invoked much more humanitarian justifications than the Gulf War of 1991 or the Iraq War of 2003. While humanitarian concerns have been put forward over centuries (Jahn, this volume), the context and content of such references have changed dramatically, given the emergence of a highly institutionalized normative international order and the evolution of human rights regimes. As, for example, Martha Finnemore[33] has shown in her analysis of 'humanitarian interventions' over time, the people in need that were to be 'protected', used to belong to particular communities such as 'white', 'Christian', 'European'—whereas the late twentieth, early twenty-first-century concepts of 'humanitarian intervention' or the 'Responsibility to Protect' have adopted a cosmopolitan outlook and reach out to all human beings, to 'saving strangers'[34] irrespective of their community belonging.

Apart from a broad array of substantial norms democratic speakers refer to, procedural arguments are fairly prominent in contemporary pre-intervention discourses: our analysis of pre-war parliamentary discourses in the Gulf War, the Kosovo

[31] Simon and Brock, above note 1, at 7; see also Vec, 'Multinormativität in der Rechtsgeschichte', in Berlin-Brandenburgische Akademie der Wissenschaften (vormals Preußische Akademie der Wissenschaften) (ed.), *Jahrbuch 2008* (2009) 155, at 162–65.

[32] For the following Geis, Müller, and Schörnig, above note 11.

[33] M. Finnemore, *The Purpose of Intervention: Changing Beliefs about the Use of Force* (2003), at 52–84.

[34] N. Wheeler, *Saving Strangers. Humanitarian Intervention in International Society* (2000).

War, and the Iraq War[35] showed that a large number of parliamentarians accept the necessity of a show of military force to back 'coercive diplomacy'.[36] They advocate the (potential) deployment of troops or military exercises close to the crisis theatre, often expecting a UN mandate to come forward in the following months or taking into consideration that such a mandate might not be attained in the end—but that 'coercive diplomacy', that is showing the hard power instruments to the opponent, is deemed necessary for now to ensure compliance by the adversary.

The other crucial and related procedural question is the legality of a potential use of force. Our case studies, though certainly not representative of all contemporary interventions, showed the high salience of international law, in particular the legitimating role of a UN Security Council (UNSC) mandate. Each democratic government felt compelled to justify its behaviour 'somehow' with reference to international law—or more precisely, with what the government framed to be the current position of international law or international state practice. For some democracies, a mandate from the UN is an absolute prerequisite. For some, it can be replaced by a common legal interpretation by the Western alliance (as offered by NATO in the Kosovo case). For a few, it seemed sufficient that a nationally accepted interpretation approves the case to be lawful. The need to make a strong legal case if the issue is disputed (such as in 1999 or 2003) was felt much more strongly by everybody else than by the United States. The role model of the United States as a world leader endowed with a special responsibility for world order, together with historic exceptionalism, suggests that the US government stands above international procedural law in its fight to enhance substantive norms.[37] The legality of going to war is thus a very important aspect in democratic publics. What that means, however, is disputed among them.

These findings suggest that the legitimacy provided by international law and the UN resonates highly within Western publics and forces both proponents and opponents of military action to address the issue of legality. But as empirical analyses show (and as is known from *constructivist* research on norms in general[38]), systems of norms such as the UN Charter can be interpreted in quite different, sometimes even opposite, ways. In addition, even if proponents understand that the given legal text as such prohibits their actions, they switch to an 'auxiliary norm' and argue that the action is enforcing the 'spirit' of the law which otherwise could not be enforced due to procedural obstructions in the UNSC. Alternatively they claim, as in the Kosovo War, that international law as it currently stands is not adapted to new challenges and that a 'creative development' of law is thus required.

In sum, liberal democracies' justifications for military interventions have been characterized by multi-normativity with humanitarian motives and international law playing particularly prominent roles.

35 For the following see Geis, Müller, and Schörnig, above note 11.

36 For an empirical evaluation see R. Art and P.M. Cronin (eds), *The United States and Coercive Diplomacy* (2007).

37 Brock, 'Between Sovereign Judgement and the International Rule of Law: The Protection of People from Mass Atrocities', in M. Albert and A.F. Lang, Jr (eds), *The Politics of International Political Theory. Reflections on the Works of Chris Brown* (2019) 87.

38 A. Wiener, *The Invisible Constitution of Politics: Contested Norms and International Encounters* (2008).

3.2 Justifying the Use of Force after '9/11'

The multi-normativity of liberal democracies' justifications for using force has been further expanded after the 2001 terrorist attacks on New York and Washington. In the wake of '9/11', there has been a renaissance of traditional concerns about national security and concomitant 'realist' frames.[39] At the same time, however, humanitarian norms have been further developed towards a 'Responsibility to Protect' (R2P), that was endorsed at the 2005 World Summit.

Katharina Emschermann's study of Western leaders' justifications of (non-)intervention in Libya in 2011 and the Syrian civil war, however, suggests a further 'complication' when democracies are confronted with today's humanitarian crises: it seems that the endorsement of 'R2P' has promoted a phenomenon that Emschermann labels 'substitute rhetoric':[40] not only leaders who *advocate* a military intervention use 'strong' rhetoric in condemning the culprits and demanding their deposition and in depicting the humanitarian disaster and expressing their empathy for the suffering people. Leaders who *reject* an intervention do the same. Hence, recourse to 'strong' moral rhetoric does not lead to the conclusion that one will plead for intervention. This rather recent phenomenon of 'substitute rhetoric' can send quite confusing signals to different audiences and weaken the effects of coercive diplomacy. Why do democratic leaders do this? They seem to attempt a tightrope walk between 'speaking out' in the face of humanitarian crises thus acknowledging that the 'international community' should not 'look away' and accept responsibility on the one hand, and on the other, the interest to avoid any commitment of their country's troops to yet another military mission in times when domestic publics have grown increasingly 'weary' of interventions.

Another striking finding of Emschermann's fine-grained analysis is the relative frequency with which democratic leaders who advocate military intervention in Libya and Syria refer to just war criteria, with regard to both *ius ad bellum* and *ius in bello* (just cause, last resort, probability of success, proportionality, right authority, and right intention).[41] In addition, also leaders who *reject* an intervention also may refer to just war criteria, for example 'just cause' and a 'right intention'.[42] The just war tradition seems to provide a perfect justificatory theory for democratic states (contra Brown, this volume). The Kosovo intervention in 1999 was another humanitarian crisis in which NATO states sought to depict their military intervention—that lacked a legal mandate by the UNSC—as fully legitimate.[43]

While the just war tradition remains important and perhaps even gained salience for democratic leaders, references to more traditional security self-interests of states provide the other salient cluster of justifications, now often framed as global campaigns

[39] On the 'war on terror', see also Michael Stohl's contribution to this volume.

[40] K. Emschermann, *A Simple No? An Analysis of (Non-)Intervention Justifications in Humanitarian Crises* (2019) (PhD thesis on file at the University of Bremen).

[41] For the just war tradition, see Anthony Lang, Jr and Chris Brown in this volume.

[42] Emschermann, above note 40, at 219.

[43] Simma, 'Die NATO, die UN und militärische Gewaltanwendung. Rechtliche Aspekte', in R. Merkel (ed.), *Der Kosovo-Krieg und das Völkerrecht* (2000) 9.

against terrorist networks or hybrid actors such as *Daesh* (ISIS) or governments suspected of sponsoring terrorist organizations. In the remainder of this section, we present some new empirical analyses of the justifications that liberal democracies use with a view to the use of force in Afghanistan and in the fight against *Daesh* in Iraq and Syria. The military mission in Afghanistan since 2001 and the military fight against *Daesh* since 2014 are particularly interesting because they include combat operations and are thus at the higher end of the spectrum of contemporary military interventions. As a consequence, they have been extensively debated in the parliaments of participating states. Furthermore, both operations have multiple purposes: both aimed—in the case of Afghanistan at least temporarily—at a military victory over a clearly defined adversary, namely the Taliban in Afghanistan and *Daesh* in Iraq and Syria. In addition, both operations are to protect and support the civilian population in the conflict zones against these adversaries. In Afghanistan, the UN International Security Assistance Force (ISAF) was deployed 'to assist the Afghan Interim Authority in the maintenance of security in Kabul and its surrounding areas, so that the Afghan Interim Authority as well as the personnel of the United Nations can operate in a secure environment'.[44] This mandate was interpreted widely to include not only the training of the Afghan army and police but also drilling wells and building schools. The Afghanistan mission is therefore often described as a state- and nation-building mission. The anti-*Daesh* intervention in Iraq was triggered by the atrocities against religious minorities and terrorist attacks in Europe. Although the intervening states refrained from state- or nation-building, they also trained local allies, such as Kurdish forces. The multi-purpose character of both missions implies that justifications can range from (collective) self-defence to humanitarian considerations.

The subsequent analysis is based on a dataset of 1065 speeches in the parliaments of Canada, Germany, and the United Kingdom.[45] This selection of countries thus includes 'militant' and 'pacifist' democracies[46] in order to cover the full spectrum of political cultures as identified by Geis, Müller, and Schörnig.[47] The United Kingdom is a typical militant democracy with a long history of interventionism whereas post-Second World War (West) Germany developed into a typical pacifist democracy with a culture of reticence[48] when it comes to the use of force; Canada occupies a 'middle ground'[49] as it has a long tradition of fighting alongside the United States and the United Kingdom (e.g. in both World Wars, the Korean War, the 1991 Gulf War, and the 1999 Kosovo campaign) but has been more selective in its participations than the United Kingdom, for example when refraining from joining the 'coalition of the willing' in Iraq 2003. Germany not only refrained from participation

[44] United Nations Security Council Resolution 1386 (2001).

[45] For a detailed introduction to the dataset, see W. Wagner, *The Democratic Politics of Military Interventions. Political Parties, Contestation and Decisions to Use Force Abroad* (2020).

[46] Müller, 'The Antinomy of Democratic Peace', 41 *International Politics* (2004) 494.

[47] Geis, Müller, and Schörnig, above note 11.

[48] T.U. Berger, *Cultures of Antimilitarism. National Security in Germany and Japan* (1998); J.S. Duffield, *World Power Forsaken. Political Culture, International Institutions, and German Security Policy after Unification* (1998).

[49] Geis, Müller, and Schörnig, 'Liberal Democracies as Militant "Forces for Good": A Comparative Perspective', in Geis, Müller, and Schörnig, above note 11, 307, at 327.

in 2003 but also decided against joining the 2011 intervention against Ghaddafi in Libya.

Building on the work of Geis, Müller, and Schörnig, arguments are clustered in several different frames. It should be noted, however, that each frame includes arguments that are brought forward in favour of a military mission as well as arguments that are critical thereof.

(1) Universalism/humanitarianism includes all references to the protection and well-being of non-nationals in conflict zones. Such references indicate a broad sense of community. References to human suffering, displacements, enslavement, and genocidal atrocities in the conflict zones under discussion, justifications to use force in order to protect the local population, and to support them in their right to self-determination as well as concerns about civilian casualties and collateral damage all belong into this category.
(2) National interest and security is the counterpart to the universalism/humanitarianism frame because the national community replaces the international community as a reference point for the justification of using force. Arguments that a military intervention is in a country's national interest or contributes to its national security belong into this category. If an argument explicitly mentions the term 'national interest', it clearly belongs into this category. However, references to nationals being threatened if military force is not used also apply. Often, it remains unspecified what the national interest is and why using force contributes to national security. Since 2015, the argument is occasionally made that a military intervention helps to keep refugees away from Europe (and to a lesser extent Canada).
(3) International law includes all arguments that explicitly refer to the *ius at bellum*, that is to provisions in international treaties, customary law, and jurisprudence that allow the use of armed force. The presence or absence of a UNSC mandate and its interpretation is of particular importance because such a mandate is one of two accepted exemptions to the general prohibition of using force. The other exemption is self-defence, which therefore also belongs in this category. Although arguments that a local government has demanded the intervention are relevant under international law, they are excluded from this category because they are almost always made without any explicit reference to international law.
(4) Democratic accountability refers to all arguments that treat governments as agents of the people and thus judge the use of force against the yardstick of citizens' interests and opinions. Most frequently, this is done in references to public opinion. Often, a military mission is criticized for its lack of popular support, but sometimes public opinion is also invoked in support of a mission. A less frequently made but related argument is that the costs for a mission come at the expense of more important policies, especially social policies. Finally, this frame also includes arguments that a government has a particular responsibility for its citizens in the armed forces and should therefore do everything in its power to avoid casualties amongst its own troops. This frame has a strong affinity to the theories of the Democratic Peace.

(5) Creating a strong enemy image is an established strategy to mobilize the public for the use of force. This frame includes arguments demonizing the adversary as well as those that question and criticize this.
(6) Multilateralism and alliances includes all references to alliance obligations, allies, and coalitions responsible for a military mission. Alliance obligations are understood broadly to include formal obligations (e.g. once NATO Article 5 had been invoked after the 9/11 attacks) as well as informal commitments. Multilateralism refers to arguments that other states have taken a positive stance on an intervention, which would isolate a country, which refuses to go along.
(7) National identity and role conception includes all references to a country's particular history, tradition, reputation, and identity if they are invoked in favour of or in opposition to a military intervention. National identity and role conception arguments include pleas to honour a country's tradition as well as those to learn from its troubled past and possibly failures. Such arguments are often framed as moral obligations that result from 'who we are', suggesting that the participation or non-participation in an intervention impacts on a country's identity. It is important to note that a country's national identity and role conception is by no means uncontested but instead is frequently questioned, challenged, or re-interpreted. If human rights are invoked, national identity arguments may overlap with humanitarian arguments. If references are made to (significant) others' expectations, there may be an overlap with arguments about multilateralism and alliances. What distinguishes a national identity and role conception argument are references to a country's *particular* history, tradition, reputation, and identity.

Looking at the share of each frame in relation to all frames being used by all MPs in a particular country[50] for the missions in Afghanistan and the fight against Daesh, we find the following:

The considerations of international law and of alliances, universalism/humanitarianism, and national security/interest all play a prominent role in general.[51] In contrast, references to public opinion and to the government's responsibility to protect the citizens of the country are found less frequently. However, there are differences across missions and countries. Arguments about international law were made with much higher frequency in debates about fighting *Daesh* than in those about the Afghanistan mission. This may not come as a surprise as the question whether *Daesh* could legally be targeted in Syria whose government did not invite such action has been heavily contested amongst experts and decision-makers.[52]

[50] This implies that different numbers or lengths of speeches per country in the dataset do not influence the result. Furthermore, it implies that an elaborate argument weighs heavier than a short reference. All calculations by the authors.

[51] If speeches from both missions are pooled, then 'multilateralism/alliances' accounts for 21 per cent, followed by universalism/humanitarianism (19 per cent), international law (18 per cent), and national interest/security (17 per cent) whereas enemy image (10 per cent), national identity/role conception (9 per cent), and democratic accountability (6 per cent) are less prominent.

[52] Henderson, 'Editorial Comment: The Use of Force and Islamic State', 1 *Journal on the Use of Force and International Law (JUFIL)* (2014) 209.

As regards differences between countries, our data above confirm several findings from Geis et al.: arguments about national identity and role conceptions are particularly frequent in Canada. MPs often refer to 'the proud tradition of Canadian missions'.[53] MPs present the purpose of the missions as a defence of 'basic Canadian values'.[54] Therefore, participation in the military mission is considered a reaffirmation of 'the values that define us as a country and as a community'[55] and 'a deep responsibility for me as a Canadian'.[56]

Compared to Canada, national identity and role conceptions are far less frequently invoked in Britain and in Germany. In the German debate, a German *Sonderweg* is only mentioned in a pejorative way as an undesirable self-isolation.[57] In the *Bundestag*, references to national identity and role conceptions are often about the appropriate lessons learnt from German history and the obligations arising from that. Whereas to critics, German history led to a firm commitment to antimilitarism and should thus have made participation in fighting in Afghanistan impossible,[58] for supporters, German fascism has led to an obligation to show presence when people are humiliated and tortured.[59] When human rights are brought forward as arguments for military intervention, they are almost always not presented as German arguments but as universal ones.[60]

Also confirming the study by Geis et al. (2013), our data show that German MPs keep from evoking strong enemy images when portraying the Taliban or the fighters of the so-called Islamic State. In contrast, Canadian and British MPs often refer to the Taliban as well as to *Daesh* as an 'enemy',[61] as 'evil',[62] as a 'death cult',[63] and as

[53] Steven Blaney (Conservatives) 10 April 2006. See also Prime Minister Stephen Harper (Conservatives) 17 May 2006, Tony Clement (Conservatives) 30 March 2015, Peter MacKay (Conservatives), 30 March 2015 refers to the national anthem as proof 'what we are fighting for'.

[54] Minister of National Defence Bill Graham (Liberals) 15 November 2005; see also Lois Brown (Conservatives) 26 March 2015, John Duncan (Conservatives) 30 March 2015, Joyce Murray (Liberals) and Minister of Foreign Affairs Rob Nicholson (Conservatives) 26 March 2015.

[55] Marlene Jennings (Liberals) 15 October 2001.

[56] Erin O'Toole (Conservatives) 7 October 2014.

[57] See Gernot Erler (Social Democrats) 23 March 2011.

[58] Wolfgang Gehrcke (Radical Left) 5 December 2014, Alexander Neu (Radical Left) 10 November 2016.

[59] Marieluise Beck (Greens) 15 December 2016.

[60] However, Foreign Minister Sigmar Gabriel (Social Democrats) argues that it characterizes 'us Germans' to be 'on the right side, on the side of freedom and human rights' (21 November 2017).

[61] Chris Grayling (Conservatives) 4. October 2001, Steve Double, John Glen, Sarah Wollaston (all Conservatives) all on 2 December 2015 in the UK; and John Cummins, Brian Fitzpatrick, Art Hanger, Howard Hilstrom (all Conservatives), John Harvard (Liberals) all on 15 October 2001 and Jason Kenney (Conservatives) 26 March 2015, all in Canada.

[62] Prime Minister David Cameron, Fiona Bruce, John Glen, Antoinette Sandbach (all Conservatives), Hilary Benn and Ruth Smeeth (both Labour) all on 2 December 2015, Iain Duncan Smith (Conservatives) on 4 October 2001, Charles Kennedy (Liberals) on 8 October 2001 in the UK; Minister of Foreign Affairs John Malney on 17 September 2001, Leon Benoit, Stockwell Day, Brian Fitzpatrick, Howard Hilstrom (all Conservatives) all on 15 October 2001, Peter MacKay (Conservatives) on 7 October 2014, Minister of National Defence Jason Kenney and Minister of Foreign Affairs Rob Nicholson (both Conservatives) on 26 March 2015 and Cheryl Gallant, Laurie Hawn, and Peter MacKay (all Conservatives) on 30 March 2015 in Canada.

[63] With a view to the Taliban, see Iain Duncan Smith (Conservatives) and Jim Knight (Labour) both on 8 October 2001; with a view to *Daesh* see Prime Minister David Cameron, John Glen, Andrew Murrison, Tom Tugendhat (all Conservatives) all on 2 December 2015 in the UK; and Julian Fantino (Conservatives) on 26 March 2015 and Cheryl Gallant (Conservatives) on 30 March 2015 in Canada.

'murderers'.[64] *Daesh* is also framed as 'fascist',[65] and MPs remind the audience of the torture, rapes, enslavements, beheadings, and crucifixions carried out by *Daesh*.[66] MPs highlight the enemy's disrespect for human life and the impossibility to negotiate with it.

The high level of attention to questions of multilateralism and alliances in Germany confirms studies of German political and strategic culture that emphasize Germany's aversion to act unilaterally.[67] Particularly the data for Afghanistan also suggest that controversies about universalism und humanitarianism are less frequent in Canada because the protection of non-nationals is understood to be part of Canada's international role and thus appear as national identity arguments.

There is a higher frequency of democratic accountability arguments in Canada and the United Kingdom than in Germany. This can be explained with the role of parliament and thus MPs in the respective debates. Whereas German MPs are co-decision-makers on the deployment of armed forces, British and Canadian MPs are rather mouthpieces of their constituencies that address decision-makers in government.

The death of soldiers is not glorified in the parliamentary debates under discussion here. Instead, in all three countries, MPs share the goal to keep their troops out of harm's way as much as possible. They also share a sense of being (co-)responsible for risking soldiers' lives. In Germany, this notion is linked to the *Bundeswehr* as a 'parliamentary army', which gives MPs the prime responsibility for military deployments.[68] However, in Canada and the United Kingdom, where the deployment of troops has been a royal prerogative, MPs interpret as their 'duty in the House to ensure that we never put our troops in harm's way unnecessarily or irresponsibly'.[69] They assume responsibility by stating for example that 'those deaths were avoidable and the fact that this House did not oppose the expedition into Helmand province in 2006 is responsible for them; this is not down to anybody else'.[70]

Taken together, the analyses of the two post-'9-11' military missions in Afghanistan and in Iraq and Syria show a broad range of frames to justify and criticize the use of armed force. Humanitarian considerations, that were characteristic of the liberal wars in the 1990s, still play a prominent role in parliamentary debates. However, concerns for *national* interests and security and for alliance politics are about as important as humanitarian considerations.

[64] Michal Ancram (Conservatives) and Mike O'Brien (Labour) both on 4 October 2001 in the UK; Larry Bagnell (Liberals) 15 October 2001 in Canada.

[65] Hilary Benn, Dan Jarvis, Ivan Lewis (all Labour) all on 2 December 2015, Tom Tugendhat (Conservatives) 2 December 2015, all in the UK.

[66] In Canada Minister of Foreign Affairs Rob Nicholson (Conservatives) 26 March 2015, David Anderson, John Duncan, and Peter MacKay (all Conservatives) all on 30 March 2015.

[67] J.S. Duffield, *World Power Forsaken. Political Culture, International Institutions, and German Security Policy after Unification* (1998).

[68] See Anton Hofreiter (German Greens) 4 December 2015, Hans-Christian Ströbele (German Greens) 20 February 2014.

[69] John McDonnell (Labour Party) 9 September 2010, see also Paul Flynn (Labour Party) 2 December 2015, both in the UK.

[70] Paul Flynn (Labour Party) 9 September 2010 in the UK.

4. Outlook: Contextualizing Democratic Distinctiveness in the Past and in the Future

Inspired by the Democratic Peace debate and the ensuing democratic distinctiveness research, liberal democracies' justifications for military missions have received considerable attention, especially since the end of the Cold War when 'wars of choice' became the dominant form of using armed force. In line with expectations about democratic distinctiveness, liberal norms such as the protection of human rights have been identified as prominent arguments in the justificatory discourses of liberal democracies. At the same time, however, the empirical findings reviewed above clearly show a remarkable pluralism and concomitance of justifications: humanitarian motivations are brought forward alongside traditional concerns for national security and are often seconded by arguments about international law, multilateralism, and alliances. Although '9/11' and the subsequent 'Global War on Terror' have led to a renaissance of national security arguments, humanitarian justifications have persisted and remained prominent in debates about military interventions in Afghanistan, Libya, Syria, and the fight against *Daesh*.

Another recurrent finding from empirical studies points to differences between liberal democracies. A country's political culture and national identity, informed by its position in the international system as well as by its past military failures and successes, clearly impacts on the justificatory discourse. Thus, the invocation of enemy images and references to national security are more common in more 'militant' democracies such as the United States and the United Kingdom than in more 'pacifist' grounded democracies such as the one post-Second World War Germany has developed into.

The extent to which justificatory discourses exhibit country-specific features, on the one hand, and traits that are common to liberal democracies, on the other hand, remains a question for further research. In a similar vein, only systematic comparisons with the justificatory discourses in authoritarian states—which to our knowledge have not been carried out—can reveal the degree of democratic distinctiveness in legitimizing the use of force. As a follow-up to this anthology, which includes a number of historical contributions, a comparison of today's justifications and those of former times would seem warranted as well: it would be instructive to systematically compare contemporary clusters of arguments with the justifications that have been identified in the historical studies on war manifestos; but even where prima facie certain basic similarities might exist—humanitarian concerns have been put forward over many centuries—the context and content of such references have changed dramatically, given the emergence of a highly institutionalized normative international order and the evolution of human rights regimes.

Interestingly enough, a number of researchers—who do not refer to 'Democratic Peace' literature—have argued that a 'resonance' mechanism between rulers and wider audiences has already existed for many centuries and in completely undemocratic political settings: Anuschka Tischer[71] has analysed numerous war declarations from

[71] See her chapter in this volume and in more detail her monograph *Offizielle Kriegsbegründungen in der Frühen Neuzeit. Herrscherkommunikation in Europa zwischen Souveränität und korporativem Selbstverständnis* (2012).

the Early Modern Period in Europe which show that princes, kings or queens, and other rulers took great care to justify their wars vis-à-vis quite different audiences. They referred to norms and values they considered as acceptable or shared. Tischer identifies a 'European' and 'Christian' community of values that speaks from these primary sources, but the European rulers seemed to think that their norms were in a way 'global', that is acceptable also to those 'outside' (such as 'Turks and Tatars'; 'honourable pagans'; 'most barbaric nations', etc.).[72]

In a similar vein, Oona A. Hathaway and Scott J. Shapiro have analysed more than 400 war manifestos, produced between 1492 and 1945, in order to examine changing 'just causes' over the centuries.[73] Such a manifesto is a public document, issued by a sovereign against another sovereign, which contains the reasons for going to war. Hathaway and Shapiro claim that all these manifestos were also addressed to a wider 'public':

> Manifestos matter precisely *because they are propaganda*. The function of propaganda is to persuade. We can therefore tell what reasons people usually found persuasive by examining the reasons that propaganda offered to persuade them.[74]

While some of the 'just causes' for going to war have become obsolete—such as collection of debts or enforcement of hereditary rights—a number have reappeared in new variations in justification discourses to date, such as self-defence, enforcing treaty obligations, or humanitarian considerations.

However, the assessment that war manifestos are merely 'propaganda' is an undifferentiated claim as Hendrik Simon has shown in the legal-political context of nineteenth-century Europe. European rulers sought to establish discursive legitimacy for using force, which contravenes the academic myth of a *liberum ius ad bellum* of that time. Simon rejects the notion that such justifications were mere propaganda. He rather suggests that they need to be seen in the context of a regulative order of peace established by the Concert of Europe.[75]

Studies such as those conducted by Tischer, Hathaway/Shapiro, and Simon are not related to 'Democratic Peace'/'Democratic War' research and could not be expected to be so since the regimes and rulers would not qualify as 'democratic'. However, it would make an important addition to the understanding of democratic 'distinctiveness' to contextualize Democratic Peace research within historical research of this kind. Acknowledging that 'authoritarian' leaders, too, felt the *need to justify* their wars—in contrast to Immanuel Kant's observation that such heads of states 'can decide to wage war for insignificant reasons as a kind of game for amusement' (see Kant's quote above, in section 2)—could stimulate more comparative studies across centuries and across political regime types. One of the crucial *structural* differences resides in the existence of rights and freedoms in democratic systems such as the right of democratic citizens

[72] Tischer, in this volume at 72; see also Becker Lorca in this volume.

[73] O.A. Hathaway and S.J. Shapiro, *The Internationalists and their Plan to Outlaw War* (2017), at 42f.

[74] Ibid., at 42, emphasis in original.

[75] Simon, 'The Myth of *Liberum Ius ad Bellum*. Justifying War in 19th-Century Legal Theory and Political Practice', 29 *European Journal of International Law (EJIL)* (2018) 113, at 131–34; see also Simon in this volume.

to dethrone their rulers by electoral mechanisms and not by rebellion or revolution, and in exerting the civil rights that also enable citizens to participate in a public sphere.

A final note is due on a topical question about historical changes that might impact on the future justification patterns of the use of force. Since changing domestic practices and the shaping of the international order are interlinked, it can also be expected that the current rise of populist parties will leave a mark on democracies' foreign policy discourses.[76] Whether democracies will continue to portray themselves as guardians of the liberal world order or whether the rise of populism may lead to an erosion of, for example, a quintessentially liberal argument about 'saving strangers' is an open question today. However, it is plausible that the more 'illiberal' democracies become domestically, the less they might be willing to fight for a 'liberal' order.

[76] See Hobson, 'Democratic Peace: Progress and Crisis', 15 *Perspectives on Politics* (2017) 697 and Destradi and Plagemann, 'Populism and International Relations: (Un)predictability, Personalisation, and the Reinforcement of Existing Trends in World Politics', *Review of International Studies (RIS)* (2019) 1.

17

The War on Terror and the Law of War: Shaping International Order in the Context of Irregular Violence

Michael Stohl

1. Introduction

In the afternoon of 11 September 2001, United States Secretary of Defense Donald Rumsfeld, meeting with the military and Pentagon planners, laid bare the fundamentals of the US response to the tragic events of that morning.[1] As is clear from the notes taken at that meeting, Rumsfeld sought to fulfil the political maxim often falsely attributed to Machiavelli and Churchill 'never let a good crisis go to waste'. The notes, which were made available through a freedom of information suit, indicate that even in the immediate aftermath of the tragedy, when thoughts would most likely turn to identifying those behind the perpetrators, shoring up defences against a further attack, and eliminating potential additional threats, Mr Rumsfeld had begun the process of dramatically altering the administration's conception of terrorism, its threat, and response. What follows is an exploration of how 9/11 altered the constructions of the threat of terrorism and how these constructions in turn affected arguments and justification for war, both declared and not, as well as the use of force in the context of what previously has been considered simply as counter-terrorism. The attacks, were, of course, horrendous, and as far as non-state terrorism is concerned, thankfully unprecedented and nothing of that scale has been repeated in the United States or any other nation since that day.

In the years since 11 September 2001 pundits, politicians, and scholars of terrorism and international relations routinely have declared that 9/11 'changed everything'. Writing long before that day, E.V. Walter argued that the reactions to a terrorist event are *far more important* for understanding the consequences of a terrorist attack than the original action.[2] The power of the terrorist event is shaped by how policy-makers and publics react to terrorist events and these reactions are far more important than the number of victims. This goes for the attacks of 9/11, too, though they were both horrendous and shocking. And as the last nineteen years sadly have demonstrated, the choices set in motion in the immediate aftermath of the attacks were disastrous in the

[1] J. Borger, *Blogger Bares Rumsfeld's Post 9/11 Orders* (2006), available at <https://www.theguardian.com/world/2006/feb/24/freedomofinformation.september11> (last visited 10 January 2020).

[2] E.V. Walter, *Terror and Resistance: A Study of Political Violence, With Case Studies of Some Primitive African Communities* (1969), at 5.

Michael Stohl, *The War on Terror and the Law of War: Shaping International Order in the Context of Irregular Violence* In: *The Justification of War and International Order*. Edited by: Lothar Brock and Hendrik Simon, Oxford University Press (2021).
 DOI: 10.1093/oso/9780198865308.003.0017

long run for not only the United States, but also peace and security across the Middle East and as shall be discussed below, the normative foundations of international order.

As the years have passed it has also been claimed that what changed was that which was inevitable. However, the decisions that were taken in the immediate aftermath of the attacks were not the only possible reactions. With the planning commenced by Donald Rumsfeld noted above, the decisions that followed in the next two years, such as 'declaring' the 'War on Terror', swiftly passing the Patriot Act, and subsequently declaring war against Afghanistan, were all assertive choices by the United States (joined in the international arena by most members of NATO and the United Nations or mirrored by other nations in the legislative and administrative decisions made on the domestic front). They were not inevitable or necessary. Even more consequential for the norms shaping international order was the misguided and, for many, the arguably criminal war against Iraq and its impact on politics and security beyond the Middle East. As then UN Secretary General Kofi Annan stated in 2004 'I have indicated it is not in conformity with the UN Charter, from our point of view, and from the Charter point of view it was illegal.'[3]

A core component of the new construction was the declaration of the 'war on terror'. The war on terror approach was accompanied by the securitization of domestic and international policy choices in addition to counter-terrorism. The threat of terrorism and terrorist attacks was elevated to an 'existential' threat. The rhetoric and construction of the rationales that followed reflect this change in the evaluation of the threat of terrorism. Widespread national responses to the attacks, in the context of this new 'reality' demonstrated a much increased fear of further attacks and their possible consequences. These claims and responses shattered the norms that had developed in response to the failures to prevent war over the previous century and how to treat combatants and noncombatants in the aftermath of those failures. Particularly for the Great Powers, but also, for some regional powers as well, this norm-shattering helped alter the common understanding of what war in general is about, how it is justified, and how it should be fought. That this was not inevitable may be seen in the critique written in 2003 by Harold Koh, who during the Obama administration became the Legal Adviser of the US Department of State. Mr Koh argued that Mr Bush could have employed the then existing multilateralist strategy and could have employed the symbolism of standing

> in solidarity with the world's ambassadors in front of the United Nations. He could have supported the International Criminal Court as a way of bringing the Osama bin Ladens and Saddam Husseins of the world to justice.[4]

He did not.

[3] UN News, *Lessons of Iraq War Underscore Importance of UN Charter—Annan* (2004), available at <https://news.un.org/en/story/2004/09/115352-lessons-iraq-war-underscore-importance-un-charter-annan> (last visited 10 January 2020); see also Kramer and Michalowski, 'War, Aggression and State Crime: A Criminological Analysis of the Invasion and Occupation of Iraq', 45 *The British Journal of Criminology (BJC)* (2005) 446; J. Hagan, J. Kaiser, and A. Hanson, *Iraq and the Crimes of Aggressive War: The Legal Cynicism of Criminal Militarism* (2015).

[4] H.H. Koh, *Rights to Remember* (2003), online at <https://www.icasinc.org/2003/2003l/2003lhhk.html> (last visited 10 January 2020).

2. Construction of the War on Terror and the Global Network of Terror

The war on terror approach created a frame that emphasizes attacking the perpetrators and their supporters and was meant to demonstrate not only the strength of the response to the 'existential' threat represented by the attacks, but also that the Bush administration was not going to rely on what they characterized as the soft and restrained law enforcement approach of their predecessors in the Clinton administration and before. It was intended to demonstrate strength and resolve to both domestic and international audiences but that had consequences for who and what was included within the construction of terrorism.

President Bush stated on 20 September 2001:

> Our enemy is a radical network of terrorists, and every government that supports them. Our war on terror begins with al Qaeda, but it does not end there. It will not end until every terrorist group of global reach has been found, stopped and defeated ... Every nation, in every region, now has a decision to make. Either you are with us, or you are with the terrorists. From this day forward, any nation that continues to harbor or support terrorism will be regarded by the United States as a hostile regime.[5]

There have been numerous discussion as to why the war metaphor is a counterproductive rhetorical and tactical strategy for counter-terrorism.[6] Scholars, pundits, and even serving military leaders as well as numerous retired military officers have argued that the war metaphor undermined the ability of the United States to manage the problem of terrorism. In July 2005, for example, General Richard Myers, Chairman of the Joint Chiefs of Staff, noted, 'I've objected to the use of the term "war on terrorism" before, because one—if you call it a war, then you think of people in uniform as being the solution. The long-term problem is as much diplomatic, as much economic—in fact,

[5] President George W. Bush, *Address to a Joint Session of Congress and the American People* (2001), available at <https://ia800504.us.archive.org/17/items/gwb2001-09-20.flac16/gwb2001-09-20transcript.txt> (last visited 10 January 2020). This was formalized in the new National Security Strategy statement released a few months later in 2002. 'The United States of America is fighting a war against terrorists of global reach. The enemy is not a single political regime or person or religion or ideology. The enemy is terrorism—premeditated, politically motivated violence perpetrated against innocents.' The National Security Strategy of The United States of America (2002). It should be noted that Tony Blair on 16 September 2001, had already stated 'Whatever the technical or legal issues about a declaration of war, the fact is we are at war with terrorism.' BBC News, *Britain 'at War with Terrorism'* (2001), available at <http://news.bbc.co.uk/2/hi/uk_news/politics/1545411.stm> (last visited 10 January 2020).

[6] Obviously, the shift to a war approach was not simply a metaphor. In addition to the hundreds of thousands of victims of war and violence that followed, it is important also to note the impact of the disastrous decisions on budgetary allocations both within the United States and abroad and the consequences for counter-terrorism see e.g. Jahn, 'The Tragedy of Liberal Diplomacy: Democratization, Intervention, Statebuilding (Part II)', 1 *Journal of Intervention and Statebuilding* (2007) 224; see also Jahn's contribution to this volume.

more diplomatic, more economic, more political than it is military.'[7] Unfortunately, by the time General Myers spoke out publicly, the war on terror construction was deeply imbedded in the consciousness as well as the strategy and tactics of the Bush administration.

President Barack Obama came to office in January 2009 and on his first day in office fulfilled a key campaign promise and issued a series of executive orders, the purpose of which was to distance the new administration from Mr Bush's strategy and demarcate a new approach to confronting terrorism, that is one announced as based on law enforcement principles. These included the closure of the CIA's secret prisons, requiring the CIA to abide by the same interrogation standards as the military, revoking past presidential directives that authorized abusive treatment of prisoners, and reaffirming US adherence to the Geneva Conventions. Further, over the next few weeks Mr Obama renamed the Bush Global War on Terrorism to the much less elegant and forceful sounding phrase of Overseas Contingency operations to signal the shift in military, legal, and intelligence approaches adopted. But these changes, while they challenged some of the political narrative that had been established in the immediate aftermath of 9/11, did not question the main frame, that 9/11 had changed everything. And as has been frequently commented upon from civil libertarians who were quite disappointed with the Obama justice department approach to much of the legal architecture and reasoning with respect to the Bush administration legacy, not enough was truly reconceived. As Eli Lake observed,

> The U.S. still reserves the right to hold suspected terrorists indefinitely without charge, try them via military tribunal, keep them imprisoned even if they are acquitted, and kill them in foreign countries with which America is not formally at war (including Yemen, Somalia, and Pakistan). When Obama closed the secret CIA prisons known as 'black sites,' he specifically allowed for temporary detention facilities where a suspect could be taken before being sent to a foreign or domestic prison, a practice known as 'rendition.'[8]

Nonetheless, Mr Obama was immediately attacked from the Right, most notably by former Vice President Cheney (who continued to emphasize the need for the war fighting approach) for weakening the United States and opening the 'homeland' to attack. Mr Cheney was not alone. Other Republicans maintained that he was not prepared to fight the existential threat that terrorism posed.[9]

But even more consequential, while Mr Obama challenged the war on terror metaphor, he did not challenge the 'global network of terror' metaphor which accompanied it. At the 2008 Democratic National Convention 27 August 2008, Mr Obama repeated

[7] Schmitt and Shanker, 'Washington Recasts Terror War as "Struggle"', *The New York Times*, 27 July 2005, at A1.

[8] E. Lake, *The 9/14 Presidency: Barack Obama Is Operating with the War Powers Granted George W. Bush Three Days after the 9/11 Attacks* (2010), available at <http://reason.com/archives/2010/04/06/the-914-presidency> (last visited 10 January 2020).

[9] J. Gerstein, *GOP Revives 'Soft on Terror' Charge* (2010), available at <https://www.politico.com/story/2010/01/gop-revives-soft-on-terror-charge-032250> (last visited 10 January 2020).

a tag line that he employed numerous times on the campaign trail to criticize the prosecution of the war on terror.

> You don't defeat a terrorist network by occupying Iraq. You don't protect Israel and deter Iran just by talking tough in Washington.[10]

While this is a fine critique of the conduct of the war on terror, note that it increases the number of countries Mr Bush formerly had identified as being within the network from sixty to eighty countries. After winning the presidency, but before taking the oath of office, in the wake of the Mumbai attacks in late November 2008 Mr Obama said,

> The United States must continue to strengthen our partnership with India and nations around the world to root out and destroy terrorist networks.[11]

At his inauguration Obama stated that 'our nation is at war against a far-reaching network of violence and hatred'. And subsequently, he and members of his administration continued to consistently employ the network of terror metaphor throughout his consequential first year in office and thus solidified the continuation of the construction.[12]

Thus, the war on terror narrative continued as the core political narrative for the American approach to countering terrorism. Remarkably, the Bush administration actually never clearly, consistently, or fully identified what they actually meant by a network of terror and who or what actually needed to be defeated for the 'war' to be won. Some of this was undoubtedly generated by political considerations, but much of the problem was generated by the rather cavalier use of the term network. Stohl and Stohl[13] have argued that the Bush administration's construction of the network of terror metaphor had five significant false assumptions with respect to the identification, specification, and operation of a network and it is quite clear that Mr Obama and his national security colleagues did not distance the Obama administration from the network approach of the Bush administration and that they too clearly did not recognize boundary specification, nor the local, global, emergent, adaptive structures or multiplex historically constructed relations which characterize terrorists and their organizations and what the implications were and are for understanding counterterrorism.[14] By not challenging directly the Bush concept, Mr Obama's approach undermined his efforts to end the 'war on terror' construction. And consequentially,

[10] B. Obama, *Acceptance Speech at the Democratic National Convention* (2008), available at <https://millercenter.org/the-presidency/presidential-speeches/august-28-2008-acceptance-speech-democratic-national> (last visited 10 January 2020).

[11] The Washington Post, *Obama Transition Statement on Terrorist Attacks in Mumbai* (2008), available at <http://voices.washingtonpost.com/44/2008/11/obama-transition-statement-on.html> (last visited 10 January 2020).

[12] B. Obama, *Inaugural Address (2009)*, available at <https://obamawhitehouse.archives.gov/blog/2009/01/21/president-barack-obamas-inaugural-address> (last visited 10 January 2020). For further details on members of the administration identifying the global network of terror see Stohl, 'U.S Homeland Security, the Global War on Terror and Militarism', in K. Gouliamos and C. Kassimeris (eds), *The Marketing of War in the Age of Neo Militarism* (2011) 107.

[13] Stohl and Stohl, 'Networks of Terror: Theoretical Assumptions and Pragmatic Consequences', 17 *Communication Theory* (2007) 93.

[14] Ibid., at 98.

in January 2010 Mr Obama certainly backtracked in his rhetorical framing when he said:

> We are at war. We are at war against al Qaeda, a far-reaching network of violence and hatred that attacked us on 9/11, that killed nearly 3,000 innocent people, and that is plotting to strike us again. And we will do whatever it takes to defeat them.[15]

Mr Obama employed not only the war on terror metaphor but connected Al-Qaeda in the Arabian Peninsula to Osama bin Laden and all other organizations that choose to employ the name Al-Qaeda in a geographic location, regardless of whether they had actual operational connections with Al-Qaeda or not. By framing the terrorists as all connected, he inadvertently argued that all future attacks would be seen as Mr Bush and Mr Cheney originally framed them, as connected to Al-Qaeda and 'September 11', thus making it difficult for his administration (as well as the American public) to move past the Bush framing of terrorism. The result was Mr Bush's original construction was further normalized and, in an important sense, frozen in the consciousness of the American public.[16] It also contributed to the continuation of high levels of fear of future attack found in repeated public surveys from September 2001 through 2018.[17] This was not merely an American problem. The American framing was adopted by numerous governments (when politically convenient) and some nations even competed to have 'their terrorists' included in the 'global network of terror'.[18]

3. Consequences for International Order: Moving from a Domestic Rule of Law-Based Approach to Hegemonic International Law?

As has been indicated above, Mr Bush (and Mr Cheney, the chief architect of the 9/11 response), wished to demarcate 9/11 as the justification of a new era of response to terrorism. The principles prior to 9/11 privileged the rule of law at home and put domestic law enforcement (federal and local) and principally the FBI as the locus of control in those efforts. A parallel approach characterized the US approach abroad. Notwithstanding the numerous breaches (e.g. the bombing of Libya in 1986 or the

[15] B. Obama, *Remarks by the President on Strengthening Intelligence and Aviation Security* (2010), available at <https://obamawhitehouse.archives.gov/the-press-office/remarks-president-strengthening-intelligence-and-aviation-security> (last visited 10 January 2020).

[16] For a discussion of this continuing legacy see Stohl, '"There's Only Three Things He Mentions in a Sentence—a Noun, a Verb and 9/11": Terrorism, Fear and the after, after 9/11', 12 *Critical Studies on Terrorism* (2018) 63, available at <https://doi.org/10.1080/17539153.2018.1494791> (last visited 10 January 2020). See also Jackson, 'Culture, Identity and Hegemony: Continuity and (the Lack of) Change in US Counter Terrorism Policy from Bush to Obama', 48 *International Politics* (2011) 390.

[17] See Smith, Stohl, and al-Gharbi, 'Discourses on Countering Violent Extremism: The Strategic Interplay Between Fear and Security after 9/11', 12 *Critical Studies on Terrorism* (2019) 151, available at <https://doi.org/10.1080/17539153.2018.1494793> (last visited 10 January 2020).

[18] China e.g. lobbied for ten months to have the East Turkestan Islamic Movement, known as the ETIM, added to the US list of terrorist organizations linked to Osama bin Laden's global terror network. Eckholm, 'American Gives Beijing Good News: Rebels on Terror List', *The New York Times*, 27 August 2002, at Section A, 3.

cruise missile attack on Iraq in 1993), the rhetoric and basic approach of the United States at the international level was respect for the rule of law abroad and respect for international cooperation and the institutions that had been built after the Second World War. Tams delineates the standing of the principles that were considered as in force in 1989 and argues that the mainstream approach, built upon the foundations of the UN Charter,

> . . . over time had been reinforced through landmark ICJ rulings (such as *Corfu Channel* and *Nicaragua*) and important General Assembly resolutions. Of course states, especially the more powerful, were suspicious of the limits placed upon them by the Charter and felt that the drafters had gone too far in their quest to ban force, but ideological confrontation typically precluded agreement on a more flexible re-interpretation.[19]

The Bush administration argued that existential threat of the new terrorism demonstrated by 9/11 justified rejection of the principles of counter-terrorism that had characterized US policy prior to the attack. After 9/11, Mr Cheney built his One Percent Doctrine, as Matheson has argued,[20] on a variation of Pascal's Wager. 'With a low-probability, high impact event like this ... we have to treat it as a certainty in terms of our response.'[21] The Cheney doctrine thus underlay the Bush doctrine, whose preventative war argument was first explicitly stated by Mr Bush in a commencement address at West Point 'Our security will require all Americans to be forward-looking and resolute, to be ready for preemptive action when necessary to defend our liberty and to defend our lives.'[22] This approach demarcated a new era for not only the US approach to counter-terrorism but the international community's as well, what Andreopoulos has described as Hegemonic International Law.[23]

Robert Jervis argues that the Bush doctrine has four elements:

> a strong belief in the importance of a state's domestic regime in determining its foreign policy and the related judgment that this is an opportune time to transform international politics; the perception of great threats that can be defeated only by new and vigorous policies, most notably preventive war; a willingness to act unilaterally when necessary; and, as both a cause and a summary of these beliefs, an overriding sense that peace and stability require the United States to assert its primacy in world politics.[24]

[19] Tams, 'The Use of Force against Terrorists', 20 *The European Journal of International Law (EJIL)* (2009) 359, at 364.

[20] Matheson, 'The Sublime Rhetoric of Pascal's Wager', 53 *Argumentation and Advocacy* (2017) 271, at 281.

[21] Matheson's source is R. Suskind, *The One Percent Doctrine* (2006), at 61.

[22] George W. Bush, *Graduation Speech at West Point* (2002), available at <https://georgewbush-whitehouse.archives.gov/news/releases/2002/06/20020601-3.html> (last visited 10 January 2020).

[23] Andreopoulos, 'The Resurgence of Hegemonic International Law', 12 *Global-E* (2019), available at <https://www.21global.ucsb.edu/global-e/may-2019/resurgence-hegemonic-international-law> (last visited 10 January 2020).

[24] Jervis, 'Understanding the Bush Doctrine', 118 *Political Science Quarterly (PSQ)* (2003) 365, at 365, available at <https://www.psqonline.org/article.cfm?id=14813> (last visited 10 January 2020).

The Bush administration thus challenged the established *jus ad bellum* and *jus in bello* principles that had developed with the creation of the post-Second World War international law regime. International Humanitarian Law (IHL), or *jus in bello*, is the law that governs the way in which warfare is conducted. IHL is purely humanitarian, seeking to limit the suffering caused. It is independent from questions about the justification or reasons for war, or its prevention, covered by *jus ad bellum*.

Lietzau makes clear that 'Interpreting *jus ad bellum* to permit a state to preempt an apparent but as yet unrealized hostile intent was not seriously contemplated in the course of U.N. Charter negotiations.'[25] Further development of *jus ad bellum* principles to restrain states from both preemptive and retaliatory use force with respect to terrorism in the post-war period was, ironically, advanced by Cold War politics. As Tams notes, it was the Cold War bloc intransigence which paralysed the Chapter VII collective security system from the founding of the UN until the collapse of the Soviet Union and thus

> Given this lack of consensus, it comes as no surprise that the traditional *jus ad bellum* viewed assertions of a right to use anti-terrorist force rather skeptically. In line with the dominant, restrictive, analysis of the legal rules governing resort to force, international law as at 1989 effectively ruled out the possibility that states could lawfully resort to forcible measures against terrorists based in another country.[26]

Thus, it was the prevailing opinion during this period that unless terrorist attacks were attributed to a state (through state agents or clear sponsorship) the conditions necessary for the legitimate use of force as laid out in Articles 39–42 of the UN Charter were outside the Security Council's purview. And further, that anticipatory self-defence and reprisal were still outside the scope of permissible state actions. This does not mean that there were no violations of these permissible actions, but it does mean that states and the United Nations customarily supported the principles which established that which was permissible.

> Israel's 1985 raid on the PLO Headquarters outside Tunis was 'condemn[ed] vigorously' by the Security Council, which declared it an 'act of armed aggression ... in flagrant violation of the Charter of the United Nations' and urged other states 'to take measures to dissuade Israel from resorting to such acts against the sovereignty and territorial integrity of all States'. Similarly, the United States' 1986 raid on targets in Libya, while controversially discussed by commentators, was roundly rejected by the General Assembly as 'a violation of the Charter of the United Nations and of international law.' South Africa's incursions into neighbouring states, if anything, met with stiffer resistance.[27]

[25] Lietzau, 'Old Laws, New Wars: Jus ad Bellum in an Age of Terrorism', in A. von Bogdandy and R. Wolfrum (eds), 8 *Max Planck Yearbook of United Nations Law (UNYB)* (2004) 383, at 394.

[26] Tams, above note 19, at 364.

[27] Ibid., at 367. Likewise Lietzau, above note 25, at 395, argues: 'Since the Charter's inception, the concept of peacetime reprisal (the "peacetime" modifier being used to distinguish the concept from "belligerent reprisal," a *jus in bello* concept associated with particular military actions in an extant armed conflict) has been considered inconsistent with the Charter's articulation of *jus ad bellum*.'

After the 9/11 attacks, the Bush administration argued that these principles should no longer be operative, in either the *jus ad bellum* or *jus in bello* spheres. Ratner argues that this was clear 'the very night of the attacks'.[28] President George W. Bush stated, 'We will make no distinction between the terrorists who committed these acts and those who harbor them.' And this was followed by his speech to Congress on 20 September in which he stated 'Every nation, in every region, now has a decision to make. Either you are with us, or you are with the terrorists.'[29]

In the days between the these two statements, the United Nations Security Council unanimously adopted UNSC Resolution (UNSCR) 1368[30] which called for steps to respond to the attacks and combat all forms of terrorism in accordance with the UN Charter. UNSCR 1368 was particularly significant because it linked the Article 51 right to self-defence of the Charter with the response to international terrorism. A week after Mr Bush's address to Congress the UNSC also passed UNSCR 1373, which required member states to criminalize terrorism and its financing, and provided guidelines for enhanced cooperation on law enforcement and intelligence sharing. This was the first time that a Chapter VII resolution imposed a legal duty on states to introduce specific kinds of domestic counter-terrorism legislation.[31] Numerous member states complained that Resolution 1373 overstepped the legislative authority of the UNSC by imposing its will on member states.[32]

On 7 October 2001, President George W. Bush announced that airstrikes targeting Al-Qaeda and the Taliban had begun in Afghanistan under the guise of Operation Enduring Freedom (OEF). The initial military objectives of OEF were announced by President George W. Bush in his 20 September Address to the Joint Session of Congress and included the destruction of terrorist training camps and infrastructure within Afghanistan, and the capture of Al-Qaeda leaders and military installations of the Taliban regime in Afghanistan. The US was joined in the operation by the United Kingdom and Mr Bush announced that Australia, Canada, France, and Germany had pledged forces for the war.

On 20 December 2001, responding to the request of Afghan authorities, UNSCR 1386 authorized the establishment of an International Security Assistance Force (ISAF) for Afghanistan led by NATO. ISAF's initial role was to support the Afghan Interim Authority by maintaining security in and around Kabul. ISAF's force comprised troops and assets from eighteen different countries under the command authority of the United Kingdom. NATO assumed command of the ISAF mission in August 2003 and in October 2003, ISAF received a new mandate, provided for by UNSCR 1510, to expand its area of responsibility beyond Kabul and its environs.

[28] Ratner, 'Jus ad Bellum and Jus in Bello After September 11', 96 *American Journal of International Law (AJIL)* (2002) 905.

[29] Bush, above note 5.

[30] United Nations Security Council, *Resolution 1368 (2001) Adopted by the Security Council at its 4370th Meeting, on 12 September 2001*, available at <http://unscr.com/en/resolutions/doc/1368> (last visited 10 January 2020).

[31] F.N. Aoláin, *The Massive Perils of the Latest U.N. Resolution on Terrorism* (2019), available at <https://www.justsecurity.org/64840/the-massive-perils-of-the-latest-u-n-resolution-on-terrorism/> (last visited 10 January 2020).

[32] See the references to these statements as reported by Rosand, 'The Security Council As "Global Legislator": Ultra Vires or Ultra Innovative?', 28 *Fordham International Law Journal (ILJ)* (2004) 542, at 543.

Commenting on these decisions and the respective action Thomas Franck argued that the latter had transformed the previously understood *jus ad bellum* arguments restricting the right to exercise self-defence against terrorist and other non-state attacks with the result that such assertions of self-defence rights no longer were concerned with attributing the attacks to another state and the acquiescence of the international community meant that these claims 'are no longer exceptional claims'.[33] Shortly thereafter, the United States, in the release of the 2002 National Security Strategy, 'went well beyond these claims; it famously asserted a right of pre-emptive self-defense against non-imminent threats, in particular those by terrorist organizations'.[34] Then, after the US invasion of Iraq in the following year, Franck declared a much more troubling set of conclusions.

> While a few government lawyers still go through the motions of asserting that the invasion of Iraq was justified by our inherent right of self-defense, or represented a collective measure authorized by the Security Council under Chapter VII of the Charter, the leaders of America no longer much bother with such legal niceties. Instead, they boldly proclaim a new policy that openly repudiates the Article 2(4) obligation. What is remarkable, this time around, is that once-obligatory efforts by the aggressor to make a serious effort to stretch law to legitimate state action have given way to a drive to repeal law altogether, replacing it with a principle derived from the Athenians at Melos: 'the strong do what they can and the weak suffer what they must' ... The invasion of Iraq in March of 2003, and a penumbra of policy statements made concurrently by the United States, have succeeded in changing all that. Article 2(4) has died again, and, this time, perhaps for good.[35]

It is important to emphasize Franck's concerns and therefore the consequences that follow. As Lietzau notes:

> 1.) the United Nations Security Council had time and ability to act more forcefully but elected not to; 2.) neither cross-border attack nor an attack against U.S. interests had clearly occurred; 3.) the traditionally requisite criteria for anticipatory action were not present (imminent threat with no moment for deliberation); and 4.) evidence supporting the factual predicate to justify anticipatory action was weak.[36]

4. *Jus in Bello* after the War on Terror

The *jus in bello* evaluations with respect to the impact of the constructions of terrorism and the impact of the 9/11 attacks in the context of the war on terror must take place in the context of the status that the body of *jus in bello* and *jus ad bellum* had achieved

[33] T.M. Franck, *Recourse to Force: State Action against Threats and Armed Attacks* (2002), at 64, note 5.
[34] Tams, above note 19, at 381.
[35] Franck, 'What Happens Now? The United Nations after Iraq', 97 *American Journal of International Law (AJIL)* (2003) 607, at 608, 610. See also the contribution by Thilo Marauhn to this book.
[36] Lietzau, above note 25, at 421.

before the attacks. The Geneva Conventions of 1949, and their 1977 Additional Protocols, along with the UN Convention against Torture, the Universal Declaration of Human Rights, and the International Covenant on Civil and Political Rights are generally viewed as the core documents of IHL. As Bassiouni reminds us

> Clearly, the more legal instruments that exist to evidence the condemnation and prohibition of a particular crime, the better founded the proposition that the crime has risen to the level of *jus cogens*. The second consideration is the number of states that have incorporated the given proscription in their national laws. This is particularly true with respect to the military laws of 188 states that embody the normative proscriptions and prescriptions of the Geneva Conventions of August 12, 1949; 147 states for Additional Protocol 2 and 139 for Additional Protocol II.[37]

In short, before September 11 the prevailing view was that the Geneva Conventions and the additional protocols, which addressed problems arising from insurgent groups and civil wars had achieved what Meron described as 'near universal acceptance of the Conventions and their secure integration into the international system … and most of their provisions are recognized as customary law, and often as *jus cogens* or peremptory norms'.[38] This view is emphatically supported by Brooks

> Some of the norms articulated in the Universal Declaration, the ICCPR, and other human rights law treaties such as the Genocide and Torture Conventions are generally accepted as having risen to the status of jus cogens, 'peremptory norms' of international law that are binding on all states at all times, and that states cannot barter away via treaty or other agreement. Although there is some disagreement over precisely which international human rights norms have attained *jus cogens* status, most commentators (including the drafters of the Restatement (Third) of the Foreign Relations Law of the United States) agree that at a minimum, the norms against genocide; slavery; murder and 'disappearance' of individuals; torture or other cruel, inhuman, or degrading treatment or punishment; prolonged arbitrary detention; and systematic racial discrimination are peremptory in nature. These norms are binding at all times, whether in emergency situations or not, and whether in armed conflicts or not.[39]

Nonetheless, the Bush administration chose to violate its legal responsibilities.[40] And while it is true that 'human rights law permits limited derogation in times of

[37] Bassiouni, 'International Crimes: Jus Cogens and Obligatio Erga Omnes', 59 *Law and Contemporary Problems* (1996) 63, at 70, available at <https://scholarship.law.duke.edu/lcp/vol59/iss4/6> (last visited 10 January 2020).

[38] Meron, 'The Geneva Conventions and Public International Law: British Foreign and Commonwealth Office Conference Commemorating the 60th Anniversary of the 1949 Geneva Conventions', 91 *International Review of the Red Cross* (2009) 619, at 619, 625, available at <https://www.icrc.org/en/doc/assets/files/other/irrc-875-geneva-convention-int-law.pdf> (last visited 10 January 2020).

[39] Brooks, 'War Everywhere: Rights, National Security Law, and the Law of Armed Conflict in the Age of Terror', 1125 *Georgetown Law Faculty Publications and Other Works* (2004) 675, at 749, available at <https://scholarship.law.georgetown.edu/facpub/1125/> (last visited 10 January 2020).

[40] The United States indicated that the Conventions applied in principle to the conflict with the Taliban as a de facto state party, though without conferring prisoner of war (POW) status on its members under

emergency, it also outlines core rights that cannot be eliminated regardless of the nature of the threat or the existence or nonexistence of an armed conflict'.[41] From the start, beginning with the initial phases of conflict in Afghanistan, as many have noted[42] the Bush administration argued that since Al-Qaeda was not a state actor and thus not a party to the Geneva Conventions they did not apply. Unfortunately, while the United States initiated the post-9/11 assault on the applicability of IHL, it was not only in the United States that activities under the cover of UNSCR 1373 led to violations of human rights that were detailed by the UN Counterterrorism Committee,

> Under the alleged auspices of counterterrorism, China brutally repressed Uighur Muslims, Russia cracked down on Chechen separatists, and the Uzbek government targeted (PDF) its political opposition.... The report went on to address continuing concerns of torture, detention practices, misuse of emergency laws, and extrajudicial executions.[43]

Arguments for security, of political fears that the appearance of not having acted with 'strength', or 'resolve' to protect their state, also led many Western countries to abandon their duties with respect to human rights abroad as well as at home.[44] One significant breach of norms is encapsulated in the term 'extraordinary rendition'. Much of the details of what we know about extraordinary rendition comes from the work of Blakeley and Raphael of The Rendition Project. Their initial work, expanded through collaboration with Reprieve, the Bureau of Investigative Journalism and the European Parliament and Council of Europe, details what happened to the 119 men detained by the CIA, often with substantial assistance from British intelligence services, including 'the location(s) and time frame(s) of each instance of secret detention, along with the conditions and treatment to which each prisoner was subjected'.[45] It is a depressing and sobering reminder of the fragility of even 'peremptory norms' in times of fear.[46] We now know that at least fifty-four nations participated in some manner

Geneva Convention III (GC III). Nor did the Administration find GC III to be applicable at all to what it considered a separate conflict with Al-Qaeda, a non-state actor. And the brief application of Geneva Convention IV (GC IV) to civilians went largely ignored. Fischer, 'Applicability of the Geneva Conventions to "Armed Conflict" in the War on Terror', 30 *Fordham International Law Journal* (2006) 509, at 510, Article 4.

[41] Brooks, above note 39, at 684.

[42] See e.g. de Nevers, 'The Geneva Conventions and New Wars', 12 *Political Science Quarterly (PSQ)* (2006) 369, available at <http://www.jstor.org/stable/20202724>; Weissbrodt and Bergquist, 'Methods of the "War on Terror"', 16 *Minnesota Journal of International Law (MJIL)* (2007) 371, available at <http://scholarship.law.umn.edu/faculty_articles/249> (last visited 10 January 2020).

[43] Council on Foreign Relations, 'The Global Regime for Terrorism. Issue Brief', *Report by the International Institutions and Global Governance Program* (2011), available at <https://www.cfr.org/report/global-regime-terrorism> (last visited 10 January 2020).

[44] Goderis and Versteeg, 'Human Rights Violations after 9/11 and the Role of Constitutional Constraints', 41 *The Journal of Legal Studies* (2012) 131, at 134.

[45] Raphael, Black, and Blakeley, 'CIA Torture Unredacted: An Investigation into the CIA Torture Programme', *The Rendition Project*, at 4, available at <https://www.therenditionproject.org.uk/documents/RDI/190710-TRP-TBIJ-CIA-Torture-Unredacted-Full.pdf> (last visited 10 January 2020).

[46] See Weissbrodt and Bergquist; above note 42, at 379f.: '... international human rights law protects individuals in all circumstances, even those not governed by humanitarian law. Hence, even if there is no conflict or occupation, individuals are still protected by human rights law. Extraordinary rendition violates

in the Extraordinary Rendition Program and that the program had not ended well into the Obama administration, despite the announcements regarding the return to respect for international law that was proclaimed on Mr Obama's first day in office.[47]

Another troubling area of violation is connected to the use of drones. While the technologies that make drones available as a military weapon are not in themselves responsible for the violations of international law[48] that have occurred with respect to the killing of unarmed citizens and non-combatants, the use of drones lowers both the financial cost of the attacks and the oversight of the activities and has enabled the escalation of targeted killings. The United States considers 'all military-age males [killed] in a strike zone' to be 'combatants ... unless there is explicit intelligence posthumously proving them innocent',[49] without any post-attack verification attempts. As a consequence, the United States have come to claim only a very small number of civilian deaths along with their drone attacks. Data collected by the Bureau of Investigative Journalism Drone Wars project and the Air Wars project estimate that between 2004 and 2018, drone strikes in Pakistan, Afghanistan, Yemen, Somalia, Iraq, and Syria have killed at least 6878 civilians.[50]

But as with targeted killings in general, drone attacks are not limited to areas of active hostilities. The Bush administration interpreted the authorization of military force passed by Congress to mandate the hunt of Al-Qaeda members and their allies everywhere.[51] And, as indicated above, since Obama affirmed and extended the construction of the Al-Qaeda global network of terror, the consequence is that almost

numerous international human rights standards, including the International Covenant on Civil and Political Rights, the Convention and Protocol Relating to the Status of Refugees, the Convention against Torture, and the Vienna Convention on Consular Relations.' See also Borelli, 'Extraordinary Rendition, Counter-Terrorism and International Law', in B. Saul (ed.), *Research Handbook on International Law and Terrorism* (2014) 361, available at <https://www.ssrn.com/abstract=2432083> (last visited 10 January 2020).

[47] Afghanistan, Albania, Algeria, Australia, Austria, Azerbaijan, Belgium, Bosnia-Herzegovina, Canada, Croatia, Cyprus, the Czech Republic, Denmark, Djibouti, Egypt, Ethiopia, Finland, Gambia, Georgia, Germany, Greece, Hong Kong, Iceland, Indonesia, Iran, Ireland, Italy, Jordan, Kenya, Libya, Lithuania, Macedonia, Malawi, Malaysia, Mauritania, Morocco, Pakistan, Poland, Portugal, Romania, Saudi Arabia, Somalia, South Africa, Spain, Sri Lanka, Sweden, Syria, Thailand, Turkey, United Arab Emirates, United Kingdom, Uzbekistan, Yemen, and Zimbabwe Open Society Foundations (OSF), *Globalizing Torture: CIA Secret and Extraordinary Rendition* (2013), available at <https://www.justiceinitiative.org/publications/globalizing-torture-cia-secret-detention-and-extraordinary-rendition>. See also C. Whitlock, 'Renditions Continue under Obama, Despite Due-Process Concerns', *The Washington Post*, 1January 2013, available at <https://www.washingtonpost.com/world/national-security/renditions-continue-under-obama-despite-due-process-concerns/2013/01/01/> (last visited 10 January 2020).The British government was also in violation of the norms. See the compilation of evidence provided by Bellaby, 'Extraordinary Rendition: Expanding the Circle of Blame in International Politics', 22 *The International Journal of Human Rights* (2018) 574, at 597, available at <https://doi.org/10.1080/13642987.2017.1397634> (last visited 10 January 2020).

[48] In armed conflicts that—based on IHL—allow for the use of deadly force against the enemy, armed drones are just another instrument of warfare. Thimm, 'From Exception to Normalcy: The United States and the War on Terrorism', 7 *SWP Research Paper* (2018), at 24f.

[49] Becker and Shane, 'A Measure of Change. Secret "Kills List" Proves a Test of Obama's Principles and Will', *New York Times* (2012), available at <http://www.nytimes.com/2012/05/29/world/obamas-leadership-in-war-on-al-qaeda.html?r=0> (last visited 10 January 2020).

[50] Blakeley, 'Drones, State Terrorism and International Law', 11 *Critical Studies on Terrorism* (2018) 321, at 322, available at <https://doi.org/10.1080/17539153.2018.1456722> (last visited 10 January 2020).

[51] Thimm, above note 48, at 24–25.

anywhere may be considered the 'battlefield'. A leaked US Department of Justice memo further indicates that this may also include US citizens.[52] While some thought that Mr Trump's 'America First' policy would reduce US involvement in foreign war, since taking office, Mr Trump has dramatically increased the number of drone strikes, cancelled the Obama era rules of reporting civilian deaths (even if there was great criticism of how the military counted civilians), and transferred drone attacks to the CIA which had previously been responsible for the surveillance function and thus enabling the evasion of existing Pentagon reporting requirements.

5. Conclusion

9/11 altered the constructions of the threat of terrorism and how these constructions in turn affected arguments and justification for war, both declared and not, and the use of force in the context of what previously has been considered simply as counter-terrorism. It shattered well-established norms of international law and, particularly for the Great Powers, but also some regional powers as well, common understandings of what war in general is about and against whom it may be fought, when it is justified, and how it should be fought.

The world community crossed the Rubicon when it did not forcibly object to US choices in its response to the 9/11 attacks. Ratner introduces the 'Eiffel Tower Factor' and argues that it is inconceivable that any other major power would have not acted as did the United States.[53] Paraphrasing Caesar's alleged statement, the die has been cast.[54] As argued above, there was scant opposition to US actions in commencing the war in Afghanistan, and while there was initial opposition to the war in Iraq, there has been greater acquiescence than one would have expected to the continuing operations in both countries.

> In the early years of the global war on terror, US methods were strongly criticized by Europe's governments. This has now largely ceased. It is not clear whether this is due to the fact that Europeans are now increasingly experiencing terrorist attacks of their own, whether they became less critical out of understanding or sympathy for the Obama administration, or whether they have simply become accustomed.[55]

[52] 'However, a US Department of Justice memo that was leaked in February 2013 indicated that the killing of US citizens was deemed lawful if they pose an "imminent threat" of violent attack and capture is not feasible.' Blakeley, above note 50, at 329.

[53] 'It is simply unimaginable that France, Russia, China, or India, each of which spars with the United States on numerous foreign policy issues, would have responded otherwise had Al Qaeda crashed planes into the Eiffel Tower, the Kremlin, the Forbidden City, or the Taj Mahal. In such a situation, the view of state responsibility proffered by the ICJ, the ICTY, and the ILC—and by many international lawyers—becomes, in effect, instantly anachronistic as a limitation on self-defense.' Ratner, above note 28, at 919. And indeed, even stalwart critics of US policy such as Richard Falk appear to agree: 'There were no credible alternatives to war ...' R. Falk, *Appraising the War Against Afghanistan,* available at <http://www.ssrc.org/septll> (last visited 10 January 2020).

[54] Caesar stated '*alea iacta est*'—the die is cast. As reported by Suetonius, *The Lives of the Caesars*, available at <http://penelope.uchicago.edu/Thayer/E/Roman/Texts/Suetonius/12Caesars/Julius*.html> (last visited 10 January 2020).

[55] Thimm, above note 48, at 38; see also Ratner, above note 28, at 910: 'If state B protests A's action, it might object to that conduct as either a legal or a policy (i.e., a nonlegal) matter; but the absence of protest—where

It is long past time to reconsider how the construction of terrorism as an existential threat has shattered the long, hard development of *jus ad bellum* and *jus in bello* norms and law. New efforts are required to rein in or prevent the next set of abuses when the inevitable next 'existential' threat occurs. Recognizing the mistakes of the global response to terrorism during the past two decades, and that they were not inevitable, is a good place to begin the hard work that will be required.

state A's act is of a major, even constitutive nature, affecting all states—would strongly suggest that B endorsed, or at least did not object to, both the legal and the nonlegal bases for A's actions.'

18
'We Are Going to War.' Narratives of Self-Defence and Responsibility in Afghanistan War Documentaries

Axel Heck and Gabi Schlag

1. Introduction

Western states are involved in nearly every trans- and international conflict currently going on. But as almost all of these wars are taking place in the Global South they come with substantially higher costs for the people living in those countries than for Western societies and forces. In Afghanistan, for example, the Watson Institute at Brown University speaks of 147,000 people who were killed since 2001, more than 38,000 civilians among them.[1] Afghan President Ashraf Ghani confirmed that since 2015 more than 28,000 service members lost their lives in the wake of the insurgency against the government and the coalition.[2] In contrast, the casualties of the intervening coalition forces add up to ca. 3,500 dead soldiers since 2001 (Operation Enduring Freedom and ISAF combined).

Nevertheless, though interveners suffer less than the intervened, troop deployment is always a highly contested issue, especially in the so-called post-heroic societies of Europe and North America, where the numbers of soldiers killed or wounded in action evoke scepticism about the necessity of joining a war. According to a report of the Pew Research Centre published in 2019 only about 36 per cent of the people in the United States confirm that the war in Afghanistan was 'worth fighting'. Among veterans, about 38 per cent respondents agree with this statement.[3] After the terror attacks on 11 September 2001, the approval rates to the US engagement in Afghanistan were close to 90 per cent. As these numbers illustrate, going to war and sending troops overseas is a controversial political issue—especially in democracies where such decisions must be justified vis-à-vis a critical electorate.[4]

As constructivist and post-structuralist research in International Relations (IR) has shown, wars and military interventions are made possible and sustained by discursive

[1] <https://watson.brown.edu/costsofwar/costs/human/civilians/afghan> (last visited 30 December 2019).

[2] <https://www.theguardian.com/world/2018/dec/04/us-army-chief-says-death-toll-of-afghan-security-forces-is-unsustainable> (last visited 30 December 2019).

[3] <https://www.pewresearch.org/fact-tank/2019/07/10/majorities-of-u-s-veterans-public-say-the-wars-in-iraq-and-afghanistan-were-not-worth-fighting/> (last visited 30 December 2019).

[4] A. Geis, H. Müller, and N. Schörnig, *The Militant Face of Democracy: Liberal Forces for Good* (2013); A. Geis, L. Brock, and H. Müller (eds), *Democratic Wars: Looking at the Dark Side of Democratic Peace* (2006).

Axel Heck and Gabi Schlag, *'We Are Going to War.' Narratives of Self-Defence and Responsibility in Afghanistan War Documentaries* In: *The Justification of War and International Order.* Edited by: Lothar Brock and Hendrik Simon, Oxford University Press (2021). © the many contributors. DOI: 10.1093/oso/9780198865308.003.0018

practices.[5] In this context, research at the intersection of IR and media studies remind us that the mediatization and visual representation of war informs and shapes narratives of justification.[6] To better understand how military engagements and troop deployments are made possible, these narratives play a crucial role and must be critically scrutinized. Narratives of justification are more than mere propaganda; they cannot be reduced to flimsy reasons given by political leaders for why a military intervention seems necessary. Justification narratives reveal sedimented beliefs and ethical considerations held to be true and right within society. They can be found in many spheres of societal discourses: in parliaments, political speeches, and newspapers, but also in social media, culture, and art.[7] These justification narratives represent the contested assumptions and ideas about the legitimacy of the use of force as a constituent element of political order but also about the righteousness of the use of force and of the accompanying military sacrifice, that is to fight and to die for a noble cause.

In this contribution, we will discuss how documentary films that depict soldiers fighting at the war-front in Afghanistan craft and perform justification narratives of war and how they relate to the normativity of global order. As Hendrik Simon and Lothar Brock state in the introduction to this volume, debates on the justification of war have often been rather abstract, neglecting the mundane practices of how agents on the ground carry out, justify, and criticize wars. If it is true that 'war is also about the dominant ways of seeing that shape how we understand war', as Rens van Munster argues, films and documentaries can be considered as an important site of justification and critique.[8]

Therefore, we will take a closer look at two documentary films about the Afghanistan war, *Restrepo* (D: Sebastian Junger/Tim Hetherington, 2010) and its sequel *Korengal* (D: Sebastian Junger, 2014), both critically acclaimed and internationally awarded documentaries which focus on the experiences of US soldiers deployed to Afghanistan. These documentaries offer rare opportunities to see how the 'managers of violence'[9] act in combat and how they reflect upon the normativity of their

[5] D. Campbell, *Writing Security: United States Foreign Policy and the Politics of Identity* (1992); Doty, 'Foreign Policy as Social Construction: A Post-Positivist Analysis of US Counterinsurgency Policy in the Philippines', 37 *International Studies Quarterly* (1993) 297; L. Hansen, *Security as Practice: Discourse Analysis and the Bosnian War* (2013); M. Finnemore, *The Purpose of Intervention* (2004).

[6] Heck and Schlag, 'Securitizing Images: The Female Body and the War in Afghanistan', 19 *European Journal of International Relations* (2013) 891.

[7] Ibid.; Heck, 'Analyzing Docudramas in International Relations: Narratives in the Film A Murderous Decision', 18 *International Studies Perspectives* (2017) 365; Crilley, 'Like and Share Forces: Making Sense of Military Social Media Sites', in L. J. Shepherd and C. Hamilton (eds), *Understanding Popular Culture and World Politics in the Digital Age* (2016) 67; Geis and Schlag, '"The Facts Cannot Be Denied": Legitimacy, War and the Use of Chemical Weapons in Syria', 7 *Global Discourse* (2017) 285; Schlag, 'Representing Torture in Zero Dark Thirty (2012): Popular Culture as a Site of Norm Contestation', *Media, War & Conflict* (2019), doi: 1750635219864023; J. Schuff and M. Seel (eds), *Erzählungen und Gegenerzählungen: Terror und Krieg im Kino des 21. Jahrhunderts* (2016); A. Danchev, *On Art and War and Terror* (2009); Danchev and Lisle, 'Introduction: Art, Politics, Purpose', 35 *Review of International Studies* (2009) 775; on justifications and propaganda see Simon and Brock, Introduction in this volume.

[8] van Munster, 'Inside War: Counterinsurgency and the Visualization of Violence', in C. Sylvest and R. van Munster (eds), *Documenting World Politics* (2015) 128, at 114. J. Schuff and M. Seel (eds), *Erzählungen und Gegenerzählungen: Terror und Krieg im Kino des 21. Jahrhunderts* (2016).

[9] Samuel, 'The Soldier and the State: The Theory and Politics of Civil-Military Relations', 163 *Cambridge: Belknap Press of Harvard University Press* (1957) 163.

doings. The central question we want to address in this chapter is how soldiers experience war and justify their actions, how they reflect upon official political justifications, and whether and how soldiers' doings and sayings are linked to ideas concerning the global normative order.

In a nutshell, we make two arguments. Firstly, and in line with Brock and Simon, it seems essential to us to acknowledge that discourses on the justification of war do not only materialize in official documents like speeches and declarations of war or political arenas such as the United Nations or national parliaments. We argue that the experience of those who carry out the ethically and emotionally ambivalent business of war and suffer from the traumatic consequences of violence opens up a new perspective on the issue of justification. This is the experience of the ordinary infantrymen, so-called 'grunts'. By making these soldiers (mostly men), their bodies, and emotions visible and giving voice to them, documentary films represent a subjective position that is often neglected and marginalized in IR and studies on the justification of war. Secondly, we argue that documentaries like *Restrepo* and *Korengal* create a publicly shared imaginary of war. Therefore, we are interested in the normative and ethical considerations of the soldiers represented in these films as part of the process of the justification of war and the construction of world order.

This chapter will proceed as follows. First, we will briefly discuss questions of ethics and justifications of war. Second, we turn to the two documentaries and the 'grunts perspective' as a marginalized position in (Western) discourses on the justification of war. We will show how the soldiers in the film experience and reflect their deployment and how they think about the legitimacy of the mission. We elucidate our argument by analysing several scenes in which the interviewees articulate the aim and purpose of their mission. Finally, we turn to the more general question as to what kind of normative order these documentaries reproduce. We show how the micro-cosmos of the soldier's life is informed by two justification narratives, that is killing the 'bad guys' and bringing 'relief' to the 'good guys'. In the concluding section, we argue that the micro-cosmos represented in *Restrepo* and *Korengal* mirrors a global, yet ambiguous normative order oscillating between self-defence *and* global responsibility.

2. Ethics, Legitimacy, and the Justification of War

War loosely defined as the collective use of force between at least two (state) parties remains an ethically contested choice under any condition.[10] With the prohibition of traditional war-making as an assumed right of states in the Kellogg-Briand-Pact followed by the prohibition of the unilateral use of force in the UN Charter and corresponding

[10] For the contentedness of its meaning and the changing nature of war, see Y. Boyer, J. Lindley-French, and L. Freedman, 'Defining War', available at <https://www.oxfordhandbooks.com/view/10.1093/oxfordhb/9780199562930.001.0001/oxfordhb-9780199562930-e-2> (last visited 30 December 2019); Singer, 'The Correlates of War Project: Interim Report and Rationale', 24 *World Politics* (1972) 243; Bluhm and Geis, 'Den Krieg überdenken. Kriegstheorien und Kriegsbegriffe in der Kontroverse? ein Tagungsbericht', 11 *Zeitschrift für Internationale Beziehungen* (2004) 419. On the concept of new wars see M. Kaldor, *New and Old Wars: Organised Violence in a Global Era* (2013). For a post-colonial critique of common definitions of war, see Barkawi, 'Decolonising War', 1 *European Journal of International Security* (2016) 199.

regional provisions, serious attempts to ban war from international politics have been undertaken. When it comes to the justification of war, Hendrik Simon shows that war has always been normatively problematic and, therefore, in need of justification, although justifications might vary and change over time and differ between states and cultures.[11] While it might be a relevant question, whether these justifications—or reasons given by generals and politicians—are trustworthy, references to normative reasoning can be performative and consequential beyond the intentions and motivations of those who wage wars. The avoidance of civilian casualties, for example, has become a norm regardless of the strategies that military leaders identify as most successful.[12]

The rights and wrongs of war, that is *the jus ad bellum* as well as the *jus in bello*, are issues of constant debate in theory and practice. While the practices of war-making are restricted by international humanitarian law, these legal provisions are often translated into more specific Codes of Conduct and Field Manuals that direct the combat behaviour of soldiers. As history shows, there might be situations where the use of collective violence is necessary and legitimate to prevent or fight a bigger evil such as mass atrocities.[13] However, a state's (or government's and their political leaders') decision to use force accepts that people are most likely killed and that their right to live is irreversibly violated. Maja Zehfuss, for example, has recently argued that justifications of war are based on ethical considerations and claims 'to make the world a better place for others'. Paradoxically, such justifications of war finally 'justify what they seek to prevent'.[14]

Therefore, the justification of war involves ethical issues. Ethics, Kimberly Hutchings reminds us, encompasses two meanings.[15] First, ethics refers to beliefs, values, and principles about what is right or wrong. Second, ethical thinking is also about understanding the ground and nature of these beliefs, values, and principles, that is why so many people believe that it is unethical to kill a human being but not an animal. Ethics and war, however, seem to be strange bedfellows.[16] The number of publications on the theory and practice of war is voluminous. These publications include philosophical, ethical, legal, historical, and political perspectives on the question of whether war is ever or only under specific conditions the right or necessary thing to do.[17] While pacifists mostly reject the notion that fighting wars can ever be ethically or morally justified, IR scholarship is mainly concerned with limiting and restraining war, 'both on what we can legitimately fight for and on how we may

[11] Simon, 'The Myth of *Liberum Ius ad Bellum*: Justifying War in 19th-Century Legal Theory and Political Practice', 29 *European Journal of International Law* (2018) 113.

[12] Walzer, 'The Triumph of Just War Theory (and the Dangers of Success)', in R. Wertheimer (ed.), *Empowering Our Military Conscience: Transforming Just War Theory and Military Moral Education* (2017) 27.

[13] Brock, 'Between Sovereign Judgment and the International Rule of Law: The Protection of People from Mass Atrocities', in M. Albert and A.F. Lang, Jr (eds), *The Politics of International Political Theory: Reflections on the Works of Chris Brown* (2019) 87.

[14] M. Zehfuss, *War and the Politics of Ethics* (2018), at 73..

[15] K. Hutchings, *Global Ethics* (2018)..

[16] H. Frowe, *The Ethics of War and Peace* (2015); G. M. Reichberg, H. Syse, and E. Begby, *The Ethics of War: Classical and Contemporary Readings* (2006).

[17] A.F. Lang, Jr, C. O'Driscoll, and J. Williams, *Just War* (2013); A. Bellamy, 'Just Wars: From Cicero to Iraq' (2006); N. Rengger, *Just War and International Order* (2013) .

legitimately fight'.[18] Its most prominent tradition, just war theory, though, was never a purely academic project of contemplation. In the introduction to the first edition of his seminal study on *Just and Unjust Wars* published in 1977, Michael Walzer explicates his motivation to write this book:

> 'In those years of angry controversy, I promised myself that one day I would try to set out the moral argument about war in a quiet and reflective way. I still want to defend (most of) the particular arguments that underlay our opposition to the American war in Vietnam, but also and more importantly, I want to defend the business of arguing, as we did and as most people do, in moral terms.'[19]

Given the longstanding tradition of just war theory and the (formal and informal) regulations of warfare, the justification of war in democracies is often portrayed and studied as an ethical choice of governments and parliaments.[20] How soldiers experience war and justify their doings in ethical and moral terms, however, mostly remains invisible and unacknowledged in IR.[21] If normative orders are by definition orders of justification and critique, as Klaus Günther and Rainer Forst argue, then understanding the formation and re-/production of such orders requires studying the sayings and doings of those agents who are directly affected by the policies. Because justification narratives are socially and historically constructed, it is essential how they are used and whether this usage changes over time.[22] Justification narratives are situated practices, conceptualized as sayings and doings, that justify (or criticize) rules and norms. These rules and norms constitute specific political and social relations, which form more general discursive orders. Thus, orders of justification and critique are constituted by sedimented discourses that are constantly re-made by story-telling and image-making.[23]

In recent years, IR scholars have turned more attention to the cultural site of politics, in particular, how popular culture and mass media construct, represent, and memorize wars.[24] It is a widespread conviction nowadays that the meaning of

[18] Rengger, 'On the Just War Tradition in the Twenty-First Century', 78 *International Affairs* (2002) 353, at 354.

[19] M. Walzer, *Just and Unjust wars: A Moral Argument with Historical Illustrations* (1977).

[20] Geis, Brock, and Müller, above note 4; Geis, Müller, and Schörnig, above note 4; see also Brown in this volume.

[21] For micro-oriented perspectives on soldiers' experience, narration, and representation of war, see Steele, 'Ideals that Were Really Never in our Possession': Torture, Honor and US Identity', 22 *International Relations* (2008) 243. McSorley, 'Helmetcams, Militarized Sensation and 'Somatic War', 5 *Journal of War & Culture Studies* (2012) 47; Pomarede, 'Normalizing Violence through Front-Line Stories: The Case of American Sniper', 4 *Critical Military Studies* (2018) 52.

[22] R. Forst, *Normativity and Power* (2017).

[23] R. Forst and K. Günther (eds), *Die Herausbildung normativer Ordnungen* (2011).

[24] Ebbrecht, 'Docudramatizing History on TV German and British Docudrama and Historical Event Television in the Memorial Year 2005', 10 *European Journal of Cultural Studies* (2007) 35; Baer, 'Consuming History and Memory through Mass Media Products', 4 *European Journal of Cultural Studies* (2001) 491; G.R. Edgerton and P.C. Rollins, *Television Histories: Shaping Collective Memory in the Media Age* (2001); Hoskins, 'Anachronisms of Media, Anachronisms of Memory: From Collective Memory to a New Memory Ecology', in M. Neiger, O. Meyers, and E. Zandberg (eds), *On Media Memory* (2011) 278; Heck, 'Analyzing Docudramas in International Relations: Narratives in the Film A Murderous Decision', 18 *International Studies Perspectives* (2017) 365.

war is shaped discursively as well as visually.[25] War photography has been the most powerful genre that defined the public image of war and war-making, at least for North American and European societies.[26] More recently, soldiers take pictures of their deployment and share these images on social media.[27] The same holds to be true for fictional movies that are grounded in real war settings, most prominently the Vietnam war, which was also declared to be the first 'television war'.[28] Regarding the genre of documentary films, Casper Sylvest and Rens van Munster argue that 'documentaries provide powerful lenses on reality through which students can explore events and theories' and that 'documentaries actively inform public opinion and the background beliefs people take for granted and act upon'.[29] We would go a step further and argue that based on their appeal of authenticity and *Realism*, documentaries are an important site of the politics of justification and contestation of the use of force.

Although styles and techniques of documentaries may differ widely, typical political documentaries assemble first-order representations, that is real-time footage from an event, with additional data like speeches, interviews, and charts. Documentaries often build on two frames of time: first, on the experienced time of the event and second, on narrative time where agents who were involved in the events tell their stories. Given these two timeframes, documentaries can be authentic and realistic in style and scope, but they are never objective or normatively neutral. In our case, *Restrepo/Korengal* claim to offer an authentic, realistic, and nearly unfiltered view on war as an existential and traumatic experience of soldiers. While both documentaries visualize the 'micro-cosmos of combat and soldier life',[30] both films create spaces for the soldiers to reflect upon their deployment. The films give a voice to the soldiers; they listen to their stories and take their interpretations and reflections seriously. However, both films are made not by the soldiers but by the directors who decided what sceneries would be included and how these scenes were cut and edited with additional footage. It is up to the audience to bring the experienced time and the narrated time together as a second-order representation and to 'read' these documentaries.

[25] Griffin, 'Media Images of War', 3 *Media, War & Conflict* (2010) 7; Campbell, 'Representing Contemporary War', 17 *Ethics & International Affairs* (2003) 99.

[26] Hariman and Lucaites, 'Public Identity and Collective Memory in US Iconic Photography: The Image of "Accidental Napalm"', 20 *Critical Studies in Media Communication* (2003) 35; Campbell, 'Atrocity, Memory, Photography: Imaging the Concentration Camps of Bosnia–The Case of ITN versus Living Marxism, Part 1', 1 *Journal of Human Rights* (2002) 1; C. Brothers, *War and Photography: A Cultural History* (1997).

[27] Although the soldier as an author of war images is not a new phenomenon, digitization makes it possible to share these images in real time and globally. Body cameras are also deployed to provide new imagery from the combat area. Kennedy, 'Soldier Photography: Visualising the War in Iraq', 35 *Review of International Studies* (2009) 817.

[28] A. Auster and L. Quart, *How the War was Remembered: Hollywood & Vietnam* (1988); A. Hoskins, *Televising War: From Vietnam to Iraq* (2004); McMahon, 'Contested Memory: The Vietnam War and American Society, 1975–2001', 26 *Diplomatic History* (2002) 159; D. Ryan, *US Collective Memory, Intervention and Vietnam* (2007); Carvalho, 'War Hurts: Vietnam Movies and the Memory of a Lost War', 34 *Millennium—Journal of International Studies* (2006) 951.

[29] van Munster and Sylvest, 'Documenting International Relations: Documentary Film and the Creative Arrangement of Perceptibility', 16 *International Studies Perspectives* (2015) 229.

[30] van Munster, above note 8, at 115.

3. The Discursive Power of 'Reactive Observationalism'—Bringing the 'Grunts' (Back) In

Restrepo and its sequel *Korengal* were released in 2010 (*Restrepo*) and 2014 (*Korengal*). Both films are based on original video-footage and interviews with soldiers of the US army directed by photojournalist Tim Hetherington and author Sebastian Junger. Hetherington and Junger joined the Second Platoon, Battle Company, of the 173rd Airborne Brigade, which was deployed to the Korengal Valley in Afghanistan. They visited the Battle Company in the Korengal valley as embedded journalists five times and stayed for several weeks. According to Sebastian Junger, who also published a book about his experiences in Afghanistan titled *War*, they were dependent on the support of the US military providing food, shelter, transportation, and protection. But Junger makes clear that 'I was never asked—directly or indirectly—to alter my reporting in any way or to show the contents of my notebooks or my cameras.'[31]

Both films were critically acclaimed, reviewed, and are considered milestones in contemporary war documentary—especially in the genre of so-called 'grunt documentaries'.[32] They provide a boots-on-the-ground perspective on the Afghanistan war based on the subjective empowerment of soldiers and their experience of the deployment. For van Munster, *Restrepo* is 'X-raying counterinsurgency and inspiring audiences to ask macro-questions about the nature and purpose of war.'[33] What makes these films special is that in contrast to many other documentaries in the tradition of New Journalism, which became popular in the wake of the Vietnam war reporting, Junger and Hetherington do not contextualize the events and incidents.[34] The film-makers' perspective in *Restrepo/Korengal* is known as 'reactive observationalism' which is a fundamentally different approach to war reporting as it was common sense in New Journalism. According to Cawley, Hetherington's and Junger's style is

> . . . characterized by fly-on-the-wall cinematography, testimony, and overheard exchange rather than expository or narrative modes, minimal production values, and stylistic elements, and the absence of a political register or motive, among other features intended to produce an objective experience for viewers.[35]

Indeed, Junger and Hetherington are nearly absent in both films. There is no voice-over and only a few comments which explain what is happening. But the film-makers never judge the events, decisions, or portrayed people. Junger argues that the political

[31] S. Junger, *War* (2010), at 9.

[32] Aufderheide, 'Your Country, my Country: How Films about the Iraq War Construct Publics', 48 *Framework: The Journal of Cinema and Media* (2007) 56.

[33] van Munster, above note 8, at 115. For academic contributions that discuss *Restrepo* and/or *Korengal*, see Cawley, 'Who Is the Soldier? Documenting American "Grunts" from Dispatches to Restrepo', *Journal of American Studies* (2019) 1; Little, '"This Is Reality" Restrepo and the Depiction of War', 156 *The RUSI Journal* (2011) 34; Doherty, 'Korengal', 101 *Journal of American History* (2014) 1024; van Munster, above note 8.

[34] Cobley, 'Narrating the Facts of War: New Journalism in Herr's "Dispatches" and Documentary Realism in First World War Novels', 16 *The Journal of Narrative Technique* (1986) 97; A. Greenbaum, *The Tropes of War: Visual Hyperbole and Spectacular Culture* (2015); Cawley, above note 33.

[35] Cawley, above note 33, at 24.

circumstances and normative justifications of the war play only a marginal role for his film-making if they matter at all.[36] Whether directors agree or disagree, film-making always relies on a narrative structure and a point-of-view. A film is the results of multiple decisions made by the directors, cameramen, producers, and editors about which scenes will be part of the film and which ones will be not, how a scene is shot and composes, and how the final material is cut and edited, including sound, voice, and if applicable music.

As the film-makers are embedded journalists with the US army, they take in the soldiers' perspective. Insurgent fighters and Taliban are almost absent and have no individual voice in both films. One could argue that they 'communicate' through military signals like Improvised Explosive Devices (IEDs), gunfire, and Rocket Propelled Grenades (RPGs). It is only in *Korengal* that the film-makers provide some footage showing how insurgent fighters carry heavy weapons across the mountain trails. Civilians are visible only when they are picked up by US soldiers as suspects or during the *Shura*, a weekly meeting between the US forces and the elders of the villages in the valley. Aside from these scenes, civilians and Afghans have no (individual) voice, their stories and feelings do not matter for the documentary. Hence, the film does not provide a balanced nor comprehensive picture of the war but a US-centred perspective on the daily life of the deployed 'grunts'.

The documentary operates on two levels of time and narration, as mentioned. The first level covers actual happenings in the Korengal Valley at an outpost called Restrepo, named after Private Juan S. Restrepo, who was wounded during an ambush in the Korengal Valley and died on his way to the hospital at the age of twenty. The film-makers accompany the soldiers in the camp, on patrols, and meetings with the local population. Moreover, the film confronts the spectator with disturbing images of combat, including shootouts and the death of Restrepo. On the second level, the film-makers have conducted interviews with the soldiers in which they reflect on their time in Afghanistan and how the deployment changed their life. These interview scenes are a mixture of interpretation, testimony, reflection, and therapy. Some of the interviews took place during the deployment, but most of them were conducted afterward. The interviews are directed in the typical genre style mostly as a medium-shot where the person sits in front of a black background. However, sometimes the camera takes close-ups of the faces. This builds a very intimate, even intrusive relationship with the audience.[37]

4. Justifying the War—Soldier's Subjectivity and the Representation of the Afghanistan Mission

Both films, *Restrepo* and *Korengal*, show the ambivalences between the everyday, mundane, often boring life at the camp and the adrenaline-dosed combat action with heavy shootings. The documentaries also illustrate essential practices of soldering,

[36] Junger, above note 31.

[37] Many interviews as well as some scenes of the films are available on YouTube. Here, we will cite from the original films but check, if the quote or scene is available and provide the link accordingly.

like caring about your gun, cheering about the (likely) killing of the enemy, sensorial training, stitching tattoos, playing cards, or mourning the fallen soldiers.

Moreover, the documentaries reproduce gendered stereotypes of physical strength, masculinity, and brotherhood but also of emotional trauma, sadness, and agony. The disturbingly realistic *mise-en-scene* reminds us of the psychological and physical impact of war on the soldiers as human beings resulting in depression, suicidal tendencies, heavy injuries, and death. In the following section, we focus on two narratives that inform the micro-cosmos of the soldier's experience: 'killing the bad guys' and 'bringing relief'.

At the beginning of *Korengal*, Captain Kearney explicitly refers to the aims of their mission in his post-deployment interview:[38]

> I was wondering why I was in the Korengal. But then, you get told that the bigger mission is that you are not in the Korengal to hunt the bad guys, you are there so the bad guys come to you and you kill 'em. Why are we in Afghanistan, we are there to kill the bad guys. Why are we still there in Afghanistan and Iraq? If you listen to President Bush, we are there because that's where the bad guys are. The US hasn't been hit since 9/11 because we are over here, and they are coming to us now.

This statement resembles the official proclamations and common beliefs reiterated by the Bush and Obama administrations as well as many policy experts in the United States (see also Michael Stohl's chapter in this volume). The underlying strategic rationale seems to be that by bringing the battle to the enemy, people at home can live in peace. Soldiers like Kearney and his platoon are serving people at home at the distant frontline 7,000 miles away. While the attacks on 11 September 2001, hit US American soil and brought violence to US citizens at home, with the war in Afghanistan fighting 'returns' to where it came from, Kearney suggests. Now, the 'bad guys' 'are coming to us', as he says, instead of coming after the innocent people in the United States. This narrative came under fire since President Trump took office in 2016. He repeatedly claimed to pull out troops from Afghanistan the sooner, the better, but the decision to finally do so was postponed after peace negotiations with the Taliban failed.

While it is affirmed by the captain that killing 'bad guys' is one part of the mission, the invisibility of the enemy is a central motif in *Restrepo* and *Korengal*. One cannot see and talk to the Taliban and Al Qaida fighters, you just can hear them as they shoot at you. Sergeant O'Byrne explains that 'during the firefights, you never saw the enemy ... it was like we're fighting ghosts. We know the person is out there, but you cannot see him. Whenever we get a kill and see it, it was a good day.'[39] Thus, killing a ghost has been the main challenge of the deployment, which re-invokes images of a clandestine terrorist network whose members are champions in hiding while US soldiers are visible, present, and thus constantly a target.

This invisibility of the combatant reinforces the approach of both documentaries, which focus on US soldiers, their actions, experience, beliefs, and feelings. It is mainly about 'us', not them. Taliban and Al Qaida fighters are only referred to as the 'bad guys'

[38] <https://www.youtube.com/watch?v=xdVasN0vJ1c> (last visited 30 December 2019).
[39] <https://www.youtube.com/watch?v=nQTwLjObbH8> (last visited 30 December 2019).

which implies that the US soldiers are the 'good guys'. Naming combatants 'bad guys' though simplifies the relationship and defines the other side in terms of moral inferiority. From the perspective of the soldiers, they do not qualify as combatants and soldiers. They are not military equals. The label 'bad guys' trivializes the relationship between the conflicting parties to a black-and-white scheme.

The disrespect for Afghans continues when Captain Kearney speaks about the people living in the Korengal Valley in general. In his post-deployment interview, he says:

> Do I respect them? I do not respect the Korengalis. They are a bunch of liars, and they didn't want us. They didn't want our help. Are there individuals in the Korengal valley I do respect? Yes, I believe they want it better for their people, they want it better for their families, but they are caught between a rock and a hard place.

Although his statement seems to differentiate between those who accept the 'help' and those who are 'liars', Kearney sympathizes only with those who accept his presence. His respect depends on the willingness of the people to share his norms and values of a better life for their families.

On deployment, Kearney explains to the camera that the idea is to convince the people in the valley to push out the insurgents. This approach, however, is the most difficult part of the mission due to 'family ties and religious ideals'. He has to admit that 'where the road ends is where the Taliban begin'.

Captain Kearney's already cited statement about the purpose of the mission, however, continues:[40]

> We are in the Korengal to be able to take the fight away from the cities so you can bring some economic growth and prosperity to these people, to show them that there is something greater out there you can strive to do, or you can live a better life not necessarily a western culture where you need to do MTV, and you got a McDonalds and WalMart but just something better.

Although Kearney acknowledges that a better life may not necessarily imply a 'western culture', he refers to common symbols of US wealth and prosperity. Like an aid worker, US troops are stationed in the rural and remote valley close to the Afghan people. While their living conditions and purpose differ fundamentally, the troops share the location with the inhabitants. Common motifs of aid and development policies are invoked here that aim at relief for the local population.[41]

Finally, Kearney emphasizes that there is 'something greater out there you can strive to do' illustrating the idealistic overtones that some soldiers voice in *Restrepo* and *Korengal*. His deployment in the Korengal Valley is more than a military enterprise but a service to create 'something better' for humankind. Staff Sergeant Rice supports this last point when he says '(w)e are serving a purpose besides just killing the bad

[40] <https://www.youtube.com/watch?v=xdVasN0vJ1c> (last visited 30 December 2019).
[41] M. Duffield, *Development, Security and Unending War* (2007).

guys. We are helping the country; we are helping the local populace with a lot of things they never had.'[42]

In his post-deployment interview, Sergeant Major Caldwell explains the goal of the mission in similar terms:

> What were we doing in the Korengal? Our purpose in the Korengal—they had a road, and the intent was this road to go through the Korengal and out to the Chowkay Valley to be able to connect the, uh, connect the locals where they can have an easier route up toward the Pech River Valley. And our job was there to try to sustain, you know, the security for the personnel to allow them to build that road.

As we can see here, for Caldwell, the reasons why he is stationed in the Korengal Valley is to help and support the people there. This is made visible by scenes where Kearney and his team meet with the valley elders in the weekly *Shura* to explain the purpose of the mission. According to Kearny, the Korengal Valley is about to have a paved road in the next couple of years, which brings prosperity and 'makes you guys richer and makes you guys more powerful'. Kearney, though, offers a deal. To make his partners richer and more powerful, he demands their cooperation with the government.

The US army has applied its 'winning the hearts and minds'-approach since the Vietnam war.[43] While many critics dismiss it as cynical or inefficient rhetoric, it is an important reference point for the soldiers on the ground. Building trusted networks and influencing the calculated self-interest of the local population is key to Kearney and his troop in the Korengal Valley, too. However, counter-insurgency does not work because it falsely assumes that the interests of the people can be altered to fit the interests of the occupying troops. It is the opposite of listening to the needs and fears of the valley's inhabitants.

Given the idealistic overtones of the mission articulated by several soldiers, the ongoing challenges to enhance economic growth and building a road can create a source for disappointment and disillusion, in the worst case, traumatic experience of failure. The fact that not all Afghan people were welcoming US troops and the aims of US politics, Specialist Steiner expresses his disappointment:

> Hearts and minds go out of the window when you see a guy shooting at you, and he puts his wife and kids in front of him knowing full that we won't shoot back... He obviously doesn't have a heart if he's willing to sacrifice his wife and kids just to shoot a couple of rounds at us.[44]

Steiner perceives Afghan people as dishonest, disrespectful, and hypocritical when he refers to 'the guy that shakes our hand takes the ten bags of rice we gave him for his family and the school supplies and shoots an RPG at us ...fuck his heart, fuck his mind'. His emotional statement illustrates that he is disappointed because Afghan

[42] <https://www.youtube.com/watch?v=nZC4K1UtDVo> (last visited 30 December 2019).
[43] J.A. Nagl et al., *The US Army/Marine Corps Counterinsurgency Field Manual* (2008).
[44] <https://www.youtube.com/watch?v=ik9dVd5IutM> (last visited 30 December 2019).

people do not confirm his identity as the 'good guy'. He seems to be morally injured by the fact that his help, service, and duty is not wanted.

5. Conclusion: Reconstructing a Global Normative Order between Self-Defence and Responsibility

Justification narratives of war are not only found in official speeches and documents of so-called high politics. In contrast, we focus on popular representations of war, namely war documentaries, which shed light on how common infantrymen experience war at the front. This move to the micro-cosmos of the 'managers of violence' not only brings in the grunt's perspective on the justification of war but illustrates how a global order is represented and negotiated when soldiers reflect upon their doings and the purposes of a mission in the Korengal Valley.

Restrepo and *Korengal*, though, remind us that documentaries can represent a highly ambiguous imaginary of war. Both films illustrate that soldiers engage with the political justification narratives of the Afghanistan war rather superficially. They are mostly concerned with the implementation of the mission on the ground. As *Restrepo* and *Korengal* show, winning the hearts and minds of the local population and building a road is a much more complicated endeavour than shooting at and killing the enemy.

Therefore, the micro-cosmos of the 'grunts' represented in the two documentaries mirrors a global, highly ambiguous normative order which oscillates between self-defence *and* responsibility for helping people to work for a better life. As the soldiers are killing the 'bad guys' in defence of themselves and the nation, traditional security conceptions constitute an order that legitimizes the use of force for self-defence on the ground and as a means to secure national interests and identities in general. Kill or be killed is the slogan of such an order. As soldiers clarify in their interviews, self-defence is not an unregulated business. Captain Kearney emphasizes that the doings of soldiers are limited to preserve a positive self-image. Hence, it is self-defence governed by the norms of 'being civilized' and of adhering to humanitarian law. The mission, though, is not only about self-defence but intends to bring relief to the people in the Korengal Valley and thus fulfils the promises of the liberal world order. We need to see that the soldiers not only take care of each other but also claim to have a responsibility for the local population. As an ethical and moral concept, responsibility requires a response to the material and normative demands of others.

The documentaries show that soldiers want to take responsibility but may fail due to the impossibility of seeing the needs and fears of the people in the Korengal Valley. While some soldiers critically reflect on the limits of their mission and acknowledge that Afghan people have their own voice, they remain a closed US ex-pat community by definition. The global order reproduced by these narratives and images, thus, is highly contested. Often portrayed as opposing and incommensurable narratives of justification, self-defence and responsibility, killing and building roads, dualistic world views and universal normative claims may go hand in hand as *Restrepo* and *Korengal* make visible. Hence, the meaning of 'we are going to war', as Private First Class Restrepo bellows at the very beginning of the documentary, remains ambivalent.

19
Justifying Interventions—The Case of ECOWAS in Liberia

Nina Wilén

1. Introduction

Two decades into the new millennium it is clear that 'benign' interventions with the aim to stabilize conflict-ridden or troubled states are here to stay. The UN, as the world's largest international organization, stands for most of these interventions with over 90,000 peacekeepers deployed worldwide. Yet regional organizations, alliances, and different types of ad hoc coalitions have sailed up alongside the UN as new actors with the aim to enforce international order and stabilize troubled areas, mostly in Africa. At times these interventions have stretched (un)written rules and norms. The UN's authorization of an ad hoc coalition's intervention in Libya for example, was breaking new ground (and even breaking formal norms) as the situation in Libya marked the first time the Security Council authorized the use of force for human protection purposes against the wishes of a functioning state, thus straying from the UN's sacred norm of consent.[1]

In December 2015, as the crisis in Burundi developed, the African Union's Peace and Security Council also threatened to intervene against the wishes of the Burundian government to protect civilians and re-establish security.[2] These examples show that the line between internal and international affairs is rather thin and that, under these conditions, the notion of sovereignty is under constant pressure to be reinterpreted in order to widen the space for external efforts to stabilize conflict-ridden states. This begs the question of how such 'benign' interventions, which challenge the norm of sovereignty, are justified discursively in relation to an international normative order based on the very same norm of sovereignty.

This chapter explores the relationship between sovereignty, intervention, and the international normative order, by examining how ECOWAS, an African regional organization, justified its intervention in Liberia's civil war as it referred to different international norms. More precisely, it critically analyses the discourses related to ECOWAS' decision to intervene in Liberia in 1990, and the responses the justifications provoked from the UN and the United States, both major international actors with an interest in the conflict. The analysis focuses on how international actors at a certain

[1] Bellamy and Williams, 'The New Politics of Protection? Côte d'Ivoire, Libya and the Responsibility to Protect', 87 *International Affairs* (2011) 825.

[2] Wilén and Williams, 'The African Union and Coercive Diplomacy: the Case of Burundi', 56 *Journal of Modern African Studies* (2018) 673.

Nina Wilén, *Justifying Interventions—The Case of ECOWAS in Liberia* In: *The Justification of War and International Order*. Edited by: Lothar Brock and Hendrik Simon, Oxford University Press (2021). © the many contributors.
DOI: 10.1093/oso/9780198865308.003.0019

moment in time are simultaneously influenced by, and influence, a specific international normative order through their political practices (see also Simon and Brock, Introduction in this volume).

The underlying argument put forward here suggests that ECOWAS' intervention and the justifications which accompanied it were clearly influenced by an international normative order where sovereignty is the key constitutive norm and non-intervention the main regulative norm. Yet the fact that ECOWAS also violated the very same norms through their intervention created a precedent for future regional interventions and thus also influenced the international normative system.

The chapter is organized as follows. A section 2 introduces the theoretical framework, consisting of three analytical concepts: *interpretive community*, *domestic community*, and the *logic of representation*. A section 3 provides a brief background to Liberia's civil war and the regional organization ECOWAS to contextualize the following analysis. A critical discourse analysis of ECOWAS' justifications of the intervention as well as the responses it evoked follows in section 4. Finally, in the concluding part, a discussion is provided of how the justifications for the intervention and the responses they got were influencing and got influenced by a specific international normative order.

2. Normative International Order Based on Sovereignty

There is no doubt that sovereignty, whether an institution, a norm, a collection of norms, a principle, a rule or a right,[3] has been, and continues to be, fundamental for upholding today's international order. There is also no doubt that trying to define sovereignty generates a certain amnesia about its historical and culturally specific character and hence, freezes it in time.[4] Yet most authors agree that the core features of sovereignty imply that 'there is a final and absolute political authority in the political community... and no final and absolute authority exists elsewhere'.[5] This is thus a form of legitimization of the exercise of power.[6] Its absolute character notwithstanding, sovereignty is also undergoing endless modifications, negotiations, and challenges[7] and thus has always been changing. In spite of the fact that the meaning and practice of sovereignty has varied historically over time, it has always been claimed as an absolute

[3] See e.g. F.H. Hinsley, *Sovereignty* (2nd ed., 1986); Walker, 'State Sovereignty and the Articulation of Political Space/Time', 20 *Journal of International Studies* (1991) 445 ; Werner and De Wilde, 'The Endurance of Sovereignty', 7 *European Journal of International Relations* (2001) 283; Sørensen, 'Sovereignty: Change and Continuity in a Fundamental Institution', 47 *Political Studies* (1999) 590; Finnemore and Sikkink, 'International Norm Dynamics and Political Change', 52 *International Organization* (1998) 887, at 891.

[4] See e.g. Biersteker and Weber, 'The Social Construction of State Sovereignty', in T.J. Biersteker and C. Weber (eds), *State Sovereignty as Social Construct* (1996) 4; Walker above note 3, at 450.

[5] Hinsley, above note 3, at 1.

[6] De Wilde and Werner, above note 3, at 290.

[7] G. Simpson, *Great Powers and Outlaw States: Unequal Sovereigns in the International Legal Order* (2004), at 11; Biersteker and Weber, above note 4, at 11; Bartelson, 'The Concept of Sovereignty Revisited', 17 *The European Journal of International Law* (2006) 463 ; De Wilde and Werner, above note 3, at 283–313; Lake, 'The New Sovereignty in International Relations', 5 *International Studies Review* (2003) 303, at 308; Onuf, 'Intervention for the Common Good', in G.M. Lyons and M. Mastanduno (eds), *Beyond Westphalia: State Sovereignty and International Intervention* (1995) 43.

condition.[8] Sovereignty is therefore closely related to the regulative rule of the non-intervention principle, according to which states should refrain from interfering in other states' internal affairs. Even this rule has changed significantly over time and has been harshly contested during the last few years in the context of humanitarian interventions.[9] Together, sovereignty and non-intervention have constituted the core constitutive and regulative norms which have structured the international normative order over the past centuries. It is therefore against these two norms that legitimation practices and justifications for interventions are to be analysed.

In order to analyse justifications for interventions it is necessary to examine the discursive and political practices which take place between international actors. Three analytical concepts are used here to do this: *interpretive community*, *domestic community*, and the *logic of representation*. An *interpretive community* is a community created around a political or diplomatic action with the aim of interpreting and/or judging that action according to prevailing norms.[10] For example, when a country intervenes in another state, it usually justifies its action vis-à-vis a specific actor in the international community, such as the UN, the EU, or its close neighbours. To put it differently, either implicitly or explicitly, justifications of interventions address a specific interpretive community that then reacts to the action. The creation of an *interpretive community* occurs in the process of the actual justifications and contestations and is therefore something that is mutually constructed by both the acting actor and the receiving actor (see also the concept of 'communication communities' in Simon's and Brock's Introduction and Tischer's chapter in this volume). In the issue area dealt with here, the idea of an interpretive community represents the instance in which a state or a group of states reacts to an intervention by opening up a political discourse on its admissibility and consequences.[11] The act can either be understood as legitimate or illegitimate depending on the community interpreting the action, which is why the actors have an interest in addressing a community that is likely to accept its deeds.[12] This is to say that the *interpretive community* is not necessarily bipartisan. However, it evaluates the practices of states or of other organizations in the light of generally accepted norms.

The concept of *domestic community* is connected to the people in a state and asserts that, in modern international relations, the people are the true subjects of sovereignty. Exactly who these people are remains however a question to be resolved. Like the *interpretive community*, the *domestic community* is never really fixed—it is continually constructed and reconstructed in a discourse. The boundary of who is in and who is out is a function of the state's discursive authority in ambiguous situations to impose

[8] Lake, above note 7, at 308.

[9] Biersteker, 'State, Sovereignty and Territory', in W. Carlsnaes, T. Risse, and B.A. Simmons (eds), *Handbook of International Relations* (2002), at 163; Sørensen, 'Sovereignty: Change and Continuity in a Fundamental Institution', in R. Jackson (ed), *Sovereignty at the Millennium* (1999) 168, at 175.

[10] The notion of interpretive community is both borrowed from C. Weber, *Simulating Sovereignty, Intervention, the State and Symbolic Exchange* (1995), at 4f., and from S. Fish, *Is There a Text in the Class? The Authority of Interpretive Communities* (1980).

[11] Weber, above note 10, at 4f.

[12] Ibid., at 15; see also Delcourt, 'Pre-emptive Action in Iraq: Muddling Sovereignty and Intervention?', 20 *Global Society* (2006), at 54–56.

fixed and stable meanings, and thus separate the insiders from the outsiders.[13] State practices are often justified by indicating that they act on behalf of the *domestic community*, or at least of a part of the domestic community.[14] If the state loses this control over who it represents and can no longer make credible claims as to who its citizens are, there is a risk of competing claims to sovereignty by intervening powers, acting on behalf of those whom they declare to stand for the people.[15] Of interest here is the actor's designation of this community, which defines on whose behalf the action or the position is taken. There are hence competing claims to speak for the sovereign and the issue of who speaks for whom and in what context, becomes a crucial element in an analysis of how sovereignty is interpreted.

Finally, for the *domestic community* to function as a tool in discourses it needs to be understood as representing something more than just a constructed concept. A government represents its people; its domestic community, through political channels. This symbolic representation can be tied to the *logic of representation*.[16] According to this concept, language is always tied to an empirical referent that is the basis for speech. For the *logic of representation* to work it is necessary for the signifier to refer to the signified.[17] Applying this to the idea of a domestic community it is possible to see the community as the signified to which a state government / signifier refers when it wants to demonstrate the source of its sovereign authority.[18] However, as the boundaries of the domestic community are unstable, the question of who constitutes the domestic community remains an issue that needs to be solved for the logic of representation to function. If there is no actual domestic community that can function as the signified, it must be produced.[19] The intervening agent is then referring to parts of a domestic community or is simply constructing a domestic community that fits into the legitimizing discourse to justify the action.[20] All three analytical tools introduced in this section are bound together by inter-subjectivity understood as a mutual understanding of practices and interactions.[21] In the following section, the context for ECOWAS' intervention in Liberia will be presented by referring to these analytical tools.

3. Liberia's Civil War and ECOWAS' Decision to Intervene

The Liberian civil war erupted when Charles Taylor, a former official in the government of the dictator Samuel Doe, launched an insurgence from Côte d'Ivoire to Liberia

[13] Doty, 'Sovereignty and the Nation: Constructing the Boundaries of National Identity', in T.J. Biersteker and C. Weber (eds), *State Sovereignty as Social Construct* (1996) 121, at 122.
[14] Weber, above note 10, at 13.
[15] Biersteker, above note 9, at 168f.; Delcourt, above note 12, at 47–67; Weber, above note 10, at 14.
[16] Weber, above note 10, at 6f.
[17] F. de Saussure, *Cours de linguistique générale* (1972) 97; Weber, above note 10, at 6f.
[18] Weber, above note 10, at 6f.
[19] Ibid., at 7f.
[20] For an example of the problem of deciding who the domestic community is, see Delcourt, above note 12, at 52–54.
[21] See e.g. Adler, 'Seizing the Middle Ground: Constructivism in World Politics', 3 *European Journal of International Relations* (1997) 319, at 327.

on Christmas Eve 1989.[22] Shortly thereafter, the rebel movement, National Patriotic Front of Liberia (NPFL), grew in numbers with Liberians who had been disaffected with the extreme authoritarian and brutal Doe regime. Taylor's movement soon split into two groups after internal disputes, creating an Independent National Patriotic Front of Liberia (INPFL) led by Prince Johnson[23] while the third major faction to the conflict was the Armed Forces of Liberia (AFL), elements of the national army who supported the Doe regime.[24]

A few months into the conflict, when it was clear that neither the US nor the UN or the Organization of African Unity (OAU) were going to intervene,[25] the West-African organization ECOWAS with Nigeria in the lead, took the disputed decision to deploy a monitoring force, ECOMOG (ECOWAS Ceasefire Monitoring Group), to Liberia to put an end to the conflict and rescue foreign nationals from the member states that were trapped in Liberia.[26] Despite their initial objective as a neutral peacekeeping force, ECOMOG soon became one of the warring factions as it cooperated from the start with both the INPFL and AFL against Taylor.[27] Yet already in September 1990, Taylor had taken over most of Liberia and over the following years, up to the elections in 1997 Taylor was in charge of approximately 90 per cent of Liberia.[28] A UN Observer Mission (UNOMIL) of approximately 400 observers was deployed in 1993 to support ECOMOG. However, despite the international presence, the period from 1990 until the elections in 1997 saw the state basically collapse, with warring factions multiplying despite several ceasefire agreements and fourteen peace accords.[29]

The fragile peace that came with the end of the elections in 1997 lasted only for a few years. Taylor's regime quickly became known for its brutalities and abuse of children as soldiers. Corruption remained widespread and the country did not experience the expected upsurge in the economy. Shortly after the elections, new factions started to evolve with former participants from the previous rebel groups and in 2000–01 the conflict started over again. In August 2003, after increasing internal and external pressure, Charles Taylor finally stepped down to give room for a transitional government.

[22] Bøås, 'The Liberian Civil War: New War/Old War?', 19 *Global Society* (2005) 73, at 80; Howe, 'Lessons of Liberia: ECOMOG and Regional Peacekeeping', 21 *International Security* (1996) 145, at 149.

[23] Vogt, 'The Involvement of ECOWAS in Liberia's Peacekeeping', in E. Keller and D. Rothchild (eds), *Africa and the New International Order* (1996) 149, at 165.

[24] Howe, above note 22, at 149.

[25] Ibid., at 150; Wippman, 'Enforcing the Peace: ECOWAS and the Liberian Civil War', in L. Fisher Damrosch (ed.), *Enforcing Restraint: Collective Intervention in Internal Conflict* (1993) 156, at 181.

[26] See e.g. Vogt, above note 23; F. Aboagye and A.M.S. Bah (eds), *A Tortuous Road to Peace, The Dynamics of Regional, UN and International Humanitarian Interventions in Liberia* (2005); Gershoni, 'War without End and an End to a War: The Prolonged Wars in Liberia and Sierra Leone', 40 *African Studies Review* (1997) 55; M. Déme, *Law Morality and International Armed Intervention: The United Nations and ECOWAS in Liberia* (2005); Mortimer, 'ECOMOG, Liberia and Regional Security in West Africa', in Keller and Rothchild, above note 23, 149; K. van Walraven, *The Pretence of Peace-keeping: ECOMOG, West Africa and Liberia (1990–1998)* (1999).

[27] Vogt, above note 23, at 166; Mortimer, above note 26, at 152; Howe, above note 22, at 154; Wippman, above note 25, at 158.

[28] See e.g. Harris, 'From "Warlord" to "Democratic" President: How Charles Taylor Won the 1997 Liberian elections', 37 *The Journal of Modern African Studies* (1999) 431, at 432.

[29] Déme, above note 26, at 65; Mortimer, above note 26, at 155–57.

4. Examining ECOWAS' Justifications for the Intervention

ECOWAS was established in 1975 to increase the economic cooperation in the region by eliminating trade barriers and encouraging free movement of goods, persons, capital, and services.[30] The involvement of ECOWAS in the conflict was driven by the strongest member of the community, and of the region: Nigeria.[31] In May 1990, on a Nigerian initiative, ECOWAS added a new institution to its organization: the Standing Mediation Committee (SMC) entrusted with the mission of finding a settlement to the Liberian crisis.[32] This group was made up of five member states designed by the Authority of ECOWAS. In 1990, the SMC was made up of Nigeria, Ghana, Gambia, Togo, and Mali.[33] After both informal and formal talks failed between the parties to the conflict and the Doe regime still refused to resign, the decision to establish ECOMOG was taken in SMC.[34]

4.1 Internal Reasons for Intervention

The limited attention paid to Liberia's conflict by the international community spurred the regional organization to take matters in their own hands.[35] The OAU remained remarkably quiet concerning the conflict and let ECOWAS lead the different conflict resolution attempts. When the decision to intervene was taken, the OAU attended as observer and maintained the support to ECOWAS throughout the intervention. The UN Security Council thought the conflict should be solved internally and declined to even discuss the conflict until long after the decision to intervene was taken in ECOWAS.[36] This was partly due to the opposition of the Ivory Coast, Ethiopia, and Zaire, who were in the Security Council and did not want to bring the conflict up on the agenda for fear of creating a precedent.[37] There was indeed disagreement within ECOWAS as to whether an intervention was feasible and moreover legitimate. Some members accused the SMC of over-extending its mandate by establishing an intervention force, arguing that it did not have the authority to deploy a military force. Burkina Faso and Togo openly condemned the intervention as an intrusion in the internal affairs of a sovereign country with support from Ivory Coast.[38] Nigeria stood on the opposite side and was the strongest driving force for the intervention.

[30] K. van Walraven, above note 26, at 9f.

[31] S. Ellis, *The Mask of Anarchy: The Destruction of Liberia and the Religious Dimension of an African Civil War* (1999), at 31–74.

[32] Howe, above note 22, at 151;. Mortimer, above note 26, at 151; Wippman, above note 25, at 164.

[33] For more details on the constitution of ECOWAS different institutions, see van Walraven, above note 26, at 14–18.

[34] Ibid., at 15.

[35] Gershoni, above note 26, at 57; Howe, above note 22, at 151; 'Waging War to Keep the Peace: The ECOMOG Intervention and Human Rights', 5 *Africa Watch* (1993), at 5; Wippman, above note 25, at 165.

[36] Wippman, above note 25, at 159, 165.

[37] Déme, above note 26, at 12; van Walraven, above note 26, at 40.

[38] Vogt, above note 23, at 166; Déme, above note 26, at 12.

This division was, however, not only about whether the intervention was legitimate or not,[39] but also about the member states' national interests. Ivory Coast and Burkina Faso were both known supporters of Taylor's NPFL and Nigeria's President Babangida had close personal ties with President Samuel Doe at the time of the conflict's eruption.[40] In addition, Nigeria had attempted to play the role of the regional leader since the end of its own civil war and Liberia proved to be the test of this role.[41] An additional motivation for the intervention was the fear among many of the ECOWAS states that a popular overthrow of a military dictatorship in a neighbouring state might prove to be contagious and thus pose a threat to their own undemocratic regimes.[42]

4.2 Official Reasons and the Legal Basis for the Intervention

Officially, ECOWAS brought up several different justifications for the intervention at different times, which made it hard to interpret the action from a legal perspective. What also posed a problem for legal analyses was the fact that the decision to intervene was taken by the SMC, a body which had been added only months before the intervention and which allowed only five of the member states to forestall the decision which was taken on 7 August 1990.[43] The SMC group stated that it was deeply concerned about the situation in the country and the plight of foreign nationals.[44] After dwelling at length on the destruction and massacres characterizing the country, the SMC concluded:

> The result of all this is a state of anarchy and the total breakdown of law and order in Liberia. Presently, there is a government in Liberia which cannot govern and contending factions which are holding the entire population as hostage.[45]

As background for the decision by ECOWAS to intervene, the SMC expressed the fear that the situation would generate massive numbers of refugees and spill over into neighbouring countries.[46] The member states of ECOWAS

> Assume their responsibility of ensuring that peace and stability is maintained within the sub-region and in the African Continent as a whole, for they believe that the tragic situation in Liberia poses a threat to international peace and security . . . [SMC] decided to take the following immediate actions aimed at restoring peace and stability in Liberia.[47]

39 Aboagye and Bah, above note 26, at 9.
40 Déme, above note 26, at 49; Harris, above note 28, at 434; van Walraven, above note 26, at 48.
41 Africa Watch, above note 35, at 168f.; Mortimer, above note 26, at 162.
42 Wippman, above note 25, at 181.
43 Howe, above note 22, at 151; Mortimer, above note 26, at 162; van Walraven, above note 26, at 151.
44 ECOWAS Standing Mediation Committee, *Final Communiqué of the First Session,* Decision A/DEC.1/8/90 (7 August 1990).
45 ECOWAS Standing Mediation Committee, Banjul, Republic of Gambia, *Final Communiqué of the First Session* (7 August 1990).
46 Ibid., Decision A/DEC.1/8/90.
47 ECOWAS Standing Mediation Committee, above note 45.

In sum, the formal reasons given for the intervention were both focusing on humanitarian issues and on the assessment of the conflict as a threat to international peace and security, almost in the same terms as the UN Charter, Chapter VII Article 42. Indeed, the communiqué was addressed directly to the UN, calling upon the entire international community, 'and in particular the permanent members of the United Nations Security Council to support this humanitarian and political action by ECOWAS in the interest of the African people as a whole, and for the maintenance of international peace and security'.[48]

ECOWAS deliberately chose to address the UN publicly, thereby establishing whom they targeted as a potential *interpretive community*. This move was first presented as a *political* action only to be contradicted two days later in a letter to the UN Security Council by the Nigerian delegation which instead insisted on the neutrality of ECOWAS and the *non*-political character of the intervention:

> The ECOWAS Cease-fire Monitoring Group (ECOMOG) is going to Liberia first and foremost to stop the senseless killing of innocent civilian nationals and foreigners, and to help the Liberian people to restore their democratic institutions. ECOWAS intervention is in no way designed to save one part or punish another.[49]

The ECOWAS Regulations for ECOMOG also confirmed that:

> It is the duty of members of the Group to respect the laws and regulations of the Host State and to refrain from any activity of a *political character* in the Host State or other action incompatible with the international nature of their duties.[50]

There was thus an obvious contradiction in the way that ECOWAS both characterized the intervention as a 'political action' in the final communiqué and simultaneously declared that the organization was to refrain from any 'activity of a political character' as claimed in the Nigerian letter to the UN. This confusion in the description of the intervention could be due to a negligence slip on the part of ECOWAS member states in the first description, which latter was corrected in the second letter to make sure that the *interpretive community*, in the shape of the UN, was not discouraged by the 'political' character of the decision to intervene.

As to the identification of a *domestic community*, it is striking that while the documents mention the Liberian people several times in the justifications for the intervention, this could hardly be considered as reference to one people at the time of the intervention as the factions divided the people along ethnic demarcation lines. Samuel Doe, although still a president in theory, did not have any actual power and his legitimacy as a representative of the Liberian people had been in doubt ever since his violent arrival at the presidency and moreover since the fraud elections in 1985. It is also

[48] Ibid.

[49] UN Security Council, *Letter dated 9 August 1990 from the Permanent Representative of Nigeria to the United Nations addressed to the Secretary General, S/21485* (10 August 1990).

[50] Emphasis added, *ECOWAS Regulations for the Cease-fire Monitoring Group (ECOMOG) in Liberia, Republic of Gambia* (13 August 1990).

difficult to regard Charles Taylor as a true representative of the people although he did, at the time of the intervention, control most of the country.[51]

The *domestic community* remained therefore difficult to define, even more so as ECOWAS brought up different interpretations of whom the people were. On at least two occasions they referred to Liberia as in a state of anarchy with a total breakdown of law and order and with a government that could not govern.[52] This implies that there is no true referent, symbolic or political, for the people of Liberia and thus that the logic of representation was flawed. ECOWAS also claimed that different factions held the population hostage, which seems to mean that there was no such thing as a free Liberian people at the time of the decision. However, only a week after the intervention, during the National Conference convened by ECOMOG, a similar description regarding the entire population as a hostage of a government which did not govern *preceded* a statement by the Conference declaring that:

> Acting on the inherent sovereign rights of the Liberian people to make laws for their governance and alter same [sic] when their safety and happiness so require, consistent with the spirit and intent of the Constitution of Liberia ...[53]

This same declaration came with the decision to suspend parts of the Constitution to establish an interim government.[54] In this pronouncement it seems as if there *is* a free Liberian people capable of taking the decision to suspend its own Constitution in order to set up a new interim government, although in the preceding statement, the whole population is considered as having been deprived of its status as a people. Who the actual *domestic community* is and who is representing this community remains therefore vague in ECOWAS' discourses, just as the question remains unanswered on whose behalf ECOWAS was intervening.

There were hence many different reasons brought up to justify the intervention. Remarkably, what was *not* mentioned as a motivation for the intervention was President Samuel Doe's letter in which he had asked for an intervention by an ECOWAS peacekeeping force less than a month before the actual intervention.[55] However, even if this letter had been mentioned as a motivation for the intervention, it was addressed to the 'wrong' body for that kind of decision. Instead of addressing it

[51] Harris, above note 28, at 431–55.

[52] ECOWAS Standing Mediation Committee, Banjul, Republic of Gambia, *Final Communiqué of the First Session* (7 August 1990); ECOWAS Standing Mediation Committee, *Decision A/DEC.1/8/90* (7 August 1990).

[53] Interim Government of National Unity of Liberia, *Final Communiqué of the National Conference of All Liberian Political Parties, Patriotic Fronts, Interest Groups and Concerned Citizens, Republic of Gambia* (29 August 1990).

[54] M. Weller (ed.), *Regional Peace-Keeping and International Enforcement: The Liberian Crisis* (1994), at xxii.

[55] 'The Responsibility to Protect: Research Bibliography and Background', *Supplementary volume to the report of the International Commission on Intervention and State Sovereignty* (2001), available at <http://www.idrc.ca/openebooks/963-1/> (last visited 29 December 2019); 'Letter addressed by the President Doe to the Chairman and Members of the Ministerial Meeting of the ECOWAS SMC, 14 July, 1990', in Weller, above note 54.

to the Authority, the highest decision-making body in ECOWAS, Doe addressed the letter to the SMC, contrary to what is declared in the ECOWAS Protocol on Mutual Assistance in Defence (PMAD).[56] The PMAD, however, also provided a shaky basis for intervention, as interventions, according to the PMAD, only can take place in inter-state conflicts or internal ones that are sustained and maintained from outside.[57] This actually placed ECOWAS in an awkward position as on the one hand, its member states were sustaining the conflict by actively taking sides, and on the other hand, they also claimed that their support could serve as a basis for ECOWAS' intervention as a peacekeeping mission.[58]

Indeed, the legal basis for the intervention was not very solid. The ECOWAS treaty from 1975 did not provide a regional security mechanism for internal conflicts, nor did the 1978 Protocol on Non-Aggression.[59] ECOWAS intervened claiming to act as a neutral peacekeeping force, although it was clear from the beginning that this was not the case. In addition, this claim did not fall under the provision of the PMAD. To base their intervention on PMAD meant to support an existing member of the community, in this case, Doe. ECOWAS asserted that this was not the case.[60] Not only did ECOWAS assure its neutrality vis-à-vis Doe, on several occasions they also declared Liberia to be in a state of anarchy without government. This ruled out the PMAD option for justification. The decision to intervene in Liberia did therefore not have a legal basis in the ECOWAS regulation. In summary, ECOWAS's justification for intervention was based on vague and contradictory claims about who the domestic communities were and thus in whose name the intervention took place. The communities that were identified were based on a flawed logic of representation, which undermined the credibility of the statements and justifications.

5. The Reactions from the International Community

One of the main reasons for the space left for wide-ranging decisions by the regional organization was the absence of action and reaction from the international community at the time of the eruption of the conflict[61] and the intervention. Despite a clear reach-out to the UN, the international community was rather slow in constituting itself as an *interpretive community*. Its members remained vague about the admissibility and desirability of the intervention for a long time.

[56] For a more detailed explanation concerning the rules and provisions of ECOWAS on this matter, see Déme above note 26, at 60–63; van Walraven, above note 26, at 27f.

[57] See ECOWAS Protocol on Mutual Assistance in Defence; Déme above note 26, at 60–63; van Walraven, above note 26, at 27f.

[58] Both Burkina Faso, Ivory Coast, and to a certain degree Nigeria were supposedly supporting different factions in the conflict.

[59] IDRC, above note 55.

[60] Wippman above note 25, at 188.

[61] Gershoni, above note 26, at 57; Howe, above note 22, at 151; Africa Watch, above note 35, at 5; Wippman, above note 25, at 165.

5.1 The United States

Many considered the United States to be the natural peace-maker for Liberia due to the two countries' historical ties. Yet, despite calls from senior Liberians for a US intervention and an open request from President Samuel Doe, the United States argued that an African problem needed an African solution and refused to intervene.[62] The United States merely encourage ECOWAS by landing a Marine contingent outside of Monrovia's coast with the aim of protecting US and foreign nationals.[63] The United States also pushed Senegal to contribute troops to ECOMOG to avoid an all-Nigerian force.[64] This kind of involvement continued throughout the conflict even though senior US officials asserted that ECOWAS was far from neutral.[65] The support of ECOWAS proved to be ambivalent, as the United States continually refused to recognize the interim government while maintaining their support for ECOWAS.

That the United States had a problem of deciding on a representative of the people of Liberia as well became explicit during a briefing by Richard Boucher at the US State Department in September 1990:

> Q: Do we recognise anybody as the leader of the country at this point?
>
> Mr. Boucher: The situation is very confused. The two rebel groups and elements of the remaining forces of Samuel Doe are all claiming to be the legally constituted government of Liberia. In addition the All-Liberian Political Conference held under the auspices of the Economic Community of West African States in Gambia last week announced the formation of an Interim Government. We continue to be in touch with the various factions in the Liberian conflict.
>
> Q: The answer is we don't recognise anybody at this particular ...
>
> Mr. Boucher: It's not for us to choose, we've always said that.
>
> Q: Does that mean that we do not recognise a government at this point in Monrovia?
>
> Mr. Boucher: We—the question of recognition gets into complicated legal issues that we go around at this podium many many times and that I always explain in language that lawyers probably wouldn't approve because you can get a more accurate reading of this. We have an embassy there, we have an embassy in Monrovia. The question of recognition doesn't really arise when there's a change in government. We deal with governments, we work with governments as they take over if they have control of territory and the ability to enter into international obligations and things like that. At this point the situation as far as who's in charge, who's in control in Monrovia is very confused....

62 Speaking to foreign journalists in Monrovia on 11 June, President Doe had 'expressed hope that the USA will join Liberia in searching for peace in the current conflict, BBC Monitoring Report, 14 June 1990', in Weller, above note 54, at xxii; Wippman, above note 25, at 159, 165; van Walraven, above note 26, at 111.

63 President of the United States, 'Communication Concerning Armed Forces sent to Monrovia, Liberia (6 August 1990)', in Weller above note 54, at 65.

64 Mortimer above note 26, at 155; van Walraven, above note 26, at 46.

65 Africa Watch, above note 35, at 21.

> Q: You mentioned you back the ECOWAS effort. If you back the ECOWAS effort why don't you back the government that they've put forth?
>
> Mr. Boucher: We continue to support the objectives of the ECOWAS initiatives. Those objectives are a cease-fire in Liberia, the formation of an Interim Government and the holding of free and democratic elections.[66]

This discussion boils down to one of the most important questions concerning interventions in general: which faction or which party is understood as representing the sovereigns? The United States acknowledged the fact that there were at least two, possibly three, different parties to the conflict, all claiming to be the legally constituted government of Liberia. ECOWAS' attempt to set up yet another approach to represent the people of Liberia means that the claims of representation are too many and the logic of representation is muddled up. From a theoretical point of view, the United States was not an ideal part of an interpretive community: its reaction was late and ambiguous, and the involvement was weak. At the same time, their approval of the intervention and their financial help to add foreign troops gave ECOMOG more space to act independently.

5.2 The UN

The UN remained quiet about the Liberian conflict for a long time, despite several public calls and communications from ECOWAS asking for some sort of reaction. In January the Security Council finally made a statement, commending the efforts made by ECOWAS to promote peace and normalcy in Liberia and asking for all the parties to the conflict to fully cooperate with ECOWAS.[67] During the meeting in the UN Security Council, the Liberian representative raised the issue of 'reviewing and perhaps reinterpret(ing), the Charter, particularly its provision which calls for non-interference in the internal affairs of Member States', as it hampered the effectiveness of the Council to maintain its objective of international peace and security.[68] This way the question of sovereignty was put on the table. This comment did, however, not evoke a response from the Security Council. Yet, the resolution in January was the first of several statements issued by the Security Council in support of ECOWAS that later would be interpreted as *ex post facto* authorization of the intervention.[69]

It is noteworthy that the UN did not pronounce itself on ECOWAS actions until after all the warring factions had, tenuously, agreed to sign a cessation of hostilities agreement at the end of November, thereby placing ECOMOG's operations on a footing of peacekeeping by consent.[70] This is however not the same thing as a late

[66] 'United States State Department, Regular Briefing by Richard Boucher (11 September 1990)', in Weller, above note 54, at 96.

[67] *Note by the President of the United Nations Security Council, S/22133* (22 January 1991).

[68] *UN Security Council, Provisional Verbatim Record of the 2974th Meeting, S/PV.2974* (22 January 1991).

[69] The reason for the importance of this discussion is that a UN authorization of the intervention would have made the action legal and therefore also largely uncontested in the international community, Aboagye and Bah, above note 26, at 2.

[70] Weller, above note 54, at xxiv.

authorization. Some observers claim that the intervention has been interpreted as fitting under Chapter VIII in the UN Charter, classifying it as a regional organization undertaking enforcement action to address threats to international peace and security.[71] Such an action is only valid provided that the UN Security Council authorizes it,[72] which clearly was not the case at the time of the intervention. It seems instead as if the UN deliberately avoided taking a decision on the intervention until after it had taken place.[73] This way, the most difficult question remained unanswered when it would have mattered the most. Hence, the UN legitimized a peacekeeping force that was already in action. The resulting observer mission from ECOWAS was a disappointment for all parties involved.

5.3 The Actual Intervention and its Consequences

The ECOMOG troops intervened on 24 August 1990. Already before the intervention, Taylor's NPFL had clearly declared that they were against the intervention and would attack the troops.[74] With a small and badly prepared army, ECOMOG immediately stood under fire when they intervened. Without enough weapons or knowledge about the different areas in the country, ECOMOG rapidly teamed up with two of the factions in the conflict, Prince Johnson's breakaway faction from the NPFL, the INPFL (Independent National Patriotic Front of Liberia) and what was left of Doe's army, the AFL (Armed Forces of Liberia) against Taylor's NPFL.[75] This decision undoubtedly proved that their 'peacekeeping' force was not neutral. It was also a decision to regret, as the INPFL used ECOWAS' alliance with the AFL in order to capture and murder President Doe on 9 September 1990, in ECOMOG's headquarters,[76] an event which seriously put ECOMOG's capacity as a neutral peacekeeping force in doubt. Despite this failure, ECOMOG continued to cooperate with both AFL and INPFL against Taylor's NPFL.[77] The troops were poorly equipped and became soon known by the acronym 'Every Car and Moving Object Gone' for their corruption and misbehaviour in the country.[78] Despite the difficulties on the ground ECOMOG did attempt to establish an Interim Government of National Unity (IGNU) already in October 1990, which was rejected by Taylor who formed his own 'government' that had control of most of Liberia.[79]

ECOMOG's attempt to keep, or rather to enforce, peace proved to be problematic. Out of the fourteen peace accords or ceasefire agreements initiated by ECOWAS, thirteen failed. Not until 1996 did the second Abuja accord put an end to the conflict;[80]

[71] O. Corten, *Le droit contre la guerre: L'interdiction du recours à la force en droit international contemporain* (2008), at 569–79.

[72] UN Charter, Chapter VIII, Article 53, Déme, above note 26, at 45f.

[73] Wippman above note 25, at 185.

[74] Ibid., at 158.

[75] Bøås, above note 22, at 81; Howe, above note 22, at 154; Vogt, above note 23, at 166.

[76] Howe, above note 22, at 154.

[77] van Walraven, above note 26, at 54.

[78] Tuck, '"Every Car Or Movable Object Gone" The ECOMOG Intervention in Liberia', 4 *African Studies Quarterly* (2000) 1.

[79] van Walraven, above note 26, at 52.

[80] Déme, above note 26, at 65.

an end which, contrary to ECOMOG's initial objective, installed Taylor as the new president after elections in 1997.[81] The ECOMOG initiative has received mixed reviews by observers, but a large majority of the research made on the subject argues that ECOMOG was a failure that prolonged, rather than ended, the civil war in Liberia by becoming an active party to the conflict.[82]

6. Concluding Discussion—Blurring the Boundaries of Sovereignty and Intervention

The many different justifications for the intervention used by ECOWAS are evidence of how the organization tried to play into and replicate the role of the UN and its peace missions while at the same time also evoking the notion of humanitarian intervention. Firstly, ECOWAS claimed to take on the responsibility of restoring international peace and stability in the region, clearly inspired by Article 52 in Chapter VIII of the UN Charter to justify the transgression of the non-intervention principle. Secondly, by claiming to save the lives of innocent civilians, they referred to the concept of humanitarian intervention. ECOWAS thus attempted to prove that the intervention corresponded to international norms in spite of the fact that the intervention undermined one of the key norms regulating international relations: the norm of non-intervention.

The actual intervention did not live up to the rules laid down by the Charter—there was no UN authorization for it at the time of the intervention and it was not a neutral peacekeeping operation. Furthermore, it can be argued that contrary to the stated aim, the intervention actually had disastrous humanitarian consequences, given that it arguably prolonged the conflict, rather than shortened it. However, ECOWAS finally got approval for the intervention from the two different interpretive communities it addressed in the justifications: the US and the UN. Both the clear address to, and the reactions from the different interpretive communities are interesting not only because they are so vague and late in their comments but also because they encourage an intervention at a time when non-interference in domestic affairs is a prevailing norm in international relations.

This encouragement is especially surprising given ECOWAS' contradictory claims about the domestic community and the lack of a clear logic of representation that could justify an intervention on behalf of a coherent Liberian domestic community. Indeed, as the population of Liberia was divided along ethnic lines, and by supporting different armed groups at the time of the intervention, claims of intervening on behalf of *a* domestic community were clearly erroneous.

The tardiness and ambiguities of the United States' and the UN's reactions were not particularly helpful as interpretive communities. They were the most important actors from whom to get an acceptance, as they were both the most powerful (the US) and the most legitimate (the UN) actors. In Finnemore and Sikkink's words, they were the critical actors needed for the establishment of a norm.[83] Whether ECOWAS acted as

[81] Harris, above note 28, at 431–55.

[82] Vogt, above note 23; Gershoni, above note 26, at 55–76; van Walraven, above note 26; Bøås, above note 22, at 73–88.

[83] Finnemore and Sikkink, above note 3, at 901.

a 'norm entrepreneur' or not is another issue, but what is clear is that humanitarian interventions at this particular moment were not common and definitely not legitimate per se. The climate in international relations for humanitarian interventions was more sceptical than today, when the Responsibility to Protect attempts to ease the tension between sovereignty and the protection of human rights. Despite the ambiguous starting point, both interpretive communities encouraged the intervention after an initial period of hesitation. Their acceptance of the intervention meant that ECOWAS' intervention was not only shaped by the international normative order at the time, but also contributed to shaping it by creating a precedent for the toleration of intervention at the regional level under certain circumstances. This way the relationship between Chapters VII and VIII of the UN Charter was informally changed so as to open up the space for quasi-autonomous action by regional systems of collective security. Yet, the outcome left open the issue of representation. Regional organizations may be closer to the local conditions than global organizations. However, in the end, it remained an open question in whose name external actors were intervening—in the name of the international community, in the name of the regional community, or in the name of all those who constitute the collective subject of self-determination.

20

Humanitarian Intervention: Justifying War for a New International Order

Beate Jahn

1. Introduction

The concept of 'humanitarian intervention' captured the political imagination not just of academics but also of political actors and broader publics over the last two and a half decades. In a very short space of time, the term humanitarian intervention made it into the core of the academic discipline of International Relations as well as other disciplines such as International Law and Political Theory. This was all the more remarkable since the mills of change in academia tend to grind fairly slowly. Only very few concepts—another recent example being 'globalization'—manage such a meteoric rise. Similarly remarkable is the fact that the term humanitarian intervention played as important a role in public political discourse as it did in academia. This sudden prominence of the term humanitarian intervention raises the question: what was in that name? What did this term stand for that accounted for the timing and speed of its impact in academic and public discourse—as well as for its equally rapid decline today?

This question requires a historical investigation establishing the changes that provide such fruitful ground for the development and spread of the term humanitarian intervention. Section 2 of the chapter shows that at the core of the term humanitarian intervention lies the relationship between politics and morality—and that different and at times contradictory accounts of this relationship can be found in the contemporary debate on humanitarian intervention. Two different theories concerning the historical development of this relationship can be identified. One claims that the theory and practice of humanitarian intervention is made possible by a gradual universalization of moral obligations that has begun to challenge the particularist order of the international political system. The other claims that it is the historical separation of morality and politics in theory—rather than in practice—that enables the rise of the concept of humanitarian intervention. These competing accounts raise the question, in other words, whether the term humanitarian intervention stands for genuinely new policies or for old policies under a new name.

In order to answer this question, section 3 of the chapter turns to the analysis of the reflections of a number of classical authors—Francisco de Vitoria, Edmund Burke, Thomas Paine, and John Stuart Mill. It shows that classical theorists regularly contemplated the use of force in order to help citizens of other states. Furthermore, the form and content of these reflections is strongly influenced by the possibilities and limits offered within the contemporaneous international context.

Beate Jahn, *Humanitarian Intervention: Justifying War for a New International Order* In: *The Justification of War and International Order*. Edited by: Lothar Brock and Hendrik Simon, Oxford University Press (2021). © the many contributors.
DOI: 10.1093/oso/9780198865308.003.0020

There is, in conclusion, no evidence of a gradual universalization of moral obligations; instead, interventions to help citizens of other countries have been widely contemplated and practised in the past. Instead, what is new about the concept of humanitarian intervention is its counterposition of politics and morality. And this counterposition is made necessary by an international system based on the ground rule of non-intervention (especially in the political affairs of other states). By denoting a moral exception to illegal political interventions, the term serves to hide the political nature of humanitarian intervention and, ironically, contributes to the separation of politics and morality which it claims to overcome.

2. The Contemporary Debate

According to a widely accepted definition, humanitarian intervention

> is the threat or use of force across state borders by a state (or group of states) aimed at preventing or ending widespread and grave violations of the fundamental human rights of individuals other than its own citizens, without the permission of the state within whose territory force is applied.[1]

This, and similar, definitions point to the main function of the term humanitarian intervention: namely to identify an exception to the general rule of non-intervention at the core of the contemporary international system. What distinguishes humanitarian from other interventions is their moral character: the fact that they are based on the recognition of 'the common humanity that binds us all'[2] and that their aim is to realize universal moral principles in the form of individual human rights. On the basis of this moral universality, the term 'humanitarian intervention' appears to describe an inherently non-partisan and defensible act[3] which is counterposed both to interventions undertaken for partisan reasons (and thus outlawed) and to such fundamental principles of the international political and legal order as sovereignty which are explicitly designed to pursue and defend the interests of *particular* communities.

The relationship between politics and morality thus lies at the core of the concept of humanitarian intervention and it is the historical development of this relationship which is variously interpreted as providing the basis for the sudden rise to prominence of the term humanitarian intervention. One interpretation highlights the moral nature of humanitarian intervention and explicitly counterposes it to the norms and institutions of the international political order. This moral character is derived from the recognition of the moral obligation to come to the rescue of people in need—and theoretically supported by the hypothetical case of a drowning child.[4] The resulting obligation comes into conflict with the political and legal regulation of the international

[1] Holzgrefe, 'The Humanitarian Intervention Debate', in J.L. Holzgrefe and R.O. Keohane (eds), *Humanitarian Intervention. Ethical, Legal, and Political Dilemmas* (2003) 15, at 18.

[2] Tesón, 'The Liberal Case for Humanitarian Intervention', in Holzgrefe and Keohane, above note 1, 93, at 129.

[3] A. Hehir, *Humanitarian Intervention. An Introduction* (2010), at 12.

[4] Wheeler, *Saving Strangers. Humanitarian Intervention in International Society* (2000), at 49.

order by the norm of sovereignty as guaranteed in Article 2(1) UN Charter and the right of non-intervention (Article 2(4) and (7) UN Charter). Thus, the

> 'good international citizen must come to the assistance of the victims of institutionalized cruelty' but in this quest has to 'resolve the tension between legalism and progressivism in a new legal order that alters the relationship between order and justice, citizenship and humanity, and sovereignty and human rights'.[5]

Accordingly, defence of the principle of sovereignty or the right to non-intervention is presented 'as the one doctrine whose origin, design, and effect is to protect established political power and render persons defenseless against the worst forms of human evil'.[6] The failure to recognize interventions with humanitarian outcomes as such attests to the 'moral bankruptcy' of those wedded to principles of sovereignty and non-intervention,[7] and of course to the 'immoral' character of the international order based on sovereignty. What is needed, then, is a 'moral transformation' that makes governments in the West see that humanitarian intervention 'is both morally permitted and morally required'.[8] Respect for, and defence of, human rights through the legalization and practice of humanitarian intervention is thus equated with morality and counterposed to the political and legal principles of sovereignty and non-intervention as standing for particularist interests and hence immoral.

This tension between morality and politics which lies at the core of calls for the institutionalization of humanitarian intervention is seen as the result of historical development. Thus, 'non-interventionism is a doctrine of the past'[9] based on an 'insistence on sharp and morally decisive distinctions between citizens and aliens'.[10] Historically, justice was considered an 'internal norm' of specific (communication) communities[11] and force an 'external rule'—'with acts of generosity toward the foreigner embodying an exception'.[12] In addition, it is sometimes argued that 'violent conflicts', 'acts of genocide', and 'crimes against humanity' were in the past 'considered normal'.[13] These arguments assume, then, either a lack of humanitarian sensitivity in the past and/or the limitation of moral obligations to the boundaries of particular political institutions like the state.

The moral universalism embodied in the concept of humanitarian intervention is thus presented as a new development, partly motivated by the experience of the holocaust.[14] Historically, the tension between politics and morality is the result of a gradual

[5] Linklater, 'The Good International Citizen and the Crisis in Kosovo', in A. Schnabel and R. Thakur (eds), *Kosovo and the Challenge of Humanitarian Intervention: Selective Indignation, Collective Action, and International Citizenship* (2000) 482, at 493.

[6] Tesón, above note 2, at 129.

[7] Wheeler, above note 4, at 296.

[8] Ibid., at 310.

[9] Tesón, above note 2, at 128.

[10] Linklater, above note 5, at 483.

[11] See the contributions by Anuschka Tischer, Arnulf Becker Lorca, and Lauren Benton to this volume.

[12] Elshtain, 'International Justice as Equal Regard and the Use of Force', 17 *Ethics and International Affairs* (2003) 63, at 64.

[13] Müllerson, *Democracy—A Destiny of Humankind?* (2009), at 136.

[14] Wheeler, above note 4, at 302.

universalization of moral principles which was not matched by a similar development of the political and legal principles of the international order. The resulting clash between a universalist morality and a particularist political order thus requires reform of the latter 'in the name of cosmopolitan conventions whose time may have come'.[15]

And yet, what is in theory presented as an opposition between morality and politics appears in practice even to protagonists of humanitarian intervention deeply entangled. The debate shows that politics is actually taken to play a varied and indispensable role in and for humanitarian intervention. Politics, first, is seen as the root of humanitarian disasters.[16] Hence, the 'attempt to align the boundaries of the state and the boundaries of the nation' leads to ethnic cleansing;[17] regime types like 'anarchy' and 'tyranny' result in massive human rights violations;[18] and the inherently violent process of state building leads to humanitarian disasters in the post-colonial world.[19] In all these cases, the causes of systematic human rights violations are seen as 'deeply rooted in the political, economic, and social structures of societies'.[20]

The solution to such problems accordingly lies in the reconfiguration 'of political systems that violate fundamental moral principles'[21]—that is, in the establishment of alternative forms of political organization. Ethnic cleansing calls for the establishment of states 'which are more universalistic and more sensitive to cultural differences'.[22] The solution to anarchy and tyranny lies in the constitution of a liberal state.[23] The solution to the violent processes of state-building is either seen in non-intervention, thus reducing interference which might prolong the process[24] or in 'conditional independence and sub-sovereign solutions of all kinds' which involve, 'in effect, an ongoing imperial or external presence with the military or economic capacity to keep these new entities in line'.[25]

Finally, the means by which these solutions are to be implemented are also decidedly political: they lie, in accordance with the definition of humanitarian intervention, in the 'threat or use of force across state borders',[26] that is, war. And war itself is explicitly recognized as a political tool. 'The real problem is a political one and coercive force remains an extension of politics by other means'; hence many authors argue that 'calling these situations humanitarian intervention only clouds the issue'[27] and use the

[15] Linklater, above note 5, at 493.

[16] Wheeler, above note 4, at 306; Tesón, above note 2, at 96f.

[17] Linklater, above note 5, at 484.

[18] Tesón, above note 2, at 96f.

[19] Ignatieff, 'State Failure and Nation-Building', in Holzgrefe and Keohane, above note 1, 299, at 302; Ayoob, 'Humanitarian Intervention and State Sovereignty', 6 *The International Journal of Human Rights (IJHR)* (2002) 81, at 93;

[20] Wheeler, above note 4, at 306.

[21] Linklater, above note 5, at 486.

[22] Ibid., at 484.

[23] Tesón, above note 2, at 96.

[24] Ayoob, above note 19, at 94.

[25] Ignatieff, above note 19, at 305, 309; Cooper, 'The New Liberal Imperialism' (2002) at 5, available at <http://www.guardian.co.uk/world/2002/apr/07/1/print> (last visited 29 November 2019).

[26] Holzgrefe, above note 1, at 18.

[27] Elshtain, above note 12, at 68; E. Heinze, *Waging Humanitarian War. The Ethics, Law, and Politics of Humanitarian Intervention* (2009); S. Chesterman, *Just War or Just Peace? Humanitarian Intervention and International Law* (2001); Coady, 'The Ethics of Armed Humanitarian Intervention', 45 *Peaceworks* (2002), at 16; De Waal, 'No Such Thing as Humanitarian Intervention. Why We Need to Rethink the "Responsibility to Protect" in Wartime', *Harvard International Review* (21 March 2007).

term 'humanitarian war' instead. In practice, then, systematic violations of human rights are seen as rooted in particular social and political arrangements, addressing them requires alternative political arrangements, and implementing such solutions requires political means.

Politics is here integrated into the concept of humanitarian intervention, albeit functionally differentiated: morality provides the guiding principles while it is the task of politics to translate these principles into practice in particular times and places. Yet this account undermines the strict opposition of politics and morality that is necessary to establish the 'guiding' role of the latter unencumbered by the former. If political considerations and decisions are an integral part of humanitarian intervention, then the latter cannot be depicted as the moral solution to political problems as such. In other words, if both morality and politics play a crucial role in humanitarian intervention it becomes difficult to distinguish humanitarian from other forms of intervention—that is, the term humanitarian intervention does, then, not serve the function for which it was introduced, namely to clearly delineate an exception to the rule of non-intervention. And it is precisely this problem which has given rise to an extended debate on how to identify humanitarian interventions.

The challenge essentially consists in weighing moral against political motivations and considerations and it was widely taken up—without, however, resulting in a satisfactory solution. Some argue, as Hehir notes, that an intervention qualifies as humanitarian if it saves more lives than it takes.[28] Yet, such a cost-benefit analysis of lives saved and lost has two major shortcomings. The first lies in an impossibility to establish how many lives have actually been saved (which is a counterfactual and thus remains speculative). Secondly, this calculation does not take into account the political nature of the cause as well as the solution to massive human rights violations which leads Tesón to argue that 'the loss of lives is not the only indicator of the legitimacy of humanitarian intervention'; instead 'building and restoring democratic, rights-respecting institutions addresses a central cause of the problem' and should thus count as a criterion for humanitarian intervention.[29] Yet, in both these cases the humanitarian character of an intervention could only be established with hindsight—that is, when the killing is finished and bodies can be counted, or when a new political order has been established. For a decision to intervene or support an intervention, these criteria are thus not helpful. This is a problem for all consequentialist[30] definitions of humanitarian intervention.

In order to counter this problem, some authors suggest assessing the motivation of the intervener. Here, albeit very rarely, we find the suggestion that only interventions undertaken for purely altruistic reasons can count as 'humanitarian'.[31] More generally, as Hehir notes, the political considerations and interests of the intervening power are weighed against its motivation to help people in need and it is argued that the moral goal has to outweigh political interests.[32] Apart from the difficulty of finding a

[28] Hehir, above note 3, at 161, 163.

[29] Tesón, above note 2, at 117f.

[30] Wheeler, above note 4.

[31] Miller, 'Humanitarian Intervention, Altruism, and the Limits of Casuistry', 28 *Journal of Religious Ethics* (2000) 3.

[32] Hehir, above note 3, at 19.

measure by which to weigh these different motivations, this task leads straight back to the original problem: that is, it requires a clear distinction between political and moral motivations and thus assumes their separate existence albeit now within the framework of humanitarian intervention. Hence, the original problem has here simply been imported into the concept of humanitarian intervention.

This integration of politics into the concept of humanitarian intervention thus provides the basis for an alternative historical account. If politics is required in order to implement moral principles in concrete cases, then moral principles necessarily become embedded in particular political frameworks—an assumption which in turn makes the divergent historical development of politics and morality impossible (see also Simon's and Brock's Introduction and Chapter 2 by Anthony Lang, Jr in this volume). Yet, some historical change is necessary in order to explain the sudden rise of the concept of humanitarian intervention—and this change has to be identified in the gradual theoretical distinction between politics and morality.

To begin with, political institutions are indeed widely treated as embodying moral principles. Most generally,

> the state as a coercive institution is morally justifiable because in principle, it enables human beings to fulfill their potentialities, by living together according to common rules.... The nonintervention principle is therefore basic to relations between states.... There are moral reasons why a state must be recognized as having rights, in particular the right that outsiders respect its independence and boundaries.[33]

More specifically, there is widespread agreement that humanitarian interventions do not have to be undertaken if they are expected to bring 'substantial harm' to 'fellow citizens'[34] and 'states are not required to sacrifice vital interests ... for the sake of helping others' and even soldiers' lives do not have to be sacrificed 'in large numbers'.[35] There are, to be sure, disagreements on the extent of the sacrifice that can be required of fellow citizens and soldiers.[36] Yet, in all cases the obligations towards its own citizens and soldiers provide the state in general (not just the liberal or modern state) with moral standing. Hence, the universal moral obligation is explicitly recognized as not 'absolute',[37] as 'an imperfect duty'.[38]

Secondly, the debate on humanitarian intervention shows that the same moral principle can be enshrined in different political institutions and practices or, conversely, that one political institution can embody different moral principles. Thus, in the debate on humanitarian intervention, war is on the one hand offered as appropriate means to put an end to massive humanitarian disaster. On the other hand, it is argued that the destructive nature of war undermines its ability to provide the basis for

[33] Nardin, 'The Moral Basis for Humanitarian Intervention', in A.F. Lang, Jr (ed), *Just Intervention* (2003), at 20f.

[34] Elshtain, above note 12, at 74f.

[35] Wheeler, above note 4, at 49.

[36] Elshtain, above note 12, at 75; Cook, '"Immaculate War": Constraints on Humanitarian Intervention', in Lang, Jr, above note 33, at 150f.

[37] Tesón, above note 2, at 127.

[38] Nardin, above note 33, at 23.

peaceful and cooperative politics.[39] Moreover, a right to humanitarian intervention adds to the permissible causes of war and must thus be weighed against the equally moral grounds on which these causes have been strictly limited during the twentieth century.[40] 'International order and stability, international peace and security' are thus judged to be 'fundamental values' that need to be taken into account for any analysis of the costs and benefits of humanitarian wars.[41] And finally, 'humanitarian values are never under greater threat than when states get involved in wars ... War is the biggest threat to human rights' and has historically provided the framework for 'all major cases of genocide and ethnic cleansing'.[42]

What these reflections underscore is the co-constitutive nature of politics and morality which leads to the recognition that there are 'cases where whatever we do we will end up tolerating a violation of *some* fundamental rule'.[43] And, hence, the decision to be made is not one between politics and morality but requires '*moral-political* considerations'[44] or 'a mixture of principle and prudence'.[45] Consequently, what is contested in this debate is neither moral sensitivity nor substantive moral principles or the universal nature of moral obligations. Even proponents of 'humanitarian intervention' explicitly recognize that the moral judgement is essentially 'uncontroversial. For the most part, critics of humanitarian intervention do not disagree with the judgement that the situations ... that call for intervention are morally abhorrent.'[46] Indeed, 'the principles of common morality ... are in fact recognized in most communities and traditions'[47] suggesting that what is at issue in the debate on humanitarian intervention is political judgement rather than morality.

This recognition of the co-constitutive nature of politics and morality underpins an alternative historical account of the changes that have led to the recent prominence of humanitarian intervention. According to this account, moral reasoning traditionally took the form of practical political judgement[48] in the context of concrete cases—thus accounting for the intimate relationship between politics and morality. In the course of the seventeenth century, however, the role of practical political judgement began to be replaced by the elaboration of abstract logical systems of morality.

> General principles and abstract axioms were privileged over particular cases and concrete diversity, and the establishment of rules (or 'laws') that were deemed of permanent as opposed to transitory applicability came to be seen as the task of the theorist.[49]

[39] Young, 'Violence Against Power: Critical Thoughts on Military Intervention', in D.K. Chatterjee and D.E. Scheid (eds), *Ethics and Foreign Intervention* (2003) 251; Miller, 'Respectable Oppressors, Hypocritical Liberators: Morality, Intervention, and Reality', in Chatterjee and Scheid, ibid. 215, at 237.
[40] Coady, above note 27, at 17.
[41] R. Jackson, *The Global Covenant. Human Conduct in a World of States* (2000), at 291.
[42] Ibid.
[43] Tesón, above note 2, at 110.
[44] Ibid., at 12; emphasis added.
[45] Elshtain, above note 12, at 74f.; Linklater, above note 5, at 483.
[46] Tesón, above note 2, at 94.
[47] Nardin, above note 33, at 18; Bellamy, 'Ethics and Intervention: The "Humanitarian Exception" and the Problem of Abuse in the Case of Iraq', 41 *Journal of Peace Research (JPR)* (2004) 131, at 139.
[48] See also the contributions by Chris Brown and Anthony Lang, Jr to this volume.
[49] Brown, 'Selective Humanitarianism: In Defense of Inconsistency', in Chatterjee and Scheid, above note 39, 31, at 42.

In other words, 'moral reasoning became a matter of following a theoretically validated rule, rather than of making a practical judgement'.[50] This abstraction of moral reasoning from concrete political circumstances leaves the theoretical conception of morality devoid of politics and thus accounts for the tension between them in the debate on humanitarian intervention. According to this narrative, the counterposition of politics and morality has its roots in epistemic changes that do not reflect divergent historical developments of politics and morality.

The debate on humanitarian intervention shows, in sum, that central to this term is the relationship between politics and morality—and there exist two accounts of the historical development of this relationship. One holds that morality has progressively become more universal and inclusive while political institutions have remained particularist. The prominence of the term humanitarian intervention is here interpreted as the result of this progressive development of morality and it identifies the means to undermine the particularist nature of political institutions and to spread the enjoyment of human rights more widely. The other views politics and morality in practice as co-constitutive and argues that the term humanitarian intervention is the result of a historical development in which the theoretical conception of politics and morality has been separated. Here, the term humanitarian intervention serves to sell the practice as a moral act and to obscure its political dimension. Adjudicating between these historical narratives thus calls for a historical analysis.

3. The Classical Debate

Classical theorists did not use the term 'humanitarian intervention', which only gradually became common in (international) legal discourse in the nineteenth century, but was subsequently largely forgotten.[51] Nor did they develop a distinct concept of wars justified by their humanitarian goals. Nevertheless, they regularly discussed the possibility of war to punish violations of natural law, to help oppressed subjects of another ruler, or to help victims of civil war.[52] In substantive terms, these cases are comparable to those considered in contemporary debates on humanitarian intervention and I will analyse the relevant writings of Francisco de Vitoria, Edmund Burke, Thomas Paine, and John Stuart Mill with the aim to establish, in the first instance, whether they recognized a moral obligation to outsiders and, if so, how they conceived of its political realization. In addition, the writings of these four authors emanate from different periods in European history and thus provide an indication of historical change in the conception of the relationship between politics and morality in the context of 'humanitarian intervention'.[53] The classical texts here simply provide historical evidence

[50] Ibid.

[51] Rodogno, 'Humanitarian Intervention in the Nineteenth Century', in A.J. Bellamy and T. Dunne (eds), *The Oxford Handbook of the Responsibility to Protect* (2016) 19; see also the contributions by Hendrik Simon and Lauren Benton to this volume.

[52] Chesterman, above note 27, at 10–20.

[53] In addition to providing a chronological overview, this selection of authors includes 'interventionists' (Vitoria, Burke, Paine) and 'non-interventionists' (Mill) and it offers a direct debate (between Burke and Paine) that allows for an analysis of the nature of different positions.

for the existence and nature of debates on such policies of the justification of war and international order; critical engagement with their substantive positions, for lack of space, will have to wait for another opportunity.

The political event that triggered Vitoria's reflections on questions of 'humanitarian war' was the Spanish 'discovery' of America.[54] The Spanish Crown as well as a range of social groups—traders, settlers, missionaries—had an interest in extending Spanish rule over these newly discovered lands and peoples. The pursuit of these interests, however, required the moral, legal, and political justification of Spanish activities vis-à-vis the Catholic Church, other European powers, and the conscience of individuals.[55] Yet, since the very existence of the continent of America and of its peoples had been previously unknown in Europe, the Spaniards were confronted with the challenge of deciding which moral, legal, and political principles could be applied to this case. This challenge was taken up in a public debate in which Francisco de Vitoria's arguments ultimately provided the justification for Spanish rule in America.

Vitoria first established that the Amerindians were indeed human beings and then argued that as such they 'undoubtedly possessed as true dominion, both public and private, as any Christians. That is to say, they could not be robbed of their property, either as private citizens or as princes.'[56] Vitoria thus formulated an early conception of sovereignty and non-intervention which, in principle, extended to cultural and religious practices that violated, in the eyes of the Christians, the laws of nature: 'Christian princes cannot wage war on unbelievers on the grounds of their crimes against nature, any more than for other crimes which are not against nature.'[57]

Despite this robust principle of sovereignty, however, Vitoria justified the Spanish wars, and subsequent rule, in America—at least partly on 'humanitarian' grounds. The Amerindian communities were accused of practising cannibalism and human sacrifice and these practices involved harming innocent people.[58] Such harm, Vitoria argued, cannot be ignored because 'amity between men is part of natural law'.[59] The universal brotherhood of men underpins the moral right and obligation to help innocent people in need and it is therefore 'lawful to defend an innocent man even if he does not ask us to, or even if he refuses our help' and hence 'Christian princes can declare war on the barbarians'.[60] According to Vitoria's argument, the Spaniards had

[54] For a fuller discussion of Vitoria, see A. Pagden, *The Fall of Natural Man* (1982); A. Pagden, *European Encounters with the New World* (1993); and B. Jahn, *The Cultural Construction of International Relations or the Invention of the State of Nature* (2000); and A. Anghie, *Imperialism, Sovereignty, and the Making of International Law* (2005). See also Arnulf Becker Lorca's chapter and B.S. Chimni's chapter on 'humanitarianism' in this volume.

[55] See also the contributions by Anuschka Tischer and Arnulf Becker Lorca to this volume.

[56] F. de Vitoria, *Political Writings* (1991), at 250f.

[57] Ibid., at 222, 224.

[58] Ibid., at 225.

[59] Ibid., at 278f. For classical authors in general, and Vitoria in particular, it was just war theory that covered the issue of what we call humanitarian intervention today. Just war theory no doubt provides a fruitful theoretical framework for the analysis of humanitarian intervention; see Rengger, 'The Judgment of War: On the Idea of Legitimate Force in World Politics', 31 *Review of International Studies (RIS)* (2005) 143 and *Political Dilemmas* (2003) 53; Fixdal and Smith, 'Humanitarian Intervention and Just War', 42 *Mershon International Studies Review* (1998) 283. Engaging with that literature, however, would go beyond the scope of this chapter.

[60] Vitoria, above note 56, at 225.

no authority to punish violations of natural law—but they did have a moral obligation (also based on natural law) to help innocent people in need: 'the reason why the barbarians can be conquered is not that their anthropophagy and human sacrifices are against natural law, but because they involve injustice *(iniuria)* to other men'.[61]

This justification is in line with present day conceptions of 'humanitarian intervention' and the parallels extend to the limits as well as modalities of such interventions. On the one hand, 'if war is declared on the barbarians by this title (to help the victims of atrocities), it is not lawful to continue once the cause ceases, nor to seize their goods and their lands on this pretext'.[62] On the other hand, even though regime change or the imposition of a different culture or religion as such are not lawful, 'if there is no other method of ensuring safety except by setting up Christian princes over them, this too will be lawful, as far as necessary to secure that end'.[63] In short, if particular political arrangements are identified as facilitating the atrocities in question, then regime change (and concretely Spanish rule) becomes necessary and lawful in pursuit of the 'humanitarian' aim.

Vitoria's reflections clearly contradict the claim that morality was limited to, and coincided with, political boundaries in times past (or that our predecessors were lacking in moral sensitivity). Even though Vitoria provided an early and influential formulation of the concept of sovereignty and the right to non-intervention, crucially for all and not just European or Christian communities, he also subscribed to universal moral principles that trumped this particular right under certain circumstances. But it is only this political context that is capable of turning universal moral principles into actual morality, for 'in moral terms, unbelievers are *not* to be forcibly converted to the faith'; only if forcible conversion or the establishment of a Christian prince were the only means to end and prevent the harm done to innocent people, then 'in political terms . . . it seems that such compulsion would be altogether lawful'.[64] Generally, therefore, Vitoria provides a moral justification for the principle of sovereignty as the basis of the international order; and intervention is only justified where this principle is in practice abused. In other words, the morality of intervention is entirely determined by the particular political context.

Moving on to the debate between Edmund Burke and Thomas Paine on the French Revolution, we can identify continuity but also change.[65] Contrary to the claim that moral sensitivity was less developed in the past, both Burke and Paine agree that the brutalities unfolding in the course of the French Revolution are morally objectionable. The fact that the revolutionaries 'murdered their King, and imprisoned, butchered, confiscated, and banished their fellow Subjects' shocked 'the moral sentiments of all virtuous and sober minds'.[66] Indeed, that 'heads were stuck on spikes and carried

61 Ibid.
62 Ibid., at 226.
63 Ibid.
64 Ibid., at 222.
65 On Edmund Burke more generally, see J. Welsh, *Edmund Burke and International Relations* (1995); U.D. Mehta, *Liberalism and Empire. A Study in Nineteenth-Century British Liberal Thought* (1999); Hampsher-Monk, 'Edmund Burke's Changing Justification for Intervention', 48 *The Historical Journal* (2005) 65. On Paine, see Walker, 'The Forgotten Prophet: Tom Paine's Cosmopolitanism and International Relations', in 44 *International Studies Quarterly (ISQ)* (2000) 51; and Jahn, above note 54.
66 E. Burke, *Reflections on the Revolution in France* (1987), at 101, 108.

about the city' were 'outrages'.[67] Though in general recognizing the principle of sovereignty, both Burke and Paine subscribed to a universal moral law that was able to trump sovereignty. 'As to the right of men to act any where [sic] according to their pleasure, without any moral tie, no such right exists.'[68] Accordingly, both Burke and Paine justified war—albeit for opposed reasons. While Burke called on England and other European powers to declare war on France, Paine advocated French 'campaigns' in England and Germany.[69]

Yet, it is the source of these differences that is of interest for the contemporary debate on humanitarian intervention. The disagreements of these authors did not arise from different moral principles. Both Burke and Paine subscribed to the rights of life, liberty, and property[70]—or the rights of man.[71] Instead, the disagreements have their roots in different judgements on the political causes of, and solutions to, the brutalities. For Burke, the root of the problem was the revolutionary government of France, and in particular the principle of democracy:

> Of this I am certain, that in a democracy the majority of the citizens is capable of exercising the most cruel oppressions upon the minority whenever strong divisions prevail in that kind of polity, as they often must; and that oppression of the minority will extend to far greater numbers and will be carried on with much greater fury than can almost ever be apprehended from the dominion of a single scepter.[72]

In contrast, Paine argued that 'these outrages were not the effect of the principles of the Revolution, but of the degraded mind that existed before the Revolution, and which the Revolution is calculated to reform'.[73] Burke's analysis thus identifies democracy as facilitating brutal oppression and offers 'the dominion of a single scepter'—or constitutional monarchy—as a solution while Paine argues that these brutalities were the result of socialization under the *ancien regime* and offers the Revolution as solution.[74] In short, Burke and Paine do not pitch different moral norms against each other; and they do not pitch political interests against moral principles as the concept of humanitarian intervention today suggests; instead, these authors simply arrive at different political judgements on the nature of the problem as well as its solution.

Nevertheless, Burke's and Paine's debate does begin to introduce a gap between politics and morality. The formulation of abstract moral principles—the rights of man—enabled both authors to theoretically identify a particular form of government (democracy and constitutional monarchy respectively) as corresponding to these rights, as being best suited to their realization. For Burke, it is the definition of democracy as majority rule that implies the oppression of minorities; though supported by the example of the French Revolution, this is asserted as a theoretical claim

67 T. Paine, *Rights of Man, Common Sense and Other Political Writings* (1995), at 108, 110.
68 E. Burke, *The Writings and Speeches of Edmund Burke*, vol. IX (1991), at 249.
69 For Kant's perspective on these issues see the contribution by Oliver Eberl to this volume.
70 Burke, above note 68, at 238.
71 Paine, above note 67.
72 Burke, above note 66, at 110.
73 Paine, above note 67, at 110.
74 Ibid,, at 108.

with general validity. Hence, by undermining the belief in God, the rights of property, and non-democratic forms of government, the French Revolution 'violates the rights upon which … all communities are founded',[75] held Burke. Consequently, regime type becomes a legitimate concern for all members of the international system. Even if the French Revolution had so far not done any direct harm to its neighbours, such harm could be anticipated.[76] In civil society, this anticipation provides grounds for a lawsuit but

> where there is no constituted judge, as between independent states there is not, … this principle … has bestowed on the grand vicinage of Europe a duty to know, and a right to prevent, *any* capital innovation which may amount to the erection of a dangerous nuisance.[77]

This theoretical linkage of moral principles with a corresponding form of government thus led to a justification of wars in which the goal of liberating citizens in other states was inextricably linked with self-defence. Burke thus argued that

> a war to preserve national independence, property, liberty, life, and honour, from certain universal havock, is a war just, necessary, manly, pious; and we are bound to persevere in it by every principle, divine and human, as long as the system which menaces them all, and all equally, has an existence in the world.[78]

Paine follows exactly the same line of argument. Only now it is despotism which by definition violates the right to liberty—not only of the despot's own subjects but also of the citizens in other states. Unlike Burke, who had to acknowledge that revolutionary France had not (yet) done any direct harm to its neighbours, Paine could of course point to the intervention of the European monarchies in France and conclude that exporting the Revolution was justified as a war of defence: 'when France shall be surrounded with revolutions, she will be in peace and safety …'.[79] Here, too, the theoretical link between the Revolution and the realization of the rights of man led to the demand for a universalization of the Revolution. It also became a primary justification narrative of war in the political discourse of revolutionary France. Some revolutionaries increasingly aspired to a new international normative order.[80] Wars for the purpose of regime change were justified in general—and justified by the same inextricably linked goals of liberation and defence: 'with how much more glory, and advantage to itself does a nation act, when it exerts its powers to rescue the world from bondage, and to create itself friends'.[81] The abstract link between moral principles and particular

[75] Burke, above note 68, at 240–42, 252.
[76] Ibid., at 250.
[77] Ibid., at 251, emphasis added; see Iain Hampsher-Monk for a thorough discussion of the process by which Burke eventually came to settle on this particular justification for war against France (above note 65).
[78] Burke, above note 68, at 238.
[79] Paine, above note 67, at 201–02.
[80] E.J. Kolla, *Sovereignty, International Law, and the French Revolution* (2017); see also Hendrik Simon's chapter in this volume.
[81] Paine, above note 67, at 320.

regime types thus establishes a link between morality and self-interest: the rescue of others is inextricably linked to self-defence and the creation of friends while at the same time denying the moral core to alternative political projects. In other words, while one's own political judgement is equated with morality as such, alternative political judgements appear by definition as politics devoid of morality.

And this move has serious political implications for it leads Burke and Paine to justify war not just against those states in which atrocities actually take place but against all states whose form of government is associated in the abstract with the violation of the rights of man. Thus, just as Burke provides a justification for war against *any* state that introduces an 'innovation' irrespective of its performance with regard to the rights of man, Paine offers his services for a campaign against Germany without any mention of particular abuses being committed in those states at the time. Instead, the goal of such a campaign is 'the extinction of German despotism' and the establishment of 'the freedom of all Germany',[82] just as the goal of a suggested campaign against his native England was the export of the Revolution rather than the ending of any particular atrocities.[83]

In addition, the ultimate aim to generalize a particular regime type implies the need to establish appropriate rules for the international system at large. Hence, Burke argued that France has not only 'annulled all their old treaties; but they have renounced the law of nations from whence treaties have their force. With a fixed design, they have outlawed themselves, and to their power outlawed all other nations.'[84] The French identification of the rights of man with democratic government undermines the foundations of the entire international system:

> The colonies assert to themselves an independent constitution and a free trade. They must be constrained by troops. In what chapter of your code of the rights of men are they able to read that it is a part of the rights of men to have their commerce monopolized and restrained for the benefit of others? As the colonists rise on you, the Negroes rise on them. Troops again—massacre, torture, hanging! These are your rights of men![85]

Both Burke and Paine concluded, therefore, that peaceful international cooperation in general required 'a common language', 'some common recognised principle',[86] or 'a common and correspondent principle'.[87] The theoretical link between moral principles and particular political institutions did thus not only lead to the demand for the generalization of a particular regime type but also to a call for fundamental rules of the international system in line with, or conducive to, this end.

Burke and Paine recognized, however, that such a reordering of the entire international system was beyond the capabilities of the main international actors at the time. 'Distance of place', argues Burke, 'does not extinguish the duties or the rights

[82] Ibid., at 101f.
[83] Walker, above note 65, at 62, 66, 68.
[84] Burke, above note 68, at 240.
[85] Burke, above note 66, at 195.
[86] Burke, above note 68, at 340.
[87] Paine, above note 67, at 287.

of men; but it often renders their exercise impracticable.... But there are situations where this difficulty does not occur; and in which, therefore, these duties are obligatory, and these rights are to be asserted.'[88] That is, a global enforcement of the rights of men would have been 'impracticable' but within Europe the 'Law of Neighbourhood' demands the realization of these principles—and therefore war. And Paine, too, limits his advocacy of actual wars to the European theatre.

Neither Burke nor Paine, in sum, suffered from a lack of moral sensitivity just as neither of them limited moral obligations to the boundaries of the state or any other particularist institution. On the contrary, both unequivocally subscribed to a universal conception of moral obligations that cut across and trumped political boundaries—leading both authors to justify wars for the purpose of regime change. Thus, the moral principles become embedded within particular political institutions and practices. Moreover, the debate between Burke and Paine highlights that it is the political judgement that is contested, not the nature or extent of moral principles or their opposition to political interests.

And yet, unlike Vitoria for whom the justice of a war (against the Amerindians) was entirely determined by the concrete circumstances of the particular case, both Burke and Paine identify a particular regime type—democracy and constitutional monarchy respectively—*in general* as best suited to the realization of moral principles. And this abstraction from the particular case of the French Revolution leads to the demand that all states in the international system adopt that regime and a concomitant general justification of wars for regime change and a 'better' international order. Hence, the debate between Burke and Paine does attest to a process of universalization—but this is not the universalization of moral principles; rather, it is the universalization of a particular political programme which is associated with universal principles. This political programme, however, is practically pursued only within the European theatre for the simple reason that there existed no international actor at the time with sufficient power to lay down such a law for the entire world. Thus, we have a theoretical universalization of the right to intervention limited only by the lack of power to practically pursue this goal worldwide.

In contrast to Vitoria, Burke, and Paine, the writings of John Stuart Mill, finally, appear to provide some support for the assumption of a historical—rather than just theoretical—separation of politics and morality.[89] Mill, after all, famously and explicitly advocated the principle of non-intervention and thus appears to restrict moral obligations to the domestic sphere. And yet, like his predecessors, Mill did not only recognize the rights of independent political communities but explicitly also universal moral principles. Britain, he argues, only interferes in the affairs of foreign states

> in the service of others ...: to mediate in the quarrels which break out between foreign states, to arrest obstinate civil wars, to reconcile belligerents, to intercede for mild

[88] Burke, above note 68, at 250.

[89] On Mill more generally, see Mehta, above note 65; J. Pitts, *A Turn to Empire. The Rise of Imperial Liberalism in Britain and France* (2005); Jahn, 'Barbarian Thoughts: Imperialism in the Philosophy of John Stuart Mill', in 31 *Review of International Studies (RIS)* (2005) 599; and Varouxakis, 'John Stuart Mill on Intervention and Non-Intervention', 26 *Millennium (Mil)* (1997) 57.

> treatment of the vanquished, or finally, to procure the abandonment of some national crime and scandal to humanity, such as the slave-trade.[90]

For Mill, then, British foreign policy did pursue a range of moral aims and he clearly viewed the slave trade as a 'crime against humanity' and supported Britain's efforts to put an end to this crime. Indeed, Mill explicitly recognized 'universal rules of morality between man and man',[91] even while he put forward a forceful argument for the principle of non-intervention. This contradiction between the recognition of universal moral laws on the one hand and the principle of non-intervention on the other may be interpreted as reflecting a diverging historical development of morality and politics.

And yet, Mill manages to square that circle. Key to understanding Mill's principle of non-intervention is his work on representative government. Representative government, Mill argues, is *the* ideal form of government for the protection and realization of universal individual rights.[92] Just as Burke and Paine, then, Mill identifies these rights—of life, liberty, and property—unequivocally with a particular form of political organization. He recognizes, however, that this claim is contested within Europe as well as in the wider world. He thus distinguishes categorically between peoples who, in principle at least, are capable of recognizing and realizing these rights—civilized peoples—and those who in his opinion do not (yet) have this ability—barbarian peoples.[93] Barbarian peoples 'have no rights as *nation*, except a right to such treatment as may, at the earliest possible period, fit them for becoming one'.[94] Mill thus denies barbarian peoples political rights, that is the right to sovereignty and non-intervention, and argues that for barbarians, colonial or despotic government is 'the ideal rule' because it allows the civilized colonizer to prepare the indigenous population for an eventual enjoyment of the rights of life, liberty, and property.[95] Such despotic colonial rule may, however, have to be imposed by violent means since the 'barbarians' (by definition) do not recognize the beneficial nature of such colonial rule. And it is for these violent clashes between civilized and barbarian peoples that Mill reserves the term 'war'—which indicates 'conquest', 'annexation', and 'defence', that is, the denial of political rights for the entire community.[96]

This concept of war, Mill argues, does not apply to relations between civilized nations. Here

> the disputed question is that of interfering in the regulation of another country's internal concerns; the question whether a nation is justified in taking part on either side, in the civil wars or party contests of another; and chiefly, whether it may justifiably aid the people of another country in struggling for liberty; or may impose

[90] Mill, 'A Few Words on Non-Intervention', in J.M. Robson (ed.), *The Collected Works of John Stuart Mill*, vol. XXI, (1984) 109, at 111.
[91] Ibid., at 119.
[92] J.S. Mill, *On Liberty and Other Essays* (1998).
[93] Mill, above note 90, at 118.
[94] Ibid., at 119.
[95] Mill, above note 92, at 432, 453, 454. See Jahn, above note 89; Pitts, above note 89; and Mehta, above note 65, for a more thorough discussion of Mill's justification of colonialism.
[96] Mill, above note 90, at 121.

> on a country any particular government or institutions, either as being best for the country itself, or as necessary for the security of its neighbours.[97]

The term 'intervention' in Mill's usage is thus reserved for the use of force in cases where the political rights of the target are explicitly recognized. And it is the fact that civilized peoples are by definition capable of realizing the rights of life, liberty, and property through the establishment of representative government that leads Mill to argue that relations between civilized nations must be governed by the principle of non-intervention. Although at the time of his writing most European nations had not yet established representative governments, Mill argued that the institutionalization of freedom could only be effected from within. If a people has

> not sufficient love of liberty to be able to wrest it from merely domestic oppressors, the liberty which is bestowed on them by other hands than their own, will have nothing real, nothing permanent. No people ever was and remained free, but because it was determined to be so.[98]

Sufficient love of freedom can—or even must—be developed through 'an arduous struggle to become free by their own efforts' and people will only defend free institutions for which 'they have long fought' and 'made sacrifices'.[99]

In short, Mill argued that representative government conducive to the realization of universal moral rights could only be established through domestic struggle and he explicitly expected such struggles to entail, or even require, serious sacrifices which played an important role in schooling the population in the exercise of freedom and in establishing a firm commitment to free institutions. Conversely,

> a government which needs foreign support to enforce obedience from its own citizens, is one which ought not to exist; and the existence given to it by foreigners is hardly ever anything but the sympathy of one despotism with another.[100]

Hence, the principle of non-intervention simply ensures that representative government is truly representative—and it denies would-be interveners the chance of passing off their 'despotic' interference as a moral act.

This abstract argument from which Mill derives the general rule of non-intervention for relations between civilized states, however, required fine-tuning in light of concrete political realities and led Mill to formulate two exceptions to the rule of non-intervention. The first exception concerned cases in which the struggle for free institutions descended into

> a protracted civil war, in which the contending parties are so equally balanced that there is no probability of a speedy issue; or if there is, the victorious side cannot hope

[97] Ibid.
[98] Ibid., at 122.
[99] Ibid., at 123.
[100] Ibid., at 121.

> to keep down the vanquished but by severities repugnant to humanity and injurious to the welfare of the country.[101]

In such cases intervention was permitted and, Mill argued, had actually been practised quite frequently[102] and thus entered into a principle of customary international law.[103] These interventions do entail a humanitarian element in that they aim to prevent what one might call politically 'unfruitful' humanitarian costs. That is, the exception does not cover all cases of atrocities in civil wars but only those in which existing sacrifices do not contribute to a resolution of the political struggle—one way or the other. Accordingly, such interventions do not aim to establish a particular regime in the target country or to support the 'right' side in the contest—but a reconciliation on 'equitable terms of compromise'.[104]

The second exception is the case of counter-intervention. In this case, it is assumed that a foreign power is already meddling in the internal affairs of another state and thus distorts the domestic political process. In such cases, 'intervention to enforce non-intervention is always rightful, always moral, if not always prudent'.[105] Here again, the intervention is conceived as unpolitical. It does not aim to establish a particular political solution to the conflict but simply, just as in the case of protracted civil wars, to re-establish the integrity of the domestic political process.

Like his predecessors, then, John Stuart Mill was clearly sensitive to systematic brutalities like the slave trade and the humanitarian costs of civil wars and he explicitly subscribed to a universal moral law. Yet, the fact that Mill advocated the principle of non-intervention does not indicate that he restricted moral obligations to the limits of the state. On the contrary, Mill's vision of the world involves the active realization of moral rights like life, liberty, and property universally. This goal was to be achieved, however, through the institution of the state rather than through intervention. He propagates an international order in which all people are governed by the political regime—representative government or despotic colonial rule—that will ensure their enjoyment of moral rights at the earliest possible opportunity. In the case of 'civilized' peoples this requires protection of the indigenous political process through the principle of non-intervention; and in the case of 'barbarian' peoples it requires colonial rule by 'civilized' states. Far from limiting moral obligations, therefore, Mill's principle of non-intervention plays a crucial role for the establishment and protection of moral rights worldwide.

Just like Burke and Paine, nonetheless, Mill develops his position through the abstract identification of representative government as the ideal form of rule for the realization of moral principles. In practice, however, this ideal required adjustment to the

101 Ibid.

102 His examples are the intervention of the European powers in the conflict between Greek insurgents and Turkey, between Turkey and Egypt, and between Holland and Belgium. See Chesterman for a discussion of the 'humanitarian' merits of some of these cases (Chesterman, above note 27, at 28–35).

103 While there were in fact customary international legal norms on the use of force in the nineteenth century, interestingly, the very status of (humanitarian) intervention as positive law was highly controversial in political practice and international (legal) discourse, a fact which casts doubt on Mill's assumption. On legal norms regarding the use of force in the nineteenth century see Part III in this volume.

104 Mill, above note 90, at 121.

105 Ibid., at 123.

political possibilities. And here, Mill notes that while there were movements in most European nations aspiring to some form of representative government, the same was not the case in the non-European world. Unlike Burke and Paine, however, who had to restrict their political goals in practice to Europe, Mill can realistically envisage such an ordering of the entire world because the nation of Britain in the nineteenth century 'is equal to the greatest in extent of dominion, far exceeding any other in wealth, and in the power that wealth bestows'.[106] This power enabled Britain essentially to lay down the rules for the entire international system—yet not to such an extent that it could have imposed its vision by force everywhere. That was only possible in the non-European world where the task of establishing colonial rule was, in addition, shared between a number of 'civilized' states. Within Europe, Britain had the means to play a hegemonic role by reassuring other states of its respect through the principle of non-intervention, and by promising them a share in the benefits of Britain's power: according to Mill, 'this nation desires no benefit to itself at the expense of others', 'it makes no treaties stipulating for separate commercial advantages', it goes to war only in response to the 'aggressions of barbarians' and while bearing the costs of war, shares its fruits 'in fraternal equality with the whole human race'.[107] The realization of moral principles is thus here, too, circumscribed by the political possibilities and requires political judgement.

Yet, just as in the case of the French Revolution this political judgement is necessarily contested. Not only did many European states not share Mill's identification of representative government with morality; Britain was also faced by the accusation 'most widely current on the continent' that it 'meddles' in the affairs of other states motivated by 'egoism and selfishness'.[108] And it is against this challenge that Mill elaborates the moral, yet clearly contested, core of British foreign policy.

In sum, not one of the classical authors analysed here develops a concept of humanitarian intervention. Yet, this lacuna does not indicate either a poorly developed moral sensitivity or the limitation of moral obligations to the boundaries of the state. On the contrary, Vitoria, Burke, Paine, and Mill explicitly subscribe to universal moral obligations and aim to realize these moral obligations beyond state borders. The pursuit of this aim, however, requires political action and so classical authors propagate particular political practices and institutions embodying moral principles for individual states as well as the international system at large. The main differences between these political projects lie not in the substantive moral norms they embody but rather in the political imagination and relative capabilities of their protagonists—that is, in political judgement. For Vitoria, the solution to 'human rights violations' are Catholic princes, for Burke a constitutional monarchy, for Paine a revolutionary democracy, and for Mill representative government. And these solutions are imposed where possible through just wars, wars for regime change, colonial conquest—but left to the locals where necessary. Yet, the possibilities of ordering the entire international system in accordance with such political projects have clearly increased over time. While Vitoria still had to be content with the American exception and Burke and Paine could

[106] Ibid., at 111.
[107] Ibid.
[108] Ibid., at 111f.

envisage shaping the entire European continent, Mill realistically envisaged reshaping the entire international system—albeit not everywhere by means of force. Yet this process of universalization is decidedly not one of moral principles or sensitivities which are universal for all authors. Instead, it is a gradual universalization of power and of geopolitically shifting inequalities of power that appears to enable the abstraction of morality from political judgement in concrete cases and the presentation of particular political projects as embodying morality per se.

4. Conclusion

The reflections of classical theorists on war for the purpose of helping victims of abuse in foreign states illustrate, then, how morality and politics are indeed mutually constitutive. And the interventions advocated today as humanitarian are as much political projects as the wars and interventions contemplated by the classical authors. Thus, the constitution of a 'liberal state' is offered as solution to problems supposedly generated by anarchy and tyranny.[109] A more universal state is offered as solution to the particularism associated with ethnic cleansing.[110] And state building (neo-imperialist or indigenous) is offered as solution to the violence of modernization processes.[111] These interventions thus advocate a particular—broadly liberal or modern—political order as solution to the problems apparently generated by other political regimes and configurations.

And just as in classical theory, the nature of these political projects is defined and circumscribed by the political imagination and material capabilities of their protagonists. Liberal (modern, universal, European, democratic) states are seen to meet the standard for 'comparative moral reliability';[112] to have developed more universalistic forms of community;[113] and/or to have the economic, political, and military capacity 'to provide protection and a good economic environment'.[114] These communities become the 'custodians of the global human rights culture'[115] and are tasked with reforming not just the political system in particular target states but the legal norms of the entire international order.[116]

Concrete proposals to this end include 'a treaty based coalition among liberal democratic states' with the right to override a deadlocked Security Council on questions of humanitarian intervention,[117] a call for 'gradations in

[109] Tesón, above note 2, at 96.
[110] Linklater, above note 5, at 492.
[111] Ayoob, above note 19, at 93f; Ignatieff, above note 19, at 302, 313f.
[112] Buchanan and Keohane, 'The Preventive Use of Force: A Cosmopolitan Institutional Proposal', 18 *Ethics and International Affairs* (2004) 1, at 19.
[113] Linklater, above note 5, at 484.
[114] Ignatieff, above note 19, at 313f., 309; Cooper, above note 25, at 5.
[115] Linklater, above note 5, at 486.
[116] Ibid., at 493; Wheeler, above note 4, at 310; Tesón, above note 2, at 127.
[117] Buchanan and Keohane, above note 112; Buchanan, 'Reforming the International Law of Humanitarian Intervention', in Holzgrefe and Keohane, above note 1, 130, at 171; For an overview and critique of such political proposals, see Clark, 'Democracy in International Society: Promotion or Exclusion?', 37 *Millennium—Journal of International Studies* (2009) 563, and on the constitutionalization of international law, see Cohen, 'Whose Sovereignty? Empire versus International Law', 18 *Ethics and International Affairs* (2004) 1.

sovereignty',[118] right up to the establishment of a new imperialism on the grounds that 'the weak still need the strong and the strong still need an orderly world'.[119] The result of this reasoning is an international order no longer based on the principle of equality. 'State-majoritarianism, under current conditions in which many states are not democratic cannot be viewed as having the same legitimacy-conferring power as the consent of individuals.'[120] These projects aim to replace legal equality with legal inequality as the basis of the international order. The concept of humanitarian intervention thus functions 'as a doctrinal advance guard'[121] for a new international order—which has its roots in a shift in power relations as a consequence of the demise of the Soviet Union rather than in substantive moral development.

In other words, the demise of the Soviet Union provided powerful Western actors (from states through inter-governmental organizations and non-governmental organizations to their academic observers) with the opportunity of proactively spreading the political order variously identified as best suited to the protection of human rights.[122] Yet, the ground rule of the international system, especially since decolonization, explicitly forbids interference in the political constitution of other states—a ground rule which could not be challenged directly. It is this historical conjuncture which explains the timing of the rise of humanitarian intervention to prominence. The term, moreover, entails the project of a radical revolution of the ground rules of the international system at large—and it is this content which explains the speed and impact of the term amongst protagonists and critics alike. It also explains its speedy replacement by R2P (Responsibility to Protect): an attempt to codify humanitarian intervention under a new name for protagonists and its simultaneous subversion through the focus on state responsibility by critics.[123]

The term humanitarian intervention provides legitimacy for illegal interventions by distinguishing unequivocally between humanitarian, that is universal moral, and political, that is particularist and hence illegitimate and illegal, interventions. The term thus serves to hide the political nature of 'humanitarian interventions' and, conversely, functions to deny moral standing to alternative political positions and projects. This is what accusations of 'moral bankruptcy',[124] of failing to serve as 'good international citizens',[125] and of providing support for 'the worst forms of human evil'[126] are about. The discourse on humanitarian intervention thus impedes the search for solutions to concrete 'humanitarian' problems by delegitimating a wide range of potential political solutions.[127]

[118] Keohane, 'Political Authority after Intervention: Gradations in Sovereignty', in Holzgrefe and Keohane, above note 1, 275; Paris, *At War's End. Building Peace After Civil Conflict* (2004). For a discussion of the internal contradictions of this position, see Jahn, 'The Tragedy of Liberal Diplomacy. Part II', 1 *Journal of Intervention and Statebuilding* (2007) 211.

[119] Cooper, above note 25, at 5.

[120] Buchanan, above note 117, at 171.

[121] Farer, 'Humanitarian Intervention before and after 9/11: Legality and Legitimacy', in Holzgrefe and Keohane, above note 1, 53, at 55.

[122] See also the chapter by Anna Geis and Wolfgang Wagner.

[123] Bellamy, 'Responsibility to Protect or Trojan Horse? The Crisis in Darfur and Humanitarian Intervention after Iraq', 19 *Ethics and International Affairs* (2005) 31.

[124] Wheeler, above note 4, at 296.

[125] Linklater, above note 5, at 493.

[126] Tesón, above note 2, at 129.

[127] De Waal, above note 27.

In sum, the concept of humanitarian intervention rests on false premises: it assumes that there exists a historically developed gap between morality and politics which it sets out to bridge. Yet, the codification or legalization of humanitarian intervention—as it may be aspired to with the *Responsibility to Protect*[128]—does not alter 'the relationship between order and justice, citizenship and humanity, and sovereignty and human rights'[129] for the simple reason that these concepts do not constitute opposites. Order does not stand for politics without morality, nor justice for morality without politics: humanitarian interventions, after all, aim to produce order as a precondition for justice. Citizenship does not stand for a particularist political concept nor humanity for a universalist moral one: rather citizenship has a moral standing invariably acknowledged by advocates of humanitarian intervention while humanity requires political institutions such as the state to come into its own. And sovereignty is not the opposite of human rights but rather the institution through which these rights are, for the most part, realized. That is, 'the abstract attempt to apply ethics to politics, truth to power, or universality to particularity simply conjures away the very specific ways in which these apparent opposites are already mutually constitutive'.[130] And so it is with the term humanitarian intervention. Though explicitly aiming to bridge the gap between politics and morality, born of a particular political conjuncture, the term humanitarian intervention ironically contributes to their separation.

There is, then, 'no such thing as humanitarian intervention' defined as moral solution to political problems.[131] As the recent case of Syria demonstrated so clearly, moral positions were shifting according to political conditions on the ground and the 'red lines' of human rights violations were regularly crossed—partly because rigidly respecting them would have had 'immoral' implications. Taking seriously the mutually constitutive nature of politics and morality thus requires giving up the concept of humanitarian intervention—lest it continues to play not a moral but a highly ideological role

[128] Brock, 'Between Sovereign Judgment and the International Rule of Law', in M. Albert and A. F. Lang, Jr (eds): *The Politics of International Political Theory: Reflections on the Works of Chris Brown* (2019) 87.
[129] Linklater, above note 5, at 493.
[130] Walker, *After the Globe, Before the World* (2010), at 129.
[131] De Waal, above note 27.

PART VI
ALTERNATIVE PATHS
Non-Western Perspectives on the Justification of War and International Order from Past to Present

21
The Islamic Law of War and Peace and the International Legal Order: Convergence or Dissonance?

*Sohail H. Hashmi**

1. Introduction

The advent of an international legal regime governing war and peace has had an obvious and profound impact upon Muslim re-conceptualizations of relevant Islamic laws and norms. As Muslim states acceded to the League of Nations Covenant, United Nations Charter, Geneva Conventions, and other treaties that comprise this legal regime, they contributed to the claim that public international law is a universally accepted set of principles and procedures, the first such in history. This claim of universality threatened to make Islamic theories of world order, articulated in the classical jurists' works on jihad and siyar, completely obsolete. In practice, Muslim states never fully conformed to the theory even at the time of its development from the eighth through the fourteenth centuries, long before European penetration fundamentally altered their behaviour. Yet, now, with the rise of international law, the legal theory was threatened not just by desuetude, but by supersession, if not outright renunciation. International law was something that no Muslim statesman, jurist, or activist could ignore.

This chapter explores the different reactions that the rise of international law elicited among Muslim theorists. I divide the responses into three broad categories. The first group I call the assimilationists. They treat the classical theory largely as a historical and now obsolete conception of world order. The assimilationists accept the universality of public international law and argue that through the accession of Muslim states to international law, most Muslims do so as well. The second group is the accommodationists, who coalesce around the claim that while international law appropriately governs the conduct of Muslim states in international society as a whole, Islamic law could and should have a role in the mutual relations of Muslim states. In other words, they see the potential for an Islamic international law alongside public international law. Potential conflict of laws is minimal, in the accommodationist view, because the basic Islamic principles governing international relations are fundamentally compatible with those underlying modern international law. The third group is that of the

* This chapter is a revised version of my previously published essay 'Jihad and the Geneva Conventions: The Impact of International Law on Islamic Theory', in S. Hashmi (ed.), *Just Wars, Holy Wars, and Jihads: Christian, Jewish, and Muslim Encounters and Exchanges* (2014) 325.

Sohail H. Hashmi, *The Islamic Law of War and Peace and the International Legal Order: Convergence or Dissonance?* In: *The Justification of War and International Order*. Edited by: Lothar Brock and Hendrik Simon, Oxford University Press (2021).
 DOI: 10.1093/oso/9780198865308.003.0021

rejectionists, who view international law as an alien code imposed on Muslims by Europeans. The rejectionists affirm the superiority of Islamic law over public international law and call for its application by Muslim states, not just in their relations with each other, but in their relations with non-Muslim states as well.

These three categories are ideal types; there are, of course, nuances within each of these general positions, and few writers fit perfectly into one camp or the other. Still, the three groups do represent, I believe, an accurate expression of the spectrum of Muslim responses to the advent of international law on war and peace. And on this spectrum, the clear tendency of Muslim scholarship and state practice is towards the assimilationist and accommodationist positions. In other words, over the past century the broad Muslim consensus on the Islamic laws of war and peace has tended more towards convergence rather than dissonance with international humanitarian law.

2. Islamic Law of War and Peace: The Classical Theory

By the time Gratian composed the *Decretum* in the early twelfth century and gave the first full articulation of the laws of war in Christendom, the development of Islamic law had reached its golden age. Indeed, throughout the European Middle Ages, Islamic law was by far the most developed legal system on issues pertaining to war and peace in the Mediterranean littoral.

A coherent tradition of Muslim jurisprudence on war and peace began to develop in the late eighth century, following the consolidation of the Abbasid caliphate in Baghdad. In their efforts to codify *shari'a*, which they understood as a comprehensive body of divine law, out of the general precepts in the Qur'an and *sunna* (normative example) of the Prophet Muhammad, the jurists could not but include the laws of war and peace. Thus, in the major compendia of legal opinions of each of the four surviving schools of Sunni jurisprudence, as well as in the dominant Twelver Shi'i school, there are invariably sections on this subject containing detailed injunctions on the legitimate start, conduct, and conclusion of war.[1]

At the heart of the classical theory was the division of the world into two contending spheres: *dar al-Islam* (abode or territory of Islam), a unitary state comprising the community of Muslims, living by the shari'a, and led by the just ruler (*imam*). Dar al-Islam included as well non-Muslim communities, the *dhimmis* (protected people), whose separate communal laws and leaders were tolerated by the Islamic state so long as they did not challenge Muslim sovereignty. Opposed to dar al-Islam was *dar al-harb* (abode or territory of war), where Islamic law did not prevail, leading presumably to anarchy and moral corruption.

The classical jurists' attention was focused on what may be called the expansionist jihad. The imam was obliged to undertake jihad whenever the conditions of the Islamic state permitted him to reduce dar al-harb and bring its lands and peoples into dar al-Islam. This was a collective duty of the Muslim community (*fard kifaya*),

[1] For a concise overview, see Hashmi, 'War and Peace', in R. Peters and P. Bearman (eds), *Ashgate Research Companion to Islamic Law* (2014) 193.

one that required participation only from those financially and physically capable of undertaking it.

The jihad in dar al-harb, in the view of the scholars, was aimed at bringing Islam's higher civilization to those unaware of it, not territorial conquest, plunder, or forcible conversion. Accordingly, they elaborated rules on what Muslim armies may or may not do in dar al-harb. Indeed, *jus in bello* concerns received far greater elaboration than *jus ad bellum*. The basis for such moral injunctions was the Qur'an's general command, 'Do not transgress limits, for God loves not transgressors' (Q. 2:190), which was given greater specificity by the practice of the Prophet and his first four successors, the 'rightly guided' caliphs. Before the start of any attack, the enemy was to be offered the choice of accepting Islam, in which case no further action against them was permissible. If they refused, they were to be offered dhimmi status as an autonomous community within dar al-Islam. This option, deriving from Q. 9:29, initially pertained to Jews and Christians, but was steadily expanded to include Zoroastrians, Hindus, and Buddhists as the Islamic frontiers expanded. Only the polytheist Arabs who had fought so bitterly against Muhammad and the early Muslim community were excluded from the dhimmi option and forced to convert according to Q. 9:5.

In fighting the enemy, Muslim soldiers were to avoid directly targeting women and children. Some jurists included old men, peasants, hermits, merchants, the insane, and other males who do not ordinarily take part in fighting on the list of prohibited targets. According to most scholars, all able-bodied, adult males could, at the discretion of the imam, be killed, whether they were fighting or had been taken prisoner. The jurists permitted the use of all types of weapons or military tactics that were necessary to overcome the enemy, including laying siege to fortresses, firing incendiary devices, cutting off the water supply, or flooding. A few practices were categorically prohibited, based on injunctions attributed to the Prophet, including killing by mutilation or torture, burning individuals alive, and violating oaths or grants of security to soldiers or envoys.

The second type of jihad may be termed the defensive jihad. It was commonly understood that Muslims had an individual obligation (*fard 'ayn*) to defend dar al-Islam whenever it was threatened by aggression from dar al-harb. When the classical jurists were developing this theory of war and peace, Muslims were already under attack and on the defensive throughout the northern Mediterranean. Yet this type of war received little explicit attention in the chapters on jihad. Even when the Crusades forced Muslims into a protracted defensive posture, resulting in an upsurge of writings extolling the virtues of jihad, there was no clear acknowledgment that a different form of jihad was now necessary. The brief discussion that does exist focuses on the moral obligation of Muslims to fight the enemy and on whether Muslim populations who now found themselves behind enemy lines could legitimately remain within dar al-harb. The majority of jurists counselled that such Muslims should—like the Prophet and the earliest Muslims—perform a *hijra*, that is migrate out of dar al-harb and into dar al-Islam at the earliest possible occasion. The expectation was that the Islamic state would rush to their succour through a defensive jihad. Echoes of these debates would inform controversies centuries later, when in the nineteenth century European imperialism brought large Muslim populations under non-Muslim rule.[2]

[2] See R. Peters, *Islam and Colonialism: The Doctrine of Jihad in Modern History* (1979).

On the question of how a defensive jihad is to be waged, the classical jurists suggest—again, without much specificity—that various restraints imposed on the expansionist jihad are relaxed. All able-bodied Muslims, male and female, are required to rush to the defence of the Muslim victims. If some Muslims are not close to the fighting and they cannot travel to the battlefield, they are required to assist the Muslim defenders through material and moral support. Requirements relating to proper authority—that is, who may declare and under whose leadership the jihad may be fought—become more ambiguous. If the leaders of the Islamic state are incapable or unwilling to lead the defensive struggle, other Muslims must assume this responsibility. And finally, the normal constraints on how Muslims may fight to repulse the aggression are loosened under claims of necessity. This ambiguity in the classical theory gives an opening to modern-day Muslim militants to proclaim their violence as jihad.

All of the classical schools of jurisprudence accepted the validity of treaties that suspended (not eliminated) the jihad obligation. One school of Sunni jurisprudence, the Shafi'i, formalized the treaty relationship by interposing a third *dar*, *dar al-sulh* (territory of truce), between dar al-Islam and dar al-harb. The maximum duration of such a truce, according to most scholars, was ten years, although nothing prevented the imam from renewing the truce indefinitely if he deemed it in the Muslims' interest.

Whether wars within dar al-Islam constituted jihad was an issue that created ambivalence among Sunni jurists. Some preferred to treat such conflicts, including those against renegade apostates, highway robbers or pirates, and Muslim rebels as a type of police action, not jihad strictly speaking. Rebels who claimed to be upholding a different interpretation of the Qur'an were in particular given special privileges, such as the right not to be pursued and killed during a rout, and the right to retain their property if they surrendered.[3]

No such ambivalence is evident among Shi'i jurists. Jihad could be properly waged against all non-Shi'is, whether Muslim or not. The other major difference in Shi'i views on jihad was that only the righteous imam, a descendant of 'Ali, could lead the expansionist jihad. Since the line of imams ended with the disappearance of the twelfth imam in the ninth century, according to the dominant strand of Shi'ism, only a defensive jihad to repulse enemy aggression is theoretically possible.[4]

The classical theory described in broad outline above had its dissenters. Some early jurists challenged the validity of the expansionist jihad, arguing that only the defensive jihad was an obligation while the expansionist jihad was supererogatory, a praiseworthy but optional war.[5] In addition, the characteristics of dar al-Islam and dar al-harb were disputed, and writers frequently added many more types. Such terms were open to dispute because they are found nowhere in the Qur'an and *hadith* (traditions ascribed to the Prophet Muhammad), and thus were devoid of a textual 'anchor', unlike the term *jihad*. Indeed, the worldview itself that emerges out of the dar al-Islam/dar al-harb bifurcation has no clear grounding in either Qur'an or hadith.

[3] The most detailed study on this topic is K. Abou El Fadl, *Rebellion and Violence in Islamic Law* (2001).

[4] Lambton, 'A Nineteenth-Century View of Jihad', 32 *Studia Islamica* (1970) 181; Kohlberg, 'The Development of the Imami Shi'i Doctrine of Jihad', 126 *Zeitschrift der Deutschen Morgenländischen Gesellschaft* (1976) 64.

[5] Chiragh 'Ali lists several such writers, including Ibn Shubruma (d. 761) and Sufyan al-Thawri (d. 778). C. 'Ali, *A Critical Exposition on the Popular Jihad* (1885), at 134.

Yet, by the end of the classical period, the dissenters had been largely marginalized, so much so that their works are not extant.

The classical theory of a triumphant, expansionist, imperial Islam was simply that, a theory, as it was being formulated. Dar al-Islam had fragmented, with Andalusia and North Africa breaking away from the Abbasid caliphate, and the forward expansion of dar al-Islam's frontiers giving way to protracted border wars with the Byzantines, Normans, and Christian rulers of Iberia, as well as internal threats from Persian, Kurdish, Turkish, and other warlords. Yet in the centuries that followed, no formal or systematic reformulation of the classical theory was undertaken. Throughout the nineteenth and into the early twentieth century, as European imperialism advanced in Muslim territories and as Muslim minority communities developed in remote parts of these empires, Muslims were debating whether they lived in dar al-Islam or dar al-harb, what was demanded of them in one territory or the other according to the shari'a, and how they should respond to the emerging *droit public européen*, the basis of modern international law.

3. Islamic Law and International Law: Points of Agreement and Contrast

As international law evolved, Muslim involvement in its formulation was sporadic and minor until after the Second World War. No Muslim state was represented at the conference that resulted in August 1864 in the signing of the First Geneva Convention. However, the Ottoman Empire acceded within one year (in July 1865), and the Qajar government of Iran followed suit in December 1874. Ottoman and Qajar delegations were present at the 1899 and the 1907 Hague Peace Conferences. The delegates—'westernized and westernizers', as James Cockayne characterizes them—made little attempt to inject Islamic rulings on the conduct of war into the deliberations, but 'as a result of the Islamic delegations' interventions, the Hague Peace Conferences officially confirmed the principle of religious non-discrimination as a central tenet of IHL'.[6]

Rather than the substance of international humanitarian law, it was the red cross symbol adopted in 1864 that emerged as the most contentious issue for the Ottoman Empire and later other Muslim states. The Ottomans, during their war in 1876 with Serbia and later Russia, declared they would display a red crescent emblem instead of the red cross because the latter evoked the Crusades and thus was offensive to their soldiers. The Russians initially challenged the Ottomans' move but eventually agreed to respect the red crescent emblem so long as the Ottomans reciprocated for the red cross. This proved to be merely the opening round of a prolonged controversy over the emblem.

Once the red crescent had been tacitly accepted as an alternative emblem, other states moved to have their own distinctive emblems recognized as well, including Persia, which proposed a red lion with sun, based on the Qajar insignia. Realizing that a Pandora's box had been opened, the International Committee of the Red Cross

[6] Cockayne, 'Islam and International Humanitarian Law: From a Clash to a Conversation between Civilizations', 84 *International Review of the Red Cross* (2002) 597, at 608.

(ICRC) mobilized efforts, official and unofficial, to reinstate the red cross as the unique symbol of the movement. The 1906 conference that drafted the Second Geneva Convention unanimously adopted a resolution denying any religious origins or connotation to the emblem. Likewise, numerous ICRC spokesmen gave accounts of how the red cross symbol was adopted in attempts to establish its purely secular origins and significance. These efforts proved in vain. Upon the insistence of Turkey, Persia, and Egypt, the 1929 conference that drafted the Third Geneva Convention formally approved the red crescent and the red lion with sun for use only by those national societies already using them, hoping thereby to limit their use. This effort also failed. As new Muslim-majority states gained independence and acceded to the Geneva Conventions, they adopted, with very few exceptions, the red crescent symbol for their own national societies. Iran dropped the use of the red lion with sun and adopted the red crescent soon after the Iranian Revolution.[7]

Muslim engagement in efforts to broaden the scope of the laws of war would not be prominent until the negotiations for the 1977 Geneva Protocols. At that time, a number of Muslim states, motivated primarily by the Palestinian conflict with Israel, pressed for the adoption of the second protocol elaborating the rights of irregular fighters. But again they did so without recourse to any specifically Islamic injunctions.

As a result of international law's distinctively Western origins and the generally peripheral role played by Muslims in its development, international law's relationship to the jihad tradition is fundamentally different from its relationship to the just war tradition. As Ann Elizabeth Mayer observes:

> The West today deals with international law as a familiar and integral component of Western civilization that has evolved pari passu with Western culture and religion over centuries. In contrast, the Islamic doctrines of war and peace have been part of a juristic culture that has remained closer to its pre-modern roots.[8]

In other words, when Muslims engage with international law as Islamic scholars or jurists, their point of reference is the theory of world order developed more than a millennium ago that was never fully implemented, but never fully rejected or revised either. For many Muslims, therefore, international law is not an outgrowth of their own culture's previous discourses and practices; it is a rival system to the authentically Islamic system.[9]

[7] In 2005, a third protocol to the Geneva Conventions relating to the Adoption of an Additional Distinctive Emblem was approved. The move came after continuing controversy over the red cross symbol, particularly in Israel, where the emblem used is a red Shield (Star) of David. The third protocol adopts a third symbol, the red diamond, inside which distinctive national insignia may be displayed. See Bugnion, 'Towards a Comprehensive Solution to the Question of the Emblem', 85 *International Review of the Red Cross* (2003) 427; Quéguiner, 'Commentary on the Protocol Additional to the Geneva Conventions of 12 August 1949, and Relating to the Adoption of an Additional Distinctive Emblem (Protocol III)', 89 *International Review of the Red Cross* (2007) 175; J. Benthall and J.-B. Jourdan, *The Charitable Crescent: Politics of Aid in the Muslim World* (2003), at 47–53.

[8] Mayer, 'War and Peace in the Islamic Tradition and International Law', in J. Kelsay and J.T. Johnson (eds), *Just War and Jihad* (1991) 195, at 195.

[9] For a sustained argument of this view and a critique of Muslim theorists who challenge it, see Westbrook, 'Islamic International Law and Public International Law: Separate Expressions of World Order', 33 *Virginia Journal of International Law* (1993) 819.

This is not to say that the classical Islamic laws of war and peace are essentially incompatible with modern international law. Indeed, the classical theory of jihad shares many substantive points with just war theory, and inasmuch as these just war principles underlay the evolution of modern international law, the principles of jihad are also compatible with international law. But the jihad tradition evinces far less creative rethinking and adaptation in the modern period than does the just war tradition. The difference in the two traditions is seen in the fact that few modern international lawyers would feel bound by the writings of Hugo Grotius, whereas expositors of Islamic laws of war and peace ineluctably turn to the eighth-century jurist Muhammad al-Shaybani as an authority. Even those Muslims who embrace or accommodate international law cannot ignore the classical legacy. Their task is to reconcile or otherwise deal with certain basic conflicts between Islamic law and modern international law.

At the broadest conceptual level are basic epistemological differences. International law, whether it is seen as derived from natural law, as much of humanitarian and human rights law is, or from positive law, expressed in treaties among states, is firmly secular in foundation. By contrast, the shari'a was conceptualized as divine law, as revealed in God's dispensation, the Qur'an, and expounded by God's Messenger, Muhammad, through the hadith literature compiled centuries after his death. The Qur'an and hadith present no systematic or, some might argue, consistent theory of world order in general or laws of war and peace in particular. Thus it was left to the jurists to develop such a theory through interpretation of the Qur'an and the sunna of the Prophet. Some scholars were more willing than others to undertake creative legal reasoning (*ijtihad*). Nevertheless, all saw their jurisprudence (*fiqh*) not as human law-making but as an exercise in trying to discern God's law.

International law, or specifically public international law, regulates the interactions of states with each other. Islamic law, on the other hand, established rules of behaviour for the Muslim state and individuals in their interaction with non-Muslim political groups and individuals or with Muslim rebels, apostates, and criminals. It was conceived as a unilateral legal system binding only on Muslims, with no assumptions of reciprocity or accountability from those outside the Muslim *umma* (community). It was thus not 'international' law at all. Islamic law did validate diplomacy, treaties, and truces with non-Muslims, but by engaging in such activity, non-Muslims did not come within the legal framework of the shari'a, unless they agreed to dhimmi status. Those who did not accept dhimmi status remained outside of Islamic law, and both they and the Muslims were bound only by the specific provisions of their mutual agreement. If there were any assumptions at all about a 'law of peoples' on the part of Muslim jurists, they were what might be called a rudimentary notion of natural law.

The principal sources of international law are treaties and custom. The Islamic laws of war and peace are primarily 'jurists' law'; they are contained in the intellectual output of scholars working more than a millennium ago. Since custom, or state practice, beyond the traditions ascribed to the Prophet Muhammad and the first four caliphs was never accepted as a legitimate source, Islamic laws of war and peace were more theoretical than practical from their origins. The problem was compounded not just by dramatic changes in the international system and the technology of war but by the intellectual conservatism that set in after the collapse of the Abbasid Empire in the mid-thirteenth century. The generations of jurists who followed busied themselves with preserving their legal heritage rather than revising it.

This approach is evident even among twentieth-century Muslim commentators on Islamic international law. The earliest extended treatises in Western languages by Muslim authors date from the period between the two World Wars. In 1929, Najib Armanazi (d. 1968), a Syrian, published his doctoral dissertation titled *L'Islam et le droit international* (Islam and international law), written at the University of Paris.[10] The Indian scholar Muhammad Hamidullah (d. 2002), published in 1935 his doctoral dissertation submitted to the University of Bonn on neutrality in Islamic law. Hamidullah published a greatly expanded English edition of the work titled *The Muslim Conduct of State* in 1942. This book underwent several revisions in the following decades and became a standard reference for students of Islamic international law.[11] Ahmed Rechid, a Turkish legal scholar, published *L'Islam et le droit des gens* (Islam and the law of nations) in 1937.[12] Armanazi and Hamidullah draw almost exclusively on classical sources as they elucidate Islamic laws of international relations. Their references to Muslim practice are confined mainly to the era of the Prophet and the first four caliphs. Rechid, however, does devote considerable space to Ottoman practices, particularly in the field of conflict of laws, reflecting no doubt his prior experience as a legal advisor to the Turkish Ministry of Foreign Affairs.

For their part, Muslim states adapted to their changing circumstances and adopted various extra-shari'a norms in their statecraft. The Ottomans began to accommodate themselves to European custom by the mid-fifteenth century. In the twentieth century, all postcolonial Muslim states readily acceded to the modern international legal regime. All are members of the United Nations; seven are charter members, including the Kingdom of Saudi Arabia. All are signatories to the Geneva Conventions. Which leads to the fundamental question: through their practice, have not Muslim states all but renounced any place for a distinctively Islamic international law? The answer would seem to be yes, especially as far as relations between Muslim states and non-Muslim states are concerned. But this answer must be qualified by the observation that not infrequently, and especially in their mutual relations with each other, Muslim statesmen still invoke Islamic values and doctrines, including jihad, as if they are still relevant. Moreover, as I discuss at the end of this chapter, Muslim states have taken a tantalizing step towards the creation of an institution that would give substance to the claim that Islamic international law is more than just a historical artifact.

Moving from the broad conceptual level to the specific rulings on war and peace, modern Muslim commentators again face a number of important differences between Islamic law and international law. Under the legitimate grounds for war, the first issue they must grapple with is the classical theory's bifurcation of the world into the two opposing spheres of dar al-Islam and dar al-harb. Though conceived as a single political entity, dar al-Islam was not united even at the time the classical theory was being developed, and as the centuries passed, Muslim jurists found ways to rationalize the

[10] N. Armanazi, *L'Islam et le droit international* (1929).

[11] M. Hamidullah, *The Muslim Conduct of State* (5th ed., 1996).

[12] Rechid, 'L'Islam et le droit des gens', 60 *Recueil des cours de l'Académie de Droit International de la Haye* (1937) 371.

existence of separate Muslim states. Most Muslim scholars and activists are reconciled today to the legitimacy of multiple Muslim states, but the principle of state sovereignty which lies at the heart of international law remains to be fully reconciled with the Islamic ethical ideals of Muslim unity and fraternity.[13]

Dar al-harb was defined as territory where the shari'a was not enforced, and as such, the classical theorists conceived it as populated by benighted peoples. Muslims were called upon by God to bring the light of Islam to them, through peaceful proselytizing if possible, and if not, through war. If the residents of dar al-harb elected to submit to Muslim sovereignty as dhimmis, their territory was considered absorbed into dar al-Islam. The expansionist jihad to which the classical jurists devoted so much attention is an obvious and glaring problem for modern Muslim theorists who would accommodate Islamic law to international law.

Under the legitimate conduct of hostilities, Islamic injunctions do enjoin discrimination and proportionality, the twin principles at the core of international humanitarian law. But as we delve deeper into the specific opinions of the different legal schools, we find a number of provisions that dilute the force of these principles. Discrimination in war meant refraining from the intended killing of women, children, and various categories of males deemed incapable or unwilling to fight. But this notion of discrimination is not equivalent to the modern principle of non-combatant immunity. All able-bodied adult male prisoners, whether they were previously engaged in fighting or not, could be executed upon the decision of the Muslim commander, unless they had surrendered with an explicit guarantee of security (*aman*). All of the enemy population, military and civilian, could be held for ransom or sold in slavery.[14]

Classical jurists also gave Muslim commanders wide latitude in terms of the weapons and tactics they could employ in overcoming the enemy. Many of these could hardly be implemented without the expectation of large-scale civilian casualties. Discrimination and proportionality were tempered by appeals to military necessity and reciprocity.

Finally, the classical Islamic theory contains no formal provisions for punishing Muslim commanders or troops who violate restrictions on the proper conduct of jihad. Under Islamic law, a Muslim who kills a fellow Muslim is both morally and legally accountable. But a Muslim soldier who violates the rules of war against a non-Muslim enemy, either military or civilian, is at most morally accountable, meaning he will be judged and punished by God in the hereafter, not by humans in this life.[15]

These are merely some of the most prominent conflicts between the classical Islamic laws of war and modern international law. All Muslim theorists, if they are to be intellectually credible, must address these basic concerns, whether their goal is to assimilate, accommodate, or reject international law.

[13] I have discussed these issues in Hashmi, 'Islam, the Middle East, and the Pan-Islamic Movement', in B. Buzan and A. Gonzalez-Pelaez (eds), *International Society and the Middle East: English School Theory at the Regional Level* (2009) 170.

[14] Kelsay, 'Islam and the Distinction between Combatants and Non-Combatants', in J.T. Johnson and J. Kelsay (eds), *Cross, Crescent, and Sword* (1990) 197; Hashmi, 'Saving and Taking Life in War: Three Modern Muslim Views', in J. Brockopp (ed.), *The Islamic Ethics of Life: Abortion, War, and Euthanasia* (2003) 129.

[15] E. Landau-Tasseron, *"Non-Combatants" in Muslim Legal Thought* (2006), at 2–3.

4. The Assimilationists

Muslim proponents of assimilationism are few in number, at least as far as theorists are concerned. The reason may have to do with the fact that assimilationism treats the Islamic theory of world order as a historical artifact, one with no real applicability in the modern age. Pragmatically, it has no applicability because the conditions of international relations today are fundamentally different from those in which the theory was elaborated. Morally, it has no applicability because it is not grounded in either the Qur'an or sunna, but on unidentified, non-Islamic sources that influenced the classical jurists who developed it. Thus, for the assimilationists, the impetus to comment on the classical Islamic theory is not great; they assume that public international law is practically and normatively binding on Muslims. Muslim jurists who have contributed to the development of international law during the past century have most often done so without invoking Islam in their profession.

Majid Khadduri (d. 2007) is the best-known and most influential scholar of the Islamic laws of war and peace in the West. He was an Iraqi Christian, not a Muslim, but this fact seems to be unknown or unimportant to a number of Muslim and non-Muslim commentators on his work. His impact on the study of the Islamic theory of international relations is certainly comparable to that of any contemporary Muslim scholar, so he merits inclusion in this discussion. Khadduri was not only a scholar of Islamic law and philosophy, and Middle Eastern history and politics, he was also a lawyer and diplomat who participated in the drafting of the UN Charter as a member of the Iraqi delegation. He wrote a number of articles and books on Islamic theories of international relations, including the survey *War and Peace in the Law of Islam* and a translation of one of the seminal works in Islamic law, a treatise by the jurist al-Shaybani.[16] There is a paradox here: through his scholarship, Khadduri revived interest in the classical Islamic theory, among both Muslims and non-Muslims, but through his career as a diplomat and his reflections on the place of Islamic theory in the modern age, he consigned this theory largely to the past.

Two articles published by Khadduri during the 1950s, 'Islam and the Modern Law of Nations' and 'The Islamic System: Its Competition and Co-Existence with Western Systems', outline most clearly his views on the role of Islamic law in the modern age. Both essays open with the claim that Christendom and Islam had developed historically as two opposing civilizations, 'each advocating a world order sharply in conflict with the other … The world orders of the two opposing ideologies were at the outset so exclusive and inflexible that each deemed the destruction of the other as absolutely necessary for its survival.'[17] Such an intractably hostile stance could not be maintained for long, Khadduri continues, and so, 'after a long period of competition and warfare by virtue of which each came to the inevitable conclusion that its moral and political

[16] M. Khadduri, *War and Peace in the Law of Islam* (1955); *The Islamic Law of Nations: Shaybani's Siyar* (2001).

[17] Khadduri, 'The Islamic System: Its Competition and Co-Existence with Western Systems', 53 *Proceedings of the American Society of International Law* (1959) 49, at 49.

principles could not be imposed on the other', both civilizations began to adapt to the necessities of coexistence.[18]

The triumph of pragmatism over ideology, Khadduri writes, resulted in a 'thorough re-examination' by Muslims of their earlier 'legal theory of foreign relations'. He cites three changes embraced by Muslim powers, in particular the Ottoman Empire: the separation of religious doctrine from the conduct of foreign relations; acceptance of the possibility of peaceful intercourse among nations of different religions; and the territoriality of law, which undergirds the idea of sovereign states.[19] Still, despite the thorough revision in Islamic principles of statecraft, 'neither Islam nor Christendom [was yet] prepared to meet on a common ground and modify their religious principles for the purpose of developing a law of nations based on equality and reciprocity'. Had the Ottoman Empire been integrated into the European system during the formative period of the law of nations (by which Khadduri means the sixteenth century) this law might have become universal much earlier in time, he suggests.[20]

'Twentieth-century Islam found itself completely reconciled to the Western secular system', Khadduri concludes. 'The active participation of Muslim states in international councils and organizations demonstrates Islam's willingness to take active part in the promotion of international peace and security and its support of the development of the modern law of nations.'[21] In particular, jihad has become 'an obsolete weapon'.[22] By this Khadduri means the notion of expansionist jihad, for elsewhere he clarifies, 'Even those publicists who objected to the secularization of the internal law of Islam have accepted marked departures from the traditional Muslim law governing Islam's foreign relations. Almost all of them, who often invoked the *jihad* against Western encroachment on Islam, repudiated the idea that the *jihad* is offensive in character.'[23]

Khadduri ends one of his essays by acknowledging that some Muslim jurists (those I have labelled accommodationists) maintain that the modern law of nations, in order to meet the needs of 'an expanding world community', should incorporate elements of Islamic law. They often base their claims, Khadduri notes, on Article 38 of the Statute of the International Court of Justice, which 'permits the adoption of new maxims of law from the legal systems of "civilized nations"'.[24] What principles might Islamic civilization contribute to the further development of international law? Khadduri suggests two: (1) recognition of the individual as a subject of international law; and (2) moral principles should not be divorced from law, for 'the historical experiences of Islam, indeed the historical experiences of all mankind, demonstrate that any system of law, whether municipal or international, would become meaningless if divorced completely from moral principles'. He hastens to add, however, that the moral principles he has in mind are not based on specific religious doctrines. 'Religious doctrine as a basis for the conduct of the state created friction and continuous warfare with other

[18] Ibid., at 50.
[19] Ibid., at 51.
[20] Khadduri, 'Islam and the Modern Law of Nations', 50 *American Journal of International Law* (1956) 358, at 364.
[21] Khadduri, above note 17, at 51.
[22] Khadduri, above note 20, at 371.
[23] Ibid., at 370.
[24] Khadduri, above note 17, at 51f.

nations; but religion in terms of moral principles prompted the Muslims to observe humane principles embodied in their law during their hostilities with other nations.'[25]

5. The Accommodationists

The accommodationist strand of Muslim response is characterized by the acknowledgment that most aspects of the classical theory are incompatible with modern international relations, but that through renewed interpretive activity, the moral values of Islam could provide the basis for a new and distinctive Islamic international law. This Islamic law might then be enforced as a form of 'regional law' among Muslim states. A number of distinguished Muslim jurists have espoused some form of this accommodationist line, including 'Abd al-Razzaq al-Sanhuri, Subhi Mahmassani, Muhammad Abu Zahra, Muhammad Tal'at al-Ghunaimi, and Wahba al-Zuhayli. I will focus here on the arguments of the Algerian diplomat and former judge of the International Court of Justice, Mohammed Bedjaoui.

Bedjaoui contributed a lengthy analysis of the Iran-Iraq War in light of the principles of Islamic law to an anthology titled *The Gulf War of 1980–1988: The Iran-Iraq War in International Legal Perspective*.[26] He concludes this essay by asking two pertinent questions: First, does an Islamic legal order even exist today, 'when that order not only appears to have undergone no evolution since the seventh century but also transpires in practice to have found little favour with Muslim States in their respective relations'? And second, how might an Islamic legal order, 'were it to be recognized, updated and applied . . . co-exist with the universal legal order, in accordance with the relevant provisions of the United Nations Charter'?[27]

The answer Bedjaoui gives to the first question is no—Islamic law has no modern relevance. 'More particularly, Islamic international law still remains, regrettably, a law of speculation pursued for purely academic purposes.' He blames the stagnation of Islamic law partly on Western scholars and jurists, who historically denied Muslim contributions to the development of their own legal system. Thus, Islamic law was treated as an alien system of law when European states disseminated their own law of nations in the tracks of imperialism.[28] Muslim jurists by and large accepted the European interpretation of Islamic law, according to Bedjaoui, and it is these Muslims to whom he assigns principal responsibility for the 'decadence' of Islamic law. 'With a few rare exceptions, jurists from the newly independent Muslim countries have succumbed to the easy temptation of lazily imitating the West instead of meeting the more demanding needs of creativity.'[29]

But instead of resigning himself to the complete obsolescence of Islamic law, Bedjaoui declares that Muslims can and should find a place for Islamic law in international relations. He writes:

[25] Ibid., at 52.

[26] Bedjaoui, 'The Gulf War of 1980–1988 and the Islamic Conception of International Law', in I. Dekker and H. Post (eds), *The Gulf War of 1980–1988: The Iran-Iraq War in International Legal Perspective* (1992) 277.

[27] Ibid., at 293.

[28] Ibid., at 294–96.

[29] Ibid., at 296.

> 'I can restrain myself no longer from *appealing to all jurists in the Muslim world to take greater personal responsibility and mobilize more efficiently in a decisive effort of creativity, instead of confining themselves to an attitude of slavish imitation which often results in strapping lifeless artificial limbs of foreign legal origin on living human communities*' (emphasis in original).[30]

Finally, Bedjaoui turns to his second question on how a revived Islamic law might coexist with international law. He observes that by becoming parties to the UN Charter Muslim states did not foreclose the possibility of establishing 'their own regional organs within which Islamic law could have enjoyed every opportunity of a revival'. Muslim states have in fact moved to create such 'regional' organizations, the most prominent being the Organization of the Islamic Conference, founded in 1969 (renamed in 2011 as the Organization of Islamic Cooperation) and today consisting of fifty-seven members. Yet, as Bedjaoui notes, 'so far that Organization does not appear to have produced any clear undertaking to ensure the development and modernization of Muslim public international law by adapting it to the new structure of the international community and having it applied in inter-Islamic relations'.[31] Should it ever do so, Bedjaoui is confident that no fundamental incompatibility will arise between the newly framed Islamic international law and public international law. In the case of the Iran-Iraq War, for example, 'one arrives at virtually the same legal analysis ... whatever the "key" be used, whether that of public international law or that of Islamic law'.[32]

6. The Rejectionists

At the forefront of Muslim opinion rejecting international law are, of course, the spokesmen of militant Islamic groups. Not just the leaders of Al-Qaeda but most extremist groups have long expressed their contempt for the United Nations and for international law, which they dismiss as tools for Western hegemony. In 1993, the same cell that planned and executed the bombing of the World Trade Center was plotting to bomb the UN headquarters. In 2003, one of the first targets of Al-Qaeda in Iraq was the newly established UN office in American-occupied Baghdad. Bin Laden addressed international law directly in his so-called Letter to America, released on 6 October 2002. In it, he does not condemn or reject international law directly, perhaps a reflection of the Western audience to which the message is aimed. What he rails against are the double standards and hypocrisy that the United States demonstrates with respect to international law. He cites four cases: the United States touts policies to prevent nuclear proliferation, but it exempts Israel. The United States pays lip service to the UN and UN resolutions, but it supports Israel despite its refusal to comply with UN resolutions. The United States demands the censure and prosecution of individuals it labels war criminals while at the same time demanding its troops be granted immunity from prosecution. The United States claims it supports human rights while it flouts international law in its treatment of Muslim

30 Ibid.
31 Ibid., at 297.
32 Ibid., at 298f.

prisoners. 'What happens in Guantanamo is a historical embarrassment to America and its values, and it screams into your hypocritical faces: What is the value of your signature on any agreement or treaty?'[33]

Perhaps the most detailed and sophisticated treatment of the international law of war penned by a sceptical Muslim writer is that of Abu al-A'la Mawdudi (d. 1979) in *al-Jihad fi al-Islam* (Jihad in Islam). Mawdudi began this work in 1926 as a series of articles in the Urdu-language newspaper *al-Jami'yat*, and the complete book was first published in 1930.[34] At the end of his book, Mawdudi included a lengthy essay on the development and content of the modern law of war. Here, he takes the reader through the main points of the Hague Conferences of 1899 and 1907, the Covenant of the League of Nations, and the post-First World War conferences on arms control. Turning from international law to state practice, he contends that the First World War had demonstrated the abject failure of Western humanitarian principles, but in spite of the catastrophe of that conflict, Western nations showed no signs of altering their behaviour. International law, for Mawdudi, is nothing more than rhetorical camouflage for Great Power politics.[35]

Mawdudi's intent in *al-Jihad fi al-Islam* is to demonstrate the distinctiveness and superiority of the Islamic laws of jihad, which are of divine origin, over all other systems, including international law. He concludes with the following gist of his comparison between Islamic law and international law: first, international law cannot be defined as 'law' in any true sense. It is entirely subject to the whims and self-interests of states, in particular the Great Powers. Under such circumstances, how can one establish any objective standards for determining the lawfulness of state behaviour? Second, if international laws generally have little relevance to interstate behaviour, the laws of war are especially irrelevant. States constantly resort to the notion of military necessity to set aside the military conventions they have previously signed. Third, since international law is not based on transcendent moral principles, as is the case with Islamic law, and it binds only those states that consent to be bound by treaty, the laws of war are speedily cast aside when war breaks out. One state withdrawing from its treaty obligations is enough to bring about the collapse of the entire edifice of international law, as the law of reciprocity dictates that states match the deeds of their opponents lest they lose the strategic advantage. Fourth, even though Islamic law predates Western law by more than thirteen centuries, international law has not added in any notable way to the humanitarian principles Islam legislated in the conduct of war. Fifth, Western laws of war limit only the means, not the ends of war. Thus, they do not prevent wars of conquest and plunder. Islamic law, in contrast, strictly limits both the means and the ends of war.

Having charted the differences between the two legal systems in sharp relief, Mawdudi asserts: 'Compared to Western laws, Islam's law of war is more correct,

[33] O. bin Laden, *Messages to the World: The Statements of Osama bin Laden* (2005), at 169f.

[34] See S.V.R. Nasr, *Mawdudi and the Making of Islamic Revivalism* (1996), at 22f. for the background to Mawdudi's writing of *al-Jihad fi al-Islam*.

[35] A. Mawdudi, *al-Jihad fi al-Islam* (1988), at 459–591; cf. English trans. by S.R. Shah (2017), at 323–436.

beneficial, logical, and firm.'[36] He never explicitly rejects Muslim adherence to international law, but we can infer that he would ask why, after his analysis, any Muslim of sound mind would choose international law over Islamic law.

7. Conclusion: From Theory to Practice

The advent of international law has yielded divergent responses from Muslim theorists. Nevertheless, I believe the trend is overwhelmingly in favour of the assimilationist or accommodationist view. The rejectionist position is propounded by a limited number of the most conservative scholars and activists. Even rejectionist theoreticians who bitterly criticize international law in their writings fail to effect or even advocate radical changes in state policies when they are in a position to do so. The Jamaʿat-i Islami, the party that Mawdudi led for nearly thirty years in Pakistan, never demanded that their nation withdraw from the United Nations or the international legal regime in general, or the treaties specifically governing the laws of war that Pakistan had signed. The same is true of Khomeini in Iran and the Taliban leadership in Afghanistan. They may have been rejectionist theoreticians, but they were also politicians. As such they accommodated themselves to the reality that their states operated in an international political and legal system far removed from what they considered the ideal.

Despite the evidence indicating Muslims accept the authority of international law, either begrudgingly or wholeheartedly, it would be a mistake, I think, to dismiss altogether any role for Islamic values or legal principles in contemporary international relations. The most concrete manifestation of the continuing vitality of the accommodationist position is the decision by the Organization of the Islamic Conference (OIC) in 1981 to create the International Islamic Court of Justice (IICJ). This institution emerged out of popular concerns with the inability of the OIC to resolve the Iran-Iraq War. But trouble arose soon after the OIC adopted the resolution to create the court. The statute of the IICJ was not approved until 1987. According to it, the court has jurisdiction over disputes referred to it by the OIC member states. It is also to render advisory opinions, or *fatwas*, on legal questions referred to it by the other organs of the OIC. To all intents and purposes, it is to operate as an Islamic version of the International Court of Justice. The defining characteristic of this court, however, is that its sole source of law would be the shariʿa. Its eleven judges are envisioned as being the leading experts in shariʿa provisions on international law.[37]

Everything is in place for this court to begin its operations, even its headquarters in Kuwait. Yet it has never convened because the court's statute has not been ratified by the requisite two-thirds majority of OIC member states. The failure of most states to ratify, the very decision to set the ratification number so high, and the earlier disputes

[36] Ibid., at 596–600.

[37] N. Baba, *Organisation of the Islamic Conference: Theory and Practice of Pan-Islamic Cooperation* (1994), at 214f.; M. al-Midani, 'Presentation of the Islamic International Court of Justice: A Muslim Judiciary Court', available at <http://www.acihl.org/articles.htm?article_id=8> (last visited 10 January 2020).

over the statute itself all point to a stark reality of the OIC: it is an intergovernmental organization founded on the rhetoric of Islamic universalism but mired ever since in the politics of its squabbling and ideologically divided member states. So, unfortunately, because the IICJ is not likely to meet anytime soon, we lack the presence of an authoritative body that may resolve such basic questions as whether a modern Islamic law of war and peace is possible and how this law relates to public international law.

22

In the Name of State Sovereignty? The Justification of War in Russian History and the Present

Paul Robinson and Mikhail Antonov

1. Introduction

At the time of writing, at the end of 2019, the armed forces of the Russian Federation are engaged in military operations in Syria. In the past dozen years, Russia has also fought a war against Georgia and has almost certainly provided military support to the rebels of the Donetsk and Lugansk Peoples' Republics in Ukraine. Western politicians and journalists express fears of Russian aggression in the Baltic region of Europe or elsewhere, discuss with alarm the 'Gerasimov Doctrine' (named after the current Chief of the General Staff of the Russian armed forces, General Valery Gerasimov), which supposedly blurs the line between war and peace, and proclaim their outrage at what they say is excessive violence used by the Russians in Syria. In this context, understanding Russian views of when war is justified (*jus ad bellum*) and what actions are permissible during war (*jus in bello*) is a matter of some salience.

This is far from an easy task. As Russian scholar A.A. Skvortsov notes, 'A first, even superficial, glance at Russian philosophy of war notices that it doesn't exist as an integral system.'[1] 'One can't talk of a completely original Russian philosophy of war because ... it is fragmentary and incomplete', Skvortsov adds.[2] The aim of this chapter is to assemble these fragments. To this end, the chapter first reviews the history of Russian thought on the law and ethics of war, and then examines statements by contemporary Russians about the law and ethics of both war in general and the particular wars fought by the Russian Federation in the post-Soviet period. In the process the chapter will identify various trajectories of thought which have played a significant role in Russian history and which continue to influence Russia today.

2. The Ethics and Law of War in Pre-Soviet Russia

The eminent Russian and Soviet jurist Vladimir Grabar' argued that the wars between ancient Russian princes should be considered not as international conflicts but as

1 A.A. Skvortsov, *Russkaia religioznaia etika voiny XX veka* (2002), at 25.
2 Ibid., at 28.

Paul Robinson and Mikhail Antonov, *In the Name of State Sovereignty? The Justification of War in Russian History and the Present* In: *The Justification of War and International Order*. Edited by: Lothar Brock and Hendrik Simon, Oxford University Press (2021).
 DOI: 10.1093/oso/9780198865308.003.0022

disputes taking place within one big family. Consequently, one could draw no conclusions from them about international law.[3] This is indeed the case, and these conflicts had no specific legal connotations. Nevertheless, the contrast between the constant wars of the Middle Ages and the Christian principles of love could not but provoke critical philosophical reflection.

The first important philosophical work of the Russian Middle Ages, Metropolitan Hilarion's 'Sermon on Law and Grace', was delivered in the mid-eleventh century.[4] Hilarion focused his analysis on the discrepancy between the Old Testament ethics of war between God's chosen people and their enemies, and the New Testament ethics of love and mercy. In the first 'theoretical' part of his eulogy, Hilarion argued that the grace of the New Testament replaced the law of the Old Testament, so that 'the lake of the Law has dried up, but the Stream from the Gospel swelled and flowed over to all the Earth' (para. 34).

However, in the second part of his sermon Hilarion changed his perspective, and glorified the deeds and the person of Grand Prince Vladimir, praising him as 'the monarch of his land, and the lands all around he subdued: the peaceful in peace, the rebellious he quelled with the sword. Thus he lived in his days, and he tended to his land with justice, valor, and reason' (para. 43). According to Hilarion, Vladimir's greatest deeds were fighting idolatry and protecting the baptized Rus' from pagans—'since his [Vladimir's] piety was coupled with power' (para. 47). The solution elaborated by Hilarion meant that wars were religiously justified when they were waged for the protection of peace and for the expansion of Christianity.

From Hilarion onwards, Russians have on occasion assigned a positive value to war, as an activity carried out in pursuit either of God's will or some humanitarian goal. This attitude has been associated with a messianic streak in Russian thought, according to which Russia is a nation endowed with a special moral purpose. For instance, religious language played some role in the wars of Muscovite rulers, such as Vasilii III's and Ivan IV's campaigns against the Tatar city of Kazan, lending those campaigns something of a crusading quality,[5] as it did also in some earlier struggles, such as the 1242 battle on the ice against the German crusaders or the 1380 Kulikovo battle against the Mongols. Later, religious language was prompted by the messianic conception of the Russian state which, after the fall of Constantinople, remained the only Orthodox political entity in the world. This messianism is patent in the sixteenth-century theory of 'Moscow as Third Rome'[6] and would become a hallmark of some Russian approaches to matters of war.

[3] V.E. Grabar', *Materialy k istorii literatury mezhdunarodnogo prava Rossii. 1647–1917* (2005), at 1.

[4] Its full title is *Concerning the Law given by Moses and the Grace and Truth which came by Jesus Christ. And: how the Law departed, and Grace and Truth filled all the earth, and Faith spread forth to all nations, even unto our nation of Rus'. And: an encomium to our kagan Vlodimer, by whom we were baptized. And a prayer to God from all our land* [Simon Franklin, *Sermons and Rhetoric of Kievan Rus* (1991), at 3–30].

[5] Stoyanov, 'Eastern Orthodox Christianity', in G.M. Reichberg and H. Syse (eds), *Religion, War, and Ethics: A Sourcebook of Textual Traditions* (2014) 164, at 193 and 195.

[6] Toumanoff, 'Moscow the Third Rome: Genesis and Significance of a Politico-Religious Idea', 40 *The Catholic Historical Review* (1955) 411.

Argumentation based on the Holy Scriptures was common in the Russian medieval philosophy of war. Typically, one of the greatest Russian political philosophers of the Middle Ages, Maksim Grek (1480–1556), implored rulers

> to have a good council which does not incite you to have troops and weapons, but advises on peace and reconciliation with all neighbours of your God-blessed kingdom. Because God's wrath and anger is directed against those who oppose His command: He will bring down the peoples who seek wars. As to you: have peace with everyone.[7]

The most important command for rulers was to rely on justice, whereas war was deemed to be possible only in extreme necessity.[8] Land-grabbing was resolutely condemned by Grek: 'It is not good for a true Christian tsar and true leader to do injustice by seizing and stealing lands and property that belong to others.'[9]

Religious-philosophical thinking started to cede place to rational conceptualizations of war and peace in the sixteenth century when Muscovy became independent from the Golden Horde, formed itself as a unified state, and actively engaged in diplomatic communications with the West. The Acts of the first Muscovite Committee for Foreign Affairs (*Posol'skii prikaz*, created in 1549) demonstrate this change.

Professor Grabar' opined that officials (*d'iaki*) of this Committee were the first Russian international-law scholars.[10] The Committee's Acts (*Stateinye spiski*) contain instructions given to ambassadors, and their reports about missions accomplished. These instructions and reports could not avoid dealing with the philosophy of war. Their philosophy relied on three principles quite commonly held at that time: war was justified when other means of settlement brought no result; when it had a just cause; or when the adverse state committed injustice (*nepravda*). War was also to be officially declared and was not to impede commercial relations.[11] Thus, Russian ambassadors to the Polish king were instructed to admonish him that he did not want to

> do good deeds but wishes blood and mischief against peasants. But God Almighty will see from above and will distinguish between justice and injustice. He will assist justice and crush injustice and . . . therefore we are not afraid of your threats.[12]

Muscovite ambassadors systematically collected Western ideas on the philosophy of war, and Russian ideas of that time did not differ significantly from those of the West.[13]

In 1607, during the reign of Tsar Vasilii Shuiskii, one of the officials of the *Posol'skii prikaz*, Onisim Mikhailov, drafted a Military Statute (*Voennyi ustav*). Mikhailov's

[7] Grek, 'Glavy poutchitelye nachalstvuiushchim pravoverno', in M. Grek, *Sochineniia* [Works] (1860), vol. 2, at 162.

[8] Ibid., at 159.

[9] Ibid., at 172.

[10] V.E. Grabar', 'Pervaia russkaia kniga po mezhdunarodnomu pravu', 7 *Vestnik Moskovskogo universiteta* (1950) 101.

[11] Grabar', above note 3, at 9–12.

[12] Ibid., at 10.

[13] Ibid., at 13–33.

book was the first attempt to reflect on the just causes of war in a modern fashion. He avoided citing the Holy Scriptures and explicitly relied on Western literature: indeed as the subtitle of his book says, his work is a collection of extracts from 'foreign military books'.[14]

In 1621, after the end of the Years of Trouble (*smutnoe vremia*), Mikhailov revised his draft, but it remained unpublished until 1777. Mikhailov's book was possibly too progressive for a time of religious mysticism. In his history of Russian political philosophy from 1500 to 1801, Gary Hamburg notes that, 'To a degree almost unfathomable for twenty-first century secular people, Russian thinking about politics before the eighteenth century rested on religious assumptions.'[15]

From the reign of Emperor Peter the Great, however, the rational style of thinking became more firmly established. Peter and those around him certainly did not consider war and conquest reprehensible. On the contrary, war was seen as a legitimate means of pursuing Russia's interests. One of Peter's closest aides, Theophan Prokopovich (1661–1738), praised his emperor in 1722 for showcasing Russian military might to the West:

> The opinion of foreign peoples about us was that we are barbarians ... Now, those who considered us as rude are seeking our brotherhood; those who condemned us are glorifying us; those who threatened us are fearing us and trembling: they cancelled their opinions and their previous stories about us, they started to speak and write differently.[16]

In 1717, the first Russian treatise on the philosophy of war was published by Petr Shafirov. The main goal of its author, a Russian diplomat in the service of Peter the Great, was to justify Russia's waging war against Sweden in 1700 (the Great Northern War of 1700–1722). The title of his treatise was telling: *Deliberations on just causes of the war between Russia and Sweden*.[17] Shafirov relied on the works of Pufendorf, Grotius, and other Western authors, arguing that *ius gentium* (the law of nations) justified Russia's territorial claims in the Baltic region. This justification of war in the vocabulary of Western European international law underlines Peter's efforts to integrate Russia into the European normative order of the early eighteenth century.[18] As a full-standing member of the international community, Russia could demand that other countries comply with international law, and could enforce their compliance by military means if they continued to violate the rules. Shafirov argued that in so doing,

[14] *Ustav ratnykh, pushkarskikh i drugikh del, kasaiushchikhsia do voinskoi nauki*, Part 1 (1777), Part 2 (1781).

[15] G. Hamburg, *Russia's Path Towards Enlightenment: Faith, Politics, and Reason, 1500–1801* (2016), at 726.

[16] Prokopovich, 'Slovo pokhval'noe v den' rozhdestva blagorodneishego gosudaria i tsarevitcha velikogo kniazia Petra Petrovitcha', in Th. Prokopovich, *Sochineniia* (1961) 38, at 46.

[17] P.P. Shafirov, *Rassuzhdenie o zakonnykh prichinakh voiny mezhdu Shvetsiei i Rossiei: 1700–1721* (1973).

[18] See also Piirimäe, 'Russia, The Turks and Europe: Legitimations of War and the Formation of European Identity in the Early Modern Period', 11 *Journal of Early Modern History* (2007) 63; see also Tischer in this volume.

Russia acted for the sake of eternal peace, which was only possible when all states complied with international law.[19]

During the eighteenth century, Russian scholars published a number of important books on natural law. Among these, Vladimir Zolotnitskii's 1764 *Abbreviation of Natural Law* dealt specifically with the philosophy of war.[20] Relying on the authority of Western authors, Zolotnitskii presupposed the natural right of every country to self-preservation. This right implied that every state had the right to wage war if it was necessary for the 'restoration of quiet' (§ 277), and that states were not required to tolerate insults from other states for the sake of keeping peace.[21]

Catherine II's 1780 *Declaration on Maritime Armed Neutrality* likewise relied on natural law, but offered a different interpretation.[22] Rejecting England's demands that Russia block trade with the rebellious North American colonies during the American War of Independence, Russia justified its refusal by 'simple, clear and undisputable premises of natural law'. These premises not only prohibited restrictions on trade and the capture of neutral countries' ships, but also required big empires to abide by the rules of international law in wars with their colonies.

A pacifist style of thinking was common among Enlightenment thinkers, including Russians. This can be seen, for instance, in Vasilii Malinovskii's *Deliberations on Peace and War* (published in two parts in 1790 and 1798).[23] Malinovskii considered war as 'evil in itself and as the combinations of all evils of the world ..., which combines all the ferocity of animals with ability of human mind that seeks to perish people'.[24] Malinovskii even went as far as to propose creating a 'European Union', which would become a forum of European peoples for discussions of principles of international law 'on the basis of good wishes of eternal peace'.[25] In his opinion, conflicts between European peoples should be settled not through war but by decisions of an all-European court.[26]

In the beginning of the nineteenth century international law became an independent legal discipline (*pravo obshchenarodnoe*), taught at law faculties of Russian universities along with natural law. Russian legal scholars not only systematically wrote on matters of law of war, but also proposed a number of important insights for the development of international law in this respect. It remains a matter of contention whether this development was natively Russian and stemmed from its medieval legal culture, or was something artificially transplanted from the West by Peter the Great and his successors. For instance, Mikhail Taube argued that the concept of just war was Western and did not match either Byzantine or Muscovite political doctrines,

[19] For the main ideas of Shafirov's book and their importance for Russian theory of international law, see Butler, 'Trekhsotletie vykhoda v svet pervoi russkoi original'noi knigi po mezhdunarodnomu publichnomu pravu', 4 *Vestnik Universiteta imeni O.E. Kutafina* (2017) 41.

[20] *Sokrashchenie estestvennogo prava, vybrannoe iz raznykh avtorov dlia pol'zy Rossiiskogo obshchestva Vladimirom Zolotnitskim* (1764).

[21] Ibid.

[22] F.F. Martens, *Sobranie traktatov i konventsii, zakluichennykh Rossiei s inostrannymi derzhavami* (1900), vol. 9, at 307–10.

[23] V.F. Malinovskii, *Rassuzhdeniia o mire i voine* (1803).

[24] Ibid., at 1f.

[25] Ibid., at 98.

[26] Ibid., at 114.

according to which wars were justified only by the fact that they corresponded with the interests of emperors.[27]

In the mid-nineteenth century, the Crimean and Russo-Turkish Wars provided perhaps the clearest examples in Russian history of humanitarian intervention being cited as a just cause for war. For instance, in 1854, during the Crimean War, the Slavophile philosopher Aleksei Khomiakov wrote that, 'God has not given us the task of conquering far-off lands ... but he does give us the task of saving brothers who are blood of our blood and heart of our heart.'[28] It is worth noting that even in this case, humanitarian intervention was justified not in terms of defending human beings per se, but in terms of defending people whom Russians considered 'blood of our blood'. It therefore differed from the way in which humanitarian intervention is generally understood in the Western world today.

In any case, the messianic impulse in Russian philosophy was never universally shared by Russia's elites. Tsar Nicholas I, for instance, strongly condemned Pan-Slavism and its calls for military intervention in foreign countries to liberate oppressed peoples. Presaging the arguments which modern Russians use against the idea of humanitarian intervention, Nicholas claimed that the Pan-Slavs' policies undermined state sovereignty and threatened the entire international system. He complained:

> Under the guise of sympathy for the Slavic tribes supposedly suppressed in other states there is hidden the criminal thought of a union with these tribes, in spite of the fact they are subjects of neighbouring and in part allied states. And they expect to attain this goal not through the will of God, but by means of rebellious outbreaks to the detriment and destruction of Russia herself.... And if indeed, a combination of circumstances produces such a union [of the Slavs], this will mean the ruin of Russia.[29]

In this period, among those Russian philosophers who addressed the subject of war, both pacifism and militarism were rare. Instead, most followed a third trajectory—'necessary war'—maintaining that war is an evil, and never just, but that on occasion it is a lesser evil and therefore tragically necessary. The primary criterion for determining whether war is permissible is not, therefore, justice, but necessity. This understanding of necessity was different from that of Clausewitz and his followers in Germany. Unlike them, Russian thinkers did not consider war either as natural or as justifiable, even desirable, on grounds of military or political necessity.[30] Rather, the necessity of war was a matter of *ultima ratio* in ethical choice, specifically the choice between two evils. War is evil, but in cases of extreme injustice one has an ethical or even religious obligation to resort to this evil: for instance, to wage war to prevent genocide (one of the major points in the justification of the Balkan War with Turkey).

[27] *Baron M. de Taube*, 'Études sur le développement historique du droit international dans l'Europe Orientale', 11 *Recueil des Cours* (1926) 341.

[28] Stoyanov, above note 5, at 198.

[29] W.B. Lincoln, *Nicholas I: Emperor and Autocrat of All the Russias* (1978), at 164.

[30] Simon, 'The Myth of Liberum Ius ad Bellum: Justifying War in 19th-Century Legal Theory and Political Practice', 29 *European Journal of International Law* (2018) 113; and Simon in this volume.

The religious justification of violence against evil was one of the central 'accursed Russian questions'[31] put by Ivan Karamazov in Dostoevsky's *Brother Karamazov*. Karamazov rejected God, refused salvation, and submitted himself to the tyranny of the world because of a 'child's *tears*':

> While there is still time, I hasten to protect myself, and so I renounce the higher harmony altogether. It's not worth the tears of that one tortured child who beat itself on the breast with its little fist and prayed in its stinking outhouse, with its unexpiated tears to 'dear, kind God'! It's not worth it, because those tears are unatoned for. They must be atoned for, or there can be no harmony... I don't want harmony. From love for humanity I don't want it. I would rather be left with the unavenged suffering. I would rather remain with my unavenged suffering and unsatisfied indignation, even if I were wrong. Besides, too high a price is asked for harmony; it's beyond our means to pay so much to enter on it. And so I hasten to give back my entrance ticket, and if I am an honest man I am bound to give it back as soon as possible. And that I am doing. It's not God that I don't accept, Alyosha, only I most respectfully return him the ticket.[32]

In the second half of the nineteenth century, the question of war as a 'necessity' was not discussed in Russian philosophy and literature in the terms of international legal debates (is war a catalyst of progress, is perpetual legal peace possible? etc.), but rather in the terms of theodicy and religious philosophy. An example is the late nineteenth-century philosopher Vladimir Solov'ev. For Solov'ev war was an evil, but not an absolute one. He compared it to throwing a child out of a window. In normal circumstances, this would be a reprehensible act, but if the building was on fire and there was no other way of getting the child out of the building, 'this terrible action becomes not only permissible but even obligatory'.[33] The test is whether the action is a 'real necessity, the only possible one *under the given conditions*'.[34] Trying to reconcile the contradiction of the concept of 'morally justified war', he wrote that

> from the standpoint of general moral evaluation there are no two views on this subject, and cannot be: it is unanimously recognized that peace is a norm, [it] is what shall be, and war is an anomaly, is what shall not be.[35]

At the same time, Solov'ev admitted the right of states to wage a just war for the sake of peace, and that such a war could be justified in terms of political morality: 'The national idea understood as a political justice in the name of which weak and oppressed

[31] After Dostoevsky, this phrase 'prokliatye russkie voprosy' became widely used in Russian culture. According to Isaiah Berlin, it is a translation of '*die verdammten Fragen*' in Heine's poem *Zum Lazarus* (Berlin, 'The Hedgehog and The Fox', in I. Berlin, *The Proper Study of Mankind: An Anthology of Essays* (2000), at 445f.

[32] F. Dostoevsky, *The Karamazov Brothers*, translated by Constance Garnett (2007), at 267f.

[33] V. Solov'ev, *Vladimir Solov'ev's Justification of the Moral Good*, translated by Thomas Nemeth (2015), at 345.

[34] Ibid.

[35] V.S. Solov'ev, *Sochineniia* (1988), vol. 1, at 464.

peoples are defended and freed, has a high ethical standing.'[36] Thus, this concept of 'necessary war' differs significantly from the concept developed by *Clausewitzian* lawyers in the German imperial context (see Simon in this volume), where it was intended to justify virtually unrestricted warfare and also preventive/pre-emptive strikes (see Hull in this volume).

Solov'ev's philosophy had a considerable influence on Russian scholarship on war and peace in the late nineteenth century, and encouraged the view that protecting people in conflict (i.e. humanitarian intervention) was a just cause of war. For instance, Nikolai Korkunov, a renowned Russian legal philosopher, asserted that 'to abstain from interference with matters of other states would mean to remain an indifferent witness of the most horrible violence, to sacrifice the most sacred interests of mankind in favour of an abstract principle.'[37] In his opinion, war fought in response to atrocities should be justified on the basis of morality, not on the basis of the formal principles of international law.[38]

By contrast, Mikhail Kapustin attempted to balance the principles of morality and international law, considering both of them to be necessary conditions of just war:

> Self-defence and coercion constitute the indisputable right of independent states and can lead to war, which is the application of force in order to disable an adversary's capacity to resist. Such an application of force has both a moral and a legal character, and is subject to certain rules, which are set forth and guaranteed by civilization and morality.[39]

According to Kapustin, such rules were devoid of sense if they could not be enforced by military means.[40]

Vladimir Ul'ianitskii in his 1911 book on international law meticulously examined the right to military intervention, or 'international coercion', in order to prevent atrocities and to respond to mass violations of the rights of citizens of other states (evidently, having in mind Ottoman and Habsburg Empires).[41] Against this, some other Russian international law scholars stuck to pacifist positions and rejected attempts to justify wars. One such scholar was Dmitrii Kachenovskii, who criticized the idea that war could be understood as a legitimate tool of international law. In his interpretation, 'the supreme sanction of international law is neither war nor arbitrary force . . . International law is the result of worldwide citizenship; it reflects the practical ideas obtained by common life of mankind.'[42] Therefore, 'international law is the law of peace and harmony.'[43] In the same vein, Leonid Kamarovskii formulated a theory of non-interference,[44] and as early as 1881 called for the creation of an international

[36] Ibid., at 478.

[37] *Lektsii po mezhdunarodnomu pravu professora N.M. Korkunova, tsitaemye v Voenno-Iuriditseskoi Akademii v 1883–1884* (2007), at 256f.

[38] Ibid., at 257.

[39] M.N. Kapustin, *Mezhdunarodnoe pravo* (1873; 2nd ed., 2007), at 221.

[40] Ibid., at 219.

[41] V.A. Ul'ianitskii, *Mezhdunarodnoe pravo* (1911; 2nd ed., 2010).

[42] D.I. Kachenovskii, *Kurs mezhdunarodnogo prava* (1863; 2nd ed., 2007), at 52.

[43] Ibid., at 54.

[44] L.A. Kamarovskii, *Nachala nevmeshatel'stva* (1874).

court.[45] Meanwhile, Alexander Lodyzhenskii produced an extensive analysis of the normative changes in international law which would be needed in order to implement the Kantian theory of perpetual peace.[46]

Fyodor (Friedrich) Martens sought to determine practical methods of arbitration of international conflicts so as to prevent wars. His position was far from the moralism of Solov'ev school, and was based on a strict positivism which implied that only norms and principles established in international declarations and treaties were proper tools for keeping the peace (see also Simon in this volume). The only legal method of solving disputes between sovereign states was that of arbitration. This meant that replacing wars with peaceful arbitration had to be among the most important objectives in the development of international law. However, Martens saw how far humanity was from this objective, and so regretted that 'it is war and not arbitration that remains the usual way of solving of international disputes'.[47]

This pessimistic conclusion matches the general methodology of Martens' civilizational approach:

> It follows that the necessary condition for the progress of international communication is the adoption and development by states of all the main elements of European civilization and culture. The level of participation of each nation in international communication always corresponds to the level of its Enlightenment and civic virtue.[48]

Martens contributed not only to the theoretical development of international humanitarian law,[49] but also to practical measures to implement it (such as the 1868 St Petersburg Declaration and the 1874 Brussels Conference). This research was continued by his followers such Mikhail Taube, Boris Nolde, and Andrei Mandelstam.[50] With Martens, who was highly respected not only as the Crown lawyer and diplomat of the Russian Tsar, but also in Western Europe as a legal scholar, and whose work has been translated into German, English, and French, one sees a culmination of Russia's integration into the legal community of the 'civilized nations' of nineteenth-century Europe. This integration of Russian thought in Europe was to change significantly with the decline of the international order after 1918. The Russian scholarship now developed along distinctly different paths than the West.[51]

The First World War and the Russian Civil War prompted a reappraisal of the ethics of violence among theologians and philosophers. For instance, Metropolitan Antony

[45] L.A. Kamarovskii, *O mezhdunarodnom sude* (1881).

[46] A.N. Lodyzhenskii, *Projekty vechnogo mira i ikh znachenie* (1880); on Kant, see Eberl in this volume.

[47] F.F. Martens, *Sovremennoe mezhdunarodnoe pravo tsivilizovannykh narodov* (1904), at 13.

[48] Ibid., at 157; translation cited according to L. Mälksoo, *Russian Approaches to International Law* (2015), at 44.

[49] See on Martens' role in international conferences on humanitarian international law: Wehberg, 'Friedrich von Martens und die Haager Friedenskonferenzen', 20 *Zeitschrift für internationales Recht* (1910) 343; V. Pustogarov, *Our Martens. International Lawyer and Architect of Peace*, edited and translated by W.E. Butler (2000).

[50] Tolstykh, 'Mezhdunarodnoe gumanitarnoe pravo v Rossii vo vtoroi polovine 19—nachale 20 vv.', 2 Pravo i obrazovaniie (2003) 148; Didmanidze, 'Rossiiskaia nauka mezhdunarodnogo prava XVI-nachala XX v. o voine', 7 Evraziiskii iuridicheskii zhurnal (2016) 95.

[51] See Mälksoo, above note 48.

(Khrapovitskii), one of the leading members of the Russian Orthodox Church, said in 1915, in the midst of the First World War, that: 'War is an evil, but in the given case, and in the majority of Russian wars, a lesser evil than declining war and surrendering to the power of barbarians.'[52] Following the First World War, the Russian Revolution, and the Russian Civil War, Antony fled Russia and went into exile, as did many Russian philosophers, some of whom chose to write on the ethics of violence. The consensus among them coincided with the views of Solov'ev and Antony above: war is a relative, but not absolute evil, and is sometimes necessary, even obligatory. Philosopher Nikolai Berdyaev, for instance, declared that, 'The denunciation of the evil and sin of war should not be permitted to lead to absolute pacifism, to peace at any price. In the evil condition of our world, war may be the lesser of two evils.'[53] And philosopher Nikolai Lossky similarly remarked that, 'War is a great evil. But advocating pacifism ... could lead to chaotic anarchy, which is more terrible than any war.'[54]

The most sophisticated exposition of the idea of necessary war came in a book published in 1925 by Russian émigré philosopher Ivan Il'in, entitled *On Resistance to Evil by Force*. According to Il'in, the use of force could not be considered 'just' as it invariably involved injustice, such as the inadvertent killing of innocents. Justice and morality are not identical. One may do something which causes somebody an injustice, but be justified in so doing. Similarly, the use of force, even if unjust, is permitted, and even obligatory, if it is necessary to overcome evil.[55] 'The way of the sword is an unjust path', wrote Il'in,[56] adding that, 'Resistance to evil with the sword is *permissible*, not when it is possible, but when it is *necessary*.'[57] In such a case, he elaborated, the use of force is not only permissible, but mandatory. Overall, Il'in concluded, 'What the sword bearer does in the fight with evildoers is not perfect, not holy, nor just ... it is an *unsinful* (!) perpetration of injustice', because it may be 'morally and religiously necessary'.[58]

The idea of necessary war makes the justification of war seem utilitarian. As Skvortsov says, war is permitted, 'in the event that failure to engage in it will bring about even worse consequences than war'.[59] Russian thought on the ethics of war is not, however, purely utilitarian and includes something close to the just war criterion of right intention. Russians have often referred to war as 'sacred' (*sviashchennaia*). Its 'sacred' nature derives not only from the justice of its cause but also from the fact that war involves self-sacrifice for a higher good.[60] The sacred nature of war springs from selflessness of intention. Il'in developed this idea further. The unjust nature of war requires soldiers to accept that it involves 'spiritual compromise'. The moral contradictions of war, he argued, are so great that only the spiritually purest can endure them without becoming compromised. To act morally, therefore, soldiers have to undergo

[52] Stoyanov, above note 5, at 207.
[53] Ibid., at 209.
[54] Skvortsov, above note 1, at 96.
[55] For a summary of Il'in's thinking on this matter, see Robinson, 'On Resistance to Evil by Force: Ivan Il'in and the Necessity of War', 2 *Journal of Military Ethics* (2003) 145.
[56] I.A. Il'in, *O soprotivlenii zlu siloiu* (1975), at 201.
[57] Ibid., at 195.
[58] Ibid., at 190.
[59] Skvortsov, above note 1, at 96.
[60] Ibid., at 102.

continual spiritual purification and combine the qualities of the warrior and the monk.[61]

3. The Soviet Era

The Soviet approach to ethics of war was based on class theory and reformulated Russian religious messianism accordingly. It was not belief in the true religion but adherence to Marxism-Leninism and the cause of the progressive social class (the proletariat) that distinguished between just and unjust wars.

Lenin's criterion of just war was quite simple: wars conducted by socialist states were emancipatory and just, whereas wars conducted by imperialist states were predatory and unjust by default, as 'every war is intrinsically connected with the political order from which it stems'.[62] As Lenin explained, 'the politics conducted by a certain state or by a certain class within this state long before the war will be inevitably and unavoidably continued by this class during the war, only changing the form of its action'.[63]

In one of its first decrees, the Bolshevik government proclaimed that peace was its main objective in international politics.[64] This remained a key objective as long as the Soviet state was relatively weak and could not think of spreading its ideology by military means. The Soviet government not only readily ratified the Paris Kellogg-Briand Pact (27 August 1928), but also initiated several international pacifist projects such as the London Convention for the Definition of Aggression (3 July 1933).[65]

However, the Soviets conceptualized war and peace differently depending on whether they were regarding them from a legal or an ideological point of view. For the 'international law market', the Soviet government advertised the principles of non-interference and peaceful coexistence and condemned wars as a remnant of the capitalist world.[66] By contrast, narratives designed for the 'domestic market', where ideology stood above the law, portrayed interventions and wars conducted for the cause of communism as perfectly justifiable. It was argued that if the state pursued the correct policies, in line with the laws of history and the objective of the emancipation of the proletariat, then war—regardless of its methods and its victims—would be 'legal, just, sacred—not in the sense predicted by priests, but in the human sense: will be the sacred war of the oppressed against their oppressors in order to bring them down, to liberate workers from domination'.[67] This logic meant that wars of the Soviets

[61] Il'in, above note 56, at 218f.

[62] Lenin, 'Gosudarstvo i revoliutsiia', in V.I. Lenin, *Sochineniia* (1932), vol. 30, at 333.

[63] Ibid.

[64] *Dekret o mire* (28 October 1917) where war was treated as 'the greatest crime against mankind' and where all acts of aggression were condemned.

[65] Tunkin, 'Sovetskoe opredelenie agressiii v OON', 2–3 *Sovetskoe gosudarstvo i pravo* (1953) 89, at 91.

[66] A.S. Bakhov and K.A. Baginian, *Agressiia—tiagchaishee mezhdunarodnoe prestuplenie* (1955); V.A. Romanov, *Iskliuchenie voiny iz zhizni obshchestva* (1961); G.V. Sharmazanshvili, *Ot prava voiny k pravu mira* (1967); D.B. Levin, *Mezhdunarodnoe pravo i sokhranenie mira* (1971); V.I. Menzhinskii, *Neprimenenie sily v mezhdunarodnykh otnosheniiakh* (1976); N.A. Ushakov, *Nevmeshatel'stvo vo vo vnutrennie dela gosudarstv* (1971).

[67] Lenin, 'Zapugannye krakhom starogo i boriushchiesia za novoe', in V.I. Lenin, *Sobraniie sochinenii* (1974), vol. 34, at 192.

against hostile classes and imperialist states were just wars by definition. It also implied that the treatment of prisoners of war and the methods of war should be differentiated depending on the social classes that the enemy belonged to. This was the main rationale for the Soviet Union's non-adherence to Geneva Convention Relative to the Treatment of Prisoners of War (27 July 1929) and for its adoption of the Regulation on Prisoners of War (*Polozhenie o voennoplennykh*) (13 March 1931).

Stalin correspondingly differentiated between just and unjust wars by referencing their objectives: 'just war has the goal either to protect the people from external aggression and against attempts to enslave the people, or to unshackle the people from slavery of capitalism, or finally to liberate colonies and dependent countries from domination of imperialists'.[68] In accordance with this class conception, one Bolshevik propagandist wrote in 1939 that

> in the sacred war against the enemies of socialism, the Red Army, resting on the united attempts of the great Soviet people and the world proletariat, will protect peoples of the entire world from robbery, from enslavement, from serfdom, from medieval slaughters and unheard of abuse of human dignity. The Red Army will protect workers of all countries from tortures of famine and devastation, it will protect culture of mankind against barbarism and wildness.[69]

It is not by chance that in these and many other ideological statements wars conducted by the Soviet Army were characterized as 'sacred'. This language shows that the Bolshevik philosophy of war was also messianic. Thus, after the 1917 Revolution, the outward ideological forms changed, but not their contents—any war that contributed to the realization of a global project was sacralized in the same terms as the project itself.[70] For many Soviet diplomats and scholars this correlation between politics and war was the basic norm.[71]

4. The Post-Soviet Era

In the post-Soviet era, sovereignty, non-interference, and peaceful coexistence continue to be the main reference points of Russian foreign policy. The Russian authorities regularly condemn NATO's justifications of its military interventions, while insisting

[68] *Istoriia VKP(b). Kratkii kurs* (1938), 161. In the 1940s, Soviet scholars shifted their focus from class theory to the justification of wars in terms of national interest, placing the Second World War in the context of struggle of Russian people against 'barbarians' that surrounded it (Germans, Mongols, Turks, etc.): N. Vlasov, *O spravedlivykh i nespravedlivykh voinakh* (1942).

[69] Polkovoi komissar Osipov, 'Voiny spravedlivye i nespravedlivye', 17 *Propagandist i agitator RKKA* (1939) 19, at 26, available at <http://www.oldgazette.ru/lib/propagit/17/index.html> (last visited 10 January 2020).

[70] Professor Pavlenko distinguishes four such sacralized 'global projects' in Russian history: 'Holy Rus' in Kievan Russia, 'Third Rome' in Muscovite Russia, 'Orthodoxy, Autocracy and Nationality' in Imperial Russia, and 'Bright Future' in Soviet Russia, see V.B. Pavlenko, *Global'nye proekty: teoriia I praktika. Istoricheskii i soveremennyi aspekty* (2007).

[71] 'The basic principle of law is the principle of defining the importance of the political content of war and of the correlation of forces' (Tuishkevich, 'Marksistsko-leninskoe uchenie o voine i armii: istoriia razvitiia i sovremennye otsenki', 6 *Voennaia mysl'* (1991), at 5). See also *Marksizm-leninizm o voine i armii* (1968).

on sovereignty as the keystone of international law.[72] On the whole, the official narratives of the Russian government insist on settling all disputes through negotiations between sovereign states. Most contemporary Russian international law scholars follow this line, in accordance with the arguments put forward by Martens in the nineteenth century and later further developed by Soviet scholars.[73] They also distinguish between international and domestic matters, and generally abstain from criticism of other countries for their violations of *jus in bello* in civil wars. The Chechen wars played an important role in this regard,[74] with Russia ignoring Western criticisms of its actions in Chechnya by referring to arguments of sovereignty and anti-terrorism.[75]

The stress on sovereignty is, however, somewhat undermined by the fact that Russia sees itself as responsible for protecting thousands of compatriots—ethnic Russians who live on the territories of the ex-USSR. Thus, the 2010 Russian military doctrine approved the use of force for securing the interests of Russians abroad.[76] Arguments about the 'Responsibility to Protect' come in useful for justifying Russia's interference in neighbouring territories in pursuit of these interests.[77] Russia, therefore, does not shy away from arguments of humanitarian intervention when discussing its conflict with Ukraine (e.g. sending trucks with food, equipment, and other goods to Donbass across the Ukrainian border), mirroring the arguments about 'Responsibility to Protect' employed by the Western powers.[78]

This reveals two tendencies in Russian thinking. One is based on Western scholarship as well as on the practice of international law, while the other often makes use of pre-modern messianic argumentation about 'sacred wars'. Both influence discussions about *jus ad bellum* among the Russian public. This is illustrated below with several examples taken from statements of the Russian Orthodox Church, secular scholarship, and political narratives.

4.1 The Russian Orthodox Church

In a 1989 statement entitled 'Orthodox Perspectives on Justice and Peace', a meeting of Orthodox theologians in Minsk declared that:

> The Orthodox Church unreservedly condemns war as evil. Yet it also recognizes that in the defence of the innocent and the protection of one's people from unjust attack, criminal activity and the overthrowing of oppression, it is sometimes necessary, with reluctance, to resort to arms. In every case, such a decision must be taken with

[72] Zorkin, 'Apologiia Vestfalskoi sistemy', *Rossiiskaia gazeta* (22 August 2006), no. 4150.

[73] Mälksoo, above note 48, at 131–36.

[74] Sakwa, 'Chechnya: A Just War Fought Unjustly?', in B. Coppieters and R. Sakwa (eds), *Contextualizing Secession. Normative Studies in Comparative Perspective* (2003) 156.

[75] OSCE Rapporteur's Report under the Moscow Mechanism on alleged Human Rights Violations and Impunity in the Chechen Republic of the Russian Federation (20 December 2018), available at <https://www.osce.org/odihr/407402> (last visited 10 January 2020).

[76] Point 20, *Voennaia doktrina Rossiiskoi Federatsii*, approved by RF President's Decree No. 146 (5 February 2010).

[77] Ibid., at 172–82.

[78] R. Allison, *Russia, The West and Military Intervention* (2013).

> full consciousness of its tragic dimensions. Consequently, the Greek fathers of the Church have never developed a 'just war theory', preferring instead to speak of the blessings of and the preference for peace.[79]

Since the collapse of the Soviet Union, the position of the Russian Orthodox Church seems to have shifted slightly. The Church has not endorsed just war theory, but it is no longer quite so categorical in rejecting it, and indeed Church leaders have on occasion even made reference to it in a manner which suggests some approval. Most notably, a document published by the Russian Orthodox Church in 2000, entitled *The Basis of the Social Concept*, pronounced that, 'While recognizing war as evil, the Church does not prohibit her children from participating in hostilities if at stake is the security of their neighbours and the restoration of trampled justice. Then war is considered necessary though undesirable.'[80] The document then listed the *ad bellum* criteria of Western just war theory, thereby combining elements of both necessary war and just war thinking.

The head of the Russian Orthodox Church, Patriarch Kirill, has on several occasions even referred to war as 'just'. In 2002, for instance, Kirill (at the time Metropolitan of Smolensk and Kaliningrad) declared that the United States had the right to wage war in response to the 2001 terrorist attacks in New York and Washington, as the criteria for a just war laid down by St Augustine had been satisfied.[81] In January 2016, Kirill declared that the war in Syria had 'a defensive character, it is just'. He added that, 'military actions are justified when they defend mankind, society, the fatherland'. Russia had been subjected to terrorist attacks, and thus 'what is going on is responsive, defensive actions. In this sense, we can talk about a just war.'[82]

In May 2016, Kirill went even further, calling the war against terrorism in Syria 'sacred', and saying that it was being 'waged against the enemy of all humankind'.[83] He commented that

> Today, when our warriors take part in combat operations in the Middle East, we know that this is not an aggression, occupation or an attempt to impose some ideology on other peoples. . . . This is the fight against the fearsome foe that is currently not only spreading evil through the Middle East but also threatening the whole of mankind.[84]

[79] Stoyanov, above note 5, at 218.

[80] The Russian Orthodox Church, *The Basis of the Social Concept* (2000), available at <https://mospat.ru/en/documents/social-concepts/viii> (last visited 30 December 2019).

[81] Robinson, 'The Justification of War in Russian History and Philosophy', in P. Robinson (ed.), *Just War in Comparative Perspective* (2003) 62, at 62.

[82] RBK, *Patriarkh Kirill nazval operatsiiu v Sirii 'oboronitel'noi voinoi'* (2016), available at <https://www.rbc.ru/rbcfreenews/568e42249a79477603af9985> (last visited 30 December 2019).

[83] MK.ru, *Patriarkh Kirill nazval voinu v Sirii sviashchennoi* (2016), available at <http://www.mk.ru/politics/2016/05/06/patriarkh-kirill-nazval-voynu-v-sirii-svyashhennoy.html> (last visited 30 December 2019).

[84] RT, *Russian Orthodox Patriarch Declares Worldwide 'Holy War' on Terrorism* (2016), available at <https://www.rt.com/politics/342097-russian-orthodox-patriarch-declares-international/> (last visited 17 June 2018).

In October 2016, Kirill issued a clarification, however, saying, 'I never spoke of a sacred war in Syria, I spoke of a sacred war against terrorism.... My words shouldn't be interpreted in a political way.' The war against terrorism, he said, should unite the entire world against a common evil, and 'for that reason, I call this war sacred'.[85]

4.2 Secular Philosophers

Boris Kashnikov of the Higher School of Economics in Moscow remarked in 2012 that, 'Russian discourse [on just war] is only just getting started.'[86] 'Contemporary Russian discourse on the justification of war is very simplistic', he wrote, 'It is still lacking in philosophical reflection, and so can be called a "proto-discourse".'[87] Nevertheless, the past twenty years have witnessed a gradual growth of interest in the ethics of war among academics, resulting in the publication of several books devoted to the subject,[88] as well as the translation of several Western works of just war theory.[89]

On the whole, Russian scholars are critical of Western just war theory, regarding it as overly flexible, and thus subject to abuse (see also Chris Brown in this volume). In Russian eyes, trends in just war theory in recent years have served to expand the number of circumstances in which war is said to be justified, and the theory has thus become an intellectual tool to support military aggression by Western states. This kind of just war theory serves more to encourage war than to discourage it. In a 2016 article, for instance, Andrei Ivanov states that just war theory is inconsistently and selectively applied, and therefore does not actually limit war.[90] Similarly, Arseny Kuman'kov complains of the recent tendency to justify wars by appeal to universal values. This tendency, he says, is 'extremely dangerous' as it leads 'not to the limitation of war, but to completely the opposite consequences'.[91]

Boris Kashnikov, who is probably the most prolific contemporary Russian author on matters relating to the ethics of war, shares this point of view. He notes that just war theory derives from numerous, often contradictory philosophical systems, and so 'one cannot speak of a unified theory ... the same war can be evaluated completely differently by different people'.[92] In recent years, says Kashnikov, just war theory has been used to 'justify torture, the use of drones, illegitimate detention, and murder

[85] RIA Novosti, *Patriarkh nazval voinu s terrorizmom sviashchennoi* (2016), available at <https://ria.ru/religion/20161018/1479505802.html> (last visited 17 June 2018).

[86] Kashnikov, 'Kritika sovremennogo diskursa spravedlivoi voiny', 11 *Voenno-iuredicheskii zhurnal* (2012) 22, at 23.

[87] Ibid.

[88] For instance, Skvortsov, above note 1; A.D. Kuman'kov, *Sovremennye klassiki teorii spravedlivoi voiny: M. Uoltser, N. Foushun, B. Orend, Dzh. MakMakhan* (2016); B.N. Kashnikov and A.D. Kuman'kov (eds), *Etika voiny i mira: istoriia i perspektivy issledovaniia* (2016).

[89] For instance, Walzer, 'Smena rezhima i spravedlivaia voina', translated by A.D. Kuman'kov, 7 *Voenno-iuredicheskii zhurnal* (2013), at 26–32.

[90] Ivanov, 'Moral'nye dilemmy i kontseptual'nye protivorechiia teorii spravedlivoi voiny', 69 *Gramota* (2016), at 55–58.

[91] Kuman'kov, 'Printsip pravogo dela i "novye" voiny', in Kashnikov and Kuman'kov (eds), above note 88, at 174.

[92] Kashnikov, above note 86, at 28.

without trial'.[93] He therefore denounces the 'militarism of just war', saying that 'just war based on the paradigm of human rights in reality is equivalent to militarism'.[94] This leads Kashnikov to reject the idea of humanitarian intervention. By arguing in favour of such interventions, Western just war theorists have, he says, committed 'a dangerous utilitarian error', as seen by examples such as the attacks on Iraq and Libya, which caused much worse consequences than inaction would have.[95] According to Kashnikov,

> By justifying humanitarian intervention, just war theory provides the ideological basis for total, absolute, and permanent war.... Just war theory has resurrected the practice of crusading, under the flag of human rights, humanism, and universal security. Just war has ceased to mean an exceptional short-term use of force. In the current conditions, it has turned into a justification of a global order based on permanent force.[96]

The idea of humanitarian intervention threatens state sovereignty, and so undermines the existing international system, Kashnikov claims. This system, based on the United Nations, is deficient, he says, but:

> Even the current state of the international order is better than no order at all, which is what there will be if the Westphalian system is completely destroyed and the authority of the UN replaced by that of NATO or the USA.... There is no alternative to the Westphalian system. The destruction of state sovereignty can lead only to an increase in violence. We need to preserve everything which supports state sovereignty.[97]

By supporting humanitarian intervention, just war theory has become 'a weapon of war, and not a means for limiting it'.[98] In its place, Kashnikov proposes instead the idea of necessary war, citing in particular the work of Ivan Il'in.[99] 'Contemporary war cannot be just but can be necessary', writes Kashnikov, adding that,

> War shouldn't be regarded any more as something just. It is necessary only on condition ... of aggression or the physical attack of another state or internal enemy. Such an understanding of necessity precludes not only pre-emptive, but also preventive war.[100]

[93] Kashnikov, 'Mif spravedlivoi voiny kak real'nost' absoliutnogo nasiliia', 2 *Voenno-istoricheskii zhurnal* (2017) 3, at 5.
[94] Ibid., at 5–6.
[95] B.N. Kashnikov, 'Teoriia spravedlivoi voiny kak voina i spravedlivost' global'nogo mira', 3 *Voenno-istoricheskii zhurnal* (2014) 24, at 29.
[96] Ibid.
[97] Ibid., at 30.
[98] Ibid., at 31.
[99] Kashnikov, above note 86, at 29. Also, Kashnikov, above note 95, at 30.
[100] B.N. Kashnikov, 'Nasilie spravedlivosti i spravedlivost' nasiliia', 8 *Voenno-iuredicheskii zhurnal* (2013) 20, at 27.

Kashnikov concludes that necessary war 'knows only one form of justification—an immediate and direct threat to a community's political existence', and thus only one just cause—'repulsing direct aggression, self-defence. Nothing else ... can be justified.'[101]

4.3 State Actors

In recent years, various Western commentators have accused the Russian military of seeking to elide the distinction between war and peace by embracing 'a vision of total warfare that places politics and war within the same spectrum of activities'.[102] Citing what has been called the 'Gerasimov doctrine', such commentators claim that Russia is pursuing an aggressive strategy of 'hybrid warfare' which combines military action with activities such as propaganda, computer hacking, espionage, subversion, and economic warfare.

This understanding of Russian military doctrine derives from an article written by the Chief of the Russian General Staff, General Valery Gerasimov, in 2013.[103] But as Mark Galeotti (who was the first to coin the phrase 'Gerasimov doctrine') has pointed out, in his article Gerasimov was not speaking normatively—that is he was not arguing that Russia *should* elide the distinction between war and peace. Rather, he was speaking descriptively, arguing that others were already eliding the distinction in this way. From this it followed that Russia needed to find some way to defend itself against this phenomenon.[104] The idea that the 'Gerasimov doctrine' reveals an expansive new Russian approach to war does not stand up to close scrutiny. On the contrary, analysis of the statements of senior Russian politicians reveals a rather more restrictive view of war than that espoused by their Western colleagues. They are suspicious of humanitarian intervention and the 'Responsibility to Protect' and support the principle of state sovereignty. Russian political rhetoric thereby fits fairly closely with the contemporary scholarly arguments mentioned above. As Charles Ziegler notes:

> The fundamental issue for Russia is protection of state sovereignty.... preserving domestic stability has been a key security concern since the economic free-fall and societal chaos of the 1990s ... The Kremlin opposes humanitarian interventions that could lead to state fragmentation, especially in those regions close to its borders or in partner states, while remaining indifferent to or tolerating interventions designed to strengthen territorial integrity or where Russia's interests are not directly affected.[105]

[101] Kashnikov, above note 95, at 30.

[102] M.K. McKew, *The Gerasimov Doctrine* (2017), available at <https://www.politico.eu/article/new-battles-cyberwarfare-russia/> (last visited 19 June 2018).

[103] For an English translation of this article, see Gerasimov, 'The Value of Science is in the Foresight', *Military Review* (January–February 2016), at 23–29.

[104] M. Galeotti, *I'm Sorry for Creating the Gerasimov Doctrine* (2018), available at <http://foreignpolicy.com/2018/03/05/im-sorry-for-creating-the-gerasimov-doctrine/> (last visited 19 June 2018).

[105] Ziegler, 'Russia on the Rebound: Using and Misusing the Responsibility to Protect', 30 *International Relations* (2016) 346, at 348.

Having twice experienced regime change in their own country in the twentieth century (in 1917 and 1991), and having seen the chaos which ensued, Russians are reluctant to support military interventions to overthrow regimes. This reluctance has been reinforced by seeing the results of recent Western military campaigns. Vladimir Baranovsky and Anatoly Mateiko note that, 'Russia differs from the West in its understanding of how to make populations more secure. It suggests that a brutal and repressive regime can sometimes be a better option than intervention, which makes the whole country ungovernable.'[106]

This attitude helps to explain Russia's support for the government of Bashar al-Assad in Syria. Russian Foreign Minister Sergei Lavrov has emphasized the danger of state collapse, pointing out the negative humanitarian consequences of overthrowing regimes.[107] Moscow maintains that its aim in Syria is not to protect Assad but rather to prevent the collapse of the Syrian state and a consequent spread of violence throughout the region.[108] In line with this logic, Russia supported French military intervention in Mali in 2012, in order to prop up the Malian state.[109] Russia is, therefore, not opposed to intervention per se, but regards it as primarily justified when used to support the existing state system, not to undermine it.[110] In this sense, the modern Kremlin's attitude is very similar to that of Nicholas I in the first half of the nineteenth century.

Various versions of the 'Foreign Policy Concept of the Russian Federation', published in 2008, 2013, and 2016, have denounced Western attempts to enshrine military humanitarian intervention and the Responsibility to Protect as norms of international law. The 2016 version of the document, for instance, states that, 'Russia intends ... to prevent military interventions or other forms of outside interference contrary to international law, specifically the principle of sovereign equality of States, under the pretext of implementing the "responsibility to protect" concept.'[111] According to Baranovsky and Mateiko, 'Russia tends to avoid absolutist interpretations of sovereignty and ... considers that sovereignty can actually be overridden under certain circumstances ... As a rule, however, Moscow is particularly vigilant upon signs of sovereignty being downplayed or circumvented.'[112] Thus in 2012, Putin said that, 'it is definitely true' that sovereignty takes precedence over human rights.[113]

Nevertheless, Russia reserves the right to intervene militarily in other countries under certain circumstances. In August 2008, Foreign Minister Sergei Lavrov justified Russia's military action in Georgia as follows:

[106] Baranovsky and Mateiko, 'Responsibility to Protect: Russia's Approaches', 51 *The International Spectator* (2016) 49, at 58.

[107] Averre and Davies, 'Russia, Humanitarian Intervention and the Responsibility to Protect: the Case of Syria', 91 *International Affairs* (2015) 813, at 823.

[108] Ibid., at 820.

[109] Kuhrt, 'Russia, the Responsibility to Protect and Intervention', in D. Fiott and J. Koops (eds), *The Responsibility to Protect and the Third Pillar* (2015) 97, at 107.

[110] Ibid., at 106.

[111] The Ministry of Foreign Affairs of the Russian Federation, *Foreign Policy Concept of the Russian Federation* (2016), available at <http://www.mid.ru/foreign_policy/news/-/asset_publisher/cKNonkJE02Bw/content/id/25422488> (last visited 19 June 2018).

[112] Baranovsky and Mateiko, above note 106, at 52.

[113] Ibid.

> Under the Constitution [the President] is obliged to protect the life and dignity of Russian citizens ... According to our Constitution there is also a responsibility to protect ... this is not Africa to us, this is next door. This is the area where Russian citizens live. So the Constitution of the Russian Federation, the laws of the Russian Federation make it absolutely unavoidable to us to exercise responsibility to protect.[114]

Lavrov's use of the phrase 'Responsibility to Protect' is interesting considering Moscow's general dislike of the concept. But close analysis reveals that Lavrov was not justifying Russia's actions by appealing to some requirement to protect humanity per se. His interpretation was far narrower: Russia has a responsibility to protect Russian citizens.

Dmitry Medvedev, who was president of the Russian Federation at the time, justified Russia's actions in Georgia in similar terms, claiming that the Georgian army had committed 'what amounts to an act of aggression against Russian peacekeepers located there and the civilian population in South Ossetia'. Most of the South Ossetians were 'citizens of the Russian Federation', said Medvedev, and so Russia had a responsibility to defend them.[115] Medvedev concluded that Russia 'will not allow the deaths of [its] compatriots to go unpunished'.[116]

Similarly, Vladimir Putin justified sending Russian troops into Crimea in March 2014 on the grounds of the 'danger to lives of the citizens of the Russian Federation, our compatriots, and Russian military personnel'.[117] The responsibility to the people of Crimea derived from 'the common culture and history of this region and from the fact that in people's hearts and minds, "Crimea had always been and remained an inseparable part of Russia"'.[118]

Russia, therefore, is suspicious of the idea of humanitarian intervention and places a high value on state sovereignty. But it is willing to make an exception when its own citizens, or people whom it considers to be part of the 'Russian world', are believed to be under threat.

5. Conclusion: Russian Traditions between Protecting Peace and Justifying War

According to Yuri Stoyanov, although just war notions did appear in medieval Eastern Orthodox writings, 'they were not developed in any systematic fashion or integrated into a consistent theory'.[119] From the analysis in this chapter, one can, however, identify certain trajectories of thought. These include pacifism at one end of the spectrum,

114 Evans, 'Russia, Georgia and the Responsibility to Protect', 1 *Amsterdam Law Review* (2009) 25, at 25.
115 Reshetnikov, 'Intervention as Virtue, Obligation and Moral Duty: The Meaning of Russia's Rhetoric on Responsibility during the Georgian and Crimean Crises', 2 *Russian Politics* (2017) 155, at 170.
116 Ibid., at 178.
117 Ibid., at 156.
118 Ibid., at 166.
119 Stoyanov, 'Eastern Orthodox Christianity', in G.M. Reichberg and H. Syse (eds), *Religion, War, and Ethics: A Sourcebook of Textual Traditions* (2014) 164, at 187.

and something akin to a crusading tradition at the other end. Lying in between is a third trajectory, best described by the phrase 'necessary war'.[120]

The first trajectory may be seen to date back to the first saints of the Russian Orthodox Church, Princes Boris and Gleb, who refused to resist assassins sent by their brother Svyatopolk in 1015 AD. The best-known Russian proponent of non-resistance to evil was the great nineteenth-century writer Leo Tolstoy, who in his 1894 book *The Kingdom of God is Within You* argued that violence responds only to the symptoms of evil, not its root causes, and ultimately serves only to perpetuate it.[121] Although Tolstoy won some converts to his cause, his pacifist ideology never acquired mass support and had no obvious influence on Russian practice.

On the other end of the spectrum, some Russian thinkers have justified war with reference to the messianic role that is attributed to Russia in world history. Nicholas Berdyaev aptly characterized Russians as millenarian people.[122] This characterization apparently fits very well with Russia's intellectual history, with some Russian thinkers even going so far as to portray war as a positive good. For instance, the famous Russian writer Fyodor Dostoevsky in 1877 declared that, 'We need this war for ourselves ... war will clear the air we breathe and in which we have been suffocating.'[123] Dostoevsky also said that, 'War is not always a scourge; sometimes it is salvation.'[124] He continued:

> It is not war but peace, a prolonged peace, that bestializes and hardens people. A prolonged peace always gives rise to cruelty, cowardice, and coarse, bloated egoism ... a war for some noble purpose ... strengthens every soul with the awareness of self-sacrifice and the spirit of the entire nation with the awareness of the mutual solidarity and unity of all the members that constitute it.[125]

More often than viewing war as either sacred or always wrong, Russians have tended to follow a third trajectory, speaking of war as an occasionally necessary evil. This has arguably led to a more restrictive concept of *jus ad bellum* than that often propagated by Western just war theorists. Just cause in Russian thought is most often associated with defence against an immediate danger to Russia or Russian citizens. While humanitarian intervention has gained some support, war to support existing regimes has been much more likely to garner support than war to undermine them. In the post-Soviet era, even where Russian forces have intervened in other countries, as in Georgia in 2008 and Ukraine in 2014, they have refrained from overthrowing the state.

[120] For a more detailed analysis of the various trajectories of Russian thinking, see Robinson, above note 81.

[121] L.N. Tolstoy, *The Kingdom of God is Within You* (1917).

[122] N. Berdyaev, *Russian Idea* (1948), at 193. For a discussion of this particularity of Russian political thinking, see Rowley, 'Redeemer Empire: Russian Millenarianism', 5 *American Historical Review* (1999) 1582.

[123] F. Dostoevsky, *A Writer's Diary. Volume Two, 1877–1881* (1994), at 930. For more on Dostoevsky's views on war, see Nikitina, 'Nravtvennoe opravdanie voiny V.S. Solov'evym i F.M. Dostoevskim', in B.N. Kashnikov and A.D. Kuman'kov (eds), *Etika voiny i mira: istoriia i perspektivy issledovaniia* (2016), at 18–32.

[124] Dostoevsky, above note 123, at 934.

[125] Ibid., at 938f.

At the same time, while insisting on the strict observance of the rules of post-1945 international law and while condemning interference in other countries' affairs, Russia sometimes finds it appropriate to mimic Western policies, as in the case of Responsibility to Protect (cited above). When the argument of Responsibility to Protect was triggered in Georgia, the Russian foreign office and political leadership maintained that they had acted in the same way as the West had in Serbia under the banner of a 'peace enforcement operation'. Ironically, Russian diplomatic language repeated all the formulations used by Western powers for the Kosovo operation and for the attacks on Libya, Iraq, and elsewhere.[126] Later, Russian international lawyers produced a large number of legal arguments to justify the annexation of Crimea.[127] Some of these relied not on interpretations of the legal effect of territorial exchanges in the USSR (and on the nullity of Khrushchev's 1954 transfer of Crimea from Russia to Ukraine), but on an array of arguments concerning the protection of Crimeans against possible infringements of their national, cultural, and linguistic identity, and also cited other reasons previously used by Western powers in Kosovo.[128] Russia supports the Donbass rebels in Eastern Ukraine using similar logic.

This ambivalent approach underlines the paradoxes of the Russian understanding of international order and sovereignty. Russia denies the dualism of sovereignty and human rights and sees itself as the guardian of sovereignty as the core norm of the international legal order. In line with this, Russia condemns Western interference but, at the same, Russia resorts to a completely different language when it needs to justify its own interventions or to limit the sovereignty of its ex-Soviet neighbours.[129]

All this puts Russian thinking rather at odds with contemporary versions of just war theory which argue that military power may be used to promote human rights by combatting oppressive states (for the Western perspective see Geis and Wagner as well as Jahn in this volume). On the whole, Russia views force as a tool to be used in extreme circumstances to protect Russians and to preserve the existing international order. In this sense, Russia can be seen as trying to uphold the legal status quo (sovereignty), while also trying to enhance its own position (by sometimes ignoring other states' sovereignty).[130]

[126] See Russian Federation, *Legal Aspects of War in Georgia*, available at <https://www.loc.gov/law/help/legal-aspects-of-war/russian-georgia-war.php> (last visited 10 January 2020). The former representative of Russia in the UN sarcastically mentioned about Russia's operation in Georgia: 'How can we not recall the responsibility to protect that we hear so much about in the United Nations?' UN-Doc S/PV.5952 (8 August 2008), at 5.

[127] Grant, 'Annexation of Crimea', 109 *The American Journal of International Law* (2015) 68.

[128] See Roberts, 'Crimea and the South China Sea—Connections and Disconnects among Chinese, Russian, and Western International Lawyers', in A. Roberts et al. (eds), *Comparative International Law* (2018) 111, at 116ff.

[129] R. Allison, *Russia, the West, and Military Intervention* (2013), at 212ff.

[130] In Russian foreign policy, defending the status quo means, on the one hand, rejecting encroachments on its sovereignty in the 'Westphalian' sense, including criticism of its political regime and human rights protection. On the other hand, Russia insists that factual changes in the international politics after 1990 do not enable the Western powers to reconsider the rules of international law without following the legal tracks established in the post-1945 world. This is not the same as attempting to restore Russia's former standing as a superpower, as some argue: militarily and economically, Russia is simply unable to recover this standing, and its political leadership is fully aware of it.

None of the three trajectories mentioned above entirely dominates the others. Russian narratives about the justification of war oscillate pragmatically between generally favouring pacifism and sovereignty as means of preserving the status quo and, as an exception, supporting military interventions when these are required by the transient goals of Russian foreign policy. Russia's history thus provides it with a flexible set of intellectual tools to justify both war and peace.

23
China's Approach to the Use of Force: A Short Review of China's Changing Attitudes towards the Justification of Humanitarian Intervention

Manjiao Chi[*]

1. Introduction

The United Nations (UN) Charter provides the fundamental norms for states in international relations. It sets up a high threshold requirement for the international use of force through imposing a general ban on the unilateral use of force and strictly limiting the use of force in two exceptional circumstances. These restrictions were designed by the drafters of the UN Charter with interstate wars in their vision. Though humanitarian intervention is difficult, if possible at all, to be precisely defined, it has been and still is frequently used as a justification for the use of force. As humanitarian intervention typically involves the use of force by third state(s) in the host state's territory without the latter's consent for the purpose of human rights protection, it could violate the principles of sovereignty and fail to meet the requirements for invoking the exceptions under the UN Charter. It is particularly problematic since humanitarian intervention often occurs in armed conflicts within a state. Justifying humanitarian intervention under these circumstances is quite demanding. Recent state practice seems to suggest that such justification can be supported through an expansive interpretation of the UN Charter's exceptions to the use of force or invoking the Responsibility to Protect. Both approaches are contentious in international relations and law.

Since the establishment of the People's Republic of China ('PRC' or 'China') in 1949, China has been involved in a number of use-of-force events. It has not only engaged in several regional wars, such as the Korean War in the 1950s,[1] the Sino-Indian War in the 1960s,[2] and the Sino-Vietnamese War in the 1970s,[3] but also participated in

[*] Thanks are due to Ms Jian Lin, Ph.D candidate and Junior Researcher of CIELP, UIBE Law School, for her kind assistance. The opinions and errors of this paper are the author's.

[1] See e.g. J.F. Schnabel, *United Nations Army in the Korean War Policy and Direction: The First Year* (1972); Levie, 'How It All Started and How It Ended: A Legal Study of the Korean War', 35 *Akron Law Review* (2002) 205.

[2] See e.g. L. Burkitt, A. Scobell, and L.M. Wortzel (eds), *The Lessons from History: The Chinese People's Liberation Army at 75* (2003); Maxwell, 'Settlements and Disputes: China's Approach to Territorial Issues', 41 *Economic and Political Weekly* (2006) 3873.

[3] See e.g. MacDonald, 'The Law and Politics of a Norm Violation: Punitivity and the Sino-Vietnamese War of 1979', 8 *Amsterdam Law Forum* (2016) 38.

Manjiao Chi, *China's Approach to the Use of Force: A Short Review of China's Changing Attitudes towards the Justification of Humanitarian Intervention* In: *The Justification of War and International Order.* Edited by: Lothar Brock and Hendrik Simon, Oxford University Press (2021). © the many contributors. DOI: 10.1093/oso/9780198865308.003.0023

several UN peacekeeping operations,[4] and has sent navy forces for the suppression of international piracy in the Gulf of Aden and off the Somali coast in response to the calls of the Security Council.[5] Yet, China has not engaged in humanitarian intervention events. Given its recent rise, China seeks to play a leadership role in international affairs. This vision inevitably influences China's foreign policy, including its policy on the use of force and humanitarian intervention.

This chapter aims at reviewing China's position towards humanitarian intervention. Section 1 introduces the objective of the chapter. Section 2 discusses the principles of the use of force under the UN Charter and China's position on these principles. Section 3 discusses the tension between the UN Charter and humanitarian intervention, and China's shifting positions on humanitarian intervention and the Responsibility to Protect. Section 4 concludes that, though China adheres to a narrow understanding of the UN Charter principles on the use of force, its position on humanitarian intervention has experienced a gradual and slight shift. China's position is likely to shift further, given its recent initiation of the notion of a 'community of shared future for mankind'.

2. China's Attitudes towards the UN Charter Principles of the Use of Force

The UN Charter is almost universally accepted as offering a normative framework for dealing with conflict. China is a permanent member of the Security Council, which bears a primary responsibility in ensuing international peace and security.[6] Given the cornerstone role of the UN Charter in international relations and law, a study of the use of force should start from the relevant principles of the UN Charter.

For three decades since its establishment, the PRC faced profound isolation by the international community, especially by the Western industrialized countries. Yet, such isolation does not necessarily mean that the PRC is a total stranger to the UN regime. As Wang Tieya, a prominent Chinese international lawyer, has observed, 'the general principles laid down in the UN Charter have gained whole-hearted support of the PRC since its founding, even before it took its legitimate seat in the United Nations'.[7] Strong adherence to the UN Charter made good sense since it could not only enhance China's legitimacy as a player in the international community, but also helped maintain a peaceful external environment for its development.

After the PRC was restored as the legitimate representative of China in the UN in 1971,[8] it has been gradually accepted by the international community. As a permanent

[4] See generally Institute of Security and Development Policy, *China's Role in UN Peacekeeping* (March 2018), available at <http://isdp.eu/content/uploads/2018/03/PRC-Peacekeeping-Backgrounder.pdf> (last visited 10 January 2020).

[5] See e.g. Chi, 'A Note on China's Legal and Operational Responses to International Piracy', 44 *Ocean Development & International Law* (2013) 113, at 113–22.

[6] Art. 24 UN Charter.

[7] Wang, 'International Law in China: Historical and Contemporary Perspectives', 221 *Recueil des cours* (1990) 195, at 265.

[8] See GA Res. 2758 (25 October 1971).

member of the Security Council, China along with the other four permanent members has a special responsibility for the maintenance of international peace and security and for the promotion of international cooperation for common development.[9] It is important for China to promote the UN Charter as a focus for building international order. For instance, in his statement at the General Debate of the 70th Session of UN General Assembly in 2015, Chinese President Xi Jinping stated that

> The earlier generation of mankind pooled together their wisdom and adopted the *Charter of the United Nations*, laying the cornerstone of the contemporary international order, and establishing the fundamental principles of contemporary international relations ... China was the first country to put its signature on the UN Charter. We will continue to uphold the international order and system underpinned by the purposes and principles of the UN Charter.[10]

It goes without saying that adherence to the UN Charter implies acceptance of the principles of the use of force enshrined in the Charter. The founders of the UN created an elaborate scheme, a treaty-based system that requires states to impose limits on their own right to resort to force and to depend on a collective response for protection.[11] Towards this end, the UN Charter imposes a ban on the unilateral use of force. This principle is codified in many international legal instruments. It represents not only a core value of the UN Charter, but is also universally accepted as a norm of customary international law.[12]

While the UN Charter imposes a general ban on the use of force, it allows certain exceptions where a state may legally use force, that is the Security Council enforcement actions and the right to individual and collective self-defence.[13] Under the UN Charter, the Security Council is invested with the right to use of force in order to ensure international peace and security (Articles 39, 41, and 42). According to Article 39, the Security Council 'shall determine the existence of any threat to the peace, breach of the peace, or act of aggression and shall make recommendations, or decide what measures shall be taken in accordance with Articles 41 and 42, to maintain or restore international peace and security'. Article 41 further provides that the Security Council 'may decide what measures not involving the use of armed force are to be employed to give effect to its decisions, and it may call upon the Members of the United Nations to apply such measures.' In practice, a Security Council resolution that allows UN member states to use 'all necessary means' is routinely viewed as a clear indication of an authorization for the use of force.[14] Apart from collective action, Article 51 of the UN Charter allows individual states or groups of states to use force for self-defence,

[9] Xue, 'China's Open Policy and International Law', 4 *Chinese Journal of International Law* (2005) 133, at 139.

[10] Xi Jinping, *Working Together to Forge a New Partnership of Win-win Cooperation and Create a Community of Shared Future for Mankind* (29 September 2015), available at <http://www.fmprc.gov.cn/mfa_eng/wjdt_665385/zyjh_665391/t1305051.shtml> (last visited 10 January 2020).

[11] C. Gray, 'The Charter Limitations on the Use of Force: Theory and Practice', in V. Lowe et al. (eds), *The United Nations Security Council and War: The Evolution of Thought and Practice since 1945* (2008) 86, at 86.

[12] B. Simma et al. (eds), *The Charter of the United Nations: A Commentary* (3rd ed., 2015), at 204.

[13] Ibid., at 1399.

[14] See C. Gray, above note 11, at 89f.

but this right is a 'temporary right' which only exists until the Security Council has taken measures necessary to maintain international peace and security.

Though the UN Charter allows the use of force in exceptional circumstances, it is difficult to invoke these exceptions in reality. On the one hand, the Security Council authorization for the use of force is difficult to obtain, as its decision-making is highly politicized, especially among the permanent members; on the other hand, the right to self-defence under the UN Charter is also strictly construed, which makes it difficult to justify the use of force here, too. As pointed out by Hanspeter Neuhold,

> Effective enforcement action taken by the Security Council still constitutes the exception and not the rule, as the 'P5' [the five permanent members of the UN Security Council] remain divided over many issues. Some key aspects of self-defense are still contested so that it may continue to provide a convenient justification for the recourse to force. The line between laudable pressure to protect and promote common values, on the one hand, and reprehensible intervention, on the other hand, remains to be drawn.[15]

As a permanent member of the Security Council, China has veto power in authorizing the use of force. Though China has not used this power in every use-of-force event, including humanitarian intervention events, its understanding and practice of the UN Charter principles of the use of force is critical. China's principled position on the use of force under the UN Charter can be found in the Position Paper of the People's Republic of China on the United Nations Reforms issued on 7 June 2005, which states in relevant part that

> Peaceful settlement of international disputes and non-use of force in international relations is an important principle of the UN Charter and a basic norm of international law. China consistently stands for settlement of international disputes by peaceful means and opposes the threat or use of force in international relations.
>
> China resolutely adheres to the position that Article 51 should neither be amended nor reinterpreted. As pointed out, the Charter lays down explicit provisions on the use of force. Whether an urgent threat exists should be determined and handled with prudence by the Security Council in accordance with Chapter VII of the Charter and in light of the specific situation.
>
> Given the varying causes and the nature of crises, it is both unrealistic and hugely controversial to formulate a 'one-size-fits-all' rule on, or criterion for, the permission to use force. Whether to use force or not should be decided by the Security Council in light of the reality of conflicts on a case-by-case basis.
>
> The Security Council is the only body that can decide the use of force. Regional arrangements or organizations must obtain Security Council authorization prior to any enforcement action.[16]

[15] H. Neuhold, *The Law of International Conflict: Force, Intervention and Peaceful Dispute Settlement* (2015), at 218.

[16] The Permanent Mission of the People's Republic of China to the UN, *Position Paper of the People's Republic of China on the United Nations Reforms* (7 June 2005), available at <http://www.china-un.org/eng/chinaandun/zzhgg/t199101.htm> (last visited 10 January 2020).

Several observations can be drawn from China's position. First, in general, China opposes the use of force in international relations, which conforms to the UN Charter's general ban on the use of force. Second, China adopts a strict understanding of the UN Charter's principle of the use of force, opining that there are only the two exceptions to the general ban, mentioned above. Third, China argues that the exclusive power of the Security Council to decide on the use of force should be maintained and further enhanced. To some extent, China's position on the use of force reflects a strict, and perhaps rigid, reading of the UN Charter. However, this position could turn into a double-edged sword. On the one hand, it stresses the decisive role of the Security Council in determining the use of force, which potentially enhances China's power in making use-of-force decisions as a permanent member of the Security Council. On the other hand, China binds itself strictly to the Charter provisions thus accepting that its own practice in dealing with the use of force will be subject to close scrutiny by the international community.

3. China's Principled Position on Humanitarian Intervention

Humanitarian intervention by the use of force has been and continues to be a contentious issue in international law and international relations (see also Lauren Benton's, Nina Wilén's, and Beate Jahn's chapters in this volume). There is no uniform definition of humanitarian intervention,[17] and its legal status remains contentious especially with respect to customary law.[18] Thus, from a legal perspective, it is difficult to assess whether and to what extent humanitarian issues may offer a justification for the use of force. As China has not engaged in humanitarian intervention in the past, its position and practice on this issue has to be deduced from its actual responses to various humanitarian interventions.

3.1 Humanitarian Intervention and the UN Charter

Absent a uniform definition, it is generally agreed that humanitarian intervention refers to the use of force by a state or a group of states, in another state's territory, for protecting the host state's citizens from gross human rights abuses or mass atrocities.[19] These elements not only help differentiate humanitarian intervention from the two exceptions of the use of force under the UN Charter, but also puts humanitarian

[17] See e.g. T. Franck and N. Rodley, 'After Bangladesh: The Law of Humanitarian Intervention by Military Force', 67 *American Journal of International Law* (1973) 275, at 277; N. Krylov, 'Humanitarian Intervention: Pros and Cons', 17 *Loyola Los Angeles International and Comparative Law Review* (1995) 365, at 366.

[18] See generally R. Goodman, 'Humanitarian Intervention and Pretexts for War', 100 *American Journal of International Law* (2006) 107.

[19] See generally J. Davis, 'From Ideology to Pragmatism: China's Position on Humanitarian Intervention in the Post-Cold War Era', 44 *Vanderbilt Journal of Transnational Law* (2011) 217, at 221; P. Tzeng, 'Humanitarian Intervention at the Margins: An Examination of Recent Incidents', 50 *Vanderbilt Journal of International Law* (2017) 415, at 426.

intervention in direct tension with the norms of state sovereignty enshrined in the UN Charter.[20]

There have been a number of humanitarian intervention events in world history. In the pre-UN Charter era, there was no established state practice of claiming a right of humanitarian intervention to justify the use of force.[21] In the post-UN Charter era, however, military interventions for humanitarian reasons became increasingly frequent. Somehow interestingly, the intervening states seldom explicitly invoked the doctrine of humanitarian intervention to justify the use of force in these events.[22] Since around the new Millennium, humanitarian intervention has become often used as a justification for the use of force. Typical such events include NATO's military interventions in Kosovo in 1999 and in Libya in 2011, the French interventions in Cote d'Avoire in 2011 and in Mali in 2013, as well as the US-led interventions in Syria between 2014 and 2016, to list some.[23]

As humanitarian intervention involves the use of force in a state's territory, often without the consent of that state, its legality under the UN Charter could be challenged.[24] One could argue that the UN Charter's restrictions on the use of force imply that humanitarian intervention is not a valid justification of the use of force under international law, unless it constitutes self-defence or an authorized action by the Security Council.[25] However, as pointed out by V.S. Mani, the design of the Security Council is constitutionally 'incompatible' with humanitarian intervention, as the veto power of the five permanent members could make it impossible for the Security Council to authorize humanitarian intervention operations. Even when there exists an authorization, UN member states do not have a legal obligation to participate in such operations under the UN Charter.[26]

Today, while the UN Charter remains the cornerstone guide of international relations, the world has ushered in an era of global governance. The international community now witnesses an increasing number of non-state actors stepping to the international arena—this confronts the international community with common concerns of human society concerning the protection of human rights. The highly state-centric UN Charter regime is insufficiently capable to meet the humanitarian needs expressed in human society. As can be sensed from the 1994 Rwanda massacre, if the Security Council fails to promptly authorize the use of force in response to gross human rights violations within a state, this can lead to the escalation of violence into genocide.

To justify humanitarian intervention in international law and relations, various proposals have been put forward. Two proposed justifications are typical. One is to deem humanitarian intervention as a 'new' exception of the UN Charter's general ban

20 J. Davis, above note 19, at 221.

21 See V. Lowe and A. Tzanakopoulos, 'Humanitarian Intervention', in R. Wolfrum (ed.), *Oxford Public International Law* (2015).

22 P. Tzeng, above note 19, at 430.

23 Ibid.

24 See S. Chesterman, *Just War or Just Peace?: Humanitarian Intervention and International Law* (2001), at 45–47.

25 Ibid.

26 V.S. Mani, '"Humanitarian" Intervention Today', 313 *The Collected Courses of the Hague Academy of International Law* (2005), at page 23, 265f.

on the use of force which is advocated especially by liberal democracies.[27] In this regard, it has been argued that the legality of humanitarian intervention should be appraised in light of the transformation of international human rights norms since the 1960s.[28] In the ensuing decades, the increasing body of human rights and a growing number of binding human rights obligations create a situation where humanitarian intervention may be justified in extreme circumstances.[29] Under this perspective, humanitarian intervention has widened the traditional exceptions to the prohibition on the use of force under the UN Charter.[30]

The other proposed justification relies on the doctrine of the Responsibility to Protect.[31] According to the International Commission on Intervention and State Sovereignty (ICISS), 'sovereign states have a responsibility to protect their own citizens from avoidable catastrophe—from mass murder and rape, from starvation—but when they are unwilling or unable to do so, that responsibility must be borne by the broader community of states'.[32] By reconstructing sovereignty as a responsibility, as opposed to a right, this doctrine reflects a broader trend of global governance, showing a gradual pushback on the boundaries of sovereignty in regards to human security, state-building activities, and coercive intervention.[33]

3.2 China's Approach to Humanitarian Intervention

China's approach to humanitarian intervention, and to the use of force in the broader sense, is profoundly influenced by its position on the principles of sovereignty under the UN Charter. Since the establishment of the PRC, China has taken sovereignty as the most fundamental principle of international law and relations, and viewed 'independence' and 'autonomy' as the two most important elements in its foreign policy.[34] China's reliance on the principles of sovereignty is also prompted by its internal need. Especially, the Communist Party of China uses its defence of China's sovereignty to buttress its nationalist credentials and advance its broader state-building project.[35]

[27] L. Brock, 'The Use of Force in the Post-Cold War Era: From Collective Action Back to Pre-Charter Self-Defense?', in M. Bothe et al. (eds), *Redefining Sovereignty. The Use of Force after the Cold War* (2005) 21.

[28] N. Krylov, above note 17, at 380.

[29] See e.g. S. Kernot, 'Humanitarian Intervention: Human Rights Versus Humanitarian Assistance', 18 *Global Change, Peace and Security* (2006) 41; K. Annan, 'Human Rights and Humanitarian Intervention in the Twenty-First Century', in S. Power and G. Allison (eds), *Realizing Human Rights* (2000) 309; M.S. McDougal, L. Chen, and H.D. Lasswell, *Human Rights and World Public Order: The Basic Politics of an International Law of Human Dignity* (1980), at 241.

[30] P. Tzeng, above note 19, at 418.

[31] See G. Evans, 'From Humanitarian Intervention to the Responsibility to Protect', 24 *Wisconsin International Law Journal* (2006) 703, at 708.

[32] International Commission on Intervention and State Sovereignty, *The Responsibility to Protect* (2001), available at <http://responsibilitytoprotect.org/ICISS%20Report.pdf>, at VIII (last visited 10 January 2020).

[33] See generally R. Thakur, *The United Nations, Peace and Security: From Collective Security to the Responsibility to Protect* (2006).

[34] See W. Zhu, 'Responsibility to Protect: A Challenge to Chinese Diplomacy', 1 *Chinese Legal Science* (2013) 97, at 99f.

[35] See B. Gill and J. Reilly, 'Sovereignty, Intervention and Peacekeeping: The View from Beijing', 42 *Survival* (2000) 41, at 42.

The UN Charter refers to sovereign equality and non-interference into the domestic affairs of the member states.[36] In addition to the UN Charter, China further opines that the principles of sovereignty are also enshrined in the 'Five Principles of Peaceful Coexistence' ('Five Principles'), which are deemed as 'one of the major contributions made by the PRC since its establishment to the development of international law'.[37] The 'Five Principles' are not only 'closely and naturally linked with', 'inspired by', 'consistent with', and 'reflect' the UN Charter principles,[38] but also form a cornerstone of contemporary China's approach to international law and of Chinese foreign policy.[39] Several of the 'Five Principles' explicitly address the issue of sovereignty in international relations, including the principles of mutual respect for each other's territorial integrity and sovereignty, non-aggression, non-interference, equality and mutual benefit, and peaceful coexistence. Humanitarian intervention could be incompatible with these principles of sovereignty, the principle of non-interference in particular.

As China has not been involved in humanitarian intervention events as an intervening state, its position on humanitarian intervention could only be assessed from its responses to these events, especially from its voting behaviour in the Security Council. In general, China's principled position on humanitarian intervention appears cautious, but this position has undergone a gradual change over the past years.

As mentioned, China adopts a restrictive understanding of the UN Charter's constraints on the use of force. Such understanding has played an important role in shaping China's position on humanitarian intervention, which essentially brooks no exceptions for humanitarian interventions.[40] On this point, Shi Jiuyong, former Chinese judge of the International Court of Justice, opines that the justification for humanitarian intervention has much in common with the narrow approach to the interpretation of Article 2(4) of the UN Charter, which provides for the general ban on the use of force.[41] To China, humanitarian intervention does not have a legitimate political basis and violates the UN Charter and international law. It 'causes more harm than good'.[42]

China's responses to the 1999 military intervention in Kosovo by the North Atlantic Treaty Organization (NATO) offers a helpful opportunity to observe its position on humanitarian intervention. In 1998, the Kosovo Liberation Army started an armed struggle for independence from the Federal Republic of Yugoslavia, which met with

[36] Art. 2 UN Charter.

[37] T. Wang, above note 7, at 263.

[38] See ibid; T. Wang, *International Law* (1995), at 50 (original in Chinese); People's Daily, *The Five Principles of Peaceful Coexistence are the Greatest Common Divisor of International Relations* (10 June 2014), available at <http://opinion.people.com.cn/n/2014/0610/c1003-25128229.htm>l (original in Chinese) (last visited 10 January 2020); and J. Zhao, 'The Relationship between the Five Principles of Peaceful Coexistence and the UN Charter', 6 *Contemporary Law Science (Dang Dai Fa Xue)* (2014) 32 (original in Chinese).

[39] See H. Xue, 'Chinese Contemporary Perspectives on International Law: History, Culture and International Law', 355 *Recueil des cours* (2011) 41, at 75; P. Chan, 'China's Approaches to International Law since the Opium War' (2014) 27 *Leiden Journal of International Law* 859, at 892.

[40] J. Ku, *How China's Views on the Law of Jus ad Bellum Will Shape Its Legal Approach to Cyberwarfare* (2017), available at <https://www.hoover.org/sites/default/files/research/docs/ku_webreadypdf.pdf>, at 1 (last visited 10 January 2020).

[41] . Shi, 'Prohibition of Use of Force in International Law', 17 *Chinese Journal of International Law* (2018) 1, at 9.

[42] K. Zou, *China-ASEAN Relations and International Law* (2009), at 29.

military assaults from Yugoslav forces. The Security Council adopted several resolutions to address the deteriorating situation. On 31 March 1998, the Security Council, acting under Chapter VII of the UN Charter, adopted Resolution 1160, which called for a political solution and imposed an arms embargo on Yugoslavia, including Kosovo. On 23 September 1998, the Security Council adopted Resolution 1199, affirming that the deterioration of the situation in Kosovo 'constitutes a threat to peace and security in the region' and called for 'an immediate ceasefire'. Shortly after that, on 24 October 1998, the Security Council adopted Resolution 1203, characterizing the Kosovo situation as a 'continued threat to peace and security in the region'. When the fighting in Kosovo continued all the same, NATO launched a comprehensive bombing campaign against Yugoslavia starting from 24 March 1999, lasting for over two months.[43] NATO's intervention in Kosovo has never obtained authorization from the Security Council or received formal UN endorsement *ex post*.

China abstained from the voting on the above resolutions. However, during the Security Council debate on these resolutions, China has shown grave concerns and profound uneasiness. With regard to whether the situation in Kosovo constituted 'a threat to peace and security', China claimed that the characterization of the Kosovo situation as a threat to regional peace and security would result in the internationalization of a domestic secessionist movement under the guise of responding to a humanitarian crisis.[44] With regard to the question as to whether the Security Council should authorize military intervention to tackle the Kosovo situation, China argued that NATO's threats of military action constituted unlawful interference in Yugoslavia's internal affairs.[45] When NATO initiated military actions, China immediately condemned NATO for violating the UN Charter and the generally accepted norms of international law.[46] And after NATO's intervention, China further issued its denunciation, stating that the intervention has 'produced the greatest humanitarian catastrophe in post-Second-World-War Europe', and 'seriously undermined peace and stability in the Balkans'.[47]

In a sense, NATO's intervention in Kosovo not only marked 'the boldest claim to a freestanding right of humanitarian intervention to date', but also marked the high point of China's opposition to the idea of humanitarian intervention.[48] An important issue raised by the Kosovo event is the legality and justification of NATO's unilateral military invention in Kosovo under the UN Charter regime. NATO and many Western states believed that military intervention was necessary to stop the ethnic cleansing in Kosovo, and should thus be justified. China, in contrast, insisted on the UN Charter's restraints on the use of force and stressed the need to respect state sovereignty and the principle of non-interference. It has been acknowledged by some commentators that, to China, the Kosovo crisis was a 'domestic issue' of Yugoslavia, which meant that Serbia was defending its sovereignty against internal challenges, while external

[43] See generally J. Rytter, 'Humanitarian Intervention without the Security Council: From San Francisco to Kosovo—and Beyond', 70 *The Nordic Journal of International Law* (2001) 121.

[44] UN SCOR, 53d Sess., 3868th mtg at 11, UN Doc. S/PV.3868 (31 March 1998), at 3.

[45] See UN SCOR, 53d Sess., 3937th mtg at 14, UN Doc. S/PV.3937 (24 October 1998).

[46] See UN SCOR, 53d Sess., 3988th mtg at 12, UN Doc. S/PV.3988 (24 March 1999).

[47] See UN SCOR, 54th Sess., 4011th mtg at 9, UN Doc. S/PV.4011 (10 June 1999), at 8.

[48] J. Davis, above note 19, at 245.

intervention, particularly through the use of force, would amount to the 'internationalization of a domestic issue'.[49] Indeed, this position seems understandable if one takes into account China's national situation and concerns, given China's own internal challenges to sovereignty, including in Tibet, Xinjiang, and Taiwan.[50]

Another occasion to study China's position on humanitarian intervention is the East Timor event. Almost concurrent to the Kosovo event in 1999, when the majority of East Timorese voters endorsed independence in the UN-sponsored plebiscite, armed militias, backed by the Indonesian military, began a campaign of violence to reverse the decision. Quickly, the humanitarian situation in East Timor deteriorated. The Indonesian government finally gave consent to the intervention by the international community. On 15 September 1999, China supported the adoption of Security Council Resolution 1264, which authorizes the Australia-led multinational peacekeeping force, the International Force in East Timor (INTERFET), to restore peace and security in East Timor.[51]

China's responses in the Kosovo and East Timor events, though different with regard to its voting in the Security Council, reveals the essential conditions for its consent to humanitarian intervention, that is the authorization by the Security Council and the consent of the host state.[52] While these conditions were met in the East Timor event, they were not satisfied in the Kosovo event.

3.3 China's Position on the Responsibility to Protect

Since the end of the Cold War, the international community has faced growing challenges of humanitarian crises. Against such a background, the concept of Responsibility to Protect was proposed with an aim of reconceptualizing humanitarian intervention around the concept of legitimacy.[53]

In 1999, former UN Secretary-General Kofi Annan voiced support for humanitarian intervention, and proposed to redefine state sovereignty from the perspective of 'the human rights and fundamental freedoms of each and every individual as enshrined in our Charter'.[54] Later, in his report titled 'We the Peoples', Kofi Annan highlighted the dilemma of humanitarian intervention, asking: 'If humanitarian intervention is, indeed, an unacceptable assault on sovereignty, how should we respond to a Rwanda, to a Srebrenica—to gross and systematic violations of human rights that affect every precept of our common humanity?'[55] To explore this question, the ICISS was established in 2000. In 2001, the ICISS submitted a report to UN Secretary-General. This report

[49] Ibid. and UN SCOR, 53d Sess., 3868th mtg at 11, UN Doc. S/PV.3868 (31 March 1998), at 3. Above note 44.

[50] Ibid., at 247.

[51] SC Res. 1264, 15 September 1999, at para. 3.

[52] J. Davis, above note 19, at 253.

[53] See G. Evans, above note 31, at 708.

[54] K. Annan, *Balance State Sovereignty with Individual Sovereignty* (20 September 1999), available at <http://dwfed.org/pp_annan_on_sov.html> (last visited 10 January 2020).

[55] K. Annan, *We the Peoples: The Role of the United Nations in the 21st Century* (2000), available at <https://www.un.org/en/ events/pastevents/pdfs/We_The_Peoples.pdf>, at 48 (last visited 10 January 2020).

recognizes the growing need for intervention for human protection purposes over the past decades, and puts forward the concept of Responsibility to Protect:

> This report is about the so-called 'right of humanitarian intervention': the question of when, if ever, it is appropriate for states to take coercive—and in particular military—action, against another state for the purpose of protecting people at risk in that other state. At least until the horrifying events of 11 September 2001 brought to center stage the international response to terrorism, the issue of intervention for human protection purposes has been seen as one of the most controversial and difficult of all international relations questions.[56]

In the ensuing years, a number of landmark documents were adopted, mainly through UN-led efforts, which further elaborate the concept and nature of the Responsibility to Protect. These documents include, to list some, the 2004 Report of *More Secure World, Our Shared Responsibility*, the 2004 Secretary-General Report of *In Larger Freedom: Towards Development, Security and Human Rights for All*, the 2005 World Summit Outcome, and a series of Secretary-General Reports on the Responsibility to Protect between 2009 and 2015.[57] Among these documents, the 2005 World Summit Outcome seems outstanding since it clearly adopted the concept of the Responsibility to Protect, stating that

> Each individual State has the responsibility to protect its populations from genocide, war crimes, ethnic cleansing and crimes against humanity. This responsibility entails the prevention of such crimes, including their incitement, through appropriate and necessary means. We accept that responsibility and will act in accordance with it. The international community should, as appropriate, encourage and help States to exercise this responsibility and support the United Nations in establishing an early warning capability.[58]

The development of the Responsibility to Protect had a profound impact on China's position on humanitarian intervention. Originally, China was a strong objector to the notion of Responsibility to Protect. Adhering to the principles of sovereignty enshrined in the UN Charter, China argued that 'human rights are the internal affairs of a given country', and that 'sovereign equality, mutual respect for state sovereignty and noninterference in the internal affairs of others are the basic principles governing international relations today'.[59] For such reasons, China sees the Responsibility to Protect as an improper opportunity for intervention into the internal affairs of a state, and human rights protection is just an 'excuse' for the intervention. This view can be sensed from a speech of China's permanent representative to the UN in 1999, in which he stated that the Responsibility to Protect could provide the 'green light' for the strong

[56] ICISS, *The Responsibility to Protect* (December 2001), available at <http://responsibilitytoprotect.org/ICISS%20Report.pdf>, at vii (last visited 10 January 2020).
[57] A list of these documents is available at <http://www.responsibilitytoprotect.org/index.php/about-rtop/core-rtop-documents> (last visited 10 January 2020).
[58] GA Res. 60/1 (24 October 2005), at para. 138.
[59] UN GAOR, 54th Sess., 8th Plenary Meeting, UN Doc. A/54/PV.8 (22 September 1999).

and rich countries to bully the weak and poor countries, and that 'there would be no peace in the world'.[60]

However, China's position on the Responsibility to Protect has undergone gradual changes over the years. A notable development is China's adoption of the 2005 World Summit Outcome. As the Responsibility to Protect entered the international political agenda, China sensed a growing need to participate in this process, especially as an important power within the international community. Indeed, not only did China's former foreign minister, Qian Qichen, join the High-level Panel on Threats, Challenges and Change commissioned by the UN Secretary-General, but the Chinese government also clearly welcomed the establishment of the Panel and expressed confidence that the Panel would play a role in the UN reform.[61] In its Position Paper on the United Nations Reforms, China explicitly recognized the concept of the Responsibility to Protect, stating the following:

> Each state shoulders the primary responsibility to protect its own population. However, internal unrest in a country is often caused by complex factors. Prudence is called for in judging a government's ability and will to protect its citizens. No reckless intervention should be allowed.
>
> When a massive humanitarian crisis occurs, it is the legitimate concern of the international community to ease and defuse the crisis. Any response to such a crisis should strictly conform to the UN Charter and the opinions of the country and the regional organization concerned should be respected. It falls on the Security Council to make the decision in the frame of the UN in light of specific circumstances which should lead to a peaceful solution as far as possible. Wherever it involves enforcement actions, there should be more prudence in the consideration of each case.[62]

This Position Paper is a milestone concerning the shift in China's position on the Responsibility to Protect. It seems that China has since moved from a strong opponent to a cautious supporter of the Responsibility to Protect. Through a number of policy statements, China has formulated several conditions for the implementation of the Responsibility to Protect which represent a variation of the suggestions made in the original report of 2001. First, host states bear primary obligations for the protection of their citizens; second, the Responsibility to Protect only applies to four designated crimes, including genocide, ethnic cleansing, war crimes, and crimes against humanity; third, the Security Council should play a role in the implementation of the Responsibility to Protect; and fourth, states should not abuse the Responsibility to Protect.[63] Besides, it has also been argued that, as China (in line with the United States and Russia) sees the Responsibility to Protect as a *concept*, it does not recognize

[60] UN GAOR, 54th Sessi., 27th Plenary Meeting, UN Doc. A/54/PV.27 (6 October 1999), at 12.

[61] People's Daily, *Foreign Ministry: Welcome the Establishment of the High-Level Panel on UN Reform* (6 November 2003), available at <http://www.people.com.cn/GB/shizheng/1027/2173788.html> (original in Chinese) (last visited 10 January 2020).

[62] The Permanent Mission of the People's Republic of China to the United Nations, above note 16.

[63] See e.g. Y. Luo, 'The Development of "the Responsibility to Protect" and China's Positions', 35 *Journal of International Studies* (2014) 11, at 22f. (original in Chinese).

that the Responsibility to Protect constitutes a legal obligation and could thus be enforced.[64]

It should be noted that the shift of China's position on the Responsibility to Protect does not necessarily mean that China now shares the position of other countries. In fact, China remains quite cautious and vigilant towards the Responsibility to Protect when engaging in international relations. On this issue, China's responses to the intervention in Libya offers a good example. After the Arab Spring, the situation in Libya greatly deteriorated. The Security Council adopted a series of resolutions to address the Libyan situation. On 26 February 2011, the Security Council adopted Resolution 1970. This Resolution is deemed to have invoked the Responsibility to Protect, as it explicitly cites 'the Libyan authorities' responsibility to protect its population', and authorizes a series of non-military measures.[65] China voted in favour of this Resolution.

On 17 March 2011, the Security Council adopted Resolution 1973. In this Resolution, the Security Council showed grave concerns over the deteriorating situation in Libya, declared that Libya bears the primary responsibility for protecting the civilians from gross human rights violations, and authorized UN member states to 'to take all necessary measures ... to protect civilians and civilian populated areas under threat of attack in the Libyan Arab Jamahiriya'.[66] China abstained on Resolution 1973. When explaining the abstention, China's permanent representative to the UN stated nevertheless that China 'supports the Security Council in taking appropriate and necessary action to stabilize the situation in Libya and in putting an end to the acts of violence against civilians at an early date'.[67]

The adoption of Resolutions 1970 and 1973 is often viewed as a response of the Security Council to the Responsibility to Protect,[68] or even a precedent of its implementation.[69] China's abstention gave green light to the adoption of Resolution 1973, and consequently to the military intervention in Libya. In light of this, China's abstention gives rise to an interesting question, that is has China materially changed its previous position on humanitarian intervention? To put it more straightforwardly, did China come to agree that the Responsibility to Protect is a valid justification for military intervention? China did not provide a clear answer to these questions. As a matter of fact, in China's explanation of its vote in Resolution 1973, the Chinese ambassador did not refer to the Responsibility to Protect in any way, but stated that China's vote was based on the 'special circumstance in Libya'.[70] The ambassador also made clear

[64] See C. Holland, *Chinese Attitudes to International Law: China, the Security Council, Sovereignty, and Intervention* (July 2012), available at <http://nyujilp.org/wp-content/uploads/2012/07/Christopher-Holland-China-the-Security-Council-and-Intervention.pdf>, at 34 (last visited 10 January 2020).

[65] UN Security Council, *UN Security Council Resolution 1970* (26 February 2011), available at <https://undocs.org/S/RES/1970(2011)> (last visited 10 January 2020).

[66] UN Security Council, *UN Security Council Resolution 1973* (17 March 2011), available at <https://undocs.org/S/RES/1973(2011)> (last visited 10 January 2020).

[67] Permanent Mission of the People's Republic of China to the United Nations, *Explanation of Vote by Ambassador Li Baodong after Adoption of Security Council Resolution on Libya* (17 March 2011), available at <http://www.china-un.org/eng/gdxw/t807544.htm> (last visited 10 January 2020).

[68] S. Adams, *Libya and the Responsibility to Protect* (2012), available at <http://www.globalr2p.org/media/files/libyaandr2poccasionalpaper-1.pdf> (last visited 10 January 2020).

[69] See e.g. C. Powell, 'Libya: A Multilateral Constitutional Moment?', 106 *American Journal of International Law* (2012) 298, at 298.

[70] Ibid.

that 'China has serious concerns over some elements of the resolution', 'China always opposes the use of force in international relations', and that 'China has all along stressed that the action of the Security Council should comply with the Charter of the United Nations and the norms of international law'.[71] Even during the implementation of Resolution 1973, China has repeatedly reiterated its adherence to the UN Charter with regard to humanitarian intervention. Especially, on 16 June 2011, China's deputy foreign minister stated that 'the international community should promote a political solution to the issue of Libya, carefully implement the resolutions of the Security Council, refrain from taking any action beyond the authorization of the resolutions, expand humanitarian assistance to Libya and actively support the diplomatic efforts of the African Union'.[72]

As can be seen, despite its abstention on Resolution 1973, China's various ensuing explanations and statements seem to suggest that its vote was prompted by the special situation in Libya, rather than manifesting a fundamental change of its principled position on the Responsibility to Protect.

3.4 The Notion of 'Community of Shared Future for Mankind' and its Implication

China has grown to be the second biggest economy in the world, and is trying to obtain a greater say in international affairs. China's unprecedented economic rise profoundly influences its foreign policy. It has been observed that China has recently deviated from its traditional policy of 'making progress while keeping a low profile', and has begun to seek a leadership role in global governance.[73] While it is not the purpose of this chapter to deal with the issue of China's foreign policy change, it is of interest to briefly discuss the impact of China's rise on its position on humanitarian intervention.

Recently, China has raised the notion of building 'a community of shared future for mankind'. The origin of this notion, 'the community of common destiny', can be found in the 2011 White Paper of the State Council, titled 'China's Peaceful Development'.[74] In the White Paper, China stated that building a community of shared future for mankind is the 'general trend of the world and the common aspiration of all people'.[75] A typical enunciation of this notion has been made by President Xi Jinping in a statement at the general debate of the 70th Session of the UN General Assembly in 2015. In his statement, President Xi stated the following:

[71] Ibid.

[72] Permanent Mission of the People's Republic of China to the United Nations, *Address by Deputy Foreign Ministry Zhai Jun on the Issue of Libya to the United Nations Security Council* (16 June 2011), available at <http://www.china-un.org/chn/hyyfy/t831410.htm> (original in Chinese) (last visited 10 January 2020).

[73] See e.g. M. Cage, *Xi Jinping Just Made It Clear Where China's Foreign Policy Is Headed* (25 October 2017), available at <https://www.washingtonpost.com/news/monkey-cage/wp/2017/10/25/xi-jinping-just-made-it-clear-where-chinas-foreign-policy-is-headed/?noredirect=on&utm_term=.839c2ab85291> (last visited 10 January 2020).

[74] Information Office of the State Council, *China's Peaceful Development* (9 September 2014), available at <http://english.www.gov.cn/archive/white_paper/2014/09/09/content_281474986284646.htm> (last visited 10 January 2020).

[75] Ibid.

> Peace, development, equity, justice, democracy and freedom are common values of all mankind and the lofty goals of the United Nations. Yet these goals are far from being achieved, and we must continue our endeavor to meet them. In today's world, all countries are interdependent and share a common future. We should renew our commitment to the purposes and principles of the UN Charter, build a new type of international relations featuring win-win cooperation, and create a community of shared future for mankind.[76]

The notion of 'building a community of shared future for mankind' has rapidly entered into China's foreign policy-making and entered various international agendas. Especially, it has been incorporated in a few UN documents, such as a resolution on 'Social Dimensions of the New Partnership for Africa's Development', adopted at the 55th UN Commission for Social Development in 2017.[77] Despite the frequent use of the term 'building a community of shared future for mankind', it remains largely at a notional level. China as the initiator of this notion has not clarified what it means sufficiently, though President Xi enunciated several key aspects of the notion in his statement, such as political, economic, cultural, developmental, and environmental aspects. Despite its vagueness, the notion could have profound impact on China's future foreign policy-making. As implied in this notion, future international relations should be community-oriented and people-based. This further implies that China's foreign policy has in some way deviated from the traditional state-centric approach to international relations, and increasingly has become governance-oriented.

At this point of time, it remains premature to evaluate with meaningful accuracy whether and how China's foreign policy change prompted by this notion would influence its position on humanitarian intervention and the use of force. However, if China truly adopts a people-based perspective as implied in this notion, it would not be surprising to see China's understanding of the UN Charter principles on the use of force becoming more flexible, and China's position on humanitarian intervention being further softened in the future.

4. Conclusion

China's position on the use of force relies heavily on the wording of the UN Charter. As China adopts a restrictive understanding of these principles, it would be difficult for China to accept humanitarian intervention as a legitimate justification for the use of force. Yet, as demonstrated by its voting in a number of humanitarian intervention events, China's position on humanitarian intervention and the Responsibility to Protect has experienced a gradual and slight shift in the recent decades. It seems that China has shifted from a strong opponent to a cautious supporter of humanitarian

[76] Xi Jinping, *Working Together to Forge a New Partnership of Win-win Cooperation and Create a Community of Shared Future for Mankind* (29 September 2015), available at <http://www.xinhuanet.com/world/2015-09/29/c_1116703645.htm> (original in Chinese) (last visited 10 January 2020).

[77] Permanent Mission of the People's Republic of China to the United Nations, *The Notion of Forging a Community of Shared Future for Mankind First Inked in a UN Resolution*, available at <http://www.china-un.org/chn/ zgylhg/shhrq/shfz/t1437664. htm> (original in Chinese) (last visited 10 January 2020).

intervention. Despite this shift, however, China has not become a full supporter of humanitarian intervention, and its position remains different from that of many other states, especially among the liberal democracies.

Further, China's recent foreign policy change may hint that its position on humanitarian intervention could be further softened as China is preparing to play a more decisive role in international ordering. Recently, China has put forward the notion of a 'community of shared future for mankind'. Though this notion is largely undefined, it shows a clear orientation towards people, community, and governance. In a sense, this notion seems to signal a deviation from China's traditional state-centric perspective on the global normative order.

PART VII
INTERNATIONAL RULE OF LAW
Justifying, Contesting, and Perpetuating the Use of Force

24

Justified: Just War and the Ethics of Violence and World Order

Chris Brown

1. Prologue: Justified

Justified was an American TV series, aired between 2010 and 2015, which followed the exploits of US Marshall Raylan Givens in contemporary Kentucky, more specifically, Harlan County. The premise of the show, which developed from a short story by Elmore Leonard,[1] was that Harlan Country in the 2010s is, in many respects, not dissimilar from the Wild West of the 1870s and 80s, with crystal meth production standing in for cattle rustling; our hero Raylan is obliged to use deadly force with some regularity, in his own defence and in defence of others, gunfights being a feature of most episodes. Raylan kills lots of people and these killings are investigated by federal attorneys, who, understandably, are concerned that Raylan's approach to law enforcement may not be entirely by the book. This concern is not dispelled by these frequent legal post-mortems, but the latter conclude, in each case, that the killing was justified. These verdicts are not an endorsement of Raylan's methods, much less do the authorities consider it a point in his favour that he found himself so frequently having to shoot his way out of trouble—'justified' simply says that, given the circumstances and all things considered, he was entitled to act as he did.

There is a message here for those who, like me, believe that the just war tradition remains the best way to approach the ethics of force internationally, and the message is that the very term just war does the argument no favours. We should learn from this TV series and rename the position we support 'Justified War'. If we read 'justified' for 'just' some at least of the most common objections to the theory become much less compelling. Consider, for example, the doctrine of 'double effect', the principle that it may be permissible to cause a serious harm to a third party as a side effect of pursuing a good end, even if this side-effect were predictable in advance, so long as it was not directly intended. Some such principle is a necessary feature of any attempt to produce a workable ethics of violence, but to describe the result—let's say the killing of a number of innocent individuals—as 'just' sticks in the craw. To say that it is 'justified' may also be challenged, but at least with this usage there is less of an implication that the killing was a good even admirable thing, which the term 'just' does imply. That, at least, is my intuition, and what follows constitutes a defence of justified rather than just war. At times, and depending

[1] 'Fire in the Hole' in E. Leonard, *Fire in the Hole and Other Stories* (2012).

Chris Brown, *Justified: Just War and the Ethics of Violence and World Order* In: *The Justification of War and International Order*. Edited by: Lothar Brock and Hendrik Simon, Oxford University Press (2021). © the many contributors.
DOI: 10.1093/oso/9780198865308.003.0024

on context, I will use the old expression 'just war' but when I do the reader should read this on every occasion as 'justified war'.

2. Violence, Ethics, and World Order

We live in a world where the dominant normative assumption—reflected in documents such as the UN Charter, the Universal Declaration of Human Rights, and the Genocide Convention—is that violence between states is illegitimate (Article 2(4) of the UN Charter) except in very restrictive circumstances, specifically immediate self-defence (Article 51) or as mandated by the Security Council (Articles 41 and 42). This is in stark contrast to, say, the normative assumption that underpinned European international relations in the nineteenth century, which was that violence was a legitimate act of state which could be resorted to without legal sanction, always assuming certain forms were followed.[2] This permissive attitude towards the use of force between collectivities was actually typical not just of the nineteenth century international order, but of virtually all previous 'international' orders that is world orders where there was no genuine imperial centre. World empires do not accept that tributary countries have a right to make war on each other, much less on the empire itself, but world orders that have involved multiple power centres have usually adopted a permissive attitude.

That changed in 1945 when the independent countries who formed the United Nations wrote a document that abolished the right to make war at the sovereign's behest—whether some of those nations, or perhaps any of them, actually intended to follow this doctrine is another matter; the point is the law changed. However, the UN Security Council has rarely been able to act as the controller of violence the Charter seems to envisage. The veto power of the five permanent members of the Council (the P5), intended to protect their core interests, has been used much more widely to prevent the Council from acting in respect of issues where the P5 believe that they or their clients and allies have any skin in the game, which, of course, is most issues. The result has been that there have been many wars in the last seventy-five years but only two that have taken place under the authority of the UN—Korea in 1950, Iraq in 1991. In effect, the legal regime governing force is a dead letter;[3] the Security Council cannot exercise the authority given it by the Charter, and states are therefore obliged to make their own arrangements to protect themselves and advance their interests. Some aspects of contemporary international relations are governed, but not the resort to force.

[2] This position is contested by Hendrik Simon in 'The Myth of *Liberum Ius ad Bellum*: Justifying War in 19th-Century Legal Theory and Political Practice', 29 *European Journal of International Law* (2018) 113, but see e.g. the 1905–06 version of *Oppenheim's International Law* Part 2. Section 53. 'War no Illegality'; for the contestation of this position, see also the contributions by Hendrik Simon and Isabel V. Hull to Part III of this volume.

[3] The clearest statement of this position is still Glennon, 'Why the Security Council Failed', 82 *Foreign Affairs* (2003) 16.

This means that when it comes to the normative dimension of international violence, the use of force by states, we cannot outsource judgement to international lawyers. We are obliged to decide for ourselves whether a particular use of force is justified or not. Fortunately, there is a body of ethical thought that is devoted to clarifying this issue—the just war tradition (read justified war tradition). The just war tradition in its modern form has its roots in Christianity, specifically the works of Augustine and Thomas Aquinas, and some, but by no means all, of its modern adherents are Christian, but precursors of the tradition can be found in the classical world, and broadly similar approaches can be traced in most contemporary religious and ethical codes so there is no reason to think of the tradition as necessarily Eurocentric or reliant on religious sanction.[4] The tradition asks whether there are good reasons to resort to force, whether those who do so resort have a right to do so, and whether their intentions in so doing are proper; it also requires that the way in which force is resorted to should not go against the general rules of ethical conduct, that is, that it should be proportionate to the offence, protective of the innocent, and only undertaken if there is no reasonable alternative and there is a reasonable belief that it will make things better.[5] I have set this out without resort to Latin and without the usual formal terms because although the latter are designed to clarify, they are actually more likely to mislead the reader, making a set of ideas that are simply based on common sense seem unnecessarily esoteric.

Given the common-sense nature of justified war thinking, it is unsurprising that it should have a degree of recognition in contemporary international relations—thus, for example, the International Commission on Intervention and State Sovereignty (ICISS) in its report on The Responsibility to Protect (R2P) of 2001 employed the above criteria to identify circumstances where external military intervention might be justified to prevent humanitarian outrages.[6] In so doing the Commissioners did not consider that they were innovating—rather they saw this language and the conceptual apparatus it represented as the obvious way of setting out the issues. Still, for all its obviousness, the tradition is currently challenged and from two very different directions. On the one hand are those critics who argue that, contrary to its stated intentions, just/justified war thinking actually legitimizes violence. On the other hand, are those who purport to be friends of the tradition, but whose reworking thereof subverts the possibility that it might actually be applied to real-world cases. The next two sections of this chapter address the arguments of these declared critics and false friends, before the final section concludes with some thoughts on justified war and the contemporary world order.

[4] For the classical world and the just war tradition see O'Driscoll, 'Rewriting the Just War Tradition: Just War in Classical Greek Political Thought and Practice', 59 *International Studies Quarterly* (2015) 1; Valerie Morkevicius relates the tradition to Islamic and Hindu thought on war in *Realist Ethics: Just War Traditions as Power Politics* (2018); for Islamic perspectives on international law, see also Hashmi in this volume.

[5] I have set out my approach to just war in more detail in Brown, 'Just War and Political Judgement', in A. Lang, Jr, C. O'Driscoll, and J. Williams (eds), *Just War: Authority, Tradition and Practice* (2013) 23.

[6] International Commission on Intervention and State Sovereignty (ICISS), *The Responsibility to Protect* (2001), Section 4, available at <http://responsibilitytoprotect.org/ICISS%20Report.pdf> (last visited 10 January 2020).

3. The Critics of Justified War

The critics of the tradition are many and various; in what follows I will focus on the work of three critics who between them cover a great deal of ground. First, I will look at the work of N.J. Rengger whose critique is that of a disillusioned friend; I'll then go to Carl Schmitt, an overt enemy; and finish with Maya Zehfuss who stands in here for all the radical, post-positivist critics of the just war tradition. These critiques overlap to an interesting, and rather surprising, degree but they can be considered in isolation from one another, the strategy employed here.

Rengger's critique of the just war in his *Just War and International Order* (2013) is particularly compelling because he might be thought as the kind of thinker who would be sympathetic to the tradition[7]—he was a distinguished political theorist with a good knowledge of the classics, an interest in theology, and whose intellectual mentors were Michael Oakeshott and Eric Vogelin. Neither a hard-line realist nor a radical post-positivist he was, on the face of it, not the sort of writer one would expect to produce a strong critique of the just war tradition, yet that is indeed what he does. The essence of his critique is that the modern literature on the just war changes the original emphasis on restriction to one of permission—whereas originally just war thinking was designed to restrict the resort to force as far as possible and to restrict the kinds of violence that could be employed, he argues that many modern writers use the tradition to give permission for the exercise of violence. He has in mind particularly the support given by some just war thinkers to the 'War on Terror' and the intervention in Iraq in 2003—reference points here would be Jean Bethke Elshtain's *Just War Against Terror* (2003) and James Turner Johnson's *The War to Oust Saddam Hussein* (2005).[8] Both Elshtain and Johnson are writers who sit within the Christian tradition of writing on war and peace, and both, according to Rengger, have crossed a line and become advocates for war, a position that a just war theorist, again according to Rengger, ought not to occupy.

One response to this critique might be to say that while these two writers, amongst others, have used the tradition in this way, it is open to others to employ the same tradition to different effect, which was indeed done by many critics of the 2003 war. But for Rengger the permissive nature of much modern writing on the just war is no accident and not a matter of the choices made by particular writers. Instead he relates the permissive use of the just war tradition to the wider project of producing an international order based on the legalizing of international relations via such institutions as the International Criminal Court (ICC) and in turn he relates this to changes in domestic society towards a more regulated and organized domestic order, away from his favoured conception of politics as an Oakeshottian civil association. A permissive attitude to the use of force internationally runs in tandem with a permissive attitude towards the use of compulsion in support of economic and social goals domestically.

[7] N.J. Rengger, *Just War and international Order: The Uncivil Condition in World Politics* (2013).

[8] J.B. Elshtain, *Just War Against Terror: Ethics and the Burden of American Power in a Violent World* (2003); J.T. Johnson, *The War to Oust Saddam Hussein: Just War and the New Face of Conflict* (2005).

Those less attached to the Oakeshottian conception of politics will not be impressed by the argument here, but the narrower point Rengger makes needs to be answered. One possible response to his argument is that the opposition of 'restrictive' and 'permissive' on which his argument is based is false with regard to the just/justified war tradition. Just war thinking both restricts *and* permits violence. At the heart of the tradition is the notion of righting a wrong—'just cause' to use the conventional term; action *is both restricted to and permitted in* circumstances where those with authority use it with the intention of righting a wrong and with due attention to such practical considerations as the protection of the innocent, proportionality, last resort, and the prospects of success. It may or may not be the case that writers such as Elshtain and Johnson are *excessively* permissive—that is a matter for practical judgement—but that they are permissive of violence is not in itself a reason to reject their arguments.

A further point *contra* Rengger might be that, as he admits, he has no easy answers to the questions raised by the just war, no alternative formulation that will do what in his opinion the just war is failing to do. One critic who certainly does have an answer here is Carl Schmitt, whose work as a whole constitutes a critique of the just war tradition and is perhaps best summarized in his *The* Nomos *of the Earth: In the International law of the Jus Publicum Europaeum* (2003).[9] Schmitt's critique is more wide-ranging that that of Rengger, but he does have an answer of a sort to the question of the alternative—although, frustratingly for him, this answer is no longer available and can only take the form of an elegy for a lost European past.

Schmitt understands the just war as an essentially theological notion which legitimized total war. Just war theorists may have attempted to limit the role of violence in human affairs, and the Catholic Church in the Middle Ages certainly did try to use its influence to ban certain weapons and enforce truces, but these measures were always subverted by the basic logic of the just war, which, according to Schmitt, by inviting the judgement that one side in a conflict is 'just' involved identifying the other as 'unjust', with the concomitant that the unjust must be defeated whatever the cost, even if this involved using banned weapons or taking a conflict to extremes. Just war thinking justifies escalation, feeds self-righteousness, and legitimizes war—this goes beyond Rengger's critique; in this account the notion of a just war is not simply permissive of war but actively encourages it. Also, unlike Rengger, Schmitt is clear that there is an alternative framework against which just war thinking ought to be judged, and much of *The* Nomos *of the Earth* is devoted to defending this alternative.

This alternative emerged from the development of the sovereign, territorial state in Europe, which involved a spatial disposition of the Continent that undermined the jurisdiction of the Catholic Church and the Empire. This political order is no longer committed to the preservation of God's Order in the world but instead is based on Reason of State. The European princes created amongst themselves a *Jus Publicum Europaeum*, a secular legal order under which they recognized each other's rights and

[9] C. Schmitt, *The Nomos of the Earth in the International Law of the Jus Publicum Europaeum* (2003). This is the English translation—the original was published in 1950 and mostly written during the Second World War. I have written at greater length of Schmitt in 'From Humanised War to Humanitarian Intervention: Carl Schmitt's Critique of the Just War Tradition', in L. Odysseos and F. Petito (eds), *The International Thought of Carl Schmitt* (2007) 56.

interests, *within Europe* (the proviso here is crucial). Beyond the line, in the extra-European world, Europeans engage in large-scale appropriations of land, respecting neither the rights of the locals nor each other's rights, but within Europe a different *modus vivendi* was possible. In the extra-European world appalling atrocities occur which would not happen, or at least ought not to happen, in Europe. As between European rulers within Europe, war became 'bracketed'—rationalized and humanized. Rather than a divine punishment, war became an act of state. Whereas in the medieval order the enemy must necessarily be seen as unjust the new humanitarian approach to war involved the possibility of the recognition of the other as *justus hostis*, an enemy but a legitimate enemy, not someone who deserves to be annihilated, someone in whom one can recognize oneself, always a good basis for a degree of restraint. This, for Schmitt, is the great achievement of the age, and the ultimate justification for the sovereign state.

> [An] international legal order, based on the liquidation of civil war and on the bracketing of war (in that it transformed war into a duel between European states), actually had legitimated a realm of relative reason. The equality of sovereigns made them equally legal partners in war and prevented military methods of annihilation.[10]

The new thinking about war also opened up the possibility of neutrality as a legal status; since war was no longer justified in accordance with a theological judgement based on notions of good and evil, it became possible for third parties to stand aside if their interests were not engaged. Equally, the ordinary subjects of belligerent rulers need not feel obliged to become emotionally engaged in the fray; war becomes a matter for sovereigns and their servants, civil and military—the kind of wider involvement that might be appropriate to a war between good and evil becomes strictly optional. Thus was established what Schmitt clearly regarded as a kind of golden age in European international relations, a golden age that would be sabotaged in the twentieth century by the United States, with the reluctant, ambiguous, assistance of the United Kingdom—two maritime powers whose commitment to the *Jus Publicum Europaeum* was highly qualified in the case of the United Kingdom, non-existent in the case of the United States. At this point Schmitt's argument veers away from his critique of the just war and need not concern us here, but that critique stands alone and needs to be answered.

Part of the answer concerns the facts of the case. Contra Schmitt, medieval just war thinkers did not usually endorse crusades, nor did they demonize enemies in the way he suggests is inherent in the logic of a just war—obviously just war thinking could be twisted into a kind of Manichean division between good and evil but none of the central figures of the tradition fall into this trap. Again, Schmitt's picture of the bracketing of war may have been an accurate, albeit rose-tinted, description of war in Europe in the middle of the eighteenth century, but only for a short period and, even then, war was more of a disaster for those directly affected by it than Schmitt allows. But, even if one were to accept Schmitt's account of the golden age of the *Jus Publicum Europaeum* it is clear that this happy arrangement is no longer of any relevance to the twenty-first

[10] Schmitt, above note 9, at 142.

century—not only is his vision unacceptably Eurocentric, it is also based on the assumption that European power is dominant in the rest of the world. In effect, Schmitt's overall position is one of nostalgia for a world order that no longer has any relevance—even if one were to accept all his premises, which would be ill-advised. From Rengger's work we can take away the message that just/justified war thinking ought not to be excessively permissive; from Schmitt the takeaway is that just war thinkers ought not to allow their desire for justice to lead to the demonization of those deemed not to have a just cause. In both cases, this is no more than common-sense and the logic of justified force demands.

Maya Zehfuss's *War and the Politics of Ethics* (2018) is not actually oriented towards the just war tradition—instead her focus is on 'Western' war in general—but her work certainly touches on the tradition and on some of the writers who I approach in the next section of this chapter, the false friends of just war.[11] Zehfuss's central concern is with the way in which, she argues, the West has adopted the project of making the world a better place through the allegedly judicious use of violence. Western war is ethical war, no longer to be justified in terms of reason of state but by the good in terms of human rights and freedom from oppression that it is designed to produce. But, of course, war kills people, including, very often, the people it is intended to protect, and this subverts the ethical intent.

Zehfuss pulls in arguments that we have already met from Rengger and Schmitt, while giving them a twist. Rengger's concern with the permissiveness of much contemporary just war thinking is specifically endorsed by Zehfuss. Schmitt, on the other hand, does not register on her radar, which is rather strange since her critique of ethical war echoes the kind of arguments he puts forward, although her 'ethical warriors' go to considerable lengths to avoid civilian casualties and to understand their enemies, while Schmitt's crusaders take things to extreme in their war against the unjust. Part of the reluctance to engage with Schmitt may come from a fundamental disagreement with the positive side of the latter's argument—Zehfuss does not want to endorse the *Jus Publicum Europaeum* that is the ideal presented by Schmitt, but the result is that her critique of Western war does not present an alternative understanding of the role of force in human affairs.

The key here is the term 'Western' war. Like other radical critics of the wars of the last two decades such as Ken Booth whose 'Ten Flaws of Just Wars' (2000) is a brief summary of all the reasons why just war thinking is unacceptable to radicals, and indeed, like political movements such as the 'Stop the War Coalition', her focus is specifically on wars waged by the West.[12] Wars waged by non-Western states and other actors are outside the scope of her study, just as, for example, the Stop the War Coalition never critiques the war-making of Russia or Iran in Syria with anything like the intensity it devotes to opposing the occasional Western engagements in that conflict. For Zehfuss it is Western war that is the problem; it is the attempt to justify war in ethical terms that

[11] M. Zehfuss, *War and the Politics of Ethics* (2018).

[12] Booth, 'Ten Flaws of Just Wars', 4 *The International Journal of Human Rights* (2000) 314; see also, partly in response, Brown, 'A Qualified Defence of the Use of Force for Humanitarian Reasons', 4 *The International Journal of Human Rights* (2000) 282. For the Stop the War Coalition see <http://www.stopwar.org.uk> (last visited 10 January 2020).

attracts her critique—non-Western countries are, in effect, given a free pass to engage in war for traditional reasons of state presumably on the principle that killing people for political reasons is somehow less condemnable than killing them in order to make the world a better place. It is difficult to see the moral force of this position.

To summarize, critics of the just/justified war tradition argue that thinking about the use of force in line with the tradition does not actually serve the ends to which the tradition is supposed to be devoted. Far from limiting international violence, thinking in terms of the just war is in danger of having the opposite effect, being excessively permissive of the use of force, encouraging a crusading mentality, and providing ethical justifications for policies that do great harm. These are serious critiques, and they certainly identify attitudes to force that should be avoided—but they do not actually undermine the essential nature of just war thinking. In effect these critiques warn us of the dangers of misusing the idea of the just war, but the fact that it is certainly possible to misuse just war thinking is not a reason to avoid the discourse altogether.

4. New (False?) Friends of Justified War

Whereas the just war was once a rather esoteric minority interest, there are today (at least) three different conceptions of the idea that are current in contemporary discourse: first, there is what I would describe as 'neo-classical' just war thinking, drawing on the deep history of the discourse and applying it to modern conditions—this is the version of justified war defended here; then there is 'legalist' just war thinking which essentially equates the notion of a just war with that of a war fought in accordance with the Law of Armed Conflict, and the UN Charter, possibly slightly amended to allow for the possibility of humanitarian intervention; and finally there is 'revisionist' just war theory, which applies the techniques and assumptions of analytical political theory to the notion of a just war. The latter two approaches are closely connected—the legalist approach is best represented by the work of Michael Walzer who understands just war in terms of human rights, but rights as expressed via membership of a community, whereas the revisionists remove this latter restriction. From their perspective just war thinking has to be wholly consistent with a full understanding of human rights, most particularly (and obviously problematically) the right to life. The work of these thinkers has, without question, galvanized the study of justified war over the last two decades—the key question is whether their formulations have actually taken the notion too far away from its origins and undermined its capacity to actually relate to the realities of war. Are they, in other words, false friends of the justified war tradition?

Whereas the tradition presents the conventional criteria as a series of questions concerning the use of force and violence which invite the exercise of judgement—just war *theory* as espoused by the revisionists attempts to arrange these questions into a series of law-like propositions which will provide definitive answers to these questions. Early statements of this approach include David Rodin's *War and Self-Defence* (2002) which contests the notion that states have a right to defend themselves from attack, and Jeff McMahan's *The Ethics of Killing* (2002) which took war as simply one element of a larger problem about killing, but perhaps the most substantial and influential work is McMahan's *Killing in War* (2009) which contests the just war tradition on a number of points, arguing against the view that combatants should be treated as moral equals and challenging the central principle of non-combatant immunity.[13]

The central feature of this approach is a refusal to conceptualize war as a *collective* enterprise; instead, the revisionists argue, war has to be understood in terms of individual responsibility at all levels, from the high command down to the frontline soldier. As a general proposition, this stance involves replacing the idea that there is a specific body of laws associated with war—the Law of Armed Conflict—by an assertion of the universal scope and authority of International Human Rights Law. Such as substitution has dramatic consequences—in order to show how dramatic, it may be useful to set out some conventional positions which are supported by the Law of Armed Conflict and, for the most part, by arguments drawn from the just war tradition (certainly in its legalist form) but which are challenged by the revisionists.

Conventionally, both the just war tradition and the Law of Armed Conflict are concerned with and distinguish between what is usually called *ius ad bellum* (the justice of the resort to war) and *ius in bello* (the justice of the conduct of war). The division of just war thinking into these two categories is relatively recent but quite well established nowadays. A clear exposition of the distinction here can be found in Michael Walzer's highly influential *Just and Unjust Wars* (1977/2015), where he distinguishes between the 'theory of aggression' and the 'war convention'.[14] Walzer's theory of aggression argues that members of international society are entitled to defend their political and physical integrity and attacks on the same, except in very limited legal and moral circumstances, constitute the crime of aggression and justify a war of self-defence. However, the 'war convention' states that the combatants in such a war, whether on the side of the aggressor or the defender, should be treated as morally equal with the same rights.

Walzer's position is somewhat at odds with the medieval just war tradition, but it is broadly supported by modern international law. As noted above, the UN Charter outlaws the use of force (Article 2[4]) except in self-defence (Article 51) or as directed by the Security Council, and self-defence is recognized by most conventional just war thinkers as the most obvious example of a just cause for war—indeed, so obvious that it is rarely discussed. As to the 'war convention', the Law of Armed Conflict extends protection to all combatants and non-combatants irrespective of the alleged justice of their cause—the same rules apply to all. The neo-classical just war tradition is a little less committed to this principle, but there are good pragmatic and moral reasons for adhering to it. Pragmatically, it is clear that if either side in a conflict could waive the rules on the basis that their opponent was the wrongdoer, then there would be no rules, to the detriment of troops on both sides. The moral case is equally strong: most combatants in a large-scale war are likely to be conscripts, not given a choice whether or not to fight. Such soldiers are responsible for their conduct in the war, but the conventional approach acquits them of responsibility for the actual war.

[13] D. Rodin, *War and Self Defence* (2003); J. McMahan *The Ethics of Killing* (2002); and J. McMahan, *Killing in War* (2009); a good overview of the 'revisionist' position is: Lazar, 'War', in E.N. Zalta (ed.), *The Stanford Encyclopaedia of Philosophy* (2017), available at <https://plato.stanford.edu/archives/spr2017/entries/war/> (last visited 10 January 2020). See also the synthesis presented in C. Fabre, *Cosmopolitan War* (2012).

[14] M. Walzer, *Just and Unjust Wars* (1977/2015). On Walzer, see also the chapter by Anthony Lang, Jr in this volume.

Revisionist just war theory is critical of both the theory of aggression, the current version of *ius ad bellum*, and the moral equality of combatants, the current core principle of *ius in bello*, criticizing the international legal status quo and the just war tradition from an anti-collectivist position which focuses on the rights and responsibilities of the individual. David Rodin's *War and Self-Defence* presents the core argument with respect to *ius ad bellum*. States do not have an unqualified right of self-defence—such a right is conditional; only just societies have the right to defend their political and territorial integrity. Only just communities are entitled to defend themselves and even then, attacks on the sovereignty or territorial integrity of a state are not, in themselves, justifications for war—they are such if, and only if, individual rights are violated. So, for example, a bloodless invasion would not count as an act of aggression.

Jeff McMahan's *Killing in War* is the most important revisionist work on *ius in bello*. He regards all combatants (and non-combatants) as rights-bearers but posits that they hold different rights, depending on the justice of their cause—there is no moral equivalence between combatants fighting a just war and those fighting an unjust war.[15] Combatants in an unjust cause are not entitled to act aggressively or to defend themselves and the fact that they were conscripted into the army and may not share the beliefs of their leaders is irrelevant in the context of their moral responsibility. However, they are still rights bearers who possess the right to life, therefore they can only be killed in self-defence by just combatants who *do* have the right to defend themselves. The model here is very much that of domestic policing where the police have powers of arrest and may use necessary force if they are resisted; the criminal may not resist arrest but may not be subjected to violence in the event of non-resistance. Perhaps even more controversial, is the revisionist critique of the principle of non-combatant immunity set out by both McMahan and Fabre. Again, individuals are to be judged in accordance with their behaviour and those who support an unjust war, even if they do not actively participate in the fighting, may in restricted circumstances be subject to attack.

These positions are counter-intuitive, but, of course, that does not make them wrong—still there are compelling reasons to be concerned at the development of revisionist just war theory. There are two points here which are basic, and which address the underpinnings of revisionist just war theory: first, war is a collective enterprise, a *social* phenomenon; the idea that all social behaviour can be understood in terms of the behaviour of individuals is something that revisionist just war theorists share with other analytical political theorists and mainstream economics and is subject to the same critiques that those disciplines attract. Second, revisionist just war theory relies on implausible assumptions about the capacity of the theorist to make authoritative judgements as to the justice of particular causes or particular tactics. The revisionist just war theorist believes that with enough concentrated brainpower the justice of a cause can be accurately assessed. This assumes too much; human beings do not possess the means to achieve that kind of certainty. The neo-classical just war tradition, rooted in a Christian adaptation of Aristotelian *phronesis*, stresses the centrality of judgement but does not hold out the possibility of certainty. Classical just

[15] See also D. Rodin and H. Shue (eds), *Just and Unjust Warriors: The Moral and Legal Status of Soldiers* (2008).

war thinking provides questions, while revisionist just war theory purports to provide answers, often expressed with very little humility—consider, for example, McMahan's summary condemnation of Wittgenstein's 'moral stupidity' in supporting Austria-Hungary's case in the First World War.[16]

These two, as it were, generic criticisms feed into the specific content of revisionist just war theory, but an equally important point is that this approach loses contact with the realities of war and the importance of the moral framework of the soldier. Michael Walzer makes the point nicely in a 2012 online interview with Nancy Rosenblum where, contrasting their approach with his own, he remarks that for the revisionists 'the subject of just war theory is just war theory [whereas] I think the subject matter of just war theory is war'.[17] The point is that it would be next to impossible to fight and win a war that satisfied all the requirements of the revisionists. It is difficult to imagine what kind of cause would be regarded by the revisionists as fitting this bill for a just cause and assuming that, by some quirk of fate, a just cause that would satisfy the revisionists could be found and therefore a war could be justly undertaken, the kind of *in bello* restrictions that they would impose would make actually fighting the war a practical impossibility. Given an army that is unable to take the initiative, that may not close with the enemy, that is obliged to recognize the right to life of its opponents, the only possible outcome of a conflict would be a pointless stalemate—that is, if both armies were fighting with the same revisionist restraints which is an unlikely state of affairs.

The fact that it would be virtually impossible to fight and win a just war under revisionist terms is a conclusion that the revisionists themselves would probably welcome; the argument would be that if a war could not be fought justly, as they define fighting justly, it ought not to be fought. Revisionist just war theory is, when the chips are down, essentially pacifist, which loses contact with the central aim of the justified war tradition, which is to discriminate between cases—the just war is an alternative to pacifism on the one hand and a *Realpolitik* approach on the other. There is a perfectly good set of arguments in favour of pacifism, and therefore no need to approach this position from the direction of the just war tradition; better to preserve the latter for arguments in favour of discrimination. Of course, if everybody adopted revisionist ideas the problem of war would disappear, but it is clear that this is not likely to happen anytime soon.

5. Conclusion: Justified War and Contemporary World Order

There are good reasons for the widespread view that the rules-based international order, largely designed during the Cold War but re-fitted in the 1990s and 2000s, is now in real trouble, with the institutions designed to support this order failing in the

[16] McMahan, *Killing in War*, above note 13, at 1.

[17] Nancy Rosenblum, 'A Conversation with Michael Walzer' (2012), available at <https://www.youtube.com/watch?v=TvpnmmLoO38> (last visited 10 January 2020). Walzer expresses similar sentiments in the Afterword to the 2015 edition of *Just and Unjust Wars*.

task with which they have been charged. This is true in particular with respect to the central system of great power relations; the management of the latter has become increasingly difficult, to the point that it is no longer being attempted, the main post-war 'manager' of the system, the United States having, under the Trump administration, more or less given up on the task. Given this context, the collapse of the legal regime governing the use of force takes on a new significance. For most of the last twenty years, both friends and enemies of the justified war tradition have more or less assumed that the role of force in contemporary international relations is likely to involve so-called humanitarian interventions or other 'savage wars of peace'.[18] These are the Western wars that Zehfuss critiques and the wars that Rengger argues modern just war thinkers are too ready to support—and, in so far as the 'revisionist' theorists imagine their ideas could be put in practice it is in the context of the rules of engagement for conflicts such as that in Afghanistan or Syria.

These conflicts will doubtless continue, indeed are continuing not just in Afghanistan and Syria but in many areas in West Africa where French soldiers are battling the local Al-Qaeda franchises, but larger problems now loom. Great power wars are no longer unthinkable. Chinese actions in the South and East China seas and the Taiwan Strait could easily lead to conflict with the United States and its allies; *pace* current negotiations, nuclear politics in Korea could also lead to a confrontation between China and the United States; and, of course, in Europe we have already seen the use of force to change internationally agreed boundaries when Russia annexed Crimea and invaded eastern Ukraine. Threats to the independence of the Baltic Republics, (especially dangerous since, unlike the threat to Ukraine, NATO members are involved) Cold-War style probing of Western defences by Russian aircraft, and provocative boasting about Russia's new suite of offensive weapons—taken individually these steps may not amount to much but taken together they suggest that it would be unwise to dismiss the possibility of a return to large-scale warfare. And if, as is repeatedly rumoured, the United States in response to the unwillingness of European members of NATO to take the steps necessary to defend themselves, were to withdraw from the alliance, either formally or informally, it would be foolish indeed to believe that President Putin's Russia would not take advantage of the power vacuum this would create.

The prospect of great power war is, of course, truly terrible, but the very awfulness of what would be involved makes it particularly important that we—informed public opinion as well as government officials and the military—develop clear thinking about the circumstances under which a resort to force would be justified. To pick up a revisionist theme, is self-defence always a just *causus belli*? What would be a proportionate response to a Russian invasion of the Baltic Republics? Shifting to more immediately plausible scenarios, what should be the response to large-scale cyber-attacks on civilian infrastructure? What kind of rules of engagement should be imposed on militaries that might have to fight against well-organized and equipped conventional opponents whose respect for the Law of Armed Conflict could not be guaranteed?

[18] 'Savage wars of peace' comes from R. Kipling 'The White Man's Burden' and has most recently been used in M. Boot, *The Savage Wars of Peace: Small Wars and the Rise of American Power* (2002).

These are questions it would be good if we had thought through before we actually need to—indeed, if we think through these questions it maybe we can avoid ever having to answer them in the real world. And the burden of this chapter is to suggest that, *pace* critics of the tradition, and the *pace* the implausible formulations that some friends of the tradition prefer, the conventional notion of justified war is still the best place to begin with the task of grappling with these, literally, life and death questions.

25
How Many Deaths Can Article 2(4) UN Charter Die?

Thilo Marauhn

1. 'Changing Norms Governing the Use of Force by States'?

It is an honour and a challenge to take *Thomas M. Franck*'s famous essay 'Who Killed Article 2(4)?'[1] in Volume 64 of *The American Journal of International Law* as a starting point for a number of critical reflections on 'the high-minded resolve of Article 2(4)'[2] shared by the 'Allied nations gathered at San Francisco in the warming glow of victory'[3] when signing the UN Charter. It is noteworthy that *Franck* added an additional phrase to the title of his essay: 'or: Changing Norms Governing the Use of Force by States', a phrase taken up here to direct the introductory section of this chapter. As *Rosalyn Higgins* in her obituary written to commemorate *Franck* in Volume 103 of *The American Journal of International Law* perfectly illustrates, '[h]e had a talent for conjuring up alluring and irreverent titles to his many essays. He coupled this with an initial short phrase to catch the eye, framed by a colon, followed by a further phrase to explain the subject matter under analysis.'[4] The additional phrase on changing norms is essential to understand *Franck*'s argument. It is likewise essential for this chapter, which focuses on the existing and continuing tension between a normative claim and international (state) practice on the prohibition of the threat and use of force.

By way of introduction, it seems helpful to gain a few insights on the international law scholar *Thomas M. Franck* and his essay on Article 2(4). *Higgins* nicely describes *Franck*'s move from constitutional law to international law and points out that *Franck* 'was attracted above all by international institutions'.[5] She argues that he aimed at 'identifying concepts to make a more reasoned discourse leading to the taking of decisions that affect national and international communities'.[6] Likewise, he was interested in ways and means to 'make more effective the working of institutions'.[7] The United Nations was of particular interest to *Franck*, with 'the way law and politics played out within their walls'. It is very much in line with this interest that *Franck* would rather 'invoke reason and morality' than 'emphasize the technical detail of precedents or

[1] Franck, 'Who Killed Article 2(4)? Or: Changing Norms Governing the Use of Force by States', 64 *American Journal of International Law* (1970) 809.
[2] Ibid., at 809.
[3] Ibid.
[4] Higgins, 'Thomas M. Franck (1931–2009)', 103 *American Journal of International Law* 502.
[5] Ibid., at 503.
[6] Ibid.
[7] Ibid.

Thilo Marauhn, *How Many Deaths Can Article 2(4) UN Charter Die?* In: *The Justification of War and International Order.* Edited by: Lothar Brock and Hendrik Simon, Oxford University Press (2021). © the many contributors.
DOI: 10.1093/oso/9780198865308.003.0025

other complex arguments of international law'.[8] His 1995 monograph *Fairness in International Law and Institutions*[9] is most illustrative in this regard.

Convincingly thus, the title of his seminal essay links normative claim and international practice, challenging indifference and complacency towards the weakening of rules while at the same time upholding international law's normative power by illustrating the process(es) of 'changing norms'. In his essay, *Franck* argues that 'the rules prohibiting recourse to force were imperfect and, to some extent, already obsolescent by the time the Charter came into operation'.[10] But he also suggests that states, by pursuing 'their national interest through the use of force'[11] contributed to the demise of the Charter's prohibition of the threat and use of force.

Franck first addresses factors undermining Article 2(4). These include the 'invalid premise' (*Franck*) 'that the Security Council would be able to discharge its responsibility as the United Nations' principal organ for world peacekeeping'.[12] Turning to practice, he points out that 'the failure of U.N. enforcement machinery has not been occasional but endemic'.[13] Subsequently addressing Article 51 of the Charter, *Franck* underlines that 'there is usually no way for the international system to establish conclusively which state is the aggressor and which the aggrieved'.[14] Against this background he argues that the 'temptation remains what it was before Article 2(4) was conceived: to attack first and lie about it afterwards'.[15]

The second part of *Franck*'s essay discusses the effect of small-scale warfare on the prohibition. What he describes as 'rural and urban hit-and-run operations by small bands of fighters, sometimes not in uniform and often lightly armed'[16] and 'new kinds of assistance ... encouraging and aiding ... guerrilla movements'[17] does not fit 'comfortably into conventional international legal concepts and categorizations'[18] and makes the analogy to an armed attack ever more difficult.[19] *Franck* illustrates this by referring to many post-1945 examples, whether Communist governments supporting indigenous insurgents[20] or the 'prophylactic dispatch of U.S. troops to Lebanon and British forces to the Kingdom of Jordan'[21] in 1958. While he points out that 'a line of continuity runs from invasions by tanks and divisions through training, arming, sheltering and infiltrating neighboring insurgents, all the way down to hostile radio propaganda calling for revolution in a foreign country',[22] he takes up his concerns from the introductory part of the essay and identifies a procedural as well as a substantive problem, that the 'Charter provides no answer' as to how the fact of an armed attack is

[8] Ibid.
[9] T.M. Franck, *Fairness in International Law and Institutions* (1995).
[10] Franck, above note 1, 809.
[11] Ibid.
[12] Ibid., at 810.
[13] Ibid., at 811.
[14] Ibid.
[15] Ibid.
[16] Ibid., at 812.
[17] Ibid.
[18] Ibid.
[19] Ibid.
[20] Ibid., at 812f.
[21] Ibid., at 814.
[22] Ibid., at 813.

to be established,[23] and that it is close to impossible 'to define "armed attack" in a way relevant to the modern conditions of indirect, limited warfare without broadening it to the point at which disproportionate armed force can be used under the guise of self-defense against imagined or slight provocation'.[24] *Franck* eventually points out that 'clear-cut distinctions between aggression and self-defense' are 'exceedingly difficult' in light of small-scale warfare,[25] thus weakening Article 2(4) UN Charter.

Similar is the impact of nuclear warfare on the prohibition, as *Franck* illustrates in the following, vehemently questioning *Myres McDougal* who argued that Charter Article 51 should be interpreted as allowing a state to use military force when regarding 'itself as intolerably threatened by the activities of another'.[26] *Franck* argues that such a rule 'is so subjective as to be no rule at all'.[27]

The fourth part of his essay turns to regional enforcement and the Charter's Chapter VIII. *Franck* argues that the potential of regionalism 'as a legal loophole for avoiding the duty of Article 2(4) grew as more regional groupings came to see themselves as eligible'[28] to perform regional enforcement. While he considers the Charter itself as 'a compromise between universal and regional international systems',[29] he does not perceive this as automatically undermining the prohibition of the threat and use of force but he warns that 'it is important to have a clear view of which groupings of states are entitled to regard themselves as regional organizations'.[30] It is against this background that *Franck* criticizes the use of regionalism by the two superpowers to deprive the prohibition of its effectiveness within their respective spheres of influence: 'Violence may be used by the region in "self-defense" against a deviationist'.[31] This, he argues, 'is a long way from the solemn obligation of Article 2(4)'.[32]

Franck concludes that the 'prohibition against the use of force in relations between states has been eroded beyond recognition'.[33] 'What killed Article 2(4)', he argues, 'was the wide disparity between the norms it sought to establish and the practical goals the nations are pursuing in defense of their national interest'.[34] This was in 1970.

More than three decades later, in 2003, *Thomas Franck* took up his 1970 essay and raised the question who killed Article 2(4) again.[35] *Franck* takes a critical perspective on the 2003 use of force in Iraq by a US-led coalition. In this short contribution he contrasts the *Realism* of the 1970s with the optimism of the 1990s and the relapse of 2003—as he puts it.[36] In his introduction he complains that 'we now have on offer

23 Ibid., at 816.
24 Ibid.
25 Ibid., at 820.
26 'Panel: Fundamental Challenges to Legal Doctrines Affecting International Coercion: Aggression, Self-Defense, Non-Intervention, Self-Determination, Neutrality', 57 *Proceedings of the American Society of International Law* (1963) 147, at 164.
27 Franck, above note 1, at 821.
28 Ibid., at 826.
29 Ibid., at 822.
30 Ibid., at 827.
31 Ibid., at 835.
32 Ibid.
33 Ibid.
34 Ibid., at 837.
35 Franck, 'What Happens Now? The United Nations After Iraq', 97 *American Journal of International Law* (2003) 607.
36 Ibid., at 608, 609, and 610.

proposed models for interstatal relations that seem even worse than the dilapidated system to which, by 1970, state misbehaviour had reduced the postwar world'.[37] *Thomas Franck* argues that, whereas the Charter vision was lost in the 1970s 'in a miasma of so-called realpolitik', the 'model' offered in 2003 freed the remaining superpower 'from all restraints of international law'.[38] While 'unlawful recourses to force' in the period addressed by his earlier essay 'were accompanied by a fig leaf of legal justification', 'the aggressors ... have all but discarded the fig leaf'.[39]

In the course of his opinion piece, *Franck* goes on to describe the 1990s as a period during which 'the international system seemed to be moving' towards the original Charter ideas, 'with Article 2(4) miraculously reborn in a post-Cold War order underwritten by a return to the law of the long-languishing Charter'.[40] He saw 'the Security Council ... proving itself able to interpret and apply the rules in such a way as to make them responsive to new dangers'.[41] He then turns to the relapse of 2003 and discusses whether the Iraq invasion violated the Charter, carefully analysing the arguments presented by legal adviser *William Howard Taft IV* of the US Department of State[42] and by the attorney general of the United Kingdom,[43] both in support of the invasion. *Franck* by and large rejects these arguments as they 'do not fare well under close examination'[44] and he returns to the concern that the use of force against Iraq 'must be seen as either revising or undermining the provisions limiting the discretion of states to resort to force'.[45] While conceding on the one hand that, '[i]n international law, violators do sometimes turn out to be lawgivers',[46] he does not agree with such a reading of the events of 2003 but points out that the decisive question is the repudiation of the UN by powerful actors: 'But the nub of the matter is not who was right and who wrong but *who gets to decide what to do*'.[47] *Franck*'s main concern is not a re-interpretation of the prohibition of the threat and use of force but a weakening of a UN-based system for the maintenance of international peace and security. He is not attracted either by apology or utopia.[48] But he calls upon lawyers 'to stand tall for the rule of law'[49]—notwithstanding that some politicians make use of 'the present power disequilibrium'[50] to their advantage.

What is so important about *Thomas M. Franck's* two essays as to present them in a fairly extensive manner at the beginning of this chapter? Apart from the telling

37 Ibid., at 607.
38 Ibid., at 607f.
39 Ibid., at 608.
40 Ibid., at 609.
41 Ibid., at 610.
42 The position is illustrated in Taft and Buchwald, 'Preemption, Iraq, and International Law', 97 *American Journal of International Law* (2003) 557.
43 For a critical discussion, addressing the Chilcot Inquiry on the use of force against Iraq, see Green and Samuel, 'The Chilcot Report: Some Thoughts on International Law and Legal Advice', 22 *Journal of Conflict and Security Law* (2017) 333.
44 Franck, above note 35, at 614.
45 Ibid.
46 Ibid.
47 Ibid., at 616 (italics in the original).
48 Referring to Martti Koskenniemi's famous monograph on the structure of international legal argument, namely M. Koskenniemi, *From Apology to Utopia* (1989).
49 Franck, above note 35, 620.
50 Ibid.

heading of his 1970 essay, we believe that there are four aspects which deserve closer consideration in the following: *first*, looking at international relations and international law, what is the current state of the prohibition of the threat and use of force? *Second*, what is the relationship between normative claims and state practice in this particular context? *Third*, how can the detachment of international peace and security from the United Nations be prevented? And *fourth* and finally, what is the proper role for the lawyer in this particular context? *Thomas M. Franck* can provide some guidance along these four questions which will be addressed in the following sections of this chapter, both with his upright character as an academic in international law and with his recognition that international law is dynamic, not static.

2. What Is the Current State of the Prohibition of the Use of Force?

To begin with facts, the more recent trend of the use of force does not look encouraging. It is beyond the scope of this chapter to provide a full list of armed conflicts (international or internationalized)[51] occurring after the publication of *Franck*'s 2003 article. However, it is illustrative to reflect upon a number of incidents in light of Article 2(4) UN Charter.

The Iraq War, which was addressed in the 'Agora' section of *The American Journal of International Law*, lasted from 2003 until 2011, with, among others, involvement of the United States and United Kingdom.[52] The Chadian civil war (2005–2010) saw a French intervention.[53] Mention should be made of the 2006 Israel-Hezbollah war, affecting Lebanon.[54] *Operation Juniper Shield* in the trans-Sahel region, a counter-terrorism operation in the context of *Operation Enduring Freedom*, began in 2007 and is ongoing.[55] 2008 saw the Russo-Georgian war.[56] In 2011, a multi-state coalition intervened in Libya, referring to UN Security Council (SC) Resolution 1973 (2011).[57]

[51] An internationalized armed conflict is a non-international armed conflict with the armed forces of a third state intervening. For a critique of the notion of internationalized armed conflicts from the perspective of international humanitarian law, see Stewart, 'Towards a Single Definition of Armed Conflict in International Humanitarian Law', 85 *International Review of the Red Cross* (2003) 313.

[52] The section 'Agora: Future Implications of the Iraq Conflict', edited by Lori Fisler Damrosch and Bernard H. Oxman includes contributions by William H. Taft IV and Todd F. Buchwald (above note 42), John Yoo, Ruth Wedgwood, Richard N. Gardner, Richard A. Falk, Miriam Sapiro, Thomas M. Franck (above note 35), Tom J. Farer, and Jane E. Stromseth, see 97 *American Journal of International Law* (2003) 553–643. See also Weller, 'The Iraq War—2003', in T. Ruys and O. Corten (eds), *The Use of Force in International Law. A Case-Based Approach* (2018) 639; see also Michael Stohl's chapter in this volume.

[53] A 2006 incident is referred to in C.D. Gray, *International Law and the Use of Force* (3rd ed., 2008), at 89f.

[54] Tams and Brückner, 'The Israeli Intervention in Lebanon—2006', in Ruys and Corten, above note 52, 673.

[55] Joint Special Operations Task Force—Juniper Shield (JSOTF-JS), available at <https://www.globalsecurity.org/military/agency/dod/jsotf-ts.htm> (last visited 5 April 2020).

[56] Gray, 'The Conflict in Georgia—2008' in Ruys and Corten, above note 52, 712.

[57] Marauhn, 'Intervention in Support of Democratization Processes? Questions Arising from the 2011 Military Intervention in Libya', in T. Marauhn and H. Elliesie (eds), *Legal Transformation in Northern Africa and South Sudan* (2015) 37. See also Deeks, 'The NATO Intervention in Libya—2011', in Ruys and Corten, above note 52, 749.

Germany used military force to evacuate its nationals from Libya[58] but abstained when the UN SC voted on Resolution 1973. Since 2011, the Syrian civil war has seen multiple foreign actors intervening.[59] International military action against the Islamic State of Iraq and the Levant has been omnipresent since 2014.[60] In 2018 the United States, France, and the United Kingdom attacked Syria, arguably responding to the use of chemical weapons in the Syrian civil war.[61] The Kurdish-Turkish conflict, escalating in 2015, gave rise to the *Turkish Operation Peace Spring* in 2019.[62] Internal and external violence has occurred in South Sudan since its independence from Sudan in 2011.[63] Mali has been affected by a conflict with heavy international involvement since 2012.[64] 2014 saw the annexation of Crimea by the Russian Federation,[65] and the armed conflict in Donbas has not yet come to an end.[66] The ongoing Yemeni civil war, involving a Saudi-led coalition, began in 2015[67] and was preceded by a series of earlier conflicts starting in 2001 and 2004 respectively.[68] The 2019 recognition of the Israeli annexation of the Golan Heights by the United States should also be mentioned here.[69] In June 2019, Iran downed a US surveillance drone over the Strait of Hormuz,[70] and in early 2020, a US drone strike near Baghdad killed the Iranian major general *Qasem Soleimani*,[71] followed by Iranian counterattacks on US bases in Iraq.

While this is far less than a comprehensive list of (international) force applied across the globe, the above events raise questions as to the survival of Article 2(4) UN

[58] This incident is referred to in some detail in Tuzmukhamedov, 'Russian Forces in the Former USSR and Beyond' in D. Fleck (ed.), *The Handbook of the Law of Visiting Forces* (2nd ed., 2018) 686, at 699.

[59] By way of example see Lagerwall, 'Threats of and Actual Military Strikes Against Syria—2013 and 2017', in Ruys and Corten, above note 52, 828 and Aust and Payandeh, 'German Practice with Regard to the Use of Force in Syria', 61 *German Yearbook of International Law* (2018) 451.

[60] Corten, 'The Military Operations Against the 'Islamic State' (ISIL or Da'esh)—2014' in Ruys and Corten, above note 52, 873.

[61] De Guttry, 'The Western-led Military Operations in Syria in Response to the Use of Chemical Weapons', 56 *Archiv des Völkerrechts* (2018) 472.

[62] Gall and Kingsley, 'Turkish Forces Escalate Campaign in Syria Against Kurdish-Led Militia', available at <https://www.nytimes.com/2019/10/10/world/middleeast/syria-turkey-offensive.html> (last visited 5 April 2020).

[63] For an interesting account see Rossi, 'The International Community, South Sudan, and the Responsibility to Protect', 49 *New York University Journal of International Law & Politics* (2016) 129.

[64] Bannelier and Christakis, 'The Intervention of France and African Countries in Mali—2013', in Ruys and Corten, above note 52, 812; Charbonneau and Sears, 'Fighting for Liberal Peace in Mali? The Limits of International Military Intervention', 8 *Journal of Intervention and Statebuilding* (2014) 192.

[65] Grant, 'Three Years After Annexation: Of "Frozen Conflicts" and How to Characterise Crimea', 59 *German Yearbook of International Law* (2016) 49.

[66] Bodie, 'Modern Imperialism in Crimea and the Donbas', 40 *Loyola of Los Angeles International and Comparative Law Review* (2017) 267.

[67] Ruys, 'The Saudi-led Military Intervention in Yemen's Civil War—2015', in Ruys and Corten above note 52, 899.

[68] See, among others, Hughes, 'A Proxy War in Arabia. The Dhofar Insurgency and Cross-Border Raids into South Yemen', 69 *The Middle East Journal* (2015) 91.

[69] Galbraith (ed.), 'United States Recognizes Israeli Sovereignty Over the Golan Heights. Contemporary Practice of the United States Relating to International Law: State Diplomatic and Consular Relations', 113 *American Journal of International Law* (2019) 613.

[70] Marc Schack, 'Did the US Stay "Well Below the Threshold of War" With its June Cyberattack on Iran?' (EJIL Talk, 2 September 2019), available at <https://www.ejiltalk.org/did-the-us-stay-well-below-the-threshold-of-war-with-its-june-cyberattack-on-iran/> (last visited 6 April 2020).

[71] Labuda, 'The Killing of Soleimani, the Use of Force against Iraq and Overlooked Ius Ad Bellum Questions' (EJIL Talk, 13 January 2020), available at <https://www.ejiltalk.org/the-killing-of-soleimani-the-use-of-force-against-iraq-and-overlooked-ius-ad-bellum-questions/> (last visited 6 April 2020).

Charter. In the following, we will discuss three pertinent questions arising from some of the above incidents. They relate to the interpretation of the prohibition of the use of force (2.1), the extent to which reference to the right of self-defence weakens the prohibition (2.2), and to trends in respect of enforcement action under Chapter VII of the Charter (2.3).

2.1 Weakening Article 2(4) UN Charter by Extending its Scope?

The prohibition of the use of force is both a treaty-based rule laid down in Article 2(4) UN Charter and a rule of customary international law. This has been confirmed by the International Court of Justice on several occasions,[72] with international practice largely considering the two as similar or even identical in content.[73] The prohibition enjoys the status of a peremptory norm of international law (*ius cogens*),[74] and Article 2(4) UN Charter has consequently been praised as a cornerstone for the preservation of international peace and security.[75]

The argument presented here is that the status of Article 2(4) UN Charter depends on its specific scope, and that any broadening of its scope runs the risk of weakening the rule as such. To this end, Article 2(4) UN Charter must be distinguished from the prohibition of intervention and from other rules protecting a state's sovereignty. Intervention, to begin with, is 'the interference by a State in the internal or foreign affairs of another State',[76] and it is prohibited only 'when it uses methods of coercion in regard to such choices, which must remain free ones'.[77] The prohibition of the use of force is, one might argue, *lex specialis* in relation to the prohibition of intervention, as it refers to force as a particular means of coercion:

> The element of coercion, which defines, and indeed forms the very essence of, prohibited intervention, is particularly obvious in the case of an intervention *which uses force*, either in the direct form of military action, or in the indirect form of support for subversive or terrorist armed activities within another State.[78]

[72] *Military and Paramilitary Activities in and against Nicaragua (Nicaragua v. United States of America)* Jurisdiction and Admissibility, Judgment (26 November 1984) ICJ Reports 1984, 392, para. 73; *Military and Paramilitary Activities in and against Nicaragua (Nicaragua v. United States of America)* Merits, Judgment (27 June 1986) ICJ Reports 1986, 14, paras 187–190; *Legal Consequences of the Construction of a Wall in the Occupied Palestinian Territory* Advisory Opinion (9 July 2004) ICJ Reports 2004, 136, para. 87.

[73] Dörr, 'Use of Force, Prohibition of', in R. Wolfrum (ed.), *Max Planck Encyclopedia of Public International Law* (Online Edition <www.mpepil.com>, last update 2015), para. 10.

[74] Ibid., para. 32.

[75] Ibid., para. 1. As the historical contributions to this book demonstrate, the prohibition of the use of force represents the (provisional) climax of a gradual containment of the legitimate reasons for using force in international relations. Article 2(4) UN Charter was a revolutionary step, but one that had deep historical roots (see the chapters by Tischer and Eberl for early modernity and by Simon and Hull for the long nineteenth century).

[76] Kunig, 'Intervention, Prohibition of', in Wolfrum, above note 73 (Online Edition <www.mpepil.com>, last update 2008), para. 1.

[77] *Military and Paramilitary Activities in and against Nicaragua (Nicaragua v. United States of America)* Merits, Judgment (27 June 1986) ICJ Reports 1986, 14, para. 205.

[78] Ibid., italics by the author.

As Oliver Dörr points out, the

> essential feature that characterizes the prohibition of the use of force is the application of military force *as a means of coercion*, it is not the intrusion into the sovereign realm of another State, nor is it even the element of coercion as such, but only an intrusion or coercion *using the special features of military weaponry and displaying hostile intent*.[79]

Sovereignty is not just protected against forms of coercion; the protection of territorial integrity goes far beyond. In other words, a violation of the principle of sovereignty, in particular of territorial integrity, as it may be taken from 'the principle of sovereign equality of its Member States as guaranteed in Art. 2 (1) UN Charter',[80] which protects the 'supreme authority within a territory',[81] does not necessarily involve coercion or force. Hence, the prohibition of the use of force is a distinct rule, and it may be understood as coming close to a *lex specialis* in respect of the principle of sovereignty. This interplay between three different rules is important. It ensures a distinctiveness of the prohibition of force, and hence equips this rule with a particular sharpness.

It has long been accepted that 'force' refers to armed or military force.[82] The use of other means of coercion may be subject to the prohibition of intervention but not to the prohibition of the threat or use of force. Whether or not specific or new vulnerabilities could modify the interpretation of 'force' as military force is questionable. In the 1970s, with the oil crisis, there was talk about the so-called 'Arab oil weapon'[83] suggesting a broader interpretation of Article 2(4) UN Charter; a view that has never been accepted. Today, this debate has re-emerged, however, in the context of cyber operations. In this respect the question has been raised whether force refers to the means or to the effects of such operations.[84] Some have strongly argued that cyber operations can be considered as a use of force.[85] The International Law Association's Committee on the Use of Force has discussed this in its 2018 report:[86] '[C]yber operations are capable in principle of crossing the threshold into use of force', however, only if they 'take place as part and parcel of a wider kinetic attack'.[87] The 'emerging view ... that cyber operations may constitute a use of force ... if their scale and effects mirror those of

[79] Dörr, above note 73, para. 18, italics by the author.

[80] Besson, 'Sovereignty', in Wolfrum, above note 73 (Online Edition <www.mpepil.com>, last update 2011), para. 2.

[81] Ibid., para. 1.

[82] Dörr, above note 73, para. 11.

[83] On this debate see J.J. Paust (ed.), *The Arab Oil Weapon* (1977).

[84] For a discussion see Schmitt, 'Computer Network Attack and the Use of Force in International Law: Thoughts on a Normative Framework', 37 *Columbia Journal of Transnational Law* (1999) 885 and—with some different views—O'Connell, 'Cyber Security Without Cyber War', 17 *Journal of Conflict and Security Law* (2012) 187.

[85] Along these lines Schmitt, 'Below the Threshold Cyber Operations: The Countermeasure Response Option and International Law', 54 *Virginia Journal of International Law* (2014) 698, at 719f. See also International Group of Experts, *Tallinn Manual on the International Law Applicable to Cyber Warfare* (2013), at 330–48.

[86] International Law Association, Use of Force Committee (Michael Wood, Chair, and Noam Lubell Rapporteur), available at <https://www.ila-hq.org/images/ILA/DraftReports/DraftReport_UseOfForce.pdf> (last visited 6 April 2020).

[87] Ibid., at 25.

a traditional kinetic use of force'[88] is not as such convincing. It puts at risk the finely tuned differences not only in terms of rules on sovereignty, non-intervention, and non-use of force. More generally, it supports a trend to militarize international relations in matters which are non-military, *inter alia* policing or law enforcement.[89] New vulnerabilities do not justify such a broad interpretation of Article 2(4) UN Charter.

Likewise, a degree of carefulness should be applied when discussing indirect force by one state against another. There is established case law by the International Court of Justice[90] excluding mere financial support from the prohibition.[91] Whether this will increasingly come under pressure in light of counter-terrorism is difficult to assess. The international community has tightened rules on financial assistance to terrorist groups, both as a matter of treaty law and on the basis of UN SC Resolution. Whether or not such financial support could, however, be considered as indirect force within the meaning of Article 2(4) UN Charter is more than doubtful, even though the trend to consider the harbouring of terrorists as an armed attack does not make it unlikely that some states will consider other forms of support to terrorism as a violation of Article 2(4) UN Charter. In light of the distinctiveness of rules as highlighted above this might weaken the prohibition by blurring the lines of an otherwise clear category.

The sharpness of the rule becomes evident when looking at the debate about minimal uses of force. Some have suggested that this may escape the prohibition of Article 2(4) UN Charter, in other words, that there is a *de minimis* threshold.[92] Examples given include the evacuation of nationals abroad, such as the so-called *Operation Pegasus* by Germany in Libya in 2011,[93] and law enforcement operations in support of, for example, a state's fisheries jurisdiction, with, however, some contradictory case law.[94] Propositions towards a *de minimis* threshold run counter to the established broad interpretation of the prohibition of the use of force, 'to basically capture each and every form of armed force by individual States'.[95] Indeed, a broad reading of the prohibition, as has been pointed out, 'is apt to accommodate the universally acknowledged fundamental character of the prohibition'.[96] This does not automatically mean that the evacuation of nationals abroad or difficult law enforcement operations necessarily run counter to Article 2(4) UN Charter.

[88] Ibid.

[89] On non-military approaches to cybersecurity see Ney and Zimmermann, 'Cyber-Security Beyond the Military Perspective', 58 *German Yearbook of International Law* (2015) 51; Walter, 'Obligations of States Before, During and After a Cyber Security Incident', 58 *German Yearbook of International Law* (2015) 67; Dörr, 'Obligations of the State of Origin of a Cyber Security Incident', 58 *German Yearbook of International Law* (2015) 87; Brunnée and Mashel, 'Teaching an Old Law New Tricks: International Environmental Law Lessons for Cyberspace Governance', 58 *German Yearbook of International Law* (2015) 129.

[90] *Armed Activities on the Territory of the Congo (Democratic Republic of the Congo v. Uganda)* Merits, Judgment (19 December 2005) ICJ Reports 2005, 168, paras 163–65.

[91] *Military and Paramilitary Activities in and against Nicaragua (Nicaragua v. United States of America)* Merits, Judgment (27 June 1986) ICJ Reports 1986, 14, para. 228.

[92] Reference is made, in this context, to two paragraphs of the ICJ judgment of 27 June 1986, ibid., paras 191 and 195.

[93] Above note 58.

[94] The ILA Committee report, above note 86, at 5 in this context refers to the *The M/V 'SAIGA' (No. 2) (Saint Vincent and the Grenadines v. Guinea)* Judgment (1 July 1999) ITLOS Report 1999, 10, para. 155 and to the *Fisheries Jurisdiction (Spain v. Canada)* Jurisdiction of the Court, Judgment (4 December 1998) ICJ Reports 1998, 432, para. 84.

[95] Dörr, above note 73, para. 13.

[96] Ibid.

As suggested, among others, by the International Law Association's (ILA) Committee on the use of force in its 2018 report, it may be more plausible to refer to the nature of force[97] rather than to the level of force. This would preserve the categorical prohibition while distinguishing the prohibition of the use of force from other rules such as the prohibition of intervention and the prohibition of the violation of state sovereignty.

2.2 Weakening Article 2(4) UN Charter by Extending the Scope of Self-Defence?

As can be taken from *Thomas Franck*'s earlier discussions about the death of Article 2(4) UN Charter, there has always been a risk that a broad reading of Article 51 UN Charter effectively weakens Article 2(4): '... the Charter itself provided enough exceptions and ambiguities to open the rules to deadly erosion'.[98] No exception weakens a rule simply because of its existence. However, there is a risk that exceptions 'overwhelm ... the rule and transform ... the system'.[99] As *Franck*'s 1970 essay highlights, '[c]hanging circumstances of international relations, of the way nations perceive their self-interest, of strategy and tactics' may contribute to 'enlarging the exceptions to the point of virtually re-pealing the rule itself'.[100] In order to avoid this, 'there must be some machinery for determining whether (an) exception applies in particular instances'.[101]

In respect of Article 51 UN Charter, there are currently two issues which give rise to concern as to its interpretation. One of these was already present in both of *Franck*'s above-mentioned essays, namely the notion of preventive self-defence. The other relates to the question whether (transnational) armed attacks by non-state actors may justify acts of self-defence. As already mentioned above, an extensive reading of Article 2(4) UN Charter, considering certain cyber activities as 'force' might also lead to a broader reading of the notion of 'armed attack', hence leading to more instead of less unilateral force.

The discussion about preventive or anticipatory self-defence has been ongoing for a long time, as some of the historical chapters in this book underline. Reference, until today, is normally made to the so-called Caroline formula, emerging from an incident when a US steamboat was seized by a British force from Canada in 1837, set on fire, and sent over the Niagara Falls, claiming that it had assisted rebels against the British colonial government with men and supplies.[102] The formula emerged from correspondence between the British ambassador to the United States and US Secretary of State *Webster*, and it laid down criteria for the lawful exercise of preventive self-defence, claiming the need for the self-defence claimant to demonstrate that the

[97] International Law Association, above note 86, at 5.

[98] Franck, above note 1, at 809.

[99] Ibid., at 811.

[100] Ibid., at 822.

[101] Ibid., at 816.

[102] Greenwood, 'Caroline, The', in Wolfrum, above note 73 (Online Edition <www.mpepil.com>, last update 2009); Wood, 'The *Caroline* Incident—1837', in Ruys and Corten above note 52, 5; see also Hull's chapter in this volume.

> 'necessity of self-defense was instant, overwhelming, leaving no choice of means, and no moment of deliberation . . ., and that the British force, even supposing the necessity of the moment authorized them to enter the territories of the United States at all, did nothing unreasonable or excessive; since the act, justified by the necessity of self-defense, must be limited by that necessity, and kept clearly within it'.[103]

It seems that this formula has been accepted as being compatible with the UN Charter, either by referring to the 'inherent' character of the right or by integrating this into the term 'armed attack'.[104] The debate resurfaced in 2003 with the new national security strategy adopted by the United States in 2002,[105] seemingly going beyond the formula by permitting self-defence even against '"rogue States", terrorists, and weapons of mass destruction'.[106] While the United States later clarified their policy 'reaffirming the requirement of an imminent attack',[107] the debate is still ongoing. The ILA Committee on the use of force in its 2018 reports states: 'Although the matter remains unsettled, there may be reason to accept that when faced with a specific imminent armed attack based on objectively verifiable indicators, States may engage in measures to defend themselves in order to prevent the attack.'[108] The 2003–2011 Iraq war illustrates the risks arising from too broad an understanding of preventive self-defence. An attempt to reduce possible abuses while at the same time accepting the right to preventive self-defence was presented by Daniel Bethlehem in a 2012 contribution to the *The American Journal of International Law*.[109] He listed a series of principles that should be taken into account when assessing the validity of a claim to preventive self-defence. While these principles enjoy a degree of plausibility the question remains who decides, and whether or not this opens the door to increased use of force in light of new adversaries and new weapons, including terrorist and cyber activities.

Even more pressing than the debate about preventive or anticipatory self-defence is the question of self-defence against non-state actors. Several of the above-listed incidents are related to this, most evidently the international military action against the Islamic State of Iraq and the Levant since 2014.[110] It has rightly been pointed out that

103 Webster, 'Letter to Henry Stephen Fox', in K.E. Shewmaker (ed.), *The Papers of Daniel Webster: Diplomatic Papers*, vol. 1, at 1841–43 (1983), at 62.

104 Wilmshurst, 'Anticipatory Self-Defence Against Terrorists', in L.J. van den Herik and N. Schrijver (eds), *Counter-Terrorism Strategies in a Fragmented International Legal Order. Meeting the Challenges* (2013) 356, at 359.

105 The White House, *The National Security Strategy of the United States of America* (2002), available at <https://georgewbush-whitehouse.archives.gov/nsc/nss/2002/> (last visited 6 April 2020).

106 Quote taken from International Law Association, above note 86, at 13.

107 Ibid., at 13, referring to The White House, *Report on the Legal and Policy Frameworks Guiding the United States' Use of Military Force and Related National Security Operations* (2016), available at <https://www.justsecurity.org/wp-content/uploads/2016/12/framework.Report_Final.pdf> (last visited 6 April 2020).

108 International Law Association, above note 86, at 14.

109 Bethlehem, 'Self-Defense Against an Imminent or Actual Armed Attack by Nonstate Actors', 106 *American Journal of International Law* (2012) 770. See also the contributions by Stohl and Geis/Wagner to this volume and Brock, 'The Use of Force in the Post-Cold War Era: From Collective Action Back to Pre-Charter Self-Defense?', in M. Bothe, M.E. O'Connell, and N. Ronzitti (eds), *Redefining Sovereignty. The Use of Force After the Cold War* (2005) 21.

110 Waldron Davis, 'The Phantom of the Neo-Global Era. International Law and the Implications of Non-State Terrorism on the Nexus of Self-Defense and the Use of Force', in R.A. Miller et al. (eds), *Progress*

this is not a new phenomenon but that the legal debate has developed largely since the 2001 attacks against the United States.[111] Indeed, Article 51 does not specify that the armed attack must have been carried out by a state—even though the right to self-defence has long been read along these lines.[112] The 2018 report of the ILA Committee argues that '[s]elf-defence is a right triggered by an act, rather than an actor', and that 'self-defence is a right that exists in order that States are able to protect themselves when attacked from outside their borders'.[113] This perception is in contrast to earlier practice aiming at establishing a link between non-state actors and a state in order to eventually hold the state responsible. Some case law of the ICJ points this direction,[114] even though there seems to be some openness in the Court to further develop its perspective.[115] There are two problems in this context: first, there may be circumstances in which an armed attack performed by a non-state actor cannot be attributed to a state, and second, when directing acts of self-defence against a non-state actor this will most likely take place in the territory of a third state, which will then be subject to the use of force by the self-defence claimant. The two problems are at opposing ends of the debate but they illustrate that the right to self-defence might not be the most appropriate means to address violence exercised by non-state actors. In contrast to the ILA Committee's report it is not convincing to permit such self-defence and then raise the question of how to keep the act of self-defence within the boundaries of proportionality. What is at issue and remains at issue is whether reference to Article 51 in such circumstances eventually repeals Article 2(4). To add, the unable and unwilling doctrine[116] falls in between the earlier approach to establish the responsibility of the state and the more recent one to consider self-defence against non-state actors within the framework of Article 51. It further contributes to the weakening of the Charter and eventually to an ever more decentralized justification of the use of force. To take up ideas discussed by *Thomas Franck*,[117] it further contributes to depriving the Charter and the United Nations of their central role for the maintenance of international peace

in International Law (2008) 633; Barbour and Salzman, '"The Tangled Web"—The Right of Self-Defense Against Non-State Actors in the Armed Activities Case', 40 *New York University Journal of International Law & Politics* (special issue) (2008) 53; more generally: Kreß, 'Major Post-Westphalian Shifts and Some Important Neo-Westphalian Hesitations in the State Practice on the International Law on the Use of Force', 1 *Journal on the Use of Force and International Law* (2014) 11; Starski, 'Right to Self-Defense, Attribution and the Non-State Actor. Birth of the "Unable or Unwilling" Standard?', 75 *Zeitschrift für ausländisches öffentliches Recht und Völkerrecht* (2015) 455; Alvarez-Jiménez, 'From Al-Qaida in 2001 to ISIL in 2015. The Security Council's Decisions on Terrorism and Their Impact on the Right to Self-Defense Against Autonomous Non-State Actors', 26 *Minnesota Journal of International Law* (2017) 345.

[111] International Law Association, above note 86, at 14; see also the contribution by Stohl on the 'War on Terror' in this book.

[112] See, among others, Bothe, 'Terrorism and the Legality of Pre-Emptive Force', 14 *European Journal of International Law* (2003) 227.

[113] Ibid.

[114] *Legal Consequences of the Construction of a Wall in the Occupied Palestinian Territory* Advisory Opinion (9 July 2004) ICJ Reports 2004, 136, para. 139.

[115] *Armed Activities on the Territory of the Congo (Democratic Republic of the Congo v. Uganda)* Merits, Judgment (19 December 2005) ICJ Reports 2005, 168, para 2., at 158–65.

[116] Corten, 'The "Unwilling or Unable" Test—Has it Been, and Could it be, Accepted?', 29 *Leiden Journal of International Law* (2016) 777.

[117] Franck, above note 35, at 614–18.

and security, which should be read as a cornerstone of at least a minimal constitutional order at the international level.

2.3 Weakening Article 2(4) by Over-Interpretation of UN SC Authorizations?

In contrast to the decentralized character of Article 51 UN Charter, the wording of Articles 39–50 UN Charter constitutes a centralized exception to the prohibition of the use of force.[118] Chapter VII provides a legal basis for the UN Security Council to take, among others, military action (Article 42 UN Charter) if a threat to the peace, a breach of the peace, or an act of aggression exists (Article 39 UN Charter). Given Article 25 UN Charter, Member States are bound by those decisions, and they are, in the case of Article 48 UN Charter, obliged to carry them out. In contrast to the original ideas of the drafters of the Charter, the Council itself is not taking the actions envisaged in Article 42 but it authorizes Member States to perform such action. This practice means that not enforcement as such is centralized but only decision-making about enforcement.

A number of issues have been discussed over time in respect to such enforcement, in particular since 1989 when the Council indeed became more active than it had been before.[119] The end of the Cold War did not only stimulate new uses of Chapter VII but it also gave rise to a number of difficult questions related to the interpretation and handling of Chapter VII, among them the question of whether there are limits to an authorization, and how an authorization can be withdrawn.[120]

Thomas Franck, in his 2003 essay, deals with one of the arguments presented by the United States and the United Kingdom in order to justify the attack on Iraq. Referring to UN SC Resolutions 678, 687, and 1441 the argument[121] puts forward that UN SC Resolution 678 had authorized the use of force against Iraq, and that Resolution 687 imposed a series of ceasefire conditions upon Iraq. As these had been materially breached, the argument continues, Resolution 1441 'gave Iraq a "final opportunity" to comply, but stated ... that violations of the obligations ... under 1441 would constitute a further material breach',[122] and that the authorization by Resolution 678 would be 'revived by SCR 1441 (2002)'[123] if such breach occurred. *Thomas Franck* regards this

[118] R.O. Keohane, *Power and Governance in a Partially Globalized World* (2002) stresses the role of centralized enforcement in his section on the role of law (ibid., at 115–90), in particular in co-authored chapters 7 and 8 with a focus ibid., 133, at 146f. (with K.W. Abbott, A. Moravcsik, A.-M. Slaughter, and D. Snidal).

[119] See also Franck, 'The Security Council Unleashed? Post-Cold War U.N. Peacekeeping and Prospects of U.S. Participation', in M.K. Young (ed.), *Trilateral Perspectives on International Legal Issues. Relevance of Domestic Law and Policy* (1996) 475; Malone, 'The UN Security Council in the Post-Cold War World: 1987–97', 28 *Security Dialogue* (1997)393; Fassbender, 'Uncertain Steps into a Post-Cold War World', 13 *European Journal of International Law* (2002) 273.

[120] See also Šturma, 'Does the Rule of Law also Apply to the Security Council?', 32 *The Polish Yearbook of International Law* (2012) 299; Gioia, 'Decisions of the UN Security Council of Indefinite Duration', in P. Hilpold (ed.), *Das Kosovo-Gutachten des IGH vom 22. Juli 2010* (2012) 171.

[121] Franck describes the argument in his critique (above note 35), at 611–14 with further references (quoting William H. Taft IV).

[122] Ibid., at 611–12.

[123] Ibid., at 612.

reading as 'ultimately unsustainable',[124] and argues that the text of UN SC Resolution 1441 would have required a new authorization by the Security Council if force was to be used against Iraq. He states that the justification does 'not fare well under close examination' and that the attack on Iraq was an 'unauthorized recourse to force against Iraq'.[125] He rightly points out that the interpretation weakens the central peacekeeping role of the UN and eventually disposes of its decision-making powers.[126] Indeed, these developments set in motion a weakening of central decision-making, thus affecting the main difference between Chapter VII and Article 51 UN Charter.

It seems that in light of these developments there is a renewed reluctance to use Chapter VII as originally intended. The debate about UN SC Resolution 1973 further pushed this trend. As may be recalled, in response to the civil war in Libya, in March 2011, by a vote of ten in favour to none against, with five abstentions, the UN Security Council adopted Resolution 1973 (2011), establishing a no-fly zone in Libya's airspace and authorizing UN Member States to take all necessary measures to protect civilians under threat of attack in the country. What followed were military operations against Libyan government forces, initially performed by France and the United Kingdom and later headed by NATO.[127] While UN SC Resolution 1973 does not explicitly aim at regime change, it refers, if vaguely, to the *Responsibility to Protect* (*R2P*),[128] and the operations authorized did contribute *de facto* to regime change and constitutional transformation in Libya.[129] *Prima facie*, the military operations against the Libyan government forces more or less seem to follow UN SC practice. However, such an intervention in a non-international armed conflict, and the implicit reference to R2P, which eventually materialized in Libya, have led commentators not only to discuss the legality of the intervention but also to address the extent to which R2P has had and will have an impact on the mid- and long-term approach of the UN SC to the authorization of military force.

Extensive use of the authorization has generated mistrust against Western states as UN SC Resolution 1973 provided only a narrow mandate to prevent gross human rights abuses. It may be argued that the one-sided interpretation of Resolution 1973 by Western states was one of the reasons for other non-Western UN SC members to be hesitant to accept more of these. Indeed, in respect of Syria, for many reasons, including geopolitical ones, the UN SC could not agree on enforcement action to reduce the suffering of the Syrian population if not to end the war.[130]

124 Ibid., at 611.

125 Ibid., at 614.

126 Ibid., at 614–17.

127 Majinge, 'The Use of Force in International Law and the UN Security Council Resolutions 1970 & 1973 on Libya', 24 *Hague Yearbook of International Law* (2011) 153.

128 For a critical analysis see Teimouri and Subedi, 'Responsibility to Protect and the International Military Intervention in Libya in International Law: What Went Wrong and What Lessons Could Be Learnt from It?', 23 *Journal of Conflict and Security Law* (2018) 3.

129 See above Marauhn, above note 57 *passim*. Chris Brown argues the intervention *had* to result in regime change in order to fulfil the purpose of UN SC Resolution 1973; see Brown, 'The Antipolitical Theory of the Responsibility to Protect', 5 *Global Responsibility to Protect* (2013) 423, at 437; on this issue, see also Beate Jahn's and Chris Brown's chapters in this volume.

130 Kelly, 'The Downfall of the Responsibility to Protect: How the Libyan and Syrian Crises Secured the Fate of the Once-Emerging Norm', 43 *Syracuse Journal of International Law and Commerce* (2016) 381; Dembinski, 'Procedural Justice and Global Order. Explaining African Reactions to the Application of Global Protective Norms', 23 *European Journal of International Relations* (2017) 809; Deitelhoff, 'Is the R2P

2.4 Excursus: Weakening Article 2(4) by Reference to Common Goods or Humanitarian Concerns?

A noteworthy event may be taken up in this particular context even though it is not directly related to UN SC enforcement powers. On 14 April 2018, the United States, France, and the United Kingdom carried out a series of military strikes against Syria.[131] The three governments argued that these strikes were in response to chemical attacks against civilians in Douma, arguably performed by the Syrian government. These attacks can be neither justified by self-defence nor on the basis of Chapter VII. They have been considered as 'manifestly illegal'.[132] Even though neither the United States nor France sought to legally justify these strikes, they argued, among others, that the purpose was 'to establish a strong deterrent against the production, spread, and use of chemical weapons'.[133] What must be clearly stated is that there is no way to enforce a rule, namely the prohibition of chemical weapons, by individual use of force. International law, as national law, clearly distinguishes between rules and the ways and means available to enforce them.

It is interesting though that the UK government put forward a distinct justification for their participation in the strikes, arguing that '[t]he UK is permitted under international law, on an exceptional basis, to take measures in order to alleviate overwhelming humanitarian suffering. The legal basis for the use of force is humanitarian intervention.'[134] Two aspects are noteworthy in this regard: first, the United Kingdom obviously sees a need to explicitly justify such a use of force, thereby at least aiming at upholding Article 2(4); second, nevertheless such unilateral actions dramatically weaken the prohibition of the use of force, and humanitarianism (whether referring to a so-called humanitarian intervention or to the Responsibility to Protect) cannot and should never be pursued at the expense of peace.[135] Although many states most often reject such claims in most instances the simple fact that some states use this as a potential legal basis in itself is a weakening of the prohibition to use force. This is especially

Norm Failing? The Controversy about Norm Justification and Norm Application of the Responsibility to Protect', 11 *Global Responsibility to Protect* (2019) 149; on the R2P, see also the chapters by Geis/Wagner, Chi, Robinson/Antonov, and Jahn, in this volume.

[131] For a debate see O'Connell, 'Unlawful Reprisals to the Rescue against Chemical Attacks?' (EJIL Talk, 12 April 2018), available at <https://www.ejiltalk.org/unlawful-reprisals-to-the-rescue-against-chemical-attacks/> (last visited 7 April 2020); Hakimi, 'The Attack on Syria and the Contemporary Jus ad Bellum' (EJIL Talk, 15 April 2018), available at <https://www.ejiltalk.org/the-attack-on-syria-and-the-contemporary-jus-ad-bellum/> (last visited 7 April 2020); Milanovic, 'The Syria Strikes: Still Clearly Illegal' (EJIL Talk, 15 April 2018), available at <https://www.ejiltalk.org/the-syria-strikes-still-clearly-illegal/> (last visited 7 April 2020).

[132] Milanovic, ibid.

[133] Q. Jurecic (ed.), 'Transcript: President Trump's Remarks on Syria Airstrikes', available at <https://www.lawfareblog.com/transcript-president-trumps-remarks-syria-airstrikes> (last visited 7 April 2020).

[134] Prime Minister's Office, Policy Paper, 'Syria Action—UK Government Legal Position' (published 14 April 2018), available at <https://www.gov.uk/government/publications/syria-action-uk-government-legal-position/syria-action-uk-government-legal-position> (last visited 7 April 2020).

[135] On criteria see, among others, O'Connell, 'Responsibility to Peace. A Critique of R2P', 4 *Journal of Intervention and Statebuilding* (2010) 39.

due to the potential abuse of this argument in many instances. Russian involvement in Georgia and Ukraine have been mentioned in this context.[136]

3. The Importance of a Counter-Factual Normative Claim

Having illustrated that there are trends to weaken the prohibition of the use of force by extending the scope of the prohibition, by extending the scope of self-defence, and by over-interpreting UN SC resolutions, the remainder of this chapter will not pursue doctrinal arguments but rather focus on policy of law. The reason for this approach is that doctrine as such does not solve the problems identified. This has already become clear in *Thomas Franck*'s contributions. Doctrine is important but it should not lead us to ignore policy trends. Having identified trends along the lines of *Franck*'s fear that the prohibition to use force might not survive, it is all the more important to reconsider the role of international law in this particular context (section 3), to reflect on the need for a strong United Nations (section 4), and to recall the responsibility of lawyers in this discourse (section 5).

Let us briefly recall one of the most famous aphorisms in Anglo-American law, namely that 'hard cases make bad law'.[137] Is this true for all the above discussions and debates about how to interpret and how to apply the prohibition to use force? We do not think that it is. In line with *Thomas Franck*, it is also important to consider that rules should not simply be sacrificed on the altar of *realpolitik*.[138] As important as it is to take note of factual developments, and hence to be open to the environment within which international law operates, it is as important to uphold the counter-factual claim of central rules, of cornerstones, such as the prohibition of the use of force.

In order to contextualize this in respect of Article 2(4) UN Charter, it is important to recall the nature of international law. As has often been pointed out, one of the peculiar characteristics of public international law[139] is that states are not only the primary subject thereof, but they are at the same time its authors, addressees, and guardians.[140] Very much in contrast to today's national legal systems, the central feature of international law is the absence of compulsory judicial dispute settlement. While there is a growing body of rules aiming to compensate for this, such as a

[136] The Russian government has claimed humanitarian grounds and the protection of nationals abroad; for an analysis see Chatham, 'Defense of Nationals Abroad. The Legitimacy of Russia's Invasion of Georgia', 23 *Florida Journal of International Law* (2011) 76; various contributions in C. Marxsen, A. Peters, and M. Hartwig (eds), 'Symposium: "The Incorporation of Crimea by the Russian Federation in the Light of International Law"', 75 *Zeitschrift für ausländisches öffentliches Recht und Völkerrecht* (2015), at 1–231; Bothe, 'The Current Status of Crimea: Russian Territory, Occupied Territory or What?', 53 *Revue de droit militaire et de droit de la guerre* (2014) 99. See also the contribution by Robinson/Antonov to this volume.

[137] This aphorism goes back to the first half of the nineteenth century in the United States, and it has been used by Oliver Wendell Holmes Jr in a 1904 judgment in *Northern Securities Co. v. United States* 193 U.S. 197 (1904).

[138] Franck, above note 35, at 607 highlights the concerns of his earlier essay: 'The 1970 essay regretted the loss of that vision in a miasma of so-called realpolitik.'

[139] For a brilliant analysis with helpful insights see Kritsiotis, 'Public International Law and the Concept of Law Applied', 3 *Transnational Legal Theory* (2012) 189.

[140] Marauhn, 'Changing Role of the State', in D. Bodansky, J. Brunnée, and E. Hey (eds), *The Oxford Handbook of Environmental Law* (2007) 727.

partially compulsory dispute settlement mechanism included in the United Nations Convention on the Law of the Sea,[141] quasi-judicial dispute settlement within the World Trade Organization,[142] and various types of compliance control systems across different branches of international law,[143] ensuring compliance with the prohibition of the use of force still is quite basic. In effect, this means that the role of states as guardians remains central for this rule, with a lot depending on political expediency. In other words, state practice is of central importance for the prohibition of the use of force. The central role of such practice is obvious both for treaty law and for customary international law. In respect of the latter, state practice is one of the constitutive elements of law, and in respect of treaty law, as discussions within the International Law Commission have recently demonstrated,[144] subsequent state practice has remained and will remain central to the interpretations of treaties and hence to their content and scope. This does not mean that the law follows practice. But it means that just as law affects practice, practice also affects the law.

Before discussing this more specifically with regard to the prohibition of the use of force, another element of international law, illustrating the close relationship between its normative claim and its role in practice, emerges from the fact that often international law academics are (to some extent at least) also international legal practitioners. This has been brilliantly illustrated by Martti Koskenniemi in his critical historical analysis of the role of international law as a 'gentle civilizer'.[145] It would go beyond the scope of this chapter to review the overall effects of such dual role often performed by international law scholars. Suffice it to identify this as a further complication when it comes to the question of how to play out the counter-factual dimension of international law.

Recent studies on the sociology of international law have pinpointed the complexities of the interface between international law and practice. Social factors are not only 'involved in the formation and implementation of international law'[146] they also play a role when it comes to the question as to how compliance with international law can be enhanced. It has been pointed out that the broader the acceptance of rules is, the easier it is to promote compliance with them.[147] On the other hand, rules have always been drafted to respond to perceived deficiencies of the international system. The process of implementing international law is fairly well established when it comes to national implementation.[148] International implementation is less clearly defined, and has often

141 Wasum-Rainer and Schlegel, 'The UNCLOS Dispute Settlement System—Between Hamburg and The Hague', 48 *German Yearbook of International Law* (2005) 187.

142 Petersmann, 'Between "Member-Driven" WTO Governance and "Constitutional Justice": Judicial Dilemmas in GATT/WTO Dispute Settlement', 21 *Journal of International Economic Law* (2018) 103.

143 Brunnée, 'Compliance Control', in G. Ulfstein, T. Marauhn, and A. Zimmermann (eds), *Making Treaties Work. Human Rights, Environment and Arms Control* (2007) 373.

144 ILC, Fifth Report on subsequent agreements and subsequent practice in relation to interpretation of treaties, Special Rapporteur Georg Nolte (28 February 2018), UN Doc A/CN.4/715; ILC, Fifth Report on the identification of customary international law, Special Rapporteur Sir Michael Wood (14 March 2018), UN Doc A/CN.4/717.

145 M. Koskenniemi, *The Gentle Civilizer of Nations: The Rise and Fall of International Law 1870–1960* (2002).

146 M. Hirsch, *Invitation to the Sociology of International Law* (2015), at 2.

147 Ibid.

148 Ibid.

been discussed as a matter of compliance. Practice has, and this has also been taken up by *Thomas Franck* in both his essays discussed above, a strong impact on norms, and it has the potential to change them.[149] However, this is a complex process.

It does not simply mean that international law follows practice. Rather international law's initial impetus is its counter-factual claim. In other words, if one or several states act contrary to the rules of international law, their behaviour is first and foremost a violation of such rules. Rules will hardly ever change short-term. If at all, they will change mid- or long-term. This emerges from the requirements of law-making. New rules of international law do not emerge out of a violation instantly. Questions as to whose practice is relevant, how extensive and uniform such practice should be, and what the temporal element is, must be answered.[150] The burden to demonstrate this is usually higher when new rules replace previous ones than in the case of an emergence of new rules in a legal void. Furthermore, and this is particularly important with regard to Article 2(4), the institutional setting of rules in the UN framework must be borne in mind.

Against the above analysis of the various trends weakening the prohibition of the use of force it can be argued that there are many obstacles to the formation of new rules. Only few states would possibly seek to formally establish them. Many states simply do not formally argue that their behaviour is in compliance with Article 2(4), but simply remain tacit (as in the case of the Kosovo War in 1999). Others object to a particular behaviour, but not in respect of every incident. This seems to suggest that there is a weakening of the prohibition, but that the prohibition is not dead.[151] From our perception *Thomas Franck*'s argument on the killing of Article 2(4) was not meant to be a statement of the law but a warning uttered by a brilliant commentator, a warning that states should not discard one of the cornerstones of international peace and security.

4. Maintaining International Peace and Security Necessitates a Strong UN

As just argued, the institutional setting is important when it comes to an assessment of 'changing norms governing the use of force by states'. This institutional setting is the United Nations. *Thomas Franck*, in particular in his 2003 contribution, was right to argue that the primary object of attack by some states was not the prohibition of the use of force as such but rather the institutional framework, namely the United

[149] Franck, above note 35, at 614. On norms and compliance see also Checkel, 'Why Comply? Social Learning and European Identity Change', 55 *International Organization* (2001) 553; Deitelhoff, 'The Discursive Process of Legalization: Charting Islands of Persuasion in the ICC Case', 63 *International Organization* (2009) 33; Finnemore and Sikkink, 'International Norm Dynamics and Political Change', 52 *International Organization* (1998) 887; Wiener, 'Contested Compliance: Interventions on the Normative Structure of World Politics', 10 *European Journal of International Relations* (2004) 189; and Hendrik Simon's and Lothar Brock's introductory chapter in this volume.

[150] For an interesting collection of essays see B.D. Lepard (ed.), *Reexamining Customary International Law* (2017).

[151] See also Bothe, 'Has Article 2(4) Survived the Iraq War?', in M. Bothe, M.E. O'Connell, and N. Ronzitti, above note 109, at 417.

Nations.[152] It is indeed worrying that as of today there is a growing trend to circumvent the United Nations and to turn to unilateral action in matters of international peace and security. One might describe this as de-collectivizing international peace and security. In addition, it is not only unilateralism but also regionalism that partially seems to play against the United Nations.[153]

It is worthwhile to recall that '[t]he UN was not created to take mankind to heaven, but to save humanity from hell', as stated by former UN Secretary-General Dag Hammarskjöld in 1954.[154] Indeed, as the preamble of the UN Charter states, the UN was set up 'to save succeeding generations from the scourge of war' and the drafters were determined 'to unite our strength to maintain international peace and security'. Article 1 of the Charter hence names as the central purpose of the organization,

> [t]o maintain international peace and security, and to that end: to take effective collective measures for the prevention and removal of threats to the peace, and for the suppression of acts of aggression or other breaches of the peace, and to bring about by peaceful means, and in conformity with the principles of justice and international law, adjustment or settlement of international disputes or situations which might lead to a breach of the peace.

Article 2(4) immediately follows from this. The maintenance of international peace and security are central to the UN and there has not been developed any plausible or meaningful alternative to the UN.

The UN has always been under pressure from powerful or less powerful Member States and others. But so far they have been determined to preserve their central role in respect of international peace and security. This entails the need for all involved to accept compromise as a central feature of pertinent politics. Some of the cases in which force was used since 2003 have had a destabilizing effect on the UN and on the international normative order as a whole, even though some were driven by more positive motives. One of the problems the UN has been faced with is that a number of security threats have changed in nature,[155] be it that they emerge from non-state actors or originate in the domestic sphere of states or are not evidently covered by traditional concepts of the use of force. If the prohibition of the use of force is to be upheld as a meaningful rule, compromise is required by all participants. There is no alternative to a comprehensive truly universal organization, that is to the United Nations, to this end.

It is questionable whether the debates about constitutionalizing the international order have contributed to strengthening the UN or whether some states have actually

[152] Franck, above note 35, at 616.

[153] This has already been mentioned by Franck, above note 1, at 822–35. For a more recent debate see de Wet, 'The Modern Practice of Intervention by Invitation in Africa and Its Implications For the Prohibition of the Use of Force', 26 *European Journal of International Law* (2015) 979; and Kritsiotis, 'Interrogations of Consent: A Reply to Erika de Wet', 26 *European Journal of International Law* (2015) 999. See also the contribution of Nina Willén to this volume.

[154] Address by Secretary-General Dag Hammarskjöld at University of California Convocation, Berkeley, California (13 May 1954), Press Release SG/382.

[155] Still worthwhile to consider Wellens, 'The UN Security Council and the New Threats to the Peace', 8 *Journal of Conflict and Security Law* (2003) 15.

perceived this as an encroachment on their policies more generally, far beyond international peace and security. In this regard, constitutionalization potentially leads to sidestepping the UN or at least some of its organs.[156] This is, most likely, something to consider for the preservation of the central role of the UN, not to overburden the UN with too many different matters but rather sharpen its role in the maintenance of international peace and security, just as a cobbler should stick to his last. The weakening of the prohibition of the use of force illustrated above is serious, but it does not necessarily mean the killing of Article 2(4). Certain claims and behaviours could be explained as a backlash against constitutionalizing the international order. But they could also be perceived as a call to get back to normal, to concentrate on what is really decisive for the future of all of us: international peace and security.

5. The Lawyer's Responsibility

By way of conclusion, we share *Thomas Franck*'s concern that the prohibition to use force as laid down in Article 2(4) UN Charter is at risk. Too many attacks on the prohibition may lead to its unfortunate death. However, this is not without alternatives. *Franck*'s essays and this chapter should be understood as a reminder to ensure the survival of this very prohibition. There is no proper alternative—at least not if we believe that the future of this world can only be ensured by maintaining international peace and security. It is worthwhile to remind ourselves of *Franck*'s concluding section in his 2003 essay, in which he underlines the responsibility of lawyers in this particular context.[157] Few, but some have sent pertinent signals even at their own risk, including the former deputy legal adviser of the British Foreign Office who resigned in repudiation of the legal reasoning presented in support of the 2003 Iraq intervention by the then British attorney general.[158]

What does *Thomas Franck* call for? What does he consider to be 'the proper role for the lawyer?' Most importantly, as already quoted, 'to stand tall for the rule of law'.[159] This is the primary responsibility of lawyers. And the rule of law includes the prohibition to use force as laid down in Article 2(4). Lawyers should not shy away from political decision-makers. They should not shy away from the media. They should stand up and explain what the law is. This includes, speaking about 'longer-term costs' of skirting the law,[160] it includes naming breaches of the law, it necessitates defending the rule of law if states seek to subvert it.[161] And it should include our ability, even in light of perceived omnipotence of political decision-makers, 'to ask whether someday . . . we may not need the law'.[162]

156 On this risk—in both directions, in support of the UN or at its risk—see van den Herik, 'Sidestepping the Security Council—the Use of Non-UN Sanctions for UN Purposes', 50 *Revue belge de droit international* (2017) 474.

157 Franck, above note 35, at 620.

158 The resignation letter is available online at <https://www.globalpolicy.org/component/content/article/167/35802.html> (last visited 7 April 2020).

159 Franck, above note 35, at 620.

160 Ibid.

161 Ibid.

162 Ibid.

To conclude, in the words of *Thomas Franck*, lawyers 'should zealously guard their professional integrity'.[163] This is what has been and continues to be needed in order to avoid that Article 2(4) will eventually be killed. We should not simply rely on the multiple lives of cats when considering the normative power of the prohibition to use force. We should strengthen it, and remind ourselves of its historical background and current meaning.

[163] Ibid.

26

Justification and Critique: Humanitarianism and Imperialism over Time

B.S. Chimni

1. Introduction

In this essay I sketch a history of international humanitarianism that explores its complicity with the politics of imperialism. I offer an overview of the meaning and practice of humanitarianism since the sixteenth century, focusing on how it has been used to justify colonialism and neo-colonialism. I undertake this task as a step towards writing a critical global history of humanitarianism, a move crucial to developing principles and practices of humanitarianism that are not internally linked to the vision of empire.[1]

But to begin with I wish to stress that howsoever intimate the relationship of humanitarianism with imperialism it does not transform individual or collective acts of charity and sacrifice into collaboration with the politics of colonialism or empire. In exploring the relationship between humanitarianism and imperialism I therefore do not in any way wish to devalue individual or collective acts of charity.[2] All I wish to do is to draw attention to the dark side of the story of humanitarianism, in particular its roots in a questionable epistemological and historical vision, so that we can draw appropriate lessons.[3] I proceed in the following manner. *First*, I briefly look at the meaning of humanitarianism as it has evolved from colonialism to the present times from a Third World Approaches to International Law (TWAIL) perspective. *Second*,

[1] Michael Doyle defines 'empire' as 'a relationship, formal or informal, in which one state controls the effective political sovereignty of another political society. It can be achieved by force, by political collaboration, by economic, social, or cultural dependence. Imperialism is simply the process or policy of establishing or maintaining an empire', M. Doyle, *Empires* (1986), at 45. In this essay I use the words 'empire' and 'imperialism' interchangeably. On global history see generally Bentley, 'Globalizing History and Historicizing Globalization', in B.K. Gills and W.R. Thompson (eds), *Globalization and Global History* (2006) 18.

[2] See generally Haskell, 'Capitalism and the Origins of Human Sensibility, Part I', 90 *The American Historical Review* (1987) 339; Haskell, 'Capitalism and the Origins of Human Sensibility, Part II', 90 *The American Historical Review* (1987) 547; Linklater, 'Towards a Sociology of Global Morals with an "Emancipatory Intent"', 33 *Review of International Studies* (2007) 135; and K.A. Appiah, *Cosmopolitanism: Ethics in a World of Strangers* (2006).

[3] For a detailed history of modern humanitarianism see M. Barnett, *Empire of Humanity: A History of Humanitarianism* (2011). Barnett also occupies the middle ground between depicting humanitarianism as an instrument of empire and an instrument of emancipation. In his words, 'humanitarianism more closely resembles empire than many of its defenders might like, but because it is an emancipatory project, this accusation does not fit quite as well as many of its harshest critics suggest'. Ibid., at 8. The scope of 'history of humanitarianism' is vast. See generally, 'History and Humanitarianism: A Conversation', 241 *Past and Present* (2018) at e1.

B.S. Chimni, *Justification and Critique: Humanitarianism and Imperialism over Time* In: *The Justification of War and International Order*. Edited by: Lothar Brock and Hendrik Simon, Oxford University Press (2021).
DOI: 10.1093/oso/9780198865308.003.0026

I explore in its backdrop the idea of writing a critical global history of humanitarianism that would take cognizance of the non-western critique of the notion of humanitarianism. *Third*, I examine the role of humanitarianism in the era of colonialism. *Fourth*, I review the practice of contemporary humanitarianism in the matrix of key features of present world order and see how we can integrate power and principles to avoid the complicity of humanitarianism with the politics of imperialism. I conclude with some remarks on the need to redefine the Self without reference to an uncivilized Other as a precondition for a more equitable international order.

2. Meaning of Humanitarianism over Time

By 'humanitarianism' on the international plane I mean an ensemble of cultural, social, ideological, and material practices that have as their ostensible aim alleviation of human suffering or promotion of welfare of individuals, groups, and peoples of third nations that assume specific form and content in different phases of history. This definition goes beyond viewing humanitarianism as a response to individuals or groups in distress to include the larger goal of promoting human welfare, currently seen as the domain of human rights.[4] The broader understanding of humanitarianism is particularly applicable to the colonial era when the modern vocabulary of human rights was absent from international relations. But even in contemporary times, especially since the arrival of the ideology and practice of 'political humanitarianism', the worlds of humanitarianism and human rights have come to intersect in multifarious ways.

A humanitarian act is at all times the carrier of cultural meanings even as it brings material assistance and relief to peoples in distress. The cultural meaning of humanitarianism shapes and is in turn shaped by the social and political ideas of an age. While there are national variations of the idea of humanitarianism their meaning is in the final analysis constituted by a knowledge–power constellation that is a function of a particular world order.[5] Within an imperial world order, with us since the sixteenth century, the story of humanitarianism represents a 'narrative of justification' for the exercise of domination over, and infliction of violence on, other peoples and nations.[6] Albeit, the narrative of justification is framed in distinctive terms in different phases of imperialism.

The history of domination of other peoples and nations in modern times may conveniently be divided into two broad phases. In the long colonial phase, extending from 1500 to 1945, the idea of 'humanitarianism' subsumed under it all the ways through which Metropolitan powers justified the colonial project, the principal justification being the attempt to bring backward peoples and nations into civilization. In

[4] This position contrasts with the view that 'humanitarianism has a narrative of precarity and human rights progress, and these different narratives can be located in how each organizes history and gives meaning to events'. Barnett, 'Human Rights, Humanitarianism, and the Practices of Humanity', 10 *International Theory* (2018) 336.

[5] In the words of Barnett, 'humanitarianism can only be understood in relationship to the world order that constitutes it'. Barnett, 'Humanitarianism Transformed', 3 *Perspectives on Politics* (2005) 723, at 732.

[6] R. Forst, *Normativity and Power. Analyzing Social Orders of Justification* (2017); see Hendrik Simon's and Lothar Brock's introduction to this book, at 3–7.

the postcolonial period humanitarian practices have been aligned to a neo-colonial agenda. This period may again be divided into three phases.

In the first phase that roughly lasted from 1945 to 1989, humanitarianism came to be assigned a narrow and apolitical meaning. It was confined to giving 'impartial relief to victims of manmade and natural disasters'.[7] As the International Court of Justice noted in the *Nicaragua* case an 'essential feature of truly humanitarian aid' is that it is given "without discrimination" of any kind';[8] the attachment of political conditions to humanitarian acts was seen to undermine their humanitarian character.[9] In this phase, despite the adoption of a large number of human rights conventions the worlds of humanitarianisms and human rights did not seriously intersect. However, it saw liberal asylum regimes established centred around the 1951 UN Convention on the Status of Refugees and regional conventions (the Organization of African Unity (OAU) Convention on the Status of Refugees, 1969) and declarations (the Cartagena Declaration on Refugees, 1984).

In the period since 1989, humanitarianism came to be associated with a growing cosmopolitan vision that was not shy of being political. In this phase the vocabularies of humanitarianism and human rights came to be mixed. Humanitarianism became part of a broader development and security agenda.[10] It also derived its life and legitimacy from the ideas of 'democracy', 'human rights', and 'good governance'.[11] The resulting political humanitarianism meant a greatly expanded notion of humanitarianism from the one that dominated the Cold War era. It led to the dilution of the principle of impartial, neutral, and independent assistance. But paradoxically the expanded notion of humanitarianism came to coexist with a non-entrée regime against refugees that was actively constructed after it was clear that the Cold War had been won.[12]

A third phase has recently been inaugurated with the spread of nationalism and populism in both the imperial and postcolonial worlds. It has undermined the spirit of solidarity with victims of humanitarian disasters and displayed an active hostility towards migrants and asylum seekers. In this phase the cohesive imperial vision of international humanitarianism is being replaced by a politics of ad hoc humanitarianism. The trend has been consolidated in the period of Covid-19 pandemic.

Thus, over time the meaning of 'humanitarianism' has moved from being assigned a civilizational connotation in the colonial era to a constrained scope in the Cold War period to once again being attributed a cultural meaning in the immediate decades of

[7] Barnett, above note 5, at 723.

[8] *Military and Paramilitary Activities in and Against Nicaragua (Nicaragua v. United States of America)*, ICJ Reports, 1986, at para 243.

[9] S. Chesterman, *You, The People: The United Nations, Transitional Administration, and State-Building* (2004), at 187.

[10] 'The 1990s saw the concept of humanitarianism transformed, from a distinctive but narrow framework designed to mitigate the impact of war, into an organising principle for international relations, led largely by the West.' Macrae, 'Analysis and Synthesis', in J. Macrae (ed.), *The New Humanitarianisms: A Review of Trends in Global Humanitarian Action* (2002) 7, available at <https://www.odi.org/publications/243-new-humanitarianisms-review-trends-global-humanitarian-action> (last visited 1 April 2020).

[11] See also Jahn and Geis/Wagner in this volume; Anghie, 'Civilization and Commerce: The Concept of Governance in Historical Perspective', 45 *Villanova Law Review* (2000) 887.

[12] Chimni, 'The Geopolitics of Refugee Studies: A View from the South', 11 *Journal of Refugee Studies* (1998) 350.

the post-Cold War era to a fragmented approach in recent years amidst rising forces of nationalism and populism. But in each of these phases the underlying unity linking humanitarianism to the politics of imperialism has been ever present.[13] Both colonial and contemporary humanitarianism are viewed as a response to the lack of capacity of the Other for self-governance. From the Spanish colonization of Latin America in the sixteenth century[14] to the occupation of Iraq to the intervention in Libya and Syria a central narrative of justification is woven around bringing civilization to the inferior or lagging Other.[15]

In the matrix of this understanding of humanitarianism over time I turn to the idea of writing a global history of humanitarianism that takes into account the views of non-western peoples and nations.

3. Towards Critical Global History of Humanitarianism

Eric Hobsbawm, perhaps the most distinguished historian of the twentieth century, has noted how in the non-western world academic history either 'did not exist' or 'did not flourish' till the third quarter of the twentieth century.[16] Consequently world history before this period remained a history of western societies and all other societies came to be measured in relation to it.[17] Today, as non-western histories mark their presence, there is a need to move away from writing Eurocentric world histories to writing global histories that is inclusive of the past of what were deemed 'peoples without history'. There is for the first time, as Hobsbawm noted, 'an adequate framework for a genuinely global history'.[18] For non-western histories address crucial silences and distortions in the Orientalist world histories written in the western world.[19]

In other words, there are two alternative ways of thinking about and writing global histories of humanitarianism. One approach is to locate the history of humanitarianism in national histories, whose writing gained currency in Europe in the nineteenth

[13] Chesterman, above note 9.

[14] See also Becker's chapter in this book.

[15] See also Stohl's contribution to this volume, and A. Anghie, *Imperialism, Sovereignty and the Making of International Law* (2004). As Douzinas puts it in the context of the contemporary human rights movement:

> Despite differences in content, colonialism and the human rights movement form a continuum, episodes in the same drama, which started with the great discoveries of the new world and is now carried out in the streets of Iraq: bringing civilization to the barbarians. The claim to spread Reason and Christianity gave the western empires their sense of superiority and their universalizing impetus. The urge is still there; the ideas have been redefined but the belief in the universality of our worldview remains as strong as that of the colonialists.' Douzinas, 'The Many Faces of Humanitarianism', 2 *Parrhesia* (2007) 1, at 21.

Immanuel Wallerstein likewise notes in the context of humanitarian intervention: 'the intervenors, when challenged, always retort to a moral justification—natural law and Christianity in the sixteenth century, the civilizing mission in the nineteenth century, and human rights and democracy in the late twentieth and twenty-first centuries'. Cited in Douzinas, ibid.

[16] E. Hobsbawm, *Interesting Times: A Twentieth-Century Life* (2002), at 297.

[17] D. Chakrabarty, *Provincializing Europe* (2000).

[18] Hobsbawm, above note 16, at 292.

[19] See E. Said, *Orientalism* (1978); E. Said, *Culture and Imperialism* (1993).

and twentieth centuries.[20] The discrete national histories tend to extol humanitarian practices that accompanied the western colonial project and extend this narrative to postcolonial times.[21] These national histories affirm the superiority of western civilization and locate the idea of humanitarianism amidst it. It was such history when John Seeley wrote in *The Expansion of England* that 'the history of England is not in England but in America and in Asia'.[22]

The other possibility is of writing a global history of humanitarianism. A global history of humanitarianism can assume either of the following two trajectories. It may be a global history written by apologists of empire, past and present, or a global history that foregrounds the history of the struggles and emancipation of oppressed peoples from imperialist ideology and accompanying practices of humanitarianism. In other words, while the move towards global history is liberating in many ways there is no guarantee that it will not be the handmaiden of contemporary or future imperial projects. What we need then is the writing of *critical* global histories. A critical global history of humanitarianism would at least have the following four features.

First, it will call for a reordering of established historiographies in order to make space for histories that disengage from particular national, regional, ethnic, and religious traditions.[23] Writing global history should be an effort

> to find paths leading beyond notions that the world is a site divided into national spaces, beyond assumptions that European modernity is the appropriate standard for the measurement of all the world's societies, [and] beyond the various political, economic and cultural interests that have tinted but also tainted efforts to produce knowledge about the larger world.[24]

The goal of critical global history, according to another of its votaries, must thus be directed towards '... moral purposes, connected to the needs of a globalizing world'.[25] Such global history promises to 'raise awareness of the lives, achievements and sufferings of humanity as a whole'.[26] This history will seek to construct meta-narratives that will deepen our understanding of the diversities that constitute humankind and scale up our consciousness of the human condition.[27] Such history is to be distinguished from one-sided world history that has been written before.[28]

[20] Hopkins, 'The History of Globalization—and the Globalization of History?', in G. Hopkins (ed.), *Globalization in World History* (2002) 11, at 13.

[21] Lauren Benton has exemplified this in her chapter with reference to 'humanitarian interventions' in the nineteenth century.

[22] Seeley, 'The Expansion of England' (1883), available at <https://web.viu.ca/davies/H479B.Imperialism.Nationalism/Seeley.Br.Expansion.imperial.1883.htm> (last visited 1 April 2020).

[23] 'By re-mapping the geography of the subject, we hope to point the way towards a truly global history of globalization.' Hopkins, 'Introduction: Globalization-An Agenda for Historians', in Hopkins (ed.), above note 20, 1, at 2.

[24] Bentley, above note 1, at 30

[25] O'Brien, 'Historiographical Traditions and Modern Imperatives for the Restoration of Global History', 1 *Journal of Global History* (2006) 3, at 37.

[26] Ibid.

[27] Ibid.

[28] See for instance G.W.F. Hegel, *Philosophy of Right*, translated by S.W. Dyde (2005), at 200–07.

Second, it will actively discourage an understanding of humanitarianism that views it as a discrete process or occurrence. Instead, critical global history will treat the practices that constitute humanitarianism as embedded in a specific world order and understand its role in relation to it. To put it differently, a global history of humanitarianism is its history in different phases of a bourgeois world order as it has evolved since the sixteenth century. It will therefore per force involve an understanding of the history of capitalism and its internal relationship with colonialism and imperialism.[29]

Third, critical global history will explore the social, religious, and political ideas that shape the relationship between humanitarianism and evolving imperialism. Thus, for example, in terms of political ideas it has been shown that while *Liberalism* is not necessarily imperialistic 'the urge is internal to it'.[30] This inclination is traced to a conception of the political that harbours 'a deep impulse to reform the world'.[31] As Mehta notes, it explains the

> necessary tension [of the politics of empire] with other liberal notions such as tolerance, the right to representation, equality, and ... consent and sovereignty of the people. In the empire, this latent impulse—this urge to reform and progress—which otherwise so often remains obscured and contested behind a concern with rights and individual freedom, *becomes virtually determinative and singular*.[32]

The reform sentiment has, transgressing ideological boundaries, come to be embedded in the idea and working of western liberal democracies.[33]

Fourth, critical global history will follow critics of colonialism in taking seriously 'the sentiments, feelings, and attachments through which peoples are, and aspire to be, "at home"', that is, the thick social, material, and cultural practices through which peoples construct their distinctive worlds.[34] In other words, critical global history will pay heed to postcolonial histories that contend that an unguarded narrative of progress posits abstract scales of civilization from which the 'integrity of life forms' are read out.[35] The international community will then be ready to come to terms with the unfamiliar as a way of delegitimizing coercive humanitarianism.[36] In contrast, for instance, *Liberalism* in the era of British colonialism was 'derived from an epistemological view

[29] R. Luxemburg, *The Accumulation of Capital* (1951); V. Chibber, *Postcolonial Theory and the Spectre of Capital* (2017).

[30] U.S. Mehta, *Liberalism and Empire: India in British Liberal Thought* (1999), at 20. For a materialist explanation see also H. Lebovics, *Imperialism and the Corruption of Democracies* (2006), at 87–100

[31] Mehta, ibid., at 79.

[32] Ibid., at 80. Emphasis added.

[33] The complex sources of this transgression await elaboration in international relations literature. The questions are how and why did the impulse to reform travel to all Enlightenment thinking.

[34] Mehta, above note 30, at 21.

[35] Ibid., at 82. These local histories may either represent a professional history that relies on the tenets of Enlightenment western historiography and is done only in the institutional setting of the university or is popular history that has escaped its imperial reach. What kind of popular history and how it is to be evaluated are key questions that await an answer but it would certainly be a step forward in democratizing the writing of history. Chakrabarty, 'Review Essay: A Global and Multicultural "Discipline" of History?', 45 *History and Theory* (2006) 101, at 106ff.

[36] 'Unfamiliar', according to Mehta, means 'not sharing in the various ways of being and feeling that shape experience and give meaning to the communities and the individuals who constitute them—in a word, not being familiar with what was experientially familiar to others in the empire'. Mehta, above note 30, at 2.

that allowed the world to be read abstracted from its concrete aesthetic and emotive particulars and nevertheless issued in a firmness of judgment with respect to those particulars'.[37]

These features of a critical global history of humanitarianism will open up space for humanitarian practices that do not sustain the politics of imperialism. In sum, the humanitarian community needs to urgently learn from the past and come to terms with the symbiotic relationship of humanitarianism with colonialism and imperialism. I touch on this history in the next two sections.

4. Colonialism and Humanitarianism

The relationship of humanitarianism and colonialism, when written as an integral part of Orientalist and unreflective national histories is presented in sanguine terms. The anthropologist Nicholas Dirks writes (in the context of India) that 'built on fabrication, colonial history mirrors the general distortions and displacements of imperial self-representation—the use of imputed barbarism to justify, and even ennoble, imperial ambition'.[38] He notes that colonial history 'veiled its dependence on the world outside by legitimating and naturalizing empire, ultimately representing it as at best nothing more than a burden and a terrible responsibility'.[39] Thus, 'imperial history has been written in the service of empire itself'.[40]

It was Spain that first embarked on colonial conquest in the sixteenth century. The task of justifying the acquisition of colonial territories therefore first fell on its jurists. Francesco de Vitoria was foremost amongst them. He argued that Christian and civilized states had a duty to intervene in barbaric societies to end inhuman practices such as cannibalism, spread Christianity, and from that perspective protect its missionaries.[41] As Antony Anghie puts it, 'the discrepancy between the ontologically "universal" Indian and the socially, historically, "particular" Indian' was to be 'remedied by the imposition of sanctions which effect the necessary transformation'.[42] The underlying barbarian/civilized societies distinction was the preferred form of justification for colonization and violence in the next three centuries. Non-western critique of these justifications, like the ones presented by Arnulf Becker in this volume, were widely ignored.

[37] Ibid., at 83. The critics of Liberalism, such as the conservative Englishman Edmund Burke in the nineteenth century, have attempted to, on the other hand, 'facilitate an openness to the world, to its unavoidable contingencies, surprises and ambivalences.' Ibid., at 42. There is the underlying belief that, as Mehta notes, 'the world is not patterned in such a way that its emotional resonances, its passionate attachments, can be read off simply by placing a rational and temporarily progressive grid on it'. Ibid.

[38] N.B. Dirks, *The Scandal of Empire: India and the Creation of Imperial Britain* (2006), at 5.

[39] Ibid., at 332.

[40] Ibid., at 27.

[41] Bhiku Parikh writes that '"civilized" states, all of them Christian, had a duty to intervene in the internal affairs of "backward" societies to end such inhuman practices as cannibalism and human sacrifice, to spread Christianity, including preventing the harassment of Christian missionaries, and to remove obstacles to the conversion of the natives', Parikh, 'Rethinking Humanitarian Intervention', 18 *International Political Science Review* (1997) 49, at 50f.

[42] Anghie, above note 15, at 22.

Instead, the distinction between barbarian and civilized societies withstood changes in the methods of doing international law. Thus, for instance, in the literature of international law the justification survived the move from the language of natural law to that of positivism in the nineteenth century. The colonized nations were, as deficient Others, not treated as subjects of international law . European legal theory and political practice collaborated in articulating a chauvinistic ideology of civilization: Positivists, as has been pointed out,

> developed an elaborate vocabulary for denigrating non-European people, presenting them as suitable objects for conquest, and legitimizing the most extreme violence against them, all in the furtherance of the civilizing mission, the discharge of the white man's burden.[43]

At the Berlin Conference of 1884–85 Prince Bismarck declared that 'all the Governments invited share the wish to bring the natives of Africa within the pale of civilization . . . .'[44] Article 6 of the General Act of the Conference of Berlin Concerning the Congo stated in part:

> All the Powers exercising sovereign rights or influence in the aforesaid territories [the conventional Basin of the Congo] bind themselves to watch over the preservation of the native tribes, and to care for the improvement of the conditions of their moral and material well-being . . .[45]

This attempt to represent the Other as someone in need of care and protection also linked domestic and international dimensions of society in this era. In many intricate ways the internal mirrored the international and vice versa. Johanna de Groot has demonstrated how 'nineteenth century representations and discourses of sexual identity and difference drew upon and contributed to comparable discourses and representations of ethnic, "racial", and cultural identity and difference'.[46] Both women and natives were, and I quote Johanna de Groot,

> portrayed and treated as children in need of the protection and care of male/imperial authority by virtue of their weaknesses, innocence and inadequacy. The use of a parental concept of authority combined a sense of care and involvement with the subordinate sex or race as well as power and control over them, and as such was equally appropriate for the definition of the power of men over women or of dominant over subordinate races.[47]

[43] Ibid., at 38.
[44] Cited in Anghie, ibid., at 97.
[45] Ibid.
[46] Groot, '"Sex" and "Race": The Construction of Language and Image in the Nineteenth Century', in C. Hall (ed.), *Cultures of Empire: A Reader* (2000) 37, at 57.
[47] Ibid., at 37ff. Mehta notes how 'childhood is a theme that runs through the writings of British liberals in India with unerring constancy. It is the fixed point underlying the various imperial imperatives of education, forms of governance, and the alignment with progress'. Mehta, above note 30, at 31.

This mode of representation should not however be assigned a mere functional significance. It is to be viewed 'as a process of defining the self and others'.[48] The urge to reform and to care and protect went beyond finding an instrumental justification for colonialism. It was constitutive of the self-image of the colonizers.

The internal and international moment came together in colonial societies. In supporting the civilizing mission of empire in India in the early eighteenth century the argument was made, especially in the context of the practice of sati (or widow burning), that 'empire had a sacred mission to protect the women in India from a barbaric culture and religion'.[49] Indeed, the abolition of widow burning was 'perhaps the most important moral justification' for the British Empire so far as India is concerned.[50] Likewise was the attempt to get rid of *thuggee* or highway robbers that among other things impacted trade activities.[51] The provision of law and order was seen as critical in any civilized society.

But what relevance has this history today, decades after colonialism has ended? In the words of Dirks, 'the frozen paradigms of imperial history' have 'resurfaced with alarming ubiquity in recent years, in a wide range of earnest efforts to restore historical balance to the question of empire'.[52] Dirks continues, and I quote at length:

> Even as the British justified their new imperium in India through their commitment to free women from the curse of sati and protect their citizens from the threat of highway robbery, the United States has more recently represented its primary role in Afghanistan as the liberation of women from the oppression of the fundamentalist Taliban and the protection of all Afghanis from the ruthless warlords who controlled the highways and trading routes of the region. Of course, the British neglected to mention that their own presence had greatly exacerbated the importance of sati (not to mention that their general commitment to social reforms meaningful to women were inconsistently insincere), and that the ritual murder they worked so hard to suppress was in large part, where it wasn't fabricated, the direct result of early imperial conquest and occupation. That the United States neglected as well to mention its own role in establishing the Taliban and arming the regional warlords in its Cold War struggle with the Soviet Union thus had an important imperial precedent.[53]

In short, the need to go over the colonial history of humanitarianism arises from the fact that a certain commonality binds the colonial and postcolonial eras. In both eras, as Douzinas notes, the non-West is seen to combine 'the suffering mass and the radical evil-doer, the subhuman and the inhuman rolled into one'[54] providing infinite

[48] Mehta, above note 30, at 35.

[49] Dirks, above note 38, at 304.

[50] Ibid., at 305.

[51] Ibid., at 305–09.

[52] Ibid., at 333. He cites the examples of N. Fergusson, *Empire: The Rise and Demise of the British World Order and the Lessons for Global Power* (2002); and Frykenberg, 'India to 1858', in R.W. Winks (ed.), *Historiography*, vol. 5 of the Oxford History of the British Empire (1999), ch. 11.

[53] Dirks, above note 38, at 310.

[54] In the words of Douzinas, 'every human rights campaign or humanitarian intervention presupposes an element of contempt for the situation and the victims. Human rights are part of an attitude of the postcolonial world in which the "misery" of Africa is the result of its failings and corruption, its traditional attitudes and lack of modernization, its nepotism and inefficiency, in a word of its sub-humanity. We can feel

legitimacy for all western humanitarian intrusions. In the same vein Terry Nardin observes that 'most of the old justifications for empire are close to the surface in current understandings of America's mission'.[55]

5. Humanitarianism in the Era of Globalization

However, while there is a continuing relationship between humanitarianism and empire, the specific features of humanitarianism in the postcolonial period need to be delineated and understood. There is among other things a need to distinguish between the Cold War and post-Cold War eras, the latter both in its early and present phases.

In the Cold War era the sharp East–West divide made a de-politicized humanitarianism an attractive proposition. By de-coupling humanitarianism from politics its practices could be tolerated by both sides. At the same time, it allowed the West to deploy humanitarian practices to bring out the weaknesses in the Soviet political system. For example, the work of international agencies like the UNHCR, mandated as being of 'an entirely non-political character' and described as 'humanitarian and social',[56] allowed an international refugee regime constructed around a definition of 'refugee' exclusively centred on civil and political rights to embarrass the former Soviet Union. Every asylum seeker from the Soviet bloc countries granted refuge in the West was seen as voting with her feet against the socialist system. To put it differently, the Cold War explains why in this period a narrow definition of humanitarianism was adopted and humanitarian organizations, whether intergovernmental organizations (INGOs) or non-governmental organizations (NGOs, conformed to the classical principles of impartiality, neutrality, and independence.[57]

great pity for the victims of human rights abuses; but pity is tinged with a little contempt for their fickleness and passivity and huge aversion towards the bestiality of their compatriots and tormentors. We do not like these others, but we love pitying them.' See Douzinas, above note 15, at 13.

[55] Nardin, 'Humanitarian Imperialism', 19 *Ethics and International Affairs* (2005) 21, at 25:

> In the old literature of empire, humanitarianism was invoked to justify the supposed responsibility of an imperial power operating at the margins of the civilized world to uphold the standards of civilized morality by suppressing cannibalism, human sacrifice, and other barbaric practices. In today's rhetoric of empire, it is the barbarity of tyranny and terrorism that threaten these standards and that must be countered, in the name of humanity, by the exercise of imperial power. In the old literature of empire, colonial rule was rationalized as providing backward peoples the benefits of civilization: public order, public health, modern communications, economic development, and eventually constitutional rule. The new literature of empire rationalizes intervention in similar terms. Most of the old justifications for empire are close to the surface in current understandings of America's mission.

[56] See Article 2 of the Statute of the Office of the United Nations High Commissioner for Refugees, available at <https://www.unhcr.org/3b66c39e1.pdf> (last visited 1 April 2020).

[57] 'Impartiality requires that assistance be based on need and not on the basis of nationality, race, religion, gender, or political opinion. Neutrality demands that humanitarian organizations refrain from taking part in hostilities or from any action that either benefits or disadvantages the parties to the conflict. Independence demands that assistance should not be connected to any of the parties directly involved in armed conflicts or who have a stake in the outcome; accordingly, there is a general rule that agencies should either refuse or limit their reliance on government funding, especially if the donors have a stake in the outcome.' Weiss and Barnett, 'Humanitarianism: A Brief History of the Present', in T.G. Weiss and M.N. Barnett (eds), *Humanitarianism in Question: Politics, Power, Ethics* (2008), at 3.

5.1 The Neo-Liberal Global Order

With the end of the Cold War and the onset of an accelerated globalization process the ideology and practices of humanitarianism underwent change. The description of the new or political humanitarianism however needs to be preceded by the identification of key features of the international system that informed this transformation. *First*, a *global economic space* emerged in which goods, capital, and services flowed across national borders with greater ease than ever in the past. The borderless space was promoted by a transnational capitalist class (TCC) that fraction of the global capitalist class that gained most from the neo-liberal globalization process.[58] *Second*, a *global political space* was constituted in which while the principle of sovereignty remained the central organizing principle in international relations it was overcome in a variety of ways to serve powerful states; sovereignty came to be depicted as an anachronism that prevented the spread of global welfare. *Third*, a *global humanitarian space* emerged that had at its core a growing body of human rights, humanitarian, and refugee law and accompanying practices of what are professional humanitarian INGOs and NGOs. While these global spaces were intimately linked, each of the spheres possessed *a relative autonomy*. But in the final analysis the ideology and practices of humanitarianism were shaped by the objective of sustaining the neo-liberal global order. This order is being destabilized in the last few years with the emergence of the phenomena of nationalism and populism. I will have more to say on that development presently.

5.2 Emergence and Practice of Political Humanitarianism

The neo-liberal globalization of the post-Cold War era, given its harmful consequences for subaltern classes and nations, profoundly shaped the global humanitarian space. The divide between developed and developing countries grew creating conditions in the latter for internal conflicts leading to the displacement of millions of people. It profoundly impacted the world of humanitarianism.

First, when the neo-liberal globalization caused an economic or political crisis in postcolonial nations, humanitarian agencies were assigned the role of ensuring that its worst consequences were addressed through humanitarian assistance and development aid to stitch together a viable post-crisis society: humanitarian agencies were part of the global strategy to give neo-liberal imperialism a human face.[59]

Second, while there was the rapid development of international human rights, humanitarian, and refugee laws a state of exception was often decreed when it came to the peoples of the non-western world. Thus, for example, commentators have noted

[58] On TCC see B.S. Chimni, *International Law and World Order: A Critique of Contemporary Approaches* (2nd ed., 2017), at 507–09.

[59] Chimni, 'Globalization, Humanitarianism and the Erosion of Refugee Protection', 13 *Journal of Refugee Studies* (2000) 243; Chimni, 'Refugees and Post-Conflict Reconstruction: A Critical Perspective', in E. Newman and A. Schnabel (eds), *Recovering from Civil Conflict* (2002) 163; Chimni, 'The International Law of Humanitarian Intervention', in *State Sovereignty in the 21st Century* (2001) 103.

that the exclusion of the inferior Other from the application of international humanitarian laws has roots in colonial thinking.[60]

Third, neo-liberal globalization with its implications for the principle of sovereignty allowed the justification of muscular humanitarianism as in the instance of Northern Iraq (1991), Kosovo (1999), Iraq (2003), Libya (2011), and Syria (2015–).[61]

These factors cumulatively gave rise in the post-Cold War era to the emergence of the idea of political humanitarianism that refused to be limited by the classic concept of humanitarianism with its emphasis on the principles of impartiality, neutrality, and independence. International humanitarian agencies, both governmental and non-governmental, came to accept the view that it is their task to address both the causes and aftermath of a humanitarian crisis. This understanding, in the words of Barnett 'swept them … into the world of politics. Humanitarian agencies and states began to share agendas.'[62] This move of back to the future had grave implications as far as humanitarian organizations were concerned. Let me mention two.

First, humanitarian organizations became, as Donnini, Minear, and Walker note, 'a largely owned subsidiary' of dominant states, subjecting them to their political and security interests.[63] As a result the classical principles of humanitarianism were undermined in Kosovo, Afghanistan, and Iraq; humanitarian agencies were funded by the very governments whose actions were responsible for the humanitarian crisis.[64] Bilateralization of aid, earmarking of funds, and control over budgets by states were among the instruments of control.[65] A neo-liberal globalization process subjected humanitarian organizations to the logic of market leading inter alia to unhealthy competition for resources. Political humanitarianism thus undermined the independence and autonomy of humanitarian organizations and caused the shrinking of humanitarian space.

Second, humanitarian agencies neglected local voices as these often conflicted with the agenda of states that funded them. In this framework difference was viewed as a problem rather than simply a distinct mode of being. It led humanitarian agencies to

[60] Megret, 'From "Savages" to "Unlawful Combatants": A Postcolonial Look at International Humanitarian Law's "Other"', in A. Orford (ed.), *International Law and its Others* (2006) 265.

[61] See generally, 'Humanitarian Imperialism: The New Doctrine of Imperial Right', *Monthly Review* (2008) 22.

[62] Barnett, above note 5, at 724. As another author puts it: 'The blurring of the dividing line between humanitarian space and politics … constitutes the quintessential hallmark of new humanitarianism.' Yamashita, 'New Humanitarianism and Changing Logics of Political in International Relations', 43 *Millennium: Journal of International Studies* (2015) 411, at 413.

[63] Donnini, Minear, and Walker, 'Between Cooptation and Irrelevance: Humanitarian Action after Iraq', 17 *Journal of Refugee Studies* (2004) 260, at 260.

[64] '… most humanitarian actors agree that Iraq crisis has resulted in a dangerous blurring of lines between humanitarian and political action and the consequent erosion of the core humanitarian principles of the core humanitarian principles of neutrality, impartiality and independence'. Donnini, Minear, and Walker, above note 63, at 262. This was unfortunate for as Donnini, Minear, and Walker aptly point out, underlying the discarded principles of neutrality, impartiality, and independence 'lie much more nuanced issues of the duty bound nature of humanitarianism, the absolute need for trust between humanitarians, those they seek to serve and those who support them, and adherence to the principle of proportionality in assistance and engagement. Neutrality, independence and maybe even impartiality are modus operandi to support these much more fundamental ideas.' These underlying values are lost in the move to political humanitarianism. Ibid., at 267.

[65] Ibid., at 262ff., especially at 266.

accept the idea of reform from outside and as a logical extension muscular humanitarianism. This development was assisted by the doctrine of *Responsibility to Protect* (*R2P*) which rather than view armed humanitarian intervention as an exception from the principles of sovereignty and non-intervention places an obligation on the international community to intervene when the respective state fails to protect the rights of its people.[66] It is of course another matter that armed humanitarian intervention did not materialize in instances where urgently required: Rwanda being the classic case. Much of the changed approach could be traced to the desire of humanitarian agencies to be active participants in transforming the non-western world in ways that realized the western vision of good governance.

5.3 Populism, Nationalism, and Humanitarianism

The ideology and practices of humanitarianism are once again undergoing change. In recent years regressive social and political forces have attempted to harness through deploying populist rhetoric the anger of groups excluded from the benefits of neo-liberal globalization process and direct it towards some form of parochial nationalism.[67] Even liberal and left social forces have come under pressure from extreme right parties to use exclusionary ethno-nationalist vocabulary to attack traditional elites. In the matrix of the phenomenon of rising nationalism and populism the nation-state is back in prominence. Deglobalization has set in. The TCC is having to more actively compete with the national fractions of the global capitalist class that, in a bid to recover lost ground, are supporting the nationalist and populist social forces. It explains why powerful states like the United States want the globalization process and institutions to be reformed in ways that yields greater benefits to all segments of its business class. The slogan of 'America First' essentially means defining national interest not in terms of sustaining the global capitalist system but in terms of short-term benefits that can accrue.

Likewise, humanitarianism and humanitarian action are coming to be tied to narrow national interest. The open hostility to refugees and migrants in the western world is a manifestation of this new phase.[68] The same can be said of the more active opposition of United States to the International Criminal Court. As a result there is an overall decline in respecting the relative autonomy of global humanitarian space. This trend is reflected in the undermining of international humanitarian law in different conflict zones that pepper the world: Yemen, Syria, Sudan, and Democratic Republic of Congo are spaces in which humanitarian law and activities have suffered setbacks.[69]

[66] Dixon argues that liberal hawks allied with neoconservatives in the United States and the United Kingdom to form a powerful coalition to advocate for 'humanitarian intervention' to 'protect human rights' and 'prevent genocide'. It also influenced the outlook of humanitarian NGOs. Dixon, '"Endless Wars of Altruism"? Human Rights, Humanitarianism and the Syrian War', 23 *The International Journal of Human Rights* (2019) 819.

[67] C. Rovira Kaltwasser, P. Taggert, P. Ochoa Espejo, and P. Ostigy (eds), *The Oxford Handbook of Populism* (2017).

[68] To take just one data, between 2014 and 2019 18,067 individuals drowned in the Mediterranean Sea. See Missing Migrants Project, available at <https://missingmigrants.iom.int/> (last visited 1 April 2020).

[69] See UN Security Council, *Protection of Civilians in Armed Conflict: Report of the Secretary General* S/2019/373 (7 May 2019),.

There is also the phenomenon of reduced humanitarian access to people requiring humanitarian assistance.[70] Finally, there is growing scepticism of the role of international human rights law.[71] The vocabulary and institutions of human rights no longer arouse the hope of bringing about a just world order. The general crisis of multilateralism has in this period also diminished the role of the UN and its agencies in undertaking humanitarian activities.

All this means that humanitarianism in the near future will have an ad hoc quality. The impulse to reform the Other in the vocabulary of 'good governance', 'democracy', and 'human rights' will manifest only when it serves parochial interests, even if it can be readily seen that it will lead to a humanitarian crisis as in the instance of Venezuela. The new humanitarianism does not rule out use of force to defend human rights or humanitarian values but in the matrix of national rather than community interests. In other words, it can be said that no cohesive ideology of humanitarianism prevails, as it did in the early post-Cold War period which culminated in the idea of R2P. Some therefore call the new humanitarianism counter-humanitarianism.[72]

5.4 Balancing Power and Principle

What can be done to address the dark side of contemporary humanitarianism? Of course, all current humanitarian action cannot be tainted with the brush of empire. Thousands and thousands of people in distress are provided humanitarian assistance and protection every day. At the same time, it cannot be ignored that imperialism deploys humanitarianism and humanitarian organizations to serve its ends. In the face of the reality of imperialism it may not be possible to realize 'the utopian dream of an international community that would finally have integrated power and principle'.[73] But some steps forward are possible to establish *humane humanitarianism*.

First, if humane humanitarianism is to be the order of the day the international community needs to be self-conscious of its unsavoury past. This awareness can only come if appropriate lessons are drawn from a critical global history of the ideology and practices of humanitarianism.[74]

Second, the practice of humane humanitarianism requires that the impulse to reform the Other, embedded in the thinking of western liberal democracies, be attenuated. There has to be greater recognition that the reform of societies has to come from within and not without.

[70] J.D. Kurtzer, *Denial, Delay, Diversion: Tackling Access Challenges in an Evolving Humanitarian Landscape* (2019).

[71] Alston, 'The Populist Challenge to Human Rights', 9 *Journal of Human Rights Practice* (2017) 1.

[72] A. de Waal, *Mass Starvation: The History and Future of Famine* (2018).

[73] Berman, 'Intervention in a "Divided World": Axes of Legitimacy', 17 *European Journal of International Law* (2006) 743.

[74] According to Berman: 'Rather than dream of a final integration of power by principle, … we [must] continually goad power-holders by comparing them to those in the past who are now viewed as unprincipled. Depriving those with power of any secure legitimacy should spur them on to avoid words and deeds deserving of the most ignominious illegitimacy.' Ibid., at 746. It, among other things, anticipates the rejection of attempts to what Hopkins calls 'nationalizing internationalism' or 'internationalizing nationalism'. Hopkins, above note 20, at 14.

Third, it calls for the relative autonomy of the global humanitarian space to be respected. A politics of solidarity that makes a distinction between political humanitarianism and politics after humanitarianism must be promoted.[75] In other words, humanitarian organizations dispensing humanitarian assistance should adhere to the principles of classical humanitarianism. The causes of humanitarian crisis and the construction of post-conflict societies should be addressed outside that context.[76] At which point they would need to decide whether they will hitch their fate with empire or with those who are the true subjects of humanitarianism viz., the peoples of post-colonial nations.

Fourth, the international community should not, short of preventing genocide and gross violation of human rights, use war for humanitarian ends. There is a need for practicing prudent internationalism by which I mean

> the refusal to be ensnared by options for action framed by hegemonic States at times of crisis and the exercise of caution in recommending coercive international measures, especially military action, to the international community of States.[77]

The international community should 'not endorse actions that destabilize the normative consensus in the international system against the use of military action'.[78]

Fifth, the inhumane non-entrée regime created by western states to keep out refugees should be dismantled. In this regard it is important to bear in mind the history of migrations in the colonial era. Thus, for instance, 'from 1800 to 1930, 40 million Europeans migrated permanently overseas mainly to North and South America and Australia'.[79] It must also be remembered that nearly 84 per cent of refugees are hosted by developing nations.

Sixth, there should be greater South–South humanitarianism as it is likely to be less afflicted with oriental prejudices and power dynamics that characterize North–South humanitarianism. Unfortunately there is not sufficiently known about the phenomenon of South–South humanitarianism.[80]

Seventh, there is a need to find imaginative ways of dealing with the problem of funding humanitarian agencies, both governmental and non-governmental, so as to reduce their dependence on dominant states. In that circumstance alone can the

75 Kennedy, 'Reassessing International Humanitarianism: The Dark Sides', in Orford, above note 60, 131, at 152. Unfortunately, as Kennedy notes, 'there is scarcely a humanitarian practice that does not act as if governance were elsewhere—in government, statecraft, the member states, the states parties, the Security Council, the field, the headquarters, the empire.' Ibid. In reality humanitarians 'do rule', that is, 'exercise power and affect distribution among people'.

76 Chimni, 'Post-Conflict Peace-Building and the Return of Refugees: Concepts, Practices, and Institutions', in E. Newman and J. van Selm (eds), *Refugees and Forced Displacement: International Security, Human Vulnerability, and the State* (2002) 195.

77 Chimni, 'For Epistemological and Prudent Internationalism', *Harvard Human Rights Journal*, available at <http://harvardhrj.com/2012/11/for-epistemological-and-prudent-internationalism/> (last visited 1 April 2020).

78 Ibid.

79 S. Castles and M.J. Miller, *Age of Migration* (3rd ed., 2003), at 53.

80 See generally Conference Report: South-South Humanitarianism, 2014, available at <http://www.jsia.edu.in/wp-content/uploads/2019/03/south-south-humanitarianism-conference-report-jsia.pdf> (last visited 1 April 2020).

struggle for a humane humanitarian space be seriously joined by humanitarian organizations. In this regard we should, for example, talk about assessed contributions from UN member states for agencies like UNHCR.

Finally, a reform of the global humanitarian order is not possible without the reform of the global economic order. It needs to be restructured to deliver benefits to subaltern classes in both Global North and Global South.

6. Final Remarks

I turn to a few concluding remarks. Of the oft quoted words of Walter Benjamin are that 'there is no document of civilization that is not at the same time a document of barbarism'.[81] In the same vein Hendrik Simon and Lothar Brock observe in their introduction to this volume that the historical emergence of international orders of peace is shaped by violent struggles over power and normativity. Thus, any history of humanitarianism can be told both as a story of sacrifices and succour on the one hand and legitimizing domination and hegemony on the other. The ideology and practices of humanitarianism have over time been deployed by imperialism also to come to terms with anxieties and guilt that accompany doing violence to Others. Thus the role of humanitarianism is not merely functional but an integral part of a cultural and political vision that constitutes the Self through the eyes of a inferior Other. The Other can therefore be reconstituted only through redefining the Self without reference to an uncivilized counterpart. The Self–Other divide can be overcome only through deconstructing and critically understanding what philosopher Charles Taylor's calls the 'sources of the self'.

A critical global history of humanitarianism is at least one important step in this direction. It deconstructs and undermines the posited Self-Inferior Other divide by inter alia showing that the Other has often been understood outside the thick social practices that constitute its world.[82] This is not to embrace any form of cultural relativism but to advance the simple idea that the West does not necessarily mirror the non-West. It is also to accept the demonstrated reality that the non-West cannot be reformed from the outside through coercive humanitarian interventions. Where external intervention is necessary, reform can be brought about through communicative action; it rests on the ability to understand unfamiliar life worlds rather than see these always as eternal objects of reform. Once the impulse to reform is limited the possibility of distancing it from the politics of imperialism will arise. What will also need to be addressed is the economics of imperialism. The colonial project was all about commerce and exploitation. The neo-colonial agenda is the same through other means.

Each of the phases of imperialism, and sub-phases within them, gave rise to a corresponding ideology and practice of humanitarianism. The post-Cold War era saw a

[81] Benjamin, 'Theses on the Philosophy of History', in H. Arendt (ed.), *Illuminations*, translated by H. Zohn (1969) 253.

[82] That is why Mehta resists an interpretation of 'liberalism and empire' that 'focuses on contradictions'. He places 'the question of the liberal endorsement of empire as secondary to that of the liberal response to the experiences of the unfamiliar'. Mehta, above note 30, at 201.

cohesive ideology of political humanitarianism accompanying neo-liberal globalization emerge. It is today being reinterpreted in the matrix of the turn to nationalism and populism. International human rights law, humanitarian law, and refugee law are all in crises. The result is the destabilization of the idea of humanitarianism and the rise of counter-humanitarianism. The way ahead lies in restoring a genuine politics of solidarity with the subaltern peoples.[83] It does not mean that coercive humanitarianism is ruled out in all circumstances, but its use will be confined to the most urgent cases on which there is an overwhelming consensus within the international community.[84] For the rest the value of toleration must be cultivated for, as Michael Walzer has observed, toleration alone makes difference possible.[85] But unless unjust globalization, or for that matter deglobalization, is replaced by just international economic relations it will be difficult to reshape the ideology and material practices of humanitarianism.

[83] See generally UNGA, *Report of the Independent Expert on Human Rights and International Solidarity*, A/HRC/38/40 (11 April 2018).

[84] Chimni, 'Sovereignty, Rights and Armed Intervention', in H. Charlesworth and J.-M. Coicaud (eds), *Faultlines of International Legitimacy* (2010) 303.

[85] M. Walzer, *On Toleration* (First Indian Reprint, 2004), at xii.

27

The Justification and Critique of Coercion as World Order Politics

Christopher Daase and Nicole Deitelhoff

1. Introduction

The current crisis of the international liberal order requires a reorientation or at least re-evaluation of its normative foundations. In this context, we argue that the focus should be on the concept of 'legitimate coercion' rather than 'legitimate force' since in most institutions and also in international law the enforcement of norms—as the very basis of order—does not only or even primarily rely on military force but on various forms of political and economic coercion, re-establishing their validity to all norm addressees. However, there is no consensus as to when different coercive measures are legitimate or illegitimate despite numerous treaties, declarations, and resolutions trying to limit and regulate their application.[1] The chapter distinguishes various forms of coercion and reconstructs debates in International Law and International Relations with regard to their legality, legitimacy, and effectiveness. Doing so, we intend to broaden the debate on world order by redirecting the focus from the use of force to the use of less violent coercive measures as an act of communication.

The chapter starts by spelling out the normative implications of the current crisis of the international liberal order. It proceeds by discussing political options in order to save basic values and multilateral institutions of global peace and security. Doing so, it reaches back to theoretical positions of the interwar period in order to structure the issue and provide an alternative for dealing with it. Finally, we distinguish strategies of force, coercion, and sanction as a means to broaden the perspective on world order.

2. The Crisis of the International Liberal Order

When the Cold War ended, it was said that *Liberalism* had won. In his book on *The End of History* Francis Fukuyama wrote in 1989 that the triumph of the West over the Soviet Union not only sealed the historical defeat of communism but also proved the lack of alternatives to the liberal model of a well-ordered society.[2] Under this perspective the age of political ideologies was over and democracy was on the rise worldwide.[3]

[1] Rabkin and Yoo, 'A Return to Coercion. International Law and New Weapons Technologies', 42 *Hofstra Law Review* (2014) 1187; K.M. Greenhill and P. Krause (eds), *Coercion: The Power to Hurt in International Politics* (2018).

[2] F. Fukuyama, *The End of History and the Last Man* (1992).

[3] Huntington, 'Democracy's Third Wave', 2 *Journal of Democracy* (1991) 12.

Christopher Daase and Nicole Deitelhoff, *The Justification and Critique of Coercion as World Order Politics* In: *The Justification of War and International Order*. Edited by: Lothar Brock and Hendrik Simon, Oxford University Press (2021).
 DOI: 10.1093/oso/9780198865308.003.0027

Twenty-five years later the balance sheet is sobering. Between 2000 and 2015, democracy collapsed in twenty-seven countries.[4] Different democracy indices show that many authoritarian countries have become more authoritarian and many democratic countries less democratic.[5] The report of Freedom House's *Nations in Transit index* 2018 monitoring the democratic development of Central and Eastern European countries highlights that in more than half of the twenty-nine monitored countries the democracy scores had decreased or as the report puts it: 'Illiberalism established itself as the new normal in the region that stretches from Central Europa to Eurasia.'[6]

A similar picture emerges globally. Here, too, the liberal idea of an international order based on multilateral rules and institutions, free trade, and individual rights has been put on the defensive. Until recently, the narrative prevailed that the international liberal order was being undermined by emerging powers such as China or India and a reinvigorated Russia. In the meantime, however, it has become obvious that it is equally, maybe even more dramatically the United States that threaten this order, that is the very power that created and decisively shaped an international society based on liberal values, norms, and institutions for decades. Under President Trump, the United States has evolved from a guarantor power to a threat of the liberal world order. It is one thing if a normative order is criticized and challenged from outside. But it is something quite different when those who have established and maintained that order over decades turn away from it. This indeed can be described as a normative crisis.[7]

But it would be too easy to blame Donald Trump alone for the crisis of the liberal world order. Trump is just as much a consequence as a cause of the crisis. Amitav Acharya rightly points out that long before Trump became president of the United States there had been signs that free trade was diminishing, that multilateral institutions were losing their effectiveness, and that democracy was losing its appeal worldwide. Trump, he argues, only accelerated the decline of the liberal order.[8] Indeed, Trump is not a singular phenomenon. Brazil's Bolsonaro or the Philippines' Duterte do not need to shy away from a comparison with Trump when it comes to hostility to liberal values and norms. Brexit, Britain's now executed exit from the European peace and integration project, is aiming in the same direction, as are the illiberal tendencies that are evident in Poland and Hungary, the former liberal showcase countries of the 'new Central Europe', or in right-wing populist movements in France, Italy, the Netherlands, Sweden, Austria, or Germany. All these developments have in common that they are directed against the principles and institutions of an open society, free trade, freedom of movement, and international solidarity.[9]

When we speak of the international liberal order, we are referring to the institutionally entrenched modern order of international politics which Hedley Bull has

[4] L. Diamond, M.F. Plattner, and C. Walker (eds), *Authoritarianism Goes Global. The Challenge to Democracy* (2016).

[5] M.J. Mazarr et al. (eds), *Measuring the Health of the Liberal International Order* (2017).

[6] N. Shenkkan, *Confronting Illiberalism. Nations in Transit Report 2018* (2018), available at <https://freedomhouse.org/sites/default/files/FH_NationsInTransit_Web_PDF_FINAL_2018_03_16.pdf> (last visited 10 January 2020).

[7] C. Daase and S. Kroll (eds), *Angriff auf die liberale Weltordnung* (2018).

[8] Acharya, 'After Liberal Hegemony: The Advent of a Multiplex World Order', 31 *Ethics & International Affairs* (2018), 271.

[9] See in detail Mazarr, above note 5.

identified as an 'anarchic society'. Bull's concept encompasses two crucial conditions: the absence of a central ordering power, that is the 'anarchic' nature of international politics on the one hand, and the rule-based nature of a society, that is the validity of social conventions, norms, and institutions in international politics on the other.[10] We refer to this order as liberal, denoting its emphasis on multilateralism, free trade, and individual rights despite the fact that it never fully embraced nor realized these principles.[11]

Looking at these principles more closely, the crisis becomes even more tangible. With regard to multilateralism as an organizing principle of the liberal order, multilateral institutions, that is the network of international organizations, regimes, and conventions, are in a bad shape. The United Nations, after a brief interim high in the early 1990s, has once again lapsed into a state of paralysis. There is hardly any crisis in which the United Nations has not fallen short of its capabilities and aspirations, most glaringly in Syria. Clearly, this is not the fault of the UN organization, but of the member states, above all the five permanent members of the Security Council (P5), which, as during the Cold War, are blocking each other in the Security Council. During the last five years the use of the veto by the P5 was higher than at any time before the end of the Cold War.[12] Similar problems can be detected for NATO, which some consider 'brain dead' and others 'obsolete', at least for the time being. We should also mention the EU, which with Brexit has embarked on a path of orderly disintegration, but will not achieve unity until the core issues of European solidarity—migration, social union, and defence—are resolved.[13]

Institutional developments in the security field are also dramatic. Arms control as understood during the Cold War is in ruins. The INF Treaty is dead, the New START Treaty is likely to expire in 2021 without renewal. The United States and Russia cannot even agree on the classic formula that a nuclear war would endanger humanity and must be prevented at all costs. This also has implications for the multilateral nuclear non-proliferation regime. The Review Conference threatens to fail because of the dispute between nuclear-weapon states and non-nuclear-weapon states. This further weakens the entire Non-Proliferation Treaty (NPT). The other multilateral regimes are not doing much better. The Chemical Weapons Convention has been repeatedly violated by Syria while Russia has prevented Syria's condemnation by the UN Security Council. Not to mention that Russia has annexed the Crimea and is tenaciously undermining Ukraine's sovereignty.

Similarly, with multilateral trade agreements faltering or stalling, while bilateral and minilateral trade agreements are booming and bilateral trade conflicts involving punitive tariffs and duties proliferate, free trade is obviously in the defensive.

Finally, individual rights are under pressure as well: not only have individual norms such as the prohibition of torture been called into question, China and Russia use financial leverages to sideline human rights protection by the United Nations, a practice

[10] H. Bull, *The Anarchical Society: A Study of Order in World Politics* (1977).

[11] Ikenberry, 'Why the Liberal Order Will Survive', 32 *Ethics & International Affairs* (2018) 17.

[12] Bethke et al., 'Institutionelle Friedenssicherung: Vereinte Nationen unter Druck. Wo sich ein deutsches Engagement lohnt', in Bonn International Center for Conversion (BICC) et al. (eds), *Friedensgutachten* (2019) 115.

[13] D. Webber, *European Disintegration?: The Politics of Crisis in the European Union* (2018).

made possible by the financial and ideational withdrawal of the United States.[14] But also the institutions entrusted with the maintenance of individual rights, such as the International Criminal Court and the European Court of Human Rights, are under massive pressure.[15] The Responsibility to Protect, too, is in a crisis of legitimacy and is openly called into question.[16]

Two lessons, we believe, can be drawn from the above. First, history (hopefully) has no end and the liberal order is neither nationally nor internationally a given, robust acquis. Second, the liberal order will not stabilize itself out of itself, but it needs strong supporters and a coherent strategy to develop it further.

3. Political Options in Normative Crises

The current intensification of the normative crisis of international politics makes a fundamental resetting—or at least an endorsement—of the normative foundations of international order necessary. In our view, the concept of legitimate coercion should be at the centre of this process, because in many institutions and also in international law, the primary concern is not the enforcement of norms with violent means, but rather the strengthening of their binding force through the threat of sanctions.[17] The question of the extent to which coercion in this sense (i.e. as a legitimate form of threatening costs for the purpose of political control) is conceivable and desirable in the international normative order goes back to the beginning of the last century, when three legal-ethical positions emerged which, we maintain, still shape our thinking about the fate of international order.[18]

At the beginning of the twentieth century, or more precisely, under the impression of the devastating First World War, the decisive political, philosophical, and legal question was whether and how the institution of war should finally be overcome. Three positions emerged: a pacifist position, a realist position, and a liberal position. We briefly outline these positions using three reference theorists, namely Hans Wehberg, Carl Schmitt, and Hans Kelsen, in order to accentuate the political alternatives more clearly.

The *pacifist* position is best described in Hans Wehberg's 1930 lecture on the outlawing of war ('*Die Ächtung des Krieges*'). Wehberg pleads for a legal ban on war

[14] Sikkink and Schmidt, 'Breaking the Ban? The Heterogeneous Impact of US Contestation of the Torture Norm', 4 *Journal of Global Security Studies* (2019) 105; McKeown, 'Norm Regress: US Revisionism and the Slow Death of the Torture Norm', 23 *International Relations* (2009) 5; C. Lynch, *At the U.N., China and Russia Score Win in War on Human Rights* (2018), available at <https://foreignpolicy.com/2018/03/26/at-the-u-n-china-and-russia-score-win-in-war-on-human-rights/> (last visited 10 January 2020).

[15] Madsen, Cebulak, and Wiebusch, 'Backlash against International Courts: Explaining the Forms and Patterns of Resistance to International Courts', 14 *International Journal of Law in Context* (2018) 197; Arcudi, 'The Absence of Norm Modification and the Intensification of Norm Contestation: Africa and the Responsibility to Prosecute', 11 *Global Responsibility to Protect* (2019) 172.

[16] Welsh, 'Norm Robustness and the Responsibility to Protect', 4 *Journal of Global Security Studies* (2019) 53.

[17] C. Möllers, *Die Möglichkeit der Normen: über eine Praxis jenseits von Moralität und Kausalität* (2015).

[18] Peace Research Institute Frankfurt, *Coercion and Peace. PRIF's New Research Program* (2018), available at <https://www.hsfk.de/en/publications/publication-search/publication/coercion-and-peace/> (last visited 10 January 2020).

and the strongest possible restriction of the right to use of force in self-defence. However, he was aware that the mere prohibition of war alone would not lead to the disappearance of war. In his view, what was needed were law-based procedures for the resolution of disputes and a strong central power that could enforce sanctions militarily if necessary. With this legal-pacifist position, Wehberg distinguishes himself from radical pacifists who believed they could do without any means of coercion. He sees the chance to overcome war by the monopolization of coercive power within an international organization.[19]

The *realist* counter-position is exemplified by Carl Schmitt. The notorious critic of the Weimar Republic and supporter of the Nazis, is still considered one of the sharpest, but also one of the cleverest critics of liberal thinking, whether in constitutional or international law. In his 1937 essay '*Die Wende zum diskriminierenden Kriegbegriff*' ('The Turn to the Discriminatory Concept of War'), Schmitt argues against a general ban on war because it would lead back to the times of religious just war thinking by declaring the use of force to be legitimate only for one side, namely for the one that claims to represent the international community (for a critique see Hendrik Simon's contribution to this volume). The consequences, according to Schmitt, are twofold: on the one hand, the opponents in such wars would become criminals, gangsters, and pirates and would lose all rights as lawful combatants; on the other hand, the classical law of neutrality would be abolished, because in the face of a peace-breaker impartiality is no legitimate option.[20] According to Schmitt's prognosis both lead to an 'intensification of war and enmity' and to an 'ever deeper and sharper, ever more "total" distinction between friend and foe'.[21] The good intention of abolishing war would thus achieve exactly its opposite.

The *liberal* position, finally, is best grasped with reference to Hans Kelsen. As legal theorist close to social democracy, Kelsen was indeed sympathetic to the ideas of the League of Nations and normative cosmopolitanism. But like Schmitt, he criticized the prohibition of war. 'The war ban of the Briand-Kellogg Pact attempted to abolish the function of war as a decentralised sanction mechanism, but without replacing it with effective centralised sanction mechanisms.'[22] However, Kelsen draws radically different conclusions. Being a vehement advocate of modern international law, but deploring its weakness, he argues in his famous *Pure Theory of Law* ('*Reine Rechtslehre*') of 1934 that the goal of legal development must be the organizational unity of a universal legal community, that is the development of a world state. However, as long as a central monopoly on the use of force does not exist, executive power must be delegated and exercised decentrally by individual states. In his book *Peace through Law*, published in 1944, Kelsen does not shy away from returning to the doctrine of 'just war', which Schmitt is so fearful of. Kelsen writes:

[19] H. Wehberg, *Die Ächtung des Krieges, eine Vorlesung an der Haager Völkerrechtsakademie und am 'Institut universitaire de hautes études internationales'* (1930).

[20] Schmitt, *Die Wende zum diskriminierenden Kriegsbegriff* (1937), at 546–47.

[21] Ibid., at 562.

[22] Kelsen, 'Unrecht und Unrechtsfolge im Völkerrecht', 12 *Zeitschrift für Öffentliches Recht* (1932) 481; H. Kelsen, *Reine Rechtslehre. Einleitung in die rechtswissenschaftliche Problematik* (1932), at 586.

> Only when one sees in war, just as in reprisal, the reaction of law against injustice, can one see in it the starting point for a development that will gradually transform it from a means of self-help to a coercive act of central legal protection bodies.[23]

Seen in this light, the just war is not only legitimate but also legal, as it were, in anticipation of the monopoly of force to be established but not existing yet.

With Wehberg, Schmitt, and Kelsen, the normative positions of the interwar period—pacifism, *Realism*, and *Liberalism*—are well described, and it is surprising how much they still influence our thinking today. What is clear is that pacifism initially seemed to have failed in the Second World War. However, the legal pacifist position was revived in the United Nations, at least formally, by enshrining in the UN Charter an even more far-reaching ban on war than in the League of Nations Covenant or the Briand-Kellogg Pact. Nevertheless, a realist practice of international politics prevailed during the Cold War, not least based on the compromises made in the drafting of the Charter, which established a right of individual *and* collective self-defence. This opened the door to the extension of legal grounds for the use of force.[24] After the end of the Cold War, the idea of a 'New World Order' arose anew, and the UN mandated war against Iraq in 1990–91 seemed to foreshadow the monopoly of force in the sense of legal pacifism. However, it soon became clear that unanimity in the Security Council would not last long, and so subsequent wars (such as in Kosovo in 1999 or Iraq in 2003) were fought without a Security Council mandate, but instead with reference to 'just causes'—that is with recourse to the doctrine of just war.

Today, it seems, we are back on square one. Legal pacifists still argue that only the United Nations can guarantee international peace and security on a legal basis and that every use of force beyond UN mechanisms undermines international order. Both, *Realists* and *Liberals* disagree; *Realists* because they disdain international legal restrictions on national prerogatives concerning defence and security, *Liberals* because they repudiate normative shackles that prohibit legal advancement and human progress. Especially if the formal legal order is unable to work as intended, *Liberals* argue, higher principles of justice in the form of universal values and human rights come to the fore that allow, or even demand, the use of force without UN mandate.[25] Just as Kelsen did, they argue that the breach of law is cured by the intention to create a new legal order and that decentral norm-enforcement is critical for normative progress. This is exactly how the 1999 Kosovo war was justified by Jürgen Habermas 'in anticipation of a future cosmopolitan state'.[26] Here *Pacifists* and *Realists* disagree; *Pacifists* because they see no normative gain in the breach of law, but rather the breakdown of international obligations, *Realists* because they suspect the existence of global values. 'Whoever invokes

[23] Kelsen, *Peace through Law* (1944), at 594.

[24] Hurd, 'The Permissive Power of the Ban on War', 2 *European Journal of International Security* (2017) 1; I. Hurd, *How To Do Things With International Law* (2017); see also Marauhn's chapter and the editors' concluding chapter in this volume.

[25] Slaughter, 'The Real New World Order', 76 *Foreign Affairs* (1997) 183; Buchanan, 'Reforming the International Law Of Humanitarian Intervention', in J.L. Holzgrefe and R.O. Keohane (eds), *Humanitarian Intervention. Ethical, Legal, and Political Dilemmas* (2003) 130.

[26] Habermas, 'Bestialität und Humanität. Ein Krieg an der Grenze zwischen Recht und Moral', *Die Zeit* (28 April 1999).

the concept of humanity wants to cheat', Schmitt once wrote and formulates the scepticism against any norms beyond national interest that many *Realists* share.

So where do we stand or, rather, where are we going from here? The problem is that all three positions are rife with contradictions and not fully supportable. *Legal pacifism* is a noble sentiment, but it fails where positive law is still weak and does not possess the institutional basis for legitimate enforcement. *Realism* does not regard the monopolization of force at the international level as a viable option and is, consequently, equally conservative. In contrast, *Liberalism* can claim to have an idea of normative progress, though that may involve breaking the law if necessary and thus risking its weakening. For it is not a foregone conclusion that allegedly 'just wars' also change international law in such a way that what is legitimate also becomes legal, as liberal theorists like to claim (for a defence of the just war tradition see Brown's contribution to this volume).

4. Coercion and Sanction in International Relations

The focus on the legitimacy and illegitimacy of violence falls short of the mark, we believe. For all great narratives about the genesis and stabilization of political order refer in their core less to violence than to coercion as their rationale. Order *responds* to chaos and violence, and it does so by disarming, disciplining, and 'coercing' individuals to live together in peace and according to shared rules and norms. This is most evident in the case of the modern nation state as Leviathan, which according to Thomas Hobbes uses absolute power to make the individual leave the natural state and recognize law and justice. But the image of the Leviathan also underlines the ambivalent character of coercion. Coercion is not only the cause of but also a threat to political order. For where coercion is used beyond law and reason, the duty of the subjects to abide by the law expires and resistance to the then illegitimate order is justified.

Also beyond the state, legitimate coercion is a decisive component of political order. Even if there is no central power of coercion in the international system, there are manifold mechanisms of more or less decentralized coercion, for example, to settle disputes between conflict parties with robust peacekeeping missions, to verify compliance with agreements on verification mechanisms, to achieve the abandonment of activities that endanger peace and security through economic sanctions, to enforce 'weak' international law through international courts or to avert threats to peace and security by coercion short of or with military intervention. In the international system, the question of the legality, legitimacy, and effectiveness of coercion arises even more strongly than within the state.

Coercion is often understood—in a narrow sense—as the external influence on an actor under threat or use of force in order to achieve behaviour that otherwise would not come about.[27] However, the concept of coercion, which focuses on violence as a means of exerting pressure, is too narrow when we consider the everyday use of the term and the diversity of social practice. There are other kinds of measures by which

[27] T. Schelling, *Arms and Influence* (1966). D. Byman and M. Waxmann, *The Dynamics of Coercion. American Foreign Policy and the Limits of Military Might* (2002). R. Mandel, *Coercing Compliance. State-Initiated Brute Force in Today's World* (2015).

actors can be forced or nudged into doing something that they would not otherwise do: for example, psychological pressure, shaming and blaming, and the threat or implementation of non-violent sanctions such as economic boycotts, exclusion from committees and organizations, and so on. Coercion in this sense is a continuum of action through which different social costs are imposed or threatened in order to achieve a certain behaviour.[28]

While coercion can be used for purely selfish interests, it is also conceivable that coercion serves the goals of legitimate rule. Rule and power are both forms of vertical differentiation, insofar as they represent social structures of super ordination and subordination.[29] However, rule is institutionalized to a higher degree. While power, according to Max Weber, 'represents every opportunity to assert one's own will within a social relationship, even against resistance, no matter on what this opportunity is based',[30] rule is only present when power is institutionalized: 'Rule should mean the opportunity to find obedience of specific individuals for a specific demand.'[31] Only in an institutionalized context (i.e. in a context in which reasons for concrete actions are demanded and offered) does a social action, which under non-institutionalized conditions would simply be violence (in the broadest sense), become coercion, namely the exercise of power aiming at political order.[32] That is, coercion occurs—whether redeemed or not—with the claim to provide order and serve the common good. However, as Max Weber further pointed out, if coercion is to be not only legitimate but also legal, it needs an institutional basis. For Weber, law is a political order that is 'guaranteed by an apparatus of coercion'.[33] Violence—now more narrowly understood as the physical restriction of freedom—comes into play as it is exercised by a coercive apparatus serving legitimate objectives (transforming *violentia* into *potestas*). However, a certain degree of unjustified or unjustifiable violence always remains. As Derrida put it, 'behind the formal-legal institutionalization still lurks the uncheckable violence of the state'.[34] Politics, even when it is based on voluntary consent and readiness to follow, does not simply merge into 'authority', but remains dependent on coercive rule.[35] Coercion is thus never 100 per cent legitimate, as Jane Mansbridge emphasizes, and is therefore always the realm where resistance arises.[36]

[28] This understanding follows Robert Nozick who was among the first to establish a conceptual framework for investigating coercion in terms of its necessary and sufficient conditions, see Nozick, 'Coercion', in S. Morgenbesser, P. Suppes, and M. Whiten (eds), *Philosophy, Science, and Method: Essays in Honor of Ernest Nagel* (1969) 440.

[29] Daase and Deitelhoff, 'Opposition and Dissidence: Two Modes of Resistance against International Rule', 15 *Journal of International Political Theory* (2019) 11.

[30] Weber, *Wirtschaft und Gesellschaft. Grundriss der verstehenden Soziologie* (1922), at ch. 1, § 16.

[31] Ibid., at 38.

[32] C. Waldhoff, *Staat und Zwang: Der Staat als Rechtsdurchsetzungsinstanz* (2008).

[33] M. Weber, *Gesammelte Aufsätze zur Wissenschaftslehre* (1968), at 445.

[34] J. Derrida, *Zwei Essays über die Vernunft* (2003), own translation. See also Simon's and Brock's Introduction.

[35] Deitelhoff and Daase, 'Rule and Resistance in International Relations', *International Theory* (forthcoming).

[36] Mansbridge, 'Resisting Resistance', in R. Celikates, R. Kreide, and T. Wesche (eds), *Transformations of Democracy: Crisis, Protest and Legitimation* (2015) 147. See also Mansbridge, 'Taking Coercion Seriously', 3 *Constellations* (1997) 407.

Obviously, the international system lacks a coercive apparatus almost completely and international norms therefore have an enforcement deficit. This becomes particularly evident when the UN Security Council is divided and cannot agree on enforcement measures. In these situations the need for decentralized coercion to enforce the law arises or, as understood by Hans Kelsen, the need for sanctions.

4.1 Sanctions

To what extent can coercion and sanctions contribute to the establishment and maintenance of world order? For a long time, coercive diplomacy has been seen as the cheaper, lower-risk, and legally unproblematic alternative to warfare.[37] It was only with the scandalization of the humanitarian effects of economic sanctions as against Iraq in the 1990s that the effectiveness, appropriateness and, implicitly, the legitimacy of economic sanctions were questioned.[38] Since then, research on coercive diplomacy has focused on the effectiveness of political, economic, and military means to force states to comply with international norms or to change their policies. In this literature, sanctions are usually misunderstood by reducing them to their coercive character, that is to their function of changing the behaviour of states. This neglects the normative side of sanctions as they condemn an injustice (or, in the case of positive sanctions, reward correct behaviour). Therefore, sanctions are often associated with pure power politics. Karl Deutsch defined power as the 'expected ability to impose sanctions'.[39] But not every exercise of power, not even every act of coercion, is a sanction. As understood by Kelsen and in this contribution, sanctions come with the claim to enforce a general norm. Classical gunboat diplomacy, in which one state forces another state for purely selfish motives, is not a sanctions policy and not even coercion in the sense mentioned above. What is essential for sanctions is the invocation and justification of general norms, for the preservation of which political, economic, or military power is exercised. Understood this way, sanctions primarily function as an act of communication within a normative community (i.e. a community of norm makers and takers) to signal that a normative expectation is still valid and that violation of normative commitment are not accepted. Thus, the success of a sanction does not primarily reside in its ability to give rise to a change of behaviour on behalf of the addressee of a sanction but in its ability to uphold the normative expectation within the community as a whole.

The research on the normative functions, foundations, and effects of coercive diplomacy is underdeveloped. While there are clear legal and other normative criteria for the use of military force, there are no comparable criteria for the use of non-military, that is political and economic means of coercion. The UN Secretary-General already in 1993 stated that

[37] A.L. George, *Forceful Persuasion: Coercive Diplomacy as an Alternative to War* (1991, 1994). R.J. Art and P.M. Cronin, *The United States and Coercive Diplomacy* (2003).

[38] Bossuyt, 'The Adverse Consequences of Economic Sanctions on the Enjoyment of Human Rights', E/CN.4/Sub.2/2000/33 (2000). D. Cortright and G.A. Lopez, *The Sanctions Decade: Assessing UN Strategies in the 1990s* (2000).

[39] K.W. Deutsch, *The Nerves of Government* (1963), at 121–122.

> there is no clear consensus in international law on when coercive economic measures are improper, despite relevant treaties, declarations and resolutions adopted in international organizations which try to develop norms limiting the use of such measures.[40]

The UN Charter in Chapter VII provides for non-military means of peace enforcement. Sanctions imposed by the Security Council are therefore legally binding and generally considered to be legitimate. However, the question of whether individual states have the right to impose sanctions unilaterally or multilaterally remains highly disputed. It depends on how sanctions are understood and justified, whether they can be deemed to play a constructive role in the normative order of the international system or not.

Under customary international law, states can take a number of non-violent countermeasures against states that do not fulfil their treaty obligations, such as trade embargoes, freezing of assets, suspension of diplomatic or economic relations, etc. It is generally considered that such measure must (1) respond to a specific wrong, (2) be accompanied by an explanation and justification of the countermeasure, (3) be proportionate to the original wrong, and (4) if individuals are affected, be within the framework of existing human rights standards.[41] But countermeasures are not enforcement measures, that is decentral sanctions in Kelsen's sense. Whether such or similar coercive measures may also be used by states or groups of states to enforce international norms and maintain international institutions therefore remains a question in dire need of an answer.

Unfortunately, not only in the political debate, but also in the academic debate, an understanding prevails that looks at sanctions as mere instruments of coercive diplomacy. Indeed, coercive diplomacy and sanctions are often used synonymously. In this case the success of sanctions is measured by the change in behaviour of the sanctioned actor. However, sanctions and sanctioning are not at all primarily about forcing a particular behaviour on an actor.[42] Rather sanctions demonstrate approval or disapproval of a behaviour with the intention of confirming a certain norm. Therefore, 'sanctioning' can have the meaning of 'disapproval' and 'approval' at the same time, because it is not primarily concerned with the assessment of a particular action, but with the confirmation of the norm underlying that action.

A good example for this is the current debate about the 'failure' of sanctioning Hungary and Poland for domestic legal measures that run counter to Article 2 of the European Union Treaty. For both countries, the institutions of the European Union (for Poland it was the Commission in 2017, in the case of Hungary it was the European Parliament) have put in place so-called Article 7 proceedings that authorize the Council of the European Union to assess a potential violation of the rule of law guarantees and to eventually authorize sanctions. However, until to date, the behaviour

[40] Boutros-Ghali, 'Economic measures as a means of political and economic coercion against developing countries. Note by the Secretary-General', *A/48/535* (1993).

[41] I. Brownlie, *The Rule of Law in International Affairs. International Law at the Fiftieth Anniversary of the United Nations* (1998).

[42] Daase, 'Vom gerechten Krieg zum legitimen Zwang', in I.-J. Werkner and P. Rudolf (eds), *Rechtserhaltende Gewalt—zur Kriteriologie. Fragen zur Gewalt* (2019) 13.

in question is still in place or even worsened.[43] Hence, the initiation of sanctions has not resulted in Hungary or Poland refraining from continuing to pursue their norm-violating behaviour. However, it clearly signals to all EU members that their behaviour is in gross violation of core values and principles and the Union and therefore is unacceptable. Similarly, Bogdandy and colleagues argue not to focus too much on a change of behaviour in Poland (or Hungary) but rather on describing the initiation of sanctions as giving rise to a constitutional moment for the Union in which the Union needs to decide whether to become a constitutionalized community or fall back on a loose alliance.[44]

With the focus on the normative function of sanctions, the aspect of power loses importance. Powerless sanctions are also sanctions in the sense understood here. If the affirmation of a general norm is the main function of sanctions and not the change of a specific behaviour, then the success parameters of sanctions also change. Notes of protest against the treatment of prisoners by the American military are not only successful when they change the behaviour of the US military, but as also when they express a collective disapproval with reference to general norms (of the prohibition of torture, the proportionality of means, the obligations of occupying powers, etc.). Sanctions against Russia's annexation of the Crimea are not successful only when the Russian troops withdraw, but already at the moment when they express general disapproval and confirm the right to territorial integrity, the principle of secure borders, and the rule that treaties must be respected. Sanctions fail not because they do not achieve an intended change in behaviour, but when the reference to the underlying norm is not successful and the sanction is rightly considered to constitute illegitimate coercion. In this vein, sanctions work to revitalize a sense of belonging to a community of norm makers and norm takers and what it means to belong. Seen in this light, the sanctions imposed by the Trump administration after its withdrawal from the nuclear agreement with Iran are not sanctions, but coercive policies of an illegitimate nature.

If this understanding of sanctions is taken seriously, a new ethic of sanctioning as legitimate coercion could be spelled out. Sanctions would no longer be an exclusive domain of powerful states, not even of states alone, but a possibility for decentralized approval or disapproval of actions by states and citizens, (re-)instating themselves as a community of norm makers and takers.

5. Conclusion

In the normative discourse on world order so far, too much has been said about just war and too little about legitimate coercion. This has conceptually limited the possibilities of reacting to breaches of law, human rights violations, and injustices in

43 Michelot, *The 'Article 7' Proceedings Against Poland and Hungary: What Concrete Effects?* (2019), available at <https://institutdelors.eu/en/publications/__trashed/> (last visited 10 January 2020).

44 von Bogdandy et al., 'A Constitutional Moment for the European Rule of Law—Upcoming Landmark Decisions Concerning the Polish Judiciary' (2018), available at <https://papers.ssrn.com/sol3/papers.cfm?abstract_id=3199809> (last visited 10 January 2020).

international politics. There are non-violent forms of coercion with which violations of rules can be disapproved and the binding force of international norms can be increased. In order to systematically develop the possibilities of legitimate coercion, it is important to move away from a narrow understanding of sanctions. We are not claiming that this would solve all problems, especially not the problem of the legitimate use of military force. But a broad concept of sanctioning that emphasizes the normative function of approval and disapproval opens up new possibilities for international order politics.

AN ATTEMPT AT A SYNTHESIS

28

Justifications of the Use of Force as Constitutive Elements of World Order—Points of Departure, Arrivals, and Moving Destinations

Lothar Brock and Hendrik Simon

1. Points of Departure: The History of War as a History of its Justification

The contributions to this volume confirm what we have stated in our introductory chapter: the history of war can be read as a history of its justification. The question of when the use of force is justified or not has been discussed over thousands of years not only by scholars and theorists, but also by politicians and the general public. This is not to say that nothing has changed in the way in which the use of force interacts with its justification and contestation and with the formation of international order. However, change is difficult to qualify as such. Change refers to time and 'anyone who talks about time relies on metaphors such as progress and cycle, breaks and sediments'.[1] In this regard there is not only movement all through the centuries, there is also what German historian Reinhart Koselleck calls 'structural repetition' (*Wiederholungsstrukturen*).[2] This term refers to the unity of new beginnings and the persistence of established patterns in social relations. However, Koselleck can be read as saying that 'in every repetition there is rupture and therefore the possibility of something new'.[3] This is our starting point when we look at the linkage between the justification of war, its contestation, and the formation of international order.

Political, legal, and other *Realists* usually address the issue the other way around. They tend to claim that it is the structure of the international system which steers the decision on, and the role of, the use of force. From this view point, the absence of a monopoly of legitimate power as such (*Neo-Realism*),[4] or together with human nature (Morgenthau's *'classical' Realism*),[5] sets in motion a socialization process in which states learn that, under the given 'anarchy', they only have the choice either to enhance

[1] Hoffmann and Franzel, 'Introduction: Translating Koselleck', in R. Koselleck, *Sediments of Time. On Possible Histories*, translated and edited by S. Franzel and S.-L. Hoffmann (2018) ix.

[2] Koselleck, 'Wiederholungsstrukturen in Sprache und Geschichte', 57 *Saeculum* 1; see also R. Koselleck, *Begriffsgeschichten. Studien zur Semantik und Pragmatik der politischen und sozialen Sprache* (2006).

[3] Hoffmann and Franzel, above note 1.

[4] K.N. Waltz, *Theory of International Politics* (1979).

[5] H.J. Morgenthau, *Politics among Nations. The Struggle for Power and Peace* (1948).

Lothar Brock and Hendrik Simon, *Justifications of the Use of Force as Constitutive Elements of World Order—Points of Departure, Arrivals, and Moving Destinations* In: *The Justification of War and International Order*. Edited by: Lothar Brock and Hendrik Simon, Oxford University Press (2021). © the many contributors. DOI: 10.1093/oso/9780198865308.003.0028

their own power or to falter and remain string puppies in a power game run by others. While this view has been challenged all along within the social sciences, the end of the Cold War opened up a chance for alternative approaches to gain the upper hand.[6]

The quintessential response to *Realism* was formulated by Alexander Wendt at the beginning of a wave of *social constructivist* variations of Wendt's theme: 'Anarchy is what states make of it.'[7] Wendt (and *Social Constructivism*) did not reject the notion of 'anarchy' as signifying a space *sui generis* in which international relations and international law operate. What he insisted on was that this fact does not explain collective behaviour at the international level. Whether there is peace or war makes quite a difference and what makes for this difference cannot be inferred from anarchy as such but only from the politics pursued under it. This observation allows the conclusion that under 'anarchy', in principle, normative considerations and claims are as important as factors influencing the behaviour of actors as they are in a well-ordered state, though they come to bear in quite different ways. As Bear Braumoeller has stated in his recent study on peace and war: 'International political orders are, generally speaking, a lot like domestic political orders: They are an attempt to create a set of rules that everyone can live by and in so doing escape the dangers of a lawless, anarchical world.'[8]

That international political orders in this sense are a lot like domestic political orders is what makes international law (and other international norms) possible—at least in the eyes of those who assume that international law goes beyond what Ernst Fraenkel, a lawyer with an education in political science (and the founding father of political science in Germany after the Second World War), used to tell his students about international law: that it was like the music lessons in (German) primary schools—nice but not essential in the canon of things you needed to know in order to understand what was going on and what to do about it. Fraenkel was right, of course, at least in part. Depending on when and how it is invoked, reference to international law may have something ceremonial to it, something elevating like the intonation of Friedrich Schiller's '*Ode an die Freude*' in Beethoven's Symphony No. 9. In this regard, international law is nice, indeed. It can be used *to do* nice things with it, like enhancing rule-governed behaviour. But, of course, international law can also be used to do not so nice things with it, like masking arbitrariness as the rule of law.[9] Yet, whatever is being done with international law, as long as it seems worthwhile to political actors to indulge in this kind of game, it does not make much sense to ask whether international law is *really* law and as such matters. It only makes sense to ask in which way and with which results international law becomes the object of political struggle and at the

[6] Legro and Moravcsik, 'Is Anybody still a Realist?', 24 *International Security* (1999) 5.

[7] Wendt, 'Anarchy Is What States Make of It: The Social Construction of Power Politics', 46 *International Organization* (1992) 391.

[8] B. Braumoeller, *Only the Dead: The Persistence of War in the Modern Age* (2019), at 145; on the fundamental problem of force and order, see H. Kelsen, *Peace through Law* (1944), and our introductory chapter.

[9] See also I. Hurd, *How to Do Things with International Law* (2017); our introductory chapter, and below. A good example for the dual role of international law in politics is Hugo Grotius, who on the one hand can be assigned to the tradition of the *rule of law*, but on the other hand used his interpretations of law to legitimize violence and colonialism, see M.J. van Ittersum, *Profit and Principle: Hugo Grotius, Natural Rights Theories and the Rise of Dutch Power in the East Indies (1595–1615)* (2006); Koskenniemi, 'Imagining the Rule of Law: Rereading the Grotian "Tradition"', 30 *European Journal of International Law* (2019) 17; see also the chapters by Lang, Jr and Mallavarapu in this volume.

same time interferes with this struggle. As we read it, the latter is the unique feature of *all* law: it is made to be used and at the same time affects the way it is being used and it may even shape the identity of the user as a political actor.[10] This is our third point of departure in the attempt to reflect on the justification of war as it relates to the formation of international order.

2. Arrivals: The Differentiation of Justifications

When we state that those using force all through history have felt the need to justify what they were doing, we also refute the claim that early modernity witnessed a transformation of the discourse on war and international law which was consolidated in the nineteenth century and which allowed states to wage war without reference to any normative justifications.[11] The contributions to the present volume on early modernity and the nineteenth century clearly demonstrate that this claim is definitely wrong. Neither has a 'non-discriminatory concept' of war (*nicht-diskriminierender Kriegsbegriff*)[12] been accepted in early modern political practice or legal doctrine supposedly leaving just-war thinking as atavistic behind (Tischer in this volume), nor has a 'free right to go to war' prevailed in the nineteenth century political and legal discourse (Simon, Benton, and Hull in this volume).[13]

As Anuschka Tischer aptly observes in her contribution,[14] between 1492 and 1792 two narratives (which in turn were divided into many sub-justifications) were central in the European war discourse: in early modernity, war was considered 'just' if responding to a breach of law or—closely related to this—as an act of self-defence against an (unjustified) aggression. These two narratives were to remain central to the history of war justifications from past to present. With the birth of the modern discourse of war and international order in the long nineteenth century (Part III), the reasons for a legitimate war were defined much more narrowly. Unlike in early modern times now only self-defence against a direct (or imminent) aggression was considered a legitimate reason for war (Simon, Hull, and Genell/Aksakal in this volume). At the same time, however, intervention became the instrument of law enforcement for the international community (Benton and Simon in this volume). The Vienna Congress also

[10] P.J. Katzenstein (ed.), *The Culture Of National Security: Norms and Identity in World Politics* (1996); M. Koskenniemi, *From Apology to Utopia. The Structure of International Legal Argument* (2005).

[11] Q. Wright, *A Study of War* (2nd ed., 1965), at 685ff; W.G. Grewe, *The Epochs of International Law*, translated and revised by Michael Byers (2000), at 531; M. Koskenniemi, *The Gentle Civilizer of Nations. The Rise and Fall of International Law 1870–1960* (2002), at 86; S.C. Neff, *War and the Law of Nations: A General History* (2005), at 197; B. Buzan and G. Lawson, *The Global Transformation: History, Modernity and the Making of International Relations* (2015), at 86; C. Chinkin and M. Kaldor, *International Law and New Wars* (2017), at 133; for more literature see also Hendrik Simon's chapter in this volume, and particularly H. Simon, *Der Mythos vom 'freien Recht zum Krieg'. Zu einer Genealogie der modernen Kriegslegitimation* (forthcoming), chs 1 and 2.

[12] C. Schmitt, *Die Wendung zum diskriminierenden Kriegsbegriff* (2nd ed., 1988); see also C. Schmitt, *Der Nomos der Erde im Völkerrecht des Jus Publicum Europaeum* (1950), at 92.

[13] See also Simon, 'The Myth of *Liberum Ius ad Bellum*. Justifying War in 19th-Century Legal Theory and Political Practice', 29 *European Journal of International Law* (2018) 113.

[14] See also A. Tischer, *Offizielle Kriegsbegründungen in der Frühen Neuzeit: Herrscherkommunikation in Europa zwischen Souveränität und korporativem Selbstverständnis* (2012).

set in motion the modern process of embedding self-defence in international institutions and, later on, positive international law which specified the meaning of a breach of law and opened up the way to collective security and even to law enforcement. This development became manifest with the Hague Peace Conferences, the League of Nations, the Kellogg-Briand Pact, and the United Nations.

Thus, the assumption that, in certain periods of pre-modern und modern European history, there was no need for politicians to refer to norms in their decision-making on the use of force is untenable. Konrad Repgen is right when he states with reference to early modern war manifestos (*Kriegsmanifeste*) that it 'was generally agreed among the people of old Europe that only a just war should be fought'.[15] Of course, Machiavelli denied any need for the justification of the use of force claiming that what was necessary, was just.[16] But Machiavelli's attempt to systematize the *arbitrary* use of force as a mode of rule in itself constituted a kind of normative project for establishing and maintaining (domestic) political order under non-ideal conditions—that is in the absence of a functioning monopoly of legitimate power.[17] So Machiavelli, too, cannot entirely escape the interrelationship between power and norms.

Even the assumption that European states did not bother to justify the use of force in the context of colonialism and imperialism needs to be differentiated. This is shown by Arnulf Becker and Lauren Benton in their contributions to the present volume: with a view to the early modern Spanish-Amerindian encounter (Becker) and nineteenth-century (justificatory) strategies of British navy captains in one Pacific region (Benton) both authors suggest that providing for public order in the periphery of the European imperial system included justifying the use of force. This did not simply enforce the distinction between a 'civilized' Self and a 'barbarian' Other (see Mallavarapu and Chimni's chapter on humanitarianism in this volume). To paraphrase Arnulf Becker, reference to law in the colonial context provided a continuum between war and peace with plenty of space for European violence, with some space for resistance and with a certain, although limited, space for coexistence.

Contrary to a still prevailing perception,[18] the justification of the use of force was an important issue in nineteenth-century European political discourse and political decision-making. Jochen von Bernstorff has suggested to distinguish between order-related and ontological justifications of the use of force. He attributes order-related justifications (claiming to enforce international law) to the Anglo-American world whereas ontological justifications (understood as the 'right to self-preservation' of

[15] '*Es war für die Menschen in Alteuropa ziemlich unbestritten, daß nur ein gerechter Krieg geführt werden dürfe.*', Repgen, 'Kriegslegitimationen in Alteuropa. Entwurf einer historischen Typologie', 9 *Schriften des Historischen Kollegs. Vorträge* (1985), at 10. This issue has lately been taken up by Hathaway et al. in a partly quantitative study of war manifestos from the late fifteenth to the mid-twentieth century. The interesting study basically confirms the earlier findings of those who pioneered this kind of research and whose work is further specified in Parts II and III in this volume, see Hathaway et al., 'War Manifestos', 85 *University of Chicago Law Review* (2018) 1139.

[16] I. Brownlie, *International Law and the Use of Force by States* (1963), at 11.

[17] H. Münkler, *Machiavelli. Die Begründung des politischen Denkens der Neuzeit aus der Krise der Republik Florenz* (1984), at 284; Roth-Isigkeit, 'Niccolò Machiavelli's International Legal Thought—Culture, Contingency and Construction', in S. Kadelbach, T. Kleinlein, and D. Roth-Isigkeit (eds), *System, Order and International Law—The Early History of International Legal Thought* (2017) 19.

[18] For a critical examination of this perception, see Simon, above notes 11 and 13.

a nation) in his view are prevalent in the so-called late nation states, Germany and Italy in particular. According to von Bernstorff, this distinction can help to 'explain and contextualize the divergent justifications advanced by Germany and the Allies for starting and entering World War I and the toxic debate on war guilt and reparations in the interwar period'.[19]

Hendrik Simon follows a different path. He focuses on the bi-furcation of political and legal discourse on the use of force between those who picked up the ideas of the Enlightenment on organizing peace and those adhering to the anti-Enlightenment ideas about nation building as envisaged by Fichte and (according to conservative interpretations) by Hegel.[20] The former called for adapting international law to the historic necessities of world peace.[21] The latter (encouraged by the *Clausewitzian* understanding of war as the continuation of politics with extended means) confronted Kantian ideas of peace with what they considered to be the necessities of war. By the advent of the First World War, they formed a vast majority in German international legal politics. This is where Isabel Hull comes in: in her contribution to the present volume she proceeds on the assumption that the rules of the post-Napoleonic Concert which were directed against wars of aggression and conquest, according to subsequent state practice had been accepted in Europe as 'obligatory—this is as law'. This made it increasingly difficult for states to justify the use of force in intra-European conflicts. Against this background Germany's attempts to justify going to war (and violating the neutrality of Belgium and Luxemburg at the outset) as pre-emption, combined with the invocation of the 'necessities of war', did not find a relevant European audience which would have accepted these attempts to cover up aggression as genuine self-defence.[22]

Under this perspective, the First World War was a war over the viability of the normative order which had emerged since 1814/15, when, as a response to the French Revolution, the 'communication communities' of the anti-revolutionary Coalition Wars turned into the *Concert of the European Great Powers*.[23] This was not a new balance of power-system succeeding the system established in the Peace of Utrecht (1713) (Benno Teschke in this volume). Instead, for the first time a functioning international institution was created, which has been labelled the nineteenth-century 'Security Council'.[24] This international institution operated between might and right: it was

[19] von Bernstorff, 'The Use of Force in International Law before World War I: On Imperial Ordering and the Ontology of the Nation-State', 29 *European Journal of International Law* (2018) 230, at 236 f.

[20] On the bi-furcation of the debate in Germany see Simon, above notes 11 and 13. See also Simon in this volume.

[21] That especially in Germany there was no will to move towards an international law which would match the exigencies of the global community of nations was seen by Schücking as a tragedy, see Schlichtmann, 'Walther Schücking (1875–1935)—Völkerrechtler, Pazifist und Parlamentarier', 15 *Historische Mitteilungen der Ranke-Gesellschaft* (2002) 129, at 143.

[22] Ibid.;I.V. Hull, *A Scrap of Paper. Making and Breaking International Law during the Great War* (2014); Lesaffer, 'Aggression before Versailles', 29 *European Journal of International Law* (2018) 773.

[23] P.W. Schroeder, *The Transformation of European Politics, 1763–1848* (1994); M. Schulz, *Normen und Praxis. Das europäische Konzert der Großmächte als Sicherheitsrat, 1815–1860* (2009); see also H. Müller and C. Rauch (eds), *Great Power Multilateralism and the Prevention of War. Debating a 21st Century Concert of Powers* (2018).

[24] Schulz, above note 23; critically A. Zamoyski, *Rites of Peace: The Fall of Napoleon and the Congress of Vienna* (2007), at 554; and Fahrmeir, 'The Dark Side of the European Concert of Powers. Caveats to be

characterized by the invocation of normative principles[25] which were enforced against smaller powers and non-European polities (see also Benton's contribution). This demonstrated the Janus face of the *Concert*: while the legitimacy of war justifications was evaluated by the Great Powers, the *Concert* itself lacked, as James Lorimer put it in 1883, 'an element of self-control, on which even its external actions are dependent'.[26] The anti-revolutionary principles of this order—moderation and compromise as diplomatic bases of international peace—got certainly out of tune with the rampant nationalism of the late nineteenth century. However, the basic idea underlying the *Concert*—hedging the arbitrary use of force through institutionalized multilateral counsel—under different historical premises was carried into the twentieth century.[27]

The Covenant of the League of Nations did not meet the ambitions of those who had hoped for a strong answer to nationalism and militarism in Europe and beyond. Nevertheless the establishment of the League of Nations[28] in connection with the preceding Hague Peace Conferences[29] and the subsequent Kellogg-Briand Pact paid tribute to the idea of fostering peace through legalization and international organization.[30] They did so, *firstly*, by upgrading civil conflict resolution; *secondly*, by formally outlawing aggression; and *thirdly*, by tying the use of force to a formalized concept of collective security which included the idea of arms control. The latter referred particularistic interests to universal needs in an 'anarchical society' as later on conceptualized by the English School.[31] This implied that the justification of the use of force as self-defence increasingly had to respond to the requirement of international stability.

However, while these re-orientations in principle further challenged the *Clausewitzian* understanding of war as politics by extended means, the means of war themselves were extended: the invention of new weapons all through the nineteenth century culminated (for the time being) in making poisonous gas available as a weapon and in introducing aerial warfare as a new dimension of forcing one's will on the opponent or even on redefining the opponent (as an object of civilizing force). Surprisingly, the use of gas met with 'loud silence' in the nationalist European legal doctrine, as Miloš Vec brings out in his chapter. However, both, gas and aerial warfare opened up new arenas for the justification of the use of force—the introduction of gas with reference to the scope of admissible retaliation and the introduction of

Taken into Account for Successfully Managing Peace', in Müller and Rauch, above note 23; see also Simon in this volume.

[25] Vec, 'Principles in 19th Century International Law Doctrine', in L. Nuzzo and M. Vec (eds), *Constructing International Law—The Birth of a Discipline* (2012) 209.

[26] J. Lorimer, *The Institutes of the Law of Nations: A Treatise of the Jural Relations of Separate Political Communities*, vol. 1 (1883), at 57.

[27] M.M. Payk, *Frieden durch Recht? Der Aufstieg des modernen Völkerrechts und der Friedensschluss nach dem Ersten Weltkrieg* (2018).

[28] Ibid.

[29] M. Abbenhuis, C. E. Barber, and A. R. Higgins (eds), *War, Peace and International Order? The Legacies of the Hague Conferences of 1899 and 1907* (2017); J. Dülffer, *Regeln gegen den Krieg? Die Haager Friedenskonferenzen von 1899 und 1907 in der internationalen Politik* (1981).

[30] O.A. Hathaway and S.J. Shapiro, *The Internationalists: How a Radical Plan to Outlaw War Remade the World* (2017); B. Roscher, *Der Briand-Kellogg-Pakt von 1928. Der 'Verzicht auf den Krieg als Mittel nationaler Politik' im völkerrechtlichen Denken der Zwischenkriegszeit* (2004).

[31] H. Bull, *The Anarchical Society. A Study of Order in World Politics* (1977).

aerial warfare with reference to the possibility of redefining and economizing the use of force as a kind of police power ('police bombing') (Thomas Hippler in this volume). In this respect, aerial warfare in a strange way mixed with ideas about creating a global police navy as part of the organization of peace as proposed by Dutch lawyer and diplomat Cornelis van Vollenhoven.[32]

Meanwhile, the European 'semi-periphery' in the form of the Ottoman Empire was struggling hard to come to its own conclusions from its experience with the European theory and practice of justifying the use of force. In light of its nineteenth-century experience and of the Italian, British, and French methods of securing for themselves parts of the crumbling Ottoman Empire in the years before the First World War, the *High Porte*, at the beginning of the First World War, decided to go for *emancipation* from European machinations instead of for further *adaptation*. It did so by invoking self-defence. In this respect, siding with the Germans seemed their only option (Genell and Aksakal in the present volume). In Germany itself, after a long struggle over finding its place in the post-First World War world, the finally victorious National Socialists decided not to go for *emancipation* from European legal traditions but rather for *domination* of the European political order. This was done in the guise of self-defence under which National Socialists pursued a '*völkisch*' perspective on international law claiming a German right to attain an adequate '*Lebensraum*' (Lange in this volume). This claim went beyond the political revision of the European order established in Versailles though revisionism, in the 1930s, continued under the National Socialists driven in part by those who focused on regaining the former colonies. In contrast to the political elites in Imperial Germany who at the end of the nineteenth century seemed hardly able to speak 'the same legal language as the rest of Europe',[33] the pro-colonial revisionists in an increasingly professional manner now referred to international law to justify the restoration of German colonial rule. However, they blended revisionist positions in the use of force with a *völkisch* gaze which, as Lange observes, became more and more dominant in the context of the Second Word War.

Remarkably, the breakdown of the international order embodied in the League of Nations in the 1930s did not put an end to the ideas of collective security, the pacific settlement of disputes, and the prohibition of wars of aggression. Instead of discarding the idea of fusing power and principle under institutionalized multilateral cooperation, the United States and the United Kingdom agreed to advance the organization of peace by reassembling the broken pieces of the League and turn them into a more resilient international body politic. Of course, the decision to continue on the path of international organization, on which they were joined later on by the Soviet Union, was in the calculated interest of the individual members of the war alliance. However, the important point to be made here is that the Allies were convinced that their interests could best be served under a new general international organization based on power *and* principle.

[32] Van Vallenhoven was calling for a global naval police to help 'organize for peace the whole society of mankind'. Cornelis van Vollenhoven, *Grotius and Geneva. Lectures deliver at Columbia University* (1925/26), at 44; see also Schlichtmann, above note 21, at 146.

[33] Hull, above note 22, at 331.

Thus the UN Charter provided for the formal representation of the war coalition in the Security Council and prohibited categorically the unilateral use of force (see also Marauhn in this volume). For this reason, particularly in liberal historiographies and narratives of progress, the UN Charter is usually perceived as a high peak in international attempts to outlaw war. In the words of Randall Lesaffer: the 'UN Charter stands at the end of an evolution by which the rights of states to use force was progressively limited'.[34] Evidence for this may be seen in the fact that after the general prohibition of the use of force in 1945, the practice of formal war declarations has actually perished. Over the past seventy years, in thirty-six inter-state military encounters there have been only two formal declarations of war.[35] However, the question of the legitimacy of the use of force is by no means resolved: *de facto*, it remains highly contested—even though or rather because, since the end of the Second World War, classical inter-state war has become less frequent in comparison to 'intervention', 'police action', or 'peacekeeping'.[36] This is not as new as many like to think. It was also a well-known strategy even before 1945 to justify one's use of force not as unilateral war but rather as the use of force on behalf of the norms of the international community. In other words, the UN Charter certainly represents an important milestone in the legalization of international relations. But the practical effects of legalization are more limited than the conceptual progress suggests.

Furthermore, the enduring possibility to justify even the unilateral use of force after 1945 is already visible in the UN Charter itself: while Article 2(4) prohibits the use of force in general when it contradicts the aims of the Charter, Article 51 confirmed the 'inherent right' of each state to self-defence in the case of an armed attack. Yet, Article 51 was placed under Chapter VII UN Charter. In line with this stipulation the right to self-defence can only be claimed until the Security Council 'has taken measures necessary to maintain international peace and security'. This way, elements of the just-war tradition and natural law thinking were tied into a formal system of rules for dealing with conflict. The fact that self-defence was now legitimized by natural law and regulated by positive law indicates that multi-normativity remains alive even in the UN Charter. The persistence of multi-normativity (understood as a statement of fact) can be explained by the need for flexibility in coping with conflict in an 'anarchical society'.[37] But it can also be seen as pointing to loopholes in the system of collective peace-making which can be used to justify unilateral action.

Crucially, the Charter, instead of counterbalancing disruptive interests through adequate procedural provisions for collective decision-making, brought to bear the existing power disparities through the veto power of the Permanent Five. This facilitated

[34] Lesaffer, 'Too Much History: From War as Sanction to the Sanctioning of War', in M. Weller (ed.), *The Oxford Handbook of the Use of Force in International Law* (2015) 35, at 35; see also Lesaffer, above note 22; See also Peevers, 'Liberal Internationalism, Radical Transformation and the Making of World Orders', 29 *European Journal of International Law* (2018) 303.

[35] T.M. Fazal, *Wars of Law: Unintended Consequences in the Regulation of Armed Conflict* (2018), at 1; see also Fazal, 'Why States No Longer Declare War', 21 *Security Studies* (2012) 557; Hurd, 'The Permissive Power of the Ban on War', 2 *European Journal of International Security* (2016) 1; Hathaway and Shapiro, above note 30; Hathaway et al., above note 15.

[36] See also Fazal, *Wars of Law*, above note 35.

[37] Bull, above note 31.

the establishment of the UN, but it impaired its functioning as became obvious during the Cold War.[38] What is more, it remained a problem after the Cold War ended.

With a view to the 'new wars'[39] fought in 'fragile states'[40] or in former socialist countries over the reorganization of political authority after the end of the Cold War and over the structural deficiencies of the international division of labour, the protection of people from mass atrocities became a major issue of world politics. Again, these interventions are not novel per se (see also Becker's, Benton's, and Jahn's chapters): already in early modern times and particularly in the nineteenth century, wars were waged with reference to the justification narrative of protecting fellow Christians and compatriots but also of 'saving strangers'[41] as discussed in contemporary theoretical discourse.[42] The practice of interventions was closely tied up with colonialism and imperialism.[43] During the Cold War, too, unilateral interventions were justified with humanitarian concerns.[44] What was new about the situation after 1990, was the fact that the justification of interventions referred systematically to human rights with the United Nations serving temporarily as the central international authority on dealing with the new[45]—or rather: renewed—military interventionism.

The role of the UN, as a frame of reference and an actor in this context, was enhanced in two ways. *Firstly*, the Security Council agreed on strengthening the capacity of the United Nations to cope with violent conflict in the interest of the people affected. The result was a remarkable extension of the peacekeeping activities of the UN both conceptually and in numbers. *Secondly*, the Security Council on various occasions defined the violation of minority rights (Iraq 1991), the gross violation of human rights (Somalia 1992), the breakdown of public order (Bosnia 1995), and even the violent disruption of democratization (Haiti 1994) as threats to international peace, as it had done before in sanctioning Apartheid in South Africa and in enforcing the decolonization of Rhodesia. This move provided the possibility to turn humanitarian intervention into a collective security measure in accordance with Chapter VII of the UN Charter.[46]

While the UN became the central collective actor in peacekeeping and peacemaking, regional organizations also expanded their respective activities.[47] In her

[38] C. Peevers, *The Politics of Justifying Force: the Suez Crisis, the Iraq War, and International Law* (2013).

[39] M. Kaldor, *New & Old Wars: Organized Violence in a Global Era* (1999); H. Münkler, *Die neuen Kriege* (2002); Chinkin and Kaldor, above note 11.

[40] L. Brock, H.H. Holm, G. Soerensen, and M. Stohl, *Fragile States. Violence and the Failure of Intervention* (2013).

[41] N.J. Wheeler, *Saving Strangers: Humanitarian Intervention in International Society* (2000); D.J.B. Trim and B. Simms (eds), *Humanitarian Intervention: A History* (2011).

[42] S. Recchia and J.M. Welsh (eds), *Just and Unjust Military Intervention: European Thinkers from Vitoria to Mill* (2013).

[43] F. Klose, *'In the Cause of Humanity': Eine Geschichte der humanitären Intervention* (2019).

[44] Franck and Rodney, 'After Bangladesh: The Law of Humanitarian Intervention by Military Force', 67 *American Journal of International Law* (1973) 275.

[45] T. Debiel and F. Nuscheler (eds), *Der neue Interventionismus. Humanitäre Einmischung zwischen Anspruch und Wirklichkeit* (1996); Glennon, 'The New Interventionism. The Search for a Just International Law', 78 *Foreign Affairs* (1999) 2.

[46] Ronzitti, 'The Current Status of Legal Principles Prohibiting the Use of Force and Legal Justifications of the Use of Force', in M. Bothe, M.E. O'Connell, and N. Ronzitti (eds), *Redefining Sovereignty. The Use of Force after the Cold War* (2005) 91.

[47] N. Wilén, *Justifying Interventions in Africa. (De)Stabilizing Sovereignty in Liberia, Burundi and the Congo* (2012).

chapter, Nina Wilén explicates how, in 1990, ECOWAS in Liberia created a precedent for regional claims of competence for dealing with issues of peace and security. This claim was tacitly accepted by the international community which suggests that a certain pragmatism can help ease the latent tension between Chapters VII and VIII of the UN Charter. Nevertheless, as Wilén points out, it remains an open question in whose name regional organizations are intervening.

This question is even more important with regard to the (unilateral) interventions of individual states or groups of states (alliances and 'coalitions of the willing'). This type of interventions became a major issue from the mid-1990s onward when collective decisions-making in the Security Council began to stall (during and after the war in Bosnia and later on in the US-led 'war on terror' [see also Stohl in this volume]). NATO decided to intervene on its own in the case of Kosovo in 1999, as a response to the (anticipated) blockage of decision-making in the Security Council. At the same time, the United States (already under the presidency of Bill Clinton) resumed its bombing campaigns against Iraq claiming that Iraq was in material breach of Security Council Resolutions 678 and 687 of 1991. In 2003 the Bush administration and its coalition of the willing started an outright war against the regime of Saddam Hussein i.a. with the justification that the regime remained in material breach of UN Security Council resolutions, this time Resolution 1441 of 2002. Though the United States claimed to follow up on Security Council resolutions, the Security Council itself did not endorse this action.[48] In its National Security Strategy of 2002, the US government even purported to have a right to 'pre-emptive' self-defence which would allow the use force in order to avert the emergence of dangers to the national security of the United States.[49] The United States also claimed to have a right to intervene militarily in a country that was harbouring terrorist groups—an opinion which was tolerated by the UN Security Council in the case of the US response to the terrorist attacks of September 2001, but which helped to extend the meaning of self-defence and to undermine the concept of collective action as envisaged in Chapter VII UN Charter.[50] As a consequence, the focal point of justifying the use of force moved from collective security back to self-defence with the meaning of self-defence being expanded even to a point at which it could 'kill article 2(4)'.[51]

With the spreading of transnational terrorism, national security concerns blended with liberal cosmopolitan justifications of the use of force (Anna Geis and Wolfgang Wagner in this volume). New narratives of self-defence came up which reflected not

[48] Brock, 'The Use of Force in the Post-Cold War Era. From Collective Action Back to Pre-Charter Self Defence?', in Bothe et al., above note 46, at 34 f.

[49] Bothe, 'Terrorism and the Legality of Pre-emptive Force', 14 *European Journal of International Law* (2003) 227; O'Connell, 'The Myth of Pre-emptive Self-defence', *ASIL Presidential Task Force on Terrorism* (2002), at 17.

[50] Marauhn, Stohl, and Geis/Wagner in this volume. See also Byers, 'Terrorism, the Use of Force and International Law after 11 September', 51 *International and Comparative Law Quarterly* (2002) 401.

[51] Marauhn in this volume with reference to Franck, 'What Happens Now? The United Nations After Iraq', 97 *American Journal of International Law* (2003) 607. See also *US National Security Strategy 2002*; O'Connell and Welling Hall, 'Report from Columbus. Discussions on the Future of Law and Institutions on the Use of Force in a One Superpower Wold', in Bothe et al., above note 46, at 441, 446 f.; see also M. Byers (ed.), *The Role of Law in International Politics* (2000); Thilo Marauhn in this volume points to the withdrawal by the US administration of the extended meaning of 'pre-emption'/'prevention' and its return to the 'clear and present danger' formula.

only the political discourse in the hallways of High Politics but also the experience of the soldiers on the ground in Afghanistan (Michael Stohl and Axel Heck/Gabi Schlag in this volume). Self-defence as the traditional justification of the use of force was now re-invented as intervention for a new international order which would reflect the normative assets of the liberal democracies as cosmopolitan actors (Beate Jahn in this volume; see also Mallavarapu and Chimni in Part VII). In the words of Anna Geis and Wolfgang Wagner in their contribution:

> Core principles of liberal thought such as market economy, promotion of democracy, and human rights migrated into numerous international organizations and institutions beyond the liberal 'core' and thus helped to stabilize the global hegemony of 'the West'.

As Lothar Brock argues elsewhere, this development reflected a basic predicament of the UN which deepened with the breakdown of Real Socialism: the end of the Cold War did indeed enhance collective decision-making in the Security Council. Unfortunately, this effect did not last. Nevertheless, human rights and domestic order were further upgraded as objects of international concern. The normative basis for this 'new humanitarianism'[52] was laid by redefining sovereignty as responsibility of national governments vis-à-vis its own population and vis-à-vis the international community.[53] This meant that the international community was put in charge of checking on whether the member states of the UN were living up to an expanding scope of substantive norms (human rights, democracy, good governance, later on also engagement in anti-terrorism). Yet, there was no agreement on matching procedural innovations in the UN system that would mitigate the imbalances existing in the system between the veto powers of the Security Council on the one end of the spectrum, the underdogs of world society on the other. The new humanitarianism deepened the gap between the substantive norms to which all states had to adhere and the procedural provisions of the UN Charter for following up on these norms.[54] In the absence of a robust consensus among the Permanent Members of the Security Council on how to deal with conflict (after 1995), the 'new humanitarianism' helped the liberal democracies to function as 'hyper sovereign' states[55] which saw themselves as custodians of the substantive norms on which the 'empire of humanitarianism' was to be built—if necessary by the unilateral use of force according to just-war criteria.[56]

[52] Barnett, 'Humanitarianism Transformed', 3 *Perspectives on Politics* (2005) 723; see also Radice, 'Humanity in International Political Theory: Chris Brown and the Principles, Politics and Practice of Humanitarianism', in A. Lang, Jr and M. Albert (eds), *The Politics of International Political Theory. Reflections on the Works of Chris Brown* (2018) 69; A. Fischer-Lescano, *Globalverfassung. Die Geltungsbegründung der Menschenrechte im postmodernen ius gentium* (2005). For strong critique of the new humanitarianism as another camouflage of imperialism see also Chimni's chapter on humanitarianism in Part VII of this volume.

[53] F.M. Deng, S. Kimaro, T. Lyons, D. Rothchild, and I.W. Zartman, *Sovereignty as Responsibility. Conflict Management in Africa* (1996); see also Bothe et al., above note 46.

[54] Brock, 'Normative Integration und kollektive Handlungskompetenz auf internationaler Ebene', 6 *Zeitschrift für Internationale Beziehungen* (1999) 323.

[55] O'Connell, 'American Hyper-Sovereignty from Kosovo to the "Global War on Terror"', in Bothe et al., above note 46; see also Mallavarapu and Chimni in Part VII.

[56] M.N. Barnett, *Empire of Humanity: A History of Humanitarianism* (2011).

On top of this, the US response to '9/11' in the form of 'Operation Enduring Freedom' in Afghanistan was interpreted by some Western observers as a 'constitutional moment' for introducing 'civilian inviolability' as 'a globally acceptable Grundnorm' of international law.[57] This went far beyond the 'Responsibility to Protect' as accepted by the UN reform summit in 2005. According to Andreas Fischer-Lescano, the argumentation by Anne Marie Slaughter and William Burke-White amounted to a project 'where global aristocratic networks of governments could violently implement their understanding of human rights and security against the prevailing practices of other nations in every society of this world. In this case the entire world population would become the object of the military enforcement of security principles.'[58]

Summing up, we observe a differentiation of justifications for the use of force especially since the end of the Cold War which, again, widens the justification narratives referred to by Anuschka Tischer (above). In the European communication community of early modern times, there prevailed a broad understanding of what constituted a breach of right. Accordingly, justifications were relatively easy to construct: apart from self-defence, several breaches of law could be claimed to justify the resort to arms. In the nineteenth century self-defence (also against an imminent threat of aggression) remained the only legitimate reason for the unilateral use of force while in practice, however, competing justifications such as intervention or measures short of war could be used as well. Legalizations in the late nineteenth and the twentieth centuries up to the UN Charter confirmed self-defence as the most important narrative of justification but specified the breach of right which justified the use of force. The UN Charter reduces the admissible use of force to collective enforcement action (Chapter VII). At the same time it confirms the right to self-defence but attaches procedural strings to it. Nevertheless the recent justifications of international violence refer to an expansion of the notion of self-defence as problematized by Marauhn in this volume. The expansion of the notion of self-defence goes hand in hand with the amplification of substantive norms for good governance as legitimate concerns of the 'international community'. These substantive norms have been used to legitimate the unilateral use of force when consensus among the Permanent Members of the Security Council was not forthcoming. This is to say that the differentiation of justifications, while aiming at containment of arbitrariness in the use of force, also gives arbitrariness a new face as order politics.[59] The implications of this observation for international order will be discussed in the next section.

[57] Slaughter and Burke-White, 'An International Constitutional Moment', 43 *Harvard International Law Journal* (2003) 1.

[58] Fischer-Lescano, 'Redefining Sovereignty via International Constitutional Moments? The Case of Afghanistan', in Bothe et al., above note 46, at 360; Fischer-Lescano is referring to Koskenniemi, 'The Police in the Temple: Order, Justice and the UN—A Dialectical View', 6 *European Journal of International Law* (1995) 325, at 327, and to Maus, 'Menschenrecht als Ermächtigungsnormen internationaler Politik oder: der zerstörte Zusammenhang von Menschenrechten und Demokratie', in H. Brunkhorst (ed.), *Recht auf Menschenrechte* (1999) 279.

[59] Brock and Simon, 'Die Selbstbehauptung und Selbstgefährdung des Friedens als Herrschaft des Rechts. Eine endlose Karussellfahrt? (The Self-Assertion and Self-Destruction of Peace as Rule of Law. Riding a Merry-Go-Round?)', 59 *Politische Vierteljahresschrift* (2018) 269.

3. Moving Destinations: Commuting Between Sovereign Judgement and a Postcolonial Rule of Law

From the above, it can be inferred that even (or perhaps especially) in the highly political field of the use force political actors are 'justificatory beings' establishing international orders as orders of justification.[60] In such orders of justification states identify criteria for the use of force as internationally 'appropriate behaviour'.[61] This is to say that the specific instances of justifying the use of force refer to arguments which, directly or indirectly, claim meaning beyond the specific cases in which they are being raised. In this sense the communication communities emerging in early modernity around the exchange of war manifestos functioned as generators of international order (Tischer in this volume) as did the various subsequent communication communities evolving in modernity.

Order enhances security of expectations within historically specific constellations of power. Order, in turn, is enhanced by international law and other norms,[62] which reflect the historic constellation of power under which they are being generated and diffused[63]—both in theory and political practice.[64] At the same time international law and other norms stand in for universal concerns to which the dominant powers themselves refer in order to enhance their authority as guardians of international order.[65] If we were to understand the recourse to *principle* as nothing but a function of ubiquitous *power* games (i.e. as mere propaganda), all change would end up in 'more of the same' or 'structural repetition' (if under different constellations of power). International order would then oscillate in endless circles from one form of imperial rule or hegemony to the next.[66] However, as picked up from Koselleck (above), to the degree that repetition operates on the brink of rupture there is the chance of something new emerging.

Such ruptures are commonly associated with wars. Wars play, indeed, a central role in the present volume. But ruptures should not only be identified as dramatic historic events. They also occur in processes of continuous change which foster the emergence of new constellations of interests and power interacting with new ideas, concepts, and standards of adequacy.[67] This is to say that 'ruptures' may occur long before they are being brought to light through events.

[60] R. Forst, *Normativity and Power. Analyzing Social Orders of Justification* (2017); R. Forst and K. Günther (eds), *Die Herausbildung normativer Ordnungen: Interdisziplinäre Perspektiven* (2011).

[61] Jepperson, Wendt, and Katzenstein. 'Norms, Identity, and Culture in National Security', in P.J. Katzenstein (ed.), *The Culture of National Security. Norms and Identity in World Politics* (1996) 33, at 54; Finnemore and Sikkink, 'International Norm Dynamics and Political Change', 52 *International Organization* (1998) 887, at 891.

[62] Byers, above note 51.

[63] Grewe, above note 11, at 531.

[64] On minding and bridging this gap see our introductory chapter, and Simon, 'Theorising Order in the Shadow of War. The Politics of International Legal Knowledge and the Justification of Force in Modernity', in 22 *Journal of the History of International Law* (2020) 218; see also Lang, Jr in this volume.

[65] N. Rengger, *Just War and International Order. The Uncivil Condition in World Politics* (2013); M. Zürn, *A Theory of Global Governance. Authority, Legitimacy, and Contestation* (2018).

[66] U. Menzel, *Die Ordnung der Welt* (2015).

[67] Deitelhoff, 'The Discursive Process of Legalization: Charting Islands of Persuasion in the ICC Case', 63 *International Organization* (2009) 33.

For instance, as Benno Teschke argues, the Treaty of Utrecht was the single most important event in the history of early modern international re-ordering' expressing long-term changes in the constellation of power in Europe and domestic change in the United Kingdom which provided the basis for 'the first institutionalisation of a dynamic grand strategy for the re-organisation of European and colonial order executed by the first capitalist-parliamentarian state, post-1688/1707 Britain'.

The ensuing European order was characterized by the simultaneity of continued dynastic rule, limited wars (*Kabinettskriege*), rivalling colonialism, anti-dynastic and anti-colonial revolutionary aspirations as well as by intellectual attempts to make sense of what was happening and to design an international order based on reason and (scientific) knowledge. In this historical context, Immanuel Kant, for one, explored the interplay of change and repetition in the French Revolution. Kant was inspired by the French Revolution which in his opinion opened up the possibility of a fundamentally new international order. Yet, as Oliver Eberl argues in his contribution to the present volume, he did not gloss over the contradictions of the Revolution as it offered new justifications of the use of force (fighting 'unjust enemies'). According to Eberl, Kant rather accommodated this aspect of the Revolution in his construct of 'eternal peace' by combining the precept of the sovereignty of the people with that of non-intervention and international law backed by international organization and a slim version of world-citizenship.

The European world order that prevailed in the nineteenth century can hardly be seen as a mere extension of the order created in Utrecht, nor as an attempt to translate enlightened thinking like that by Kant into international politics. High Politics responded to the mixture of repetition, rupture, and long-term change in the eighteenth century with the reconstitution of dynastic legitimacy in unison with institutionalized inter-*dynastic* cooperation (the *Concert*). Inter-dynastic cooperation (in precarious concurrence with the British politics of [intra-Continental] non-intervention) helped to avoid war among the Great Powers for a long time. But it allowed 'small' wars on the periphery in the context of imperial ordering (see also Benton, Simon, and Genell/Aksakal in this volume). Furthermore, it was not flexible enough to accommodate changing relations among the member states and changes in the social and economic foundations of the *Concert*. This fostered nationalism and militarism but paradoxically also the idea of peace through law and institutionalized cooperation. Right up to the 'outbreak' of the First World War many contemporaries expected that the tensions inherent in the political order of the nineteenth century should and would be solved by cooperative diplomacy.[68] This expectation was supported by the professionalization of international law[69] as it inter alia appeared in the Hague Peace Conferences. It was repudiated by the war, but reconfirmed after it by the creation of the League as a consequence of the war.[70] Though the learning curve following the war experience remained rather flat as B.S. Chimni observes (in his contribution to Part IV of the present volume) 'the belief in "peace through law" prevailed both among leading

[68] N. Angell, *The Great Illusion: A Study of the Relation of Military Power in Nations to their Economic and Social Advantage* (1911); see also C. Clark, *The Sleepwalkers: How Europe Went to War in 1914* (2012).

[69] Koskenniemi, above note 9; Nuzzo and Vec, above note 25.

[70] Payk, above note 27.

thinkers and international lawyers in the inter-war period'—if sometimes in a paradoxical form. Here Chimni points to Hans Kelsen, the leading figure in this discourse arena, who on the one hand contended that peace can be realistically achieved only through 'a slow and steady perfection of the international legal order'[71] but who at the same time maintained that, under the conditions of anarchy, just war could be needed to keep up the possibility of achieving a peaceful order (see also Daase/Deitelhoff in this volume).

According to Chimni, the underlying premise of Kelsen's thinking was that 'the idea of law, in spite of everything [i.e., the two world wars] still seems to be stronger than any other ideology of power'.[72] Was Kelsen proven right after all when the Allies responded to the Second Word War by founding the United Nations? How far did rupture lead to overcoming mere repetition, how far did it give birth to something new?

The Second World War did not result in a final victory of *principle* over *power*. Instead the UN was to mitigate the tension between power and principle by universalizing sovereignty and combining it with the idea of a more comprehensive system of cooperation among *all* states leaving it open for the time being when the remaining colonies would join the UN. This new arrangement in the form of the *UN system* (including economic, cultural, health, and other organizations) offered a framework of reference for dealing with the tension between power and principle even during the Cold War.[73]

When the Cold War ended, this seemed to open up the political space for moving ahead and for expediting multilateral cooperation through the continuous legalization of international relations[74] and the constitutionalization of international law.[75] At the same time, for the reasons pointed out above, there was the revival of just-war thinking (Brown in the present volume; see also the comments by Lang, Jr on Brown's approach) and multi-normativity remained alive.

Those who believe that international law matters rely on the self-binding effects that go along with reference to law.[76] The proponents of just-war thinking regard such self-binding either as an illusion or as an obstacle to dealing effectively with conflict.[77] In

[71] Kelsen, above note 8, at ix.

[72] Chimni in this volume, at 252.

[73] Marauhn in this volume; Peevers, above note 38.

[74] Abbott, Keohane, Moravcsik, Slaughter, and Snidal, 'The Concept of Legalization', 54 *International Organization* (2000) 401; Deitelhoff, above note 67; H. Krieger, G. Nolte, and A. Zimmermann (eds), *The International Rule of Law: Rise or Decline?* (2019); for (international legal) historical perspectives on 'legalization', see Vec, 'Verrechtlichung internationaler Streitbeilegung im 19. und 20. Jahrhundert? Beobachtungen und Fragen zu den Strukturen völkerrechtlicher Konfliktaustragung', in S. Dauchy and M. Vec (eds), *Les conflits entre peuples. De la résolution libre à la résolution imposée* (2011) 11; Payk, 'Institutionalisierung und Verrechtlichung. Die Geschichte des Völkerrechts im späten 19. und frühen 20. Jahrhundert', 52 *Archiv für Sozialgeschichte* (2012) 861.

[75] Habermas, 'The Constitutionalization of International Law and the Legitimation Problems of a Constitution for World Society', 15 *Constellations* (2009) 444; Bogdandy, 'Constitutionalism in International Law: Comment on a Proposal from Germany', 47 *Harvard International Law Journal* (2006) 223; J. Klabbers, A. Peters, and G. Ulfstein (eds), *The Constitutionalization of International Law* (2009).

[76] F. Kratochwil, *Rules, Norms and Decisions, On the Conditions of Practical and Legal Reasoning in International Relations and Domestic Society* (1991); Risse, '"Let's Argue!": Communicative Action in World Politics', 54 *International Organization* (2000) 1; Brock, 'Frieden durch Recht?', in P. Becker, R. Braun, and D. Deiseroth, *Frieden durch Recht* (2010) 15.

[77] Brown in the present volume and Brown, 'In Response', in Lang, Jr and Albert, above note 52.

the context of the post-Cold War hegemony of the liberal democracies and the present crisis of this hegemony, this leads to two basic controversies: one on the contradiction between sovereign judgement and the rule of law,[78] another on the linkage between international law and imperialism (see Mallavarapu's chapter, and Chimni's chapter in Part VII). Our argument is that these controversies are at the centre of the question as to how the justification of the use of force and international order interact.

As to the first controversy, Chris Brown in his contribution (as before) argues that the Security Council cannot exercise the authority given to it by the charter: '[T]he legal regime governing force is a dead letter.'[79] From this Chris Brown draws the conclusion. '[W]e cannot outsource judgement to international lawyers. We are obliged to decide for ourselves whether a particular use of force is justified or not.' Anthony Lang, Jr in his contribution takes up this point. He observes that two prominent US American authors, Jean Bethke Elshtain and Michael Walzer, who both trust the sovereign judgement of the US administration more than that of the Security Council, came to contradictory positions on the legitimacy of the Iraq war of 2003: Elshtain was in favour, Walzer against. Under Brown's perspective the obvious answer to this would be that there is *good* judgement and there is *not so good* judgement and that it is the task of academic reasoning to distinguish between the two in order to advance good judgement. Nick Rengger has called this 'wishful thinking'.[80] His argument is that there is little evidence that the kinds of conditions that would enhance the possibility of (good) practical judgement exist even in the most advanced states.[81] Agreeing with Rengger, Anthony Lang, Jr writes 'that using judgement cannot be divorced from some deeper structures, structures which often emerge from historical narratives'. This implies in our view that the revival of the just-war tradition has to be seen in the concrete historical situation in which it came about: the post-Cold War struggle of the liberal democracies to free themselves from the strings attached to politics by international law under the UN system in order to be able to enforce the substantial norms advocated by the same liberal democracies (see also above). This is to say that the realm of problems calling for practical judgement expanded while the procedural provisions for dealing with contested judgement were weakened—with the unfortunate consequence of undermining the authority of the UN.

This is not to say that Chris Brown got it all wrong. Not at all: his insistence on practical judgement based on the observation of concrete political circumstances rather than abstract norms and principles is well founded. In agreement with Brown, Beate Jahn in her contribution states that 'the abstraction of moral reasoning from concrete political circumstances leaves the theoretical conception of morality devoid of politics and thus accounts for the tension between them in the debate on humanitarian intervention'. In this regard, practical judgement is part of any attempt to bring about a political order in which the arbitrary use of force is to be effectively reduced. This does

[78] Brown in this volume; Brock, 'Between Sovereign Judgement and the International Rule of Law. The Protection of People from Mass Atrocities', in Lang, Jr and Albert, above note 52.

[79] See also Franck, above note 51; Glennon, 'Why the Security Council failed', 82 *Foreign Affairs* (2003) 16; Marauhn on this issue in his contribution to the present volume.

[80] Rengger, 'Practical Judgement—Inconsistent or Incoherent?', in Lang, Jr and Albert, above note 52, at 63.

[81] Ibid.

not mean, however, that normative endeavours like the International Criminal Court or the Responsibility to Protect (R2P)[82] as such impede practical judgement.

Chris Brown and Beate Jahn are right in their sceptical look at what can be expected from the further legalization of international relations and the constitutionalization of international law.[83] Yet, we should be careful not to dump the baby with the bathwater and to proceed on the assumption that practical judgement can only be effective when it is exercised as sovereign judgment of nation states unhampered by formalized multilateralism. Since all political decisions are debatable, there is also an inevitable need for rules to deal with contested judgement. This is what the rule of law is about. The rule of law is difficult to achieve at any level of social interaction. The chances to succeed in bringing it about and keeping it up are, of course, higher within (well-functioning) states than between states. It may even be impossible to achieve in the absence of a monopoly of power. But even if this should be the case, this does not mean that there is no role for law at the international level. The present state of world affairs can be seen as demonstrating the irrelevance of international law, but it can also be understood as emphasizing the need for improving the function of international law as a normative frame of reference for practical judgement. Could a new understanding of coercion be part of the latter?

In their contribution to the present volume, Christopher Daase and Nicole Deitelhoff suggest a change of focus on sanctions to this avail. Sanctions, as understood by the two authors, should not only be valued under the aspect of changing the behaviour of the sanctioned party, but also under the aspect of confirming a normative order. They argue in their contribution:

> Sanctions demonstrate approval or disapproval of a behaviour with the intention of confirming a certain norm. Therefore, 'sanctioning' can have the meaning of 'disapproval' and 'approval' at the same time, because it is not primarily concerned with the assessment of a particular action, but with the confirmation of the norm underlying that action.

The argument of Daase and Deitelhoff has to be seen in the context of recent research on norms and the effects of their contestation.[84] The focus here is on the fate of normative orders. Contestation may lead to their confirmation and may even strengthen it; but it can also end up in their break-down.

> With the focus on the normative function of sanctions, the aspect of power loses importance.... If the affirmation of a general norm is the main function of sanctions and not the change of a specific behaviour, then the success parameters of sanctions also change. Notes of protest against the treatment of prisoners by the American military

[82] A. Bellamy and T. Dunne (eds), *The Oxford Handbook on the Responsibility to Protect* (2006).

[83] Brown, above note 77, at 249. Jahn in her contribution to this volume.

[84] Kratochwil, above note 76; Jepperson, Wendt, and Katzenstein, above note 61; Finnemore and Sikkink, above note 61; Risse, above note 76; Deitelhoff, above note 67; Brock, above note 76; Zürn, above note 65; A. Wiener, *Contestation and Constitution of Norms in Global International Relations* (2018); see also our introduction.

are not only successful when they change the behaviour of the US military, but also when they express a collective disapproval with reference to general norms.[85]

From this viewpoint it is not a mere play of words when Thilo Marauhn in his contribution comes to the conclusion that the factual contestation of Article 2(4) (which is Thomas Franck's object of concern) has weakened the norm but has not killed it. In other words, we are not at a point at which Carl Schmitt's vision of an indiscriminate notion of war finally trumps the discriminating concept of the use of force under the UN Charter. There is still room for the normative contestation of the use of force. This is fine. But is it enough? Or is it still too much in terms of forcing the minds of others?

This leads us to the second point of contention mentioned above: the linkage between international law and imperialism in discourses on the justification of war and international order. In the present volume both B.S. Chimni and Siddharth Mallavarapu emphasize that this linkage is not a matter of the past but also of the present international order. In the words of Mallavarapu (in this volume) referring to Emanuele Saccarelli and Latha Varadarajan[86]: '... [I]f you want to understand the use of force and justificatory rationales surrounding war in the contemporary epoch, there is really no escape from "a serious engagement with imperialism as an enduring historical reality".' At the same time Siddharth Mallavarapu insists that international law also bears the traces of anti-imperialist resistance. 'It is important not to view the colonized as passive putty in the hand of the colonial master as they fought back to articulate their own agency.'[87] So international law is part of the imperial project and at the same time reflects the emancipatory aspirations of the global South in discourses on hedging war and violence.

Chimni and Mallavarapu both give a voice to these aspirations in the context of postcolonial scholarship as it comes to bear in 'Third World Approaches to International Law' (TWAIL).[88] In following (and specifying) these approaches, Mallavarapu in his contribution focuses on different meanings of sovereignty as the linchpin for understanding the differences between an imperial and an emancipatory understanding of international law. His approach is mostly conceptual, not empirical. As such it helps to problematize and transcend a narrow understanding of sovereignty (as a mere bulwark against interference) by offering a deeper look into international structural or systemic curtailments of self-determination.[89] For instance, the Responsibility to Protect focuses on specific forms of large-scale harm to people without taking into consideration other kinds of mass atrocities—those of colonialism and imperialism (see also Beate Jahn's contribution).Therefore, the involvement of the politics of protection in colonialism and imperialism has to be taken into account in order to enhance practical judgement on how to deal with humanitarian issues. How can that be done? How can 'power and principles [be integrated] to avoid the complicity of humanitarianism

[85] Daase and Deitelhoff in the present volume, at 499.
[86] E. Saccarelli and L. Varadarajan, *Imperialism Past and Present* (2015), at 6.
[87] Mallavarapu in this volume, at 61; see also the contribution by Arnulf Becker.
[88] Chimni, 'Third World Approaches to International Law: A Manifesto', 8 *International Community Law Review* (2006) 3.
[89] See also Galtung's distinction between direct and structural violence, also in the form of cultural violence, see Galtung, 'Cultural Violence', 27 *Journal of Peace Research* (1990) 291.

with the politics of imperialism'? (Chimni in Part VII of this volume) This question is at the centre of Chimni's contribution. Chimni understands 'humanitarianism' as

> an ensemble of cultural, social, ideological, and material practices that have as their ostensible aim alleviation of human suffering or promotion of welfare of individuals, groups and peoples of third nations that assume specific form and content in different phases of history.[90]

This concept 'goes beyond equating humanitarianism with precarity'[91] and it certainly goes beyond equating humanitarianism with the violent protection of people from violence. Yet, as Chimni claims 'a humanitarian act is at all times the carrier of cultural meanings even as it brings material assistance and relief to peoples in distress'. He goes on:

> Within an imperial world order, with us since the sixteenth century, the story of humanitarianism represents a 'narrative of justification' for the exercise of domination over, and infliction of violence on, other peoples and nations.[92]

Historicizing his understanding of humanitarianism, Chimni distinguishes the 'civilizational' humanitarianism of the colonial period of imperialism from a post-Second World War humanitarianism which concentrated on relief operations and was guided by the principle of political neutrality. This was followed by a post-Cold War humanitarianism deriving 'its life and legitimacy from the ideas of "democracy", "human rights", and "good governance" '.[93] Today, according to Chimni, we witness an 'ad hoc humanitarianism' which reflects the rise of populism and nationalism around the world. In sum:

> over time the meaning of 'humanitarianism' has moved from being assigned a civilizational connotation in the colonial era to a constrained scope in the Cold War period to once again being attributed civilizational meaning in the immediate decades of the post-Cold War era to a fragmented approach in recent years amidst rising forces of nationalism and populism.[94]

The bottom line of all this is that 'from the Spanish colonization of Latin America in the sixteenth century to the occupation of Iraq to the intervention in Libya and Syria a central narrative of justification is woven around bringing civilization to the inferior or lagging Other'.[95] This argument brings up two more questions.

Firstly, to what extent does the argument above accommodate the fact that, as TWAIL insists, international law and other normative approaches to international order do not only reflect imperialistic designs but also bear the traces of anti-imperial

[90] Chimni in this volume, at 472.
[91] Ibid.
[92] Ibid., with reference to Barnett, above note 52; see also Becker's and Benton's chapters in this volume.
[93] Ibid., at 473, with reference to Jahn and Geis/Wagner in this volume.
[94] Ibid. See also Beate Jahn in this volume.
[95] Ibid., at 474, with references to the contributions of Geis/Wagner, Jahn, and Stohl.

struggles?[96] The *dependency* theories of the 1960s and 1970s[97] addressed this issue and brought to light that societies existing under dependency are split in themselves between centre and periphery.[98] This goes for the dominant countries as well. Imperialism not only leads to 'external' confrontations (between North and South) but also to 'internal' critique and respective struggles within the North. Under this perspective, international law also bears the traces of *intra-imperial* struggles. At the same time, and this is the important lesson which postcolonial studies have to teach, those living in the imperial centre tend to underestimate the importance of the struggles going on in the areas which are incorporated under imperial rule. In this regard, the postcolonial critique of Western ignorance, as spelled out by Chimni and Mallavarapu in their contributions, is well founded and can hardly be overtaxed as a contribution to an international order that offers new chances for hedging violence.

This brings up the *second* (our final) question. What kind of order is to be expected under the evolving multipolar and multi-normative constellations under which power interacts with principle, cosmopolitanism with nationalism, pacifism with populism, the search for identity with the lure of going with the masses, religion with a secular lifestyle? The present volume has assembled contributions from various political and cultural traditions: the 'West', the 'South' but also more specifically from China (Manjiao Chi), Russia (Paul Robinson and Mikhail Antonov), and the Islamic World (Sohail Hashmi). These contributions point to the specificities of each of the normative traditions prevailing and intersecting in the different national contexts. As presented in this volume, the non-Western traditions of thought do not rule out the possibility of a consensus under the UN Charter as a general frame of reference for justifying and hedging the use of force. But the contributions also demonstrate that, for the time being, any such consensus will continue to be of a *formal* nature stressing self-defence and non-intervention and thus confirming the international order envisaged by the UN Charter but not necessarily bringing concrete behaviour in line with it (with the possible consequence of hollowing it out even further).

For instance, Sohail Hashmi in his contribution refers to the 1981 decision of the Organization of the Islamic Conference (OIC) to create the International Islamic Court of Justice (IICJ) as a sign of Islamic accommodation to universal international law (even though the sole source of law of the IICJ would be the Shari'a): 'Everything is in place for the court to begin its operation ... Yet it never convened because the court's statute has not been ratified by the requisite two-thirds majority of the OIC member states.'[99] Harking back to the contribution by Daase and Deitelhoff, this is to say that the decision to create the court could be understood as confirming a basic consensus on the need for a law-governed international order in spite of the unwillingness of the OIC member states to follow up on this by making the Court a reality. In a positive turn, this could amount to a form of 'organized hypocrisy'[100] which keeps the door

[96] L. Eslava, M. Fakhri, and V. Nesiah (eds), *Bandung, Global History, and International Law. Critical Pasts and Pending Futures* (2017); J. von Bernstorff and P. Dann (eds), *The Battle for International Law: South-North Perspectives on the Decolonization Era* (2019).

[97] For many others: O. Sunkel and P. Paz, *El subdessarollo latino-americano y la téroia de desarrollo* (1970).

[98] Galtung, 'Eine strukturelle Theorie des Imperialismus', in D. Senghaas (ed.), *Imperialismus und strukturelle Gewalt* (1972) 29.

[99] Hashmi in this volume, at 393.

[100] S.D. Krasner, *Sovereignty. Organized Hypocrisy* (1999).

open for moving to a rule-based international order in spite of the hesitancy to do so up to now. It seems that keeping the door open for a cooperative form of dealing with conflict realistically is all that can be hoped for in the present situation (see also the contribution by Thilo Marauhn).

4. Conclusion

As already hinted above, the complex discourses on the justification of war analysed in this book confirm that justifications of war are more than 'cheap talk' without normative meaning. Even propaganda refers to an audience which constitutes itself around certain normative expectations. Whether such propaganda works is dependent on what a larger audience regards as acceptable. Justifications of violence certainly result from political power calculations. Yet, normative aspects of political action are part of such calculations. After all, reference to normative arguments (even as mere propaganda) would be meaningless—and ultimately unwise from a perspective of *realpolitik*—if norms were regarded as irrelevant for decision-making on the use force. Normative claims are crucial for the establishment of authority. And norms can only fulfil this function if states do not fundamentally question their character as norms: if states refer to norms in their efforts to legitimize violence, they are, of course, tempted to instrumentalize legal provisions for their respective purposes. However, in the words of Lothar Brock, this amounts to an effort on the part of political actors 'to make something available to themselves, the value of which for them is its unavailability'.[101] Even if a norm—as *Realists* claim—is referred to as part of power politics, this move must be accepted by the respective political actor *in principle* as binding. Otherwise, no norm can develop the desired discursive authority in a communication community—and the reference to it would be irrelevant. There would be no need for a discourse over normativity and the use of force if the latter was unconditionally allowed to every sovereign state at his own discretion, that is as a matter of sovereign (but as such not necessarily practical!) judgement.[102] Practical judgement as advocated by Chris Brown in our understanding calls for rules and norms which serve to facilitate a practical approach to dealing with conflict. Paraphrasing Geoffrey Hodgson: without such rules and norms 'all human activity would be hopeless'.[103] But without facing empirical experience rules and norms are futile.

In our introduction to this volume we asked whether it would be helpful, as Chris Brown suggests in his contribution, to return to the just-war tradition all together and be done with the analytical focus on an international rule of law. Our answer is that theorists and practitioners should continue to reflect on the *role* of law in international relations as indispensable for enabling decision-makers to navigate the tension between the legality and the morality of the use of force.[104] The issue is not

[101] Brock, above note 76, at 31; see also Payk, above note 27; Simon, *Der Mythos*, above note 11.

[102] Brock, above note 78 and above note 76.

[103] Hodgson, 'The Enforcement of Contracts and Property Rights. Constitutive versus Epiphenomenal Conceptions of Law', 13 *International Review of Sociology* 2 (2003) 375.

[104] See also Habermas, 'Bestiality and Humanity: A War on the Border between Legality and Morality', 6 *Constellations* (1999) 263. See also F. Kratochwil, *The Status of Law in World Society. Meditations on the Role and Rule of Law* (2014).

whether norms as such matter or not but rather how norms are involved in political struggle. Or even briefer: norms matter because judgement will always be contested.[105] The hope that the door to a cooperative international order supported by a basic consensus on rules and norms remains open, may be frail but it is not necessarily foolish.

105 Cf. Wiener, above note 84.

Index